MICROWAVE CAVITY ANTENNAS

The Artech House Antenna Library

Blake, L.V., *Antennas*

Gupta, K.C., and A. Banalla, eds., *Microstrip Antenna Design*

Hirsch, Herbert L., and Douglas C. Grove, *Practical Simulation of Radar Antennas and Radomes*

Kumar, A., and H.D. Hristov, *Microwave Cavity Antennas*

Law, Preston E., Jr., *Shipboard Antennas*

Li, S.T., J.C. Logan, J.W. Rockway, and D.W.S. Tam, *Microcomputer Tools for Communications Engineering*

Li, S.T., J.C. Logan, J.W. Rockway, and D.W.S. Tam, *The MININEC System: Microcomputer Analysis of Wire Antennas*

Pozar, David M., *Antenna Design Using Personal Computers*

Sletten, Carlyle J., ed., *Reflector and Lens Antennas: Analysis and Design Using Personal Computers*

Sletten, Carlyle J., ed., *Reflector and Lens Antennas: Software and Example Book*

Weiner, M.M., et al., *Monopole Elements on Circular Ground Planes*

Wolff, Edward A., *Antenna Analysis: Revised 2nd Edition*

MICROWAVE CAVITY ANTENNAS

A. Kumar
A.K. Electromagnetic, Inc., Montreal

H.D. Hristov
Higher Institute of Mechanical and Electrical Engineering, Varna

Artech House

Library of Congress Cataloging-in-Publication Data
Kumar, A. (Akhileshwar)
Microwave cavity antennas/by A. Kumar and H.D. Hristov.

p. cm.
Bibliography: p.
Includes index.
1. Microwave antennas. I. Hristov, H.D. II. Title. III. Title: Cavity antennas.
TK7871.6.K795 1989 621.38′028′3—dc19 89-217 CIP
ISBN 0-89006-334-6

ARTECH HOUSE, INC.
685 Canton Street
Norwood, MA 02062

International Standard Book Number: 0-89006-334-6
Library of Congress Catalog Card Number: 89-217

10 9 8 7 6 5 4 3 2 1

This book is dedicated to our families,
Nirmala, Anne, and Angel,
and
Tatjana and Teodora

Contents

Preface

The objective of this book is to provide a critical review of the work on microwave cavity antennas that has been done since the 1950s in both the East and West for the benefit of engineers and scientists in the fields of antennas and communication engineering. Among the available books on antennas, there was no reasonably comprehensive coverage of cavity antennas. This book attempts to fill this gap in the literature.

Chapter 1 provides an overall view of the book via a qualitative description of the types of cavity antennas that will be discussed. Chapters 2 and 3 are devoted to a comprehensive treatment of cavity antenna structures, basic excitations of generalized and cylindrical cavity antennas, application of the moment method to cavity-backed aperture, cavity-backed slot antennas, cavity-backed slot arrays, and closed-cavity antennas with translucent plane apertures. General descriptions, and experimental and theoretical considerations of cavity-backed crossed-dipole fed, helical, and spiral antennas are given in Chapter 4. In this chapter, practical applications, including multipactor breakdown of these antennas are discussed. Chapters 5 and 6 deal with the short and long backfire antennas. We have described some practical applications of these antennas in satellites, marine radars and ground terminals.

The authors, A. Kumar in particular, wish to thank AK Electromagnetic, Inc., Dollard-des-Ormeaux, Quebec, Canada for financial help during preparation of the manuscript. We also wish to thank Mr. Helmut E. Schrank of Westinghouse Electric Corporation, Baltimore, for reviewing the manuscript, and Mr. Mark Walsh, Mr. Dennis N. Ricci, and Ms. Riley Hart of Artech House, for their help during preparation of the book.

We express our sincere appreciation to the Institute of Electrical and Electronics Engineers New York, Institution of Electrical Engineers (London), Institution of Electronic and Radio Engineers (London), Microwave Exhibitions and Publishers, Ltd. Kent, Institute of Electronics and Communication Engineers of Japan (Tokyo), *Microwaves & RF, Microwave Journal, Microwave Systems News,* Gabriel Electronics, The Aerospace Corporation, Matra Espace (France), Queen Mary College

(London), EETC Coventry, USAF Cambridge Research Laboratories, National Aeronautics and Space Administration, European Space Agency, Higher Institute of Mechanical & Electrical Engineering (Varna, Bulgaria), AK Electromagnetic, Inc., Spar Aerospace Ltd. Hughes Aircraft Company, and American Institute of Aeronautics and Astronautics (Washington), Central Institute for Scientific and Technical Information (Sofia, Bulgaria), USSR Academy of Science (Moscow), Radioi Svyaz Publishing House (Moscow), Technical University of Denmark (Lyngby, Denmark), Taylor & Francis Ltd. (London), Bulgarian Academy of Science (Sofia, Bulgaria), Nauka Publishing House (Moscow), the Scientific Society for Telecommunication (Budapest, Hungary), McGraw-Hill (New York), and Pergamon Press Ltd. (Oxford, UK) for their kind permission to use a number of figures and materials from their various publications.

A. KUMAR AND H.D. HRISTOV

Chapter 1
Introduction

1.1 ANTENNA MINIATURIZATION USING RESONANT PHENOMENA

One of the basic developmental trends in electronics is miniaturization, particularly in antennas where it is of paramount significance for modern aircraft, spacecraft, and mobile radio-electronic equipment.

In antenna design, the miniaturization process has been slow to develop. The introduction of semiconductor devices and hybrid-monolithic integrated circuit construction has created a great disproportion between the dimensions of passive antennas and those of other radio-electronic circuits. Antenna miniaturization is fundamentally restricted by the inherent electrodynamic properties of the radiating systems. In addition to structural, technological, power, and other decisive factors for miniaturization in electronics, antennas are also characterized by space limitations, because an antenna's directive properties are dependent on its geometric dimensions.

Increasing the frequency, and even going to new frequency bands, is an obvious way to miniaturize an antenna. However, regardless of the electromagnetic nature of radio waves, the same antenna principles and configurations cannot be used over the entire spectrum. Any frequency increase results, in many cases, in an intolerable increase in signal loss along the propagation path. Therefore, antenna miniaturization must usually be accomplished within already utilized radio frequency bands. This requires the development of new principles, materials, and technologies.

For a constant wavelength, antenna size diminution leads to reductions in directivity, bandwidth, and efficiency. The supergain phenomenon occurs when the antenna dimensions are much smaller than the wavelength. However, supergain antennas have an insignificant bandwidth and poor efficiency, so they do not find any application in practice. Supergain apparently contradicts the common experience in which an increase in gain must be accompanied by an increase in aperture. While supergain antenna arrays may seem very attractive (for the sake of the theoretically possible high gains that can be attained, as compared with the normal gain of uniformly excited arrays), they are not desirable because of the excessively large currents that must be held within impractical tolerances. Therefore, from a practical standpoint, antenna size reduction is tolerable only to limits for which the radiation parameters have acceptable values.

Usually, antenna miniaturization is achieved in one of the following ways:

1. Insertion of dielectric or magnetic materials with high permeabilities into the antenna structure;
2. Using current stripline or microstrip technology;
3. Integration of the antenna with semiconductor devices;
4. Suitable combinations of the above three methods.

The methods indicated above are not usable at microwave frequencies when high directivity is needed from a single undersized antenna. In that case, considerable antenna size diminution can be achieved by using resonant field phenomena.

Any antenna tuned to resonate at a certain frequency, or a succession of frequencies, is called *resonant*. Resonant antennas have characteristic dimensions which are commensurable with their resonant wavelengths. All resonant antennas are conventionally divided into two basic classes:

exoresonant and endoresonant.

The outer surface of an *exoresonant antenna* is excited in resonance (for instance, a self-resonant half-wave dipole). The *endoresonant antenna* comprises a resonating or resonant cavity; thus we usually call it a "cavity-type antenna."

This book covers several cavity-type antennas: cavity-backed slot antennas and slot arrays, crossed-dipole-fed cavity antennas, cavity antennas with semitransparent apertures, short backfire antennas, long backfire antennas, beam waveguide cavity antennas, *et cetera*. Normally, they are narrow-banded directional antennas with bandwidths ranging from 0.5–2% (high-quality-factor antennas) to 10–15% (backfire antennas) and with typical gain values from 15 to 25 dB for a single-element antenna and from 20 to 35 dB for multielement antenna arrays. For the foregoing frequency, bandwidth, and gain properties, the cavity-type antennas and antenna arrays are considerably smaller in size, volume, and weight than the better known classical antennas. For example, a short backfire antenna is about one-fifth the length and half the volume of an optimum horn antenna with the same gain.

This book is the first attempt to summarize all antenna configurations of one class of miniature endoresonant antennas, referred to as *cavity-type antennas*.

1.2 CLASSIFICATION OF CAVITY-TYPE ANTENNAS

We can divide cavity-type antennas in two basic groups:

dual-reflector antennas; and other cavity-backed antennas.

Dual-reflector antennas consist of two reflectors, most frequently plane, with one-dimensional field resonance between them. One of the reflectors is much smaller than the other (backfire antennas), or else it is semitransparent (beam waveguide antennas) and acts like an end-mirror in a laser cavity.

The backfire antenna designs are termed long or short depending on the distance between the two reflectors. The *long backfire antenna* (LBFA) comprises a *surface wave* (SW) structure which is several wavelengths long. The surface wave structure connects the reflectors and is usually excited by a crossed-dipole or waveguide aperture. The dominant mode in the LBFA is simply a standing wave surrounding the surface wave structure. The latter can take many forms, but a Yagi-dipole assembly, corrugated (disk-on-rod) line, or dielectric rod are popular in practice.

In the case of the *short backfire antenna* (SBFA) the spacing between the reflectors is only half a wavelength, and the surface wave structure becomes unnecessary.

Backfire antennas (BFAs) have many varieties, as illustrated in Figure 1.1.

The *beam waveguide cavity-type antennas* (BWCAs) represent quasioptical open resonators (i.e., their dimensions are many times larger than the wavelength). The end reflector is semitransparent; for example, it can be perforated metal. As shown in Figure 1.2, BWCAs are in plane, plane-spherical, or confocal spherical configurations, depending on the profile of the reflector.

As we have already explained, dual-reflector antennas are based on open resonators. Cavity antennas, on the other hand, are more like three-dimensional cavity resonators. On one of the metal cavity walls, a multitude of small openings (slots or round holes) are made. Thus, the cavity antenna can also be considered either as an antenna array or, generally speaking, as a semitransparent aperture backed by a closed resonator cavity. Figure 1.3 shows some cavity antenna designs with perforated semitransparent apertures. Recently, there has been great interest in the ability of open cavity-backed antennas to produce circular polarization. The basic concept of the cavity-backed radiator antenna for circular polarization involves the use of a crossed-dipole feed, which is located in the aperture of a backing cavity. These antennas, shown in Figure 1.4, are gaining prominence.

1.3 CAVITY-TYPE ANTENNA PROGRESS

The LBFA dates back to the beginning of the 1960s. It was patented by Ehrenspeck in 1964 [1] under the name of "reflection antenna with dipole elements employing multiple reflection of the energy to effect increased gain." He has published a number of papers in which he describes the antenna construction, design, and experimental results characterizing the basic antenna parameters. In the first backfire antennas, uniform dipole structure (Yagi backfires) and plane disk reflectors were used. It has since been proved that LBFA gain can be increased considerably by an optimum choice of the SW phase velocity and the diameter of the large (surface wave)

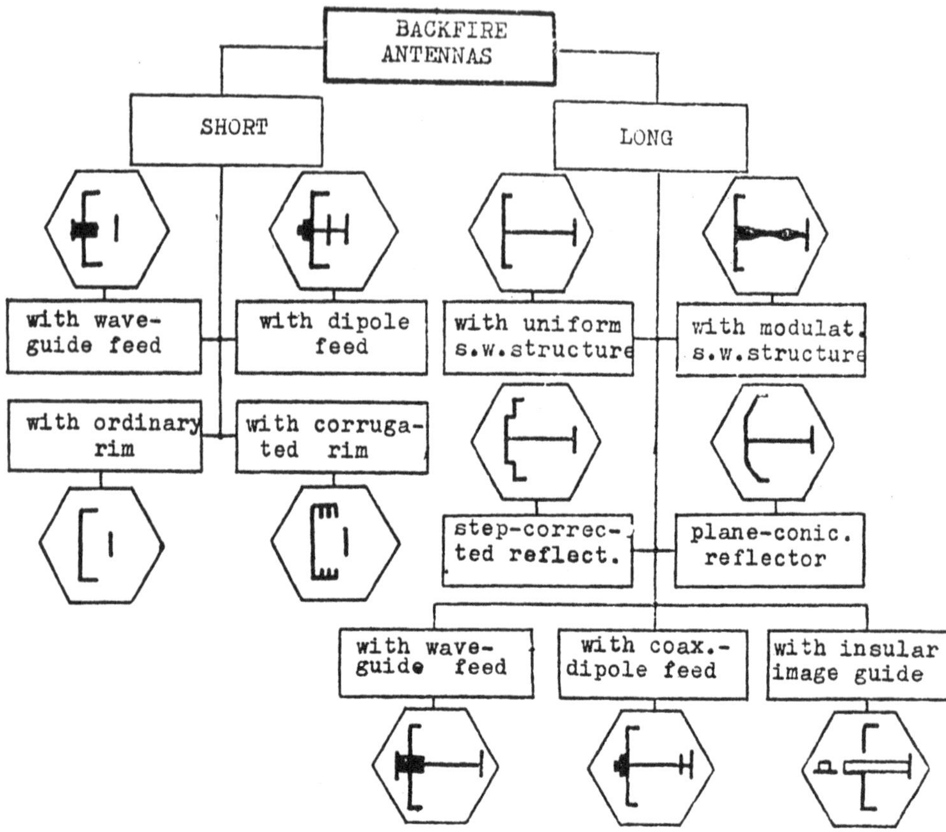

Figure 1.1 Basic varieties of long and short backfire antennas

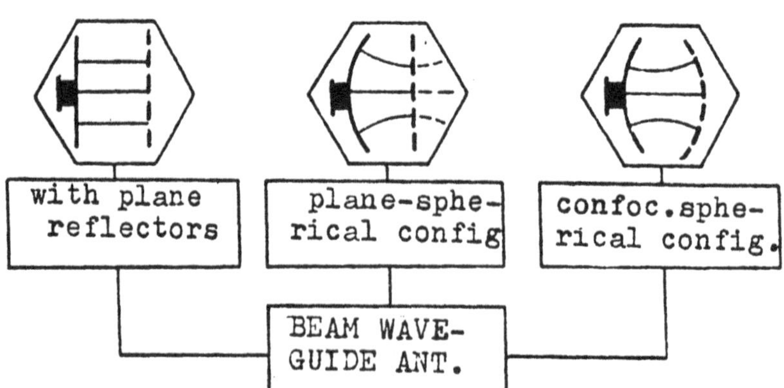

Figure 1.2 Beam waveguide quasioptical antenna configurations

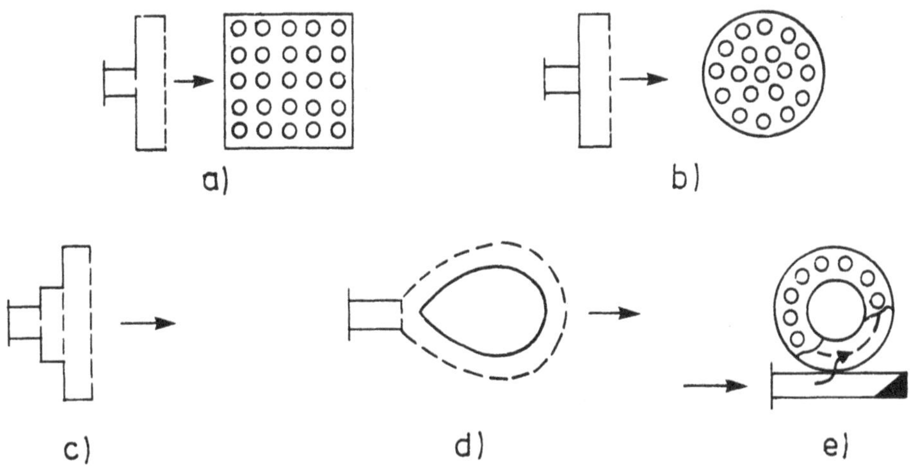

Figure 1.3 Cavity antenna designs with semitransparent radiating apertures: (a) rectangular single-cavity antenna; (b) circular single-cavity antenna; (c) two-cavity antenna; (d) double-bounded cavity antenna; (e) ring-shaped cavity antenna

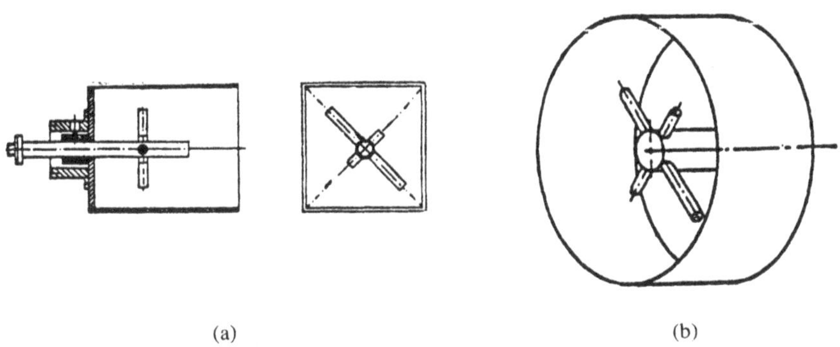

Figure 1.4 Cavity-backed open resonator antennas: (a) rectangular aperture; (b) circular aperture

reflector. (The term "surface wave reflector" was suggested by Ehrenspeck to distinguish it from the ordinary plane reflector in antenna theory.) In 1965 Zucker [2] published an approximate methodology of backfire antenna design. Ehrenspeck and his associates have developed backfire antenna construction with very good performance by modifying the large disk reflector into a *plane stepped circular* one. In principle, the feed dipole may be put in different cross section with the standing wave at maximum, but maximum gain is only reached by locating the dipole at a distance of a quarter wavelength ($\lambda/4$) from the small reflector. For higher frequencies, helical, zig-zag, meander-line, disk-on-rod (corrugated), dielectric, and metal-dielectric (Goubeau line) SW structures have been used instead of a dipole structure. A large reflector consisting of plane and conical parts (plane-conical reflector), as proposed and examined by the authors, performs a better phase correction of the field in the equivalent radiation aperture. As a result, the sidelobe levels become lower.

Tereshin and Kuznetzov [3] suggested a backfire antenna with a nonuniform, modulated dipole structure. This antenna has better directivity in comparison with an antenna having a uniform SW structure. Later, Kirov and Hristov [4] worked out a modulated, corrugated, SW structure, in which better directivity and bandwidth properties were achieved at microwave frequencies.

Since the beginning of the 1970s, LBFAs have been designed at frequencies not higher than 3–4 GHz with a coaxial dipole feed. In 1972, Hristov [5] patented a backfire antenna for centimeter waves, in which a waveguide feed of a dielectric slow wave structure is used.

During the period from 1972 to 1977, Hristov and Kumar [6] proposed a different configuration of SW structure, main reflectors, and subreflectors for LBFAs. Kumar [7] proposed a new strip subreflector for LBFAs in frequency reuse applications where a low level of cross-polarization is required between signals transmitted in the E- and H-planes. He suggested that this type of antenna may be useful in satellites, ships, aircraft, and space stations because of its structural simplicity, light weight, small size, low cross-polarization, high directivity, and symmetrical radiation pattern.

In 1978, Kumar [8] proposed a future application of LBFAs for TV reception at microwave frequencies. He also published a theoretical analysis of LBFAs containing a dielectric structure.

In 1975, Simons and Shahani [9] proposed a modification of the dielectric SW LBFA mentioned above, with an insular image guide feed compatible with integrated circuit technology for millimeter waves.

A new stage in backfire antenna development was marked by the invention of the SBFA, which was first reported by Ehrenspeck in 1965 [10]. In 1969, he received a patent for this novel BFA. However, we must point out that the idea of the SBFA and a backfire-like design had already been suggested by Trentini in 1954 [11]. A contribution to the development of the SBFA with coaxial-dipole feed had also been suggested by many researchers (Dod, Nielsen, Patel, Taylor, Jensen, Hass, Kirov, Loh, Silverberg, and others). Kooi and Leong [12] developed a SBFA with a corrugated rim to improve directivity.

In 1983, Alpha Industries of the United States developed a new application of the BFA which has performance superior to that of Cassegrain design.

A waveguide-fed SBFA was patented by Ehrenspeck and Strom in 1972 [13]. At the same time, extensive studies on waveguide-fed SBFAs were in progress in the U.K. and Eastern European countries. The SBFAs with coaxial-dipole feeds are noted for their efficiency and simple, compact design. Because of these merits, the short backfire elements are used in military and satellite communications, telemetry, and other systems. Ordinary antenna arrays, for meter and decimeter waves, consist of a large number of dipole, slot, or Yagi antennas. The substitution of such arrays by SBFAs with dipole feeds gives considerable advantage. The number of elements is reduced five to six times, while the directivity of the antenna arrays is preserved. As a result, the feed system is simplified, its losses are diminished, and gain is increased. Furthermore, the reliability of the antenna system is higher and its weight and cost are lower. In Table 1.1, the basic parameters of BFA arrays with different numbers of elements are systematized.

Table 1.1 Basic Parameters of Backfire Antenna Arrays

Antennas	*Relative antenna area* (S_A/λ^2)	*3 dB beamwidth (degrees)*		*Gain (dBi)*	*Aperture efficiency (%)*
		E-plane	*H-plane*		
1. Single-element SBFA	3.14	34	34	15.1	82
2. Two-element array	5.14	22	31	17.5	87
3. Four-element array	8.14	24	20	19.2	81
4. Eight-element array	15.9	14	15	22	80
5. Sixteen-element array	31.4	9	10	24.5	71

For comparison, the characteristics of a single-element SBFA of diameter 2λ and length (depth) 0.5λ, fed by a half-wave symmetrical dipole are presented in the table. The antennas, designated by numbers from 1 to 4, were designed and studied by Ehrenspeck and Strom in 1971 [14] at 3 GHz. The antenna shown as number 5 was designed by Hasebe and Zama in 1975 at 400 MHz.

Of special interest are the multibeam phased arrays studied by Brian and Talbot [15], and by Molker [16], which consisted of short backfire elements and were designed for mobile satellite communication and navigation systems. For example, the 18-element hexagonal array with a diameter of 2λ has a typical single beam with 3 dB width of 5°, and gain of 24 dB for L-band. Two SBFAs with multiple frequencies may be arranged coaxially, and in this case the smaller reflector of the lower frequency antenna is used as the large reflector of the higher multiple frequency antenna. This type of integrated two-frequency antenna was designed for the ground station of a telemetry satellite system.

In 1983, Horstein [17] studied the application of short backfire elements as feed elements for the cluster format for U.S. mobile communication satellite systems. He showed that the short backfire elements, placed in a resonant cup, can provide high gain with low risk.

The earliest antenna arrays consisting of waveguide-fed short backfire elements were studied at X-band almost simultaneously by Large in 1974 [18], Hristov and Urumov in 1975 [19], and Kumar in 1975 and 1980 [20,21]. Kooi and Leong [12] have worked on backfire arrays using corrugated rims at 35 GHz.

The beam waveguide cavity-type antennas are not as well studied, and we have not found them to have wide application in practice. In principle, BWCAs are fully analogous to the optical laser cavity, and they are suitable as antennas for millimeter and submillimeter waves. In 1968, Klass [22] described a beam waveguide antenna design comprised of two reflectors and a feed horn. The main reflector had a spherical shape, and the aperture reflector was a perforated (semitransparent) metal disk with a diameter of 23 cm. This antenna had a Gaussian radiation pattern, with an aperture efficiency of 80%. It was used for the remote control of an experimental helicopter model.

Another cavity-type antenna, based on a rectangular beam waveguide resonator, was proposed and studied in detail by Brauer in 1969 [23]. The semitransparent plane reflector represented a metal strip with a linear array of small holes. The beam waveguide resonator formed a Gaussian amplitude distribution in the array, and its radiation pattern had the same characteristic. The bandwidth of this antenna was 1%.

Generally, cavity antennas may be considered cavity resonators, in which the metal wall, or part of it, is semitransparent. A typical cavity antenna with a semitransparent radiating wall was proposed by Jacobsen, Andersen, and Gronlund in 1963 [24]. Its cavity resonator had rectangular shape and dimensions of $10 \times 10 \times 1.5\ \text{cm}^3$. A basic electromagnetic mode H_{101} was excited inside the cavity, and the antenna had a gain of 20 dB at 9.8 GHz.

In 1980, Paoloni [25] published a description and approximate theory of a rectangular cavity antenna with a resonant slot array on the broad wall, which was filled with a dielectric. The slots were arranged in a chessboard order and were made by using printed circuit technology. The field inside the cavity had H_{onp}-mode structure fed by a coaxial probe junction. Gorobetz and Nesterenko suggested a new type of ring-shaped cavity antenna. Essentially, it presents a slotted rectangular waveguide twisted in the form of a ring cavity resonator, and excited by another rectangular waveguide through an aperture. When unidirectional coupling between the waveguide and the ring cavity is realized, a traveling wave is obtained into the latter and the slot array radiates circularly-polarized electromagnetic waves. When there is nondirectional coupling between the ring cavity and the waveguide, a standing wave is settled into the cavity, and the radiated field has a linear polarization.

In a treatment of a general case cavity antenna with semitransparent wall, Katzenelenbaum *et al.* [26] proposed a synthesis method for finding the surface curvature and the degree of transparency that ensure a prescribed radiation pattern. A similar synthesis problem was solved by Garbacz and Pozar [27]. For a given radiation pattern, they found the proper shape of the antenna's rotary surface and the current density upon it.

Much theoretical and experimental work devoted to resonant antennas with cavity-backed semitransparent apertures has been presented recently by Hristov [28]. For example, an experimental miniature two-cavity antenna with a semitransparent aperture diameter of 20 cm and an antenna length of 4.5 cm has 3 dB beamwidth equal to 15° and sidelobe level of less than −30 dB at 10.1 GHz. The bandwidth of the antenna is about 1.5%.

Kumar [29] has used a circularly-polarized cavity-type feed for the ERS-1 shaped beam antenna at X-band. Similar types of direct radiating elements were used for the Inmarsat receiving system. The bandwidth of these antennas was not more than 3%.

REFERENCES

1. Ehrenspeck, H.W., "Reflection Antenna Employing Multiple Director Elements and Multiple Reflection of Energy to Effect Increased Gain," US Patent No. 3122745, February 1964.
2. Zucker, F.J., "The Backfire Antenna: Qualitative Approach to Its Design," *Proc. IEEE,* Vol. 53, No. 7, July 1965, pp. 746–747.
3. Tereshin, O.N., and L.N. Kuznetzov, "Backfire Antennas with Modulated Phase Velocity in Dipole Slow-wave Structure," *Radiotechnika,* Moscow, Vol. 25, No. 12, 1970 (in Russian).
4. Kirov, G.S., and H.D. Hristov, "Backfire Antenna with Modulated Disk-on-rod Surface Wave Structure," Bulgarian Patent No. 23 418, 1976.
5. Hristov, H.D., "Backfire Antenna," Bulgarian Patent No. 18 149, 1972.
6. Hristov, H.D., and A. Kumar, "X-Band Waveguide-Fed Long Backfire Antenna with Dielectric SW Structure," Research Report, Dept. of Elect. and Electron. Engineering, Queen Mary College, London University, London, August 1976.
7. Kumar, A., "Modified Backfire Antenna for Frequency Reuse Applications," Research Report, Engineering Education and Training Consultants, Coventry, U.K., Sept. 1980.
8. Kumar, A., "Backfire Antenna Aim at Direct Broadcast TV," *Microwaves,* Vol. 17, No. 4, 1978, pp. 106–112.
9. Simon, R.N. and D.T. Shahani, "Long Backfire Antenna with Insular Image Guide Feed for Dual Frequency Operation," *Electronics Letters,* Vol. 15, No. 3, February 1979, pp. 78–79.
10. Ehrenspeck, H.W., "The Short Backfire Antenna," *Proc. IEEE,* Vol. 53, No. 8, August 1965, pp. 1161–1162.
11. Trentini, G.V., "Bundelung Electrische Wellen Durch Leitsscheiben," *Z. Angew. Phys.,* Vol. 6, No. 10, 1954, pp. 462–470 (in German).
12. Kooi, P.S., and M.S. Leong, "Q-Band Short Backfire Antenna Arrays," *Proc. 10th European Microwave Conference,* Warsaw, September 1980, pp. 162–166.
13. Ehrenspeck, H.W., and J.A. Strom, "High-Frequency Waveguide Feed in Combination with Backfire Antenna," US Patent 3 774 223, November 1972.
14. Ehrenspeck, H.W., and J.A. Strom, "Short Backfire Arrays," *AFCRL Report 71-0368,* Bedford, MA, May, 1971.
15. Brian, M.T., and A.C. Talbot, "Some Characteristics of an Experimental L-Band Multibeam Phased Array Payload," *Proc. Conf. on Satellite Systems for Mobile Communications and Navigation,* IEE London, June 1983, pp. 78–82.
16. Molker, A., "High-Efficient Phased Array Antenna for Advanced Multibeam, Multiservice Mobile Communication Satellite," *Proc. Conf. on Satellite Systems for Mobile Communications and Navigation,* IEE London, June 1983, pp. 75–77.

17. Horstein, M., "Requirements for a Mobile Communications Satellite System," *Tech. Report TRW*, November 1983.
18. Large, A.C., "A Linear Array of Greatly Reduced Size and Weight, *ASWE Tech. Report TR-74-2*, Portsmouth, Hants, U.K., February 1974.
19. Hristov, H.D., and J.R. Urumov, "X-Band Backfire Antenna Arrays," *Proc. European Microwave Conf.*, Hamburg, Germany, September 1975, pp. 112–115.
20. Kumar, A., "Waveguide Fed Short Backfire Antenna," *Research Report, EETC*, Coventry, U.K., September 1975.
21. Kumar, A., "Waveguide Fed Short Backfire Antenna Array," *IEEE NTC*, Houston, Texas, 1980.
22. Klass, K.J., "Helicopter Stabilized by Microwave Beam," *Aviation Week & Space Technology*, Vol. 88, No. 17, 1968, pp. 85–87, 89.
23. Brauer, J.R.G., "A Rectangular Beam Waveguide Resonator and Antenna," *IEEE Trans. Antenna and Propagation*, Vol. AP-21, No. 3, September 1972, pp. 595–602.
24. Jacobsen, S., E. Andersen, and M. Gronlund, "An Antenna Illuminated by a Cavity Resonator," *Proc. IEEE*, Vol. 51, No. 11, November 1963, pp. 1431–1435.
25. Paoloni, F.J., "A Cavity-Backed Resonant Slot Array–Theory and Measurement," *IEEE Trans. Antenna and Propagation*, Vol. AP-28, No. 2, February 1980, pp. 259–263.
26. Katzenelenbaum, B.Z., E.N. Korshunova, L.I. Pangonis, and A.N. Sivov, "Resonant Antenna Synthesis," *Raditechnika Electronika*, USSR Academy of Science Journal, Vol. XXVIII, No. 5, pp. 854–863 (in Russian).
27. Garbacz, R.J., and D.M. Pozar, "Antenna Shape Synthesis using Characteristic Modes," *IEEE Trans. Antenna and Propagation*, Vol. AP-30, No. 5, May 1982, pp. 340–350.
28. Hristov, H.D., "Some Experiences with Resonator-Type Antennas," *IEEE Montech '86 Conference on Antennas and Communications*, Montreal, Quebec, Canada, September–October 1986.
29. Kumar, A., "Highly Shaped Beam Telemetry Antenna for the ERS-1 Satellite," *Proc. IEE*, Pt. H, Vol. 134, No. 1, February 1987, pp. 106–108.

Chapter 2
Cavity-Type Antennas

2.1 ANTENNA STRUCTURE AND BASIC PROPERTIES

It is convenient to consider the microwave cavity antenna as an electromagnetic system, consisting of conditionally-detached elements, as shown in Figure 2.1. Here $\bar{J}_a$ and $\bar{M}_a$ are equivalent electric and magnetic currents, respectively, on the aperture surface S; V_r is the resonant cavity, excited by an excitation element E_o with currents $\bar{J}_o$ and $\bar{M}_o$; M_i is a matching device between the antenna input and waveguide feeder Wg.

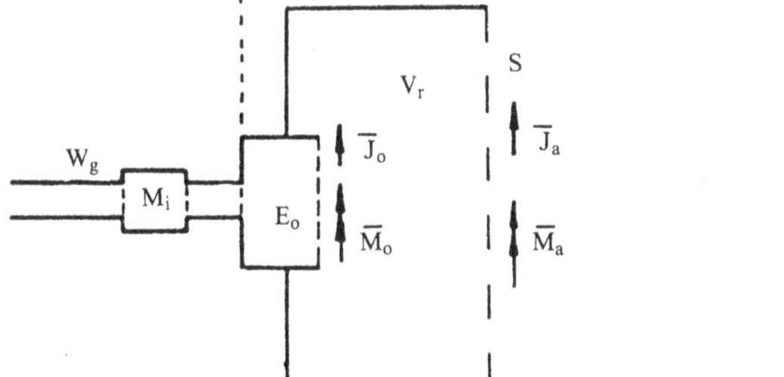

Figure 2.1 Microwave cavity antenna structure (from [1])

The set-up on Figure 2.1 is made according to the traditional approximate approach in the case of microwave antennas. The electromagnetic antenna analysis (or synthesis) is usually divided into three basic problems: outer, inner and input. The inner problem is a key one, because it is devoted to a determination of the field or current amplitude-phase distribution on the radiation aperture. On the basis of this distribution, the outer problem is solved (i.e., the far-field radiation is found). The input problem is connected to the input impedance and its matching with the waveguide characteristic impedance.

This approach is particularly suitable for cavity antennas having a high quality factor, Q_A. In that case, the inner problem is typically a resonator one, and it can be solved using the conventional eigenmode method. The transition from a closed-cavity resonator to a cavity antenna is characterized by a number of electromagnetic effects, including power radiation, cavity field perturbation, resonant frequency shifting, mode spectrum thinning.

The cavity antenna quality factor is given by:

$$1/Q_A = 1/Q_{wall} + 1/Q_{med} + 1/Q_{rad} \tag{2.1}$$

where Q_{wall}, Q_{med} and Q_{rad} are partial quality factors, connected with the metal wall power losses, cavity medium power losses, and radiation power, respectively.

In practical cavity antenna designs $1/Q_{wall} + 1/Q_{med} \ll 1/Q_{rad}$ (i.e. $Q_A \approx Q_{rad}$). The cavity antenna, as well as the real cavity resonator, has a complex resonant frequency:

$$\dot{\omega}_0 = \omega_0' + j\omega_0'' = \omega_0'\left(1 + j\frac{1}{2Q_A}\right) \tag{2.2}$$

where $\omega_0' = Re\dot{\omega}_0$ has a meaning of a real frequency and $\omega_0'' = Im\dot{\omega}_0$ is an attenuation factor.

From (2.2), we obtain

$$Q_A = \omega_0'/2\omega_0'' \tag{2.3}$$

In the perfect cavity resonator, there are constant-amplitude periodic oscillations and $\omega_0'' = 0$. When $\omega_0'' \ll \omega_0'$, the field oscillations are quasiperiodical, weakly attenuated ones.

The quality factor of the typical cavity antennas is $Q_A \geq 50$–100 (i.e., $\omega_0'' \leq 0.01$–$0.005\omega_0'$), because we assume that the radiation phenomenon does not give rise to large field perturbations. In the resonant cavity, an infinite number of characteristic modes can be excited. Their resonant frequencies form an infinite discrete spectrum. This spectrum becomes thicker with the frequency increase [2]

$$\frac{N_r}{\Delta\omega} \approx \frac{V_r}{2\pi^2 c^3}\omega^2 \tag{2.4}$$

or

$$N_r \approx 4\pi \frac{V_r}{\lambda^3} \cdot \frac{\Delta f}{f_0} \tag{2.5}$$

where $N_r/\Delta\omega$ is a number of resonant modes for a unit increment of the frequency range or frequency spectral density:

V_r = hollow cavity volume ($\varepsilon = \varepsilon_0$, $\mu = \mu_0$);
$\dfrac{2\Delta f}{f_0}$ = relative frequency band;
c = velocity of light.

The relationships shown in (2.4) and (2.5) are known in optics and quasioptics as Rayleigh-Jeans formulas. It is evident that when $V_r \gg \lambda^3$, the selective properties of the closed (three-dimensional) cavity resonators are very poor. In addition, with a frequency increase, the cavity spectrum rapidly grows thicker ($N_r/\Delta\omega$ is proportional to the frequency on a square). For comparison, we look at resonant regions with one- or two-dimensional resonances. For example, for a two-dimensional resonant region with a surface S_r, the spectral density is given by a similar formula:

$$\frac{N_r}{\Delta\omega} \approx \frac{S_r}{\pi c^2}\omega \tag{2.6}$$

Finally, for one-dimensional region with a length L_r, the spectral density does not change with frequency variation:

$$\frac{N_r}{\Delta\omega} \approx \frac{2L_r}{\pi c} = \text{const.} \tag{2.7}$$

In other words, in the last case, the resonant frequencies form an equidistant spectrum.

From the above relationships, we may conclude that, as a result of the cavity opening, frequency spectrum thinning is obtained. This effect is due to the radiation losses, which are selective ones (i.e., they are different for the various resonant modes). This is not valid for closed-cavity resonators, because the heat losses in the walls and coupling devices do not have such selective properties.

The cavity antennas and cavity resonators usually work in a single-mode regime (i.e., within a certain frequency band only one characteristic mode exists). Most often this is the *cavity fundamental mode*, which has the lowest resonant frequency.

Multimode conditions require large proportions between the cavity dimensions and operating wavelength. By choosing a suitable excitation element, only desired modes are excited. If there are some unwanted modes in the cavity, they can be suppressed by adding mode filters.

2.2 CAVITY ANTENNA EXCITATION THEORY

2.2.1 Excitation of Generalized Cavity Antenna

Let us now examine a generalized antenna model with its resonant cavity V_r confined within an arbitrarily-shaped closed surface S (see Figure 2.2). The resonant region contains a lossless medium (ε,μ) and excitation current sources ($\bar{J}_0$ = electric current density, $\bar{M}_0$ = magnetic current density). The surface S consists of the following elements: S_E = electric (metal) cavity wall; S_M = imaginary magnetic wall, and S_0 = small apertures. Electromagnetic fields excited in the volume V_r satisfy Maxwell's curl equations and the following approximate boundary conditions:

$$\bar{E}_\tau = [\bar{n}_0,\bar{E}] \approx -Z_S^E[\bar{n}_0[\bar{n}_0,\bar{H}]] \quad \text{on} \quad S_E \tag{2.8}$$

$$\bar{H}_\tau = [\bar{n}_0,\bar{H}] \approx -Y_S^M[\bar{n}_0[\bar{n}_0,\bar{E}]] \quad \text{on} \quad S_M \tag{2.9}$$

$$\bar{E}_\tau = [\bar{n}_0,\bar{E}] = \bar{E}_{\tau 0} \quad \text{in} \quad S_0 \tag{2.10}$$

where

Z_S^E = the surface impedance of the electric wall; and
Y_S^M = the surface equivalent admittance of the magnetic wall.

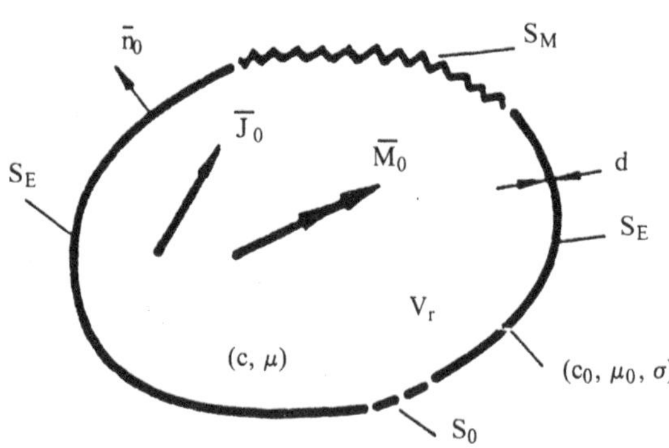

Figure 2.2 Generalized cavity antenna configuration (from [1])

The excited electromagnetic field inside the cavity can be expanded as a sum over all possible characteristic field modes, with mode indices $m = \xi, \eta, \zeta$ and $m' = \xi', \eta', \zeta'$:

$$\bar{E} = \sum_{m=1}^{\infty} A_m \bar{E}_m + \sum_{m'=1}^{\infty} A_{m'}, \bar{E}_{m'} \tag{2.11}$$

$$\bar{H} = \sum_{m=1}^{\infty} B_m \bar{H}_m + \sum_{m'=1}^{\infty} B_{m'}, \bar{H}_m \tag{2.12}$$

The expressions $[\bar{E}_m, \bar{H}_m]$ and $[\bar{E}_{m'}, \bar{H}_{m'}]$ are complete sets of solenoidal and potential field modes, respectively. For high-frequency and small cavity field perturbations, the potential modes are usually neglected.

General expressions for the mode amplitude coefficients A_m and B_m are given in [1] and [3]:

$$A_m = -\frac{Q_m^\sigma}{N_m} \cdot \frac{\dfrac{1}{\omega_m}\displaystyle\int_{V_r} \bar{J}_0 \bar{E}_m^* \mathrm{d}V + \frac{1}{\omega}\left(\int_{V_r} \bar{M}_0 \bar{H}_m^* \mathrm{d}V + \int_{S_0} [\bar{E}_{\tau 0}, \bar{H}_m^*] \mathrm{d}\bar{S}\right)}{1 + \dfrac{Q_m^\sigma}{Q_{m\Sigma}^E} + \dfrac{Q_m^\sigma}{Q_{m\Sigma}^M} + jX_{A,B}} \tag{2.13}$$

$$B_m = -\frac{Q_m^\sigma}{N_m} \cdot \frac{\dfrac{1}{\omega}\displaystyle\int_{V_r} \bar{J}_0 \bar{E}_m^* \mathrm{d}V + \frac{1}{\omega_m}\left(\int_{V_r} \bar{M}_0 \bar{H}_m^* \mathrm{d}V + \int_{S_0} [\bar{E}_{\tau}, \bar{H}_m] \mathrm{d}\bar{S}\right)}{1 + \dfrac{Q_m^\sigma}{Q_{m\Sigma}^E} + \dfrac{Q_m^\sigma}{Q_{m\Sigma}^M} + jX_{A,B}} \tag{2.14}$$

where

$$X_{A,B} = 1 + Q_m^\sigma \left[\frac{\omega}{\omega_m} - \frac{\omega_m}{\omega} - \frac{\displaystyle\int_{S_E} I_m Z_S^E |\bar{H}_{\tau m}|^2 \mathrm{d}S}{\omega_m N_m} - \frac{\displaystyle\int_{S_M} I_m Y_{S\Sigma}^M |\bar{E}_{\tau m}|^2 \mathrm{d}S}{\omega_m N_m} \right]$$

Q_m^σ, $Q_{m\Sigma}^E$ and $Q_{m\Sigma}^M$ are partial quality factors; and N_m is the modal normalization coefficient for the mth cavity mode. The quality factor Q_m^σ corresponding to the power losses into the electric (metal) wall S_E' with a conductivity σ is

$$Q_m^\sigma = \omega_m \frac{W}{P_\sigma} = \omega_m \frac{\mu \int_V |\bar{H}_m|^2 \mathrm{d}V}{\int_{S_E'} ReZ_S^\sigma |\bar{H}_{\tau m}|^2 \mathrm{d}S} \tag{2.15}$$

In (2.15) $Z_S^E = Z_S^\sigma$. If the electric wall is homogeneous and $Re\ Z_S(S_E') = R_S(S_E')$ = const., then

$$Q_m^\sigma = \frac{\omega_m N_m}{R_S^\sigma \int_{S_E'} |H_{\tau m}|^2 \mathrm{d}S} \tag{2.16}$$

By analogy, a quality factor $Q_{m\Sigma}^E$ corresponding to the radiation power through translucent electric wall (aperture) is introduced:

$$Q_{m\Sigma}^E = \frac{\omega_m N_m}{\int_{S_E''} ReZ_{S\Sigma}^E |\bar{H}_{\tau m}|^2 \mathrm{d}S} \tag{2.17}$$

If S_E'' is a perforated metal wall, the radiation aperture is equivalent to an array of small radiating holes, and in this case, the quality factor is given by the following expression:

$$Q_{m\Sigma}^E = \frac{\omega_m N_m}{\sum_{i=1}^{N} \int_{S_{Ei}''} ReZ_{S\Sigma i} |\bar{H}_{\tau m}|^2 \mathrm{d}S} \tag{2.18}$$

where

N = number of holes;
i = arbitrary hole index; and
$Z_{S\Sigma i}^E$ = equivalent surface impedance of ith array hole element.

Finally, a quality factor conforming to the radiation power through the magnetic wall is given by

$$Q^M_{m\Sigma} = \frac{\omega_m N_m}{\int_{S_M} ReY^M_{S\Sigma} |\bar{E}_{\tau m}|^2 \mathrm{d}S} \tag{2.19}$$

The total quality factor Q_{Am} is

$$\frac{1}{Q_{Am}} = \frac{1}{Q^\sigma_m} + \frac{1}{Q^E_{m\Sigma}} + \frac{1}{Q^M_{m\Sigma}} \tag{2.20}$$

The modal normalization factor is found using

$$N_m = \int_V \mu |\bar{H}_m|^2 \mathrm{d}V = \int_V \varepsilon |E_m|^2 \mathrm{d}V \tag{2.21}$$

If $N_m = 1$, the cavity field modes are called *orthonormalized*. For the resonant frequency $\omega = \omega_m$, the amplitude coefficients of the electric and magnetic fields are equal, as shown in the following equation:

$$A_m = B_m = -\frac{Q^\sigma_m}{\omega_m N_m} \cdot \frac{\int_{V_r} J_0 E^*_m \mathrm{d}V + \int_{V_r} M_0 H^*_m \mathrm{d}V + \int_{S_0} [E_{\tau 0}, H^*_m] \mathrm{d}S}{1 + Q^\sigma_m \left(\frac{1}{Q^E_{m\Sigma}} + \frac{1}{Q^M_{m\Sigma}} \right) + jX_{A,B}} \tag{2.22}$$

By equalizing $X_{A,B}$ to zero, the cavity antenna resonant frequency is obtained thusly:

$$f_m = \frac{Q^\sigma_m}{2\pi N_m} \left[\int_{S_E} ImZ^E_{S\Sigma} |\bar{H}_{\tau m}|^2 \mathrm{d}S + \int_{S_M} ImY^M_{S\Sigma} |\bar{E}_{\tau m}|^2 \mathrm{d}S \right] \tag{2.23}$$

The complex resonant frequency is

$$\dot{f}_m = f_m \left(1 + j \frac{1}{2Q_{Am}} \right) \tag{2.24}$$

If the resonant frequency of an ideal cavity is marked by f_{m0}, the frequency shifting due to the inner losses and radiation is

$$\Delta f = f_{m_0} - f_m = f_{m_0}\left(1 - \sqrt{1 - \frac{1}{4Q_{Am}^2}}\right) \tag{2.25}$$

For example, if $f_{m0} = 10$ GHz and Q_{Am} has a value of 100, from (2.25) we calculate $\Delta f = 120$ kHz.

When two modes with nearby resonant frequencies ($\omega_s \approx \omega_p$) are excited simultaneously in the antenna cavity a mutual quality factor Q_{ps} is defined and given by

$$\frac{1}{Q_{ps}} \approx \frac{1}{2}\left[\frac{1}{Q_p} + \frac{1}{Q_s} \pm \sqrt{\left(\frac{1}{Q_p} - \frac{1}{Q_s}\right)^2 + \frac{4}{Q'_{ps}}}\right] \tag{2.26}$$

where Q_p and Q_s are the respective quality factors for the two modes;

$$Q'_{ps} = -\frac{2\sqrt{M_p M_s}}{ReZ_s M_{ps}}; \quad M_{p,s} = \varepsilon \int_V \bar{E}^2_{p,s} \mathrm{d}V; \quad M_{ps} = -\oint_S \bar{H}_{p\tau} \bar{H}_{s\tau} \mathrm{d}S$$

These modes are coupled as a result of the cavity losses and radiation. With the excitation of the fundamental mode, an additional mode also arises, and the two modes have almost equal amplitudes. The presence of two near-frequency resonant modes in the cavity-type filters and antennas can be utilized for an enlargement of their amplitude-frequency characteristics. The two- or multimode field structure can also be used for an appropriate amplitude-phase aperture distribution and radiation pattern shaping.

2.2.2 Excitation of Cylindrical Cavity Antenna

The cavity modes may also be considered as a superposition of two waves traveling in opposite directions. This approach is especially suitable for electromagnetic fields in cylindrical cavity antennas, which are examined as waveguide pieces.

The cylindrical cavity antenna in Figure 2.3 is obtained from a generalized cylindrical waveguide, with a coordinate system (u,v,z). At the $z = z_1$ cross section, the waveguide is short-circuited by a metal wall S_1 and at $z = z_2$ there is a small plane reflector R, which is partially blocking the aperture S_2 (backfire antenna). The aperture surface can also be made frequency-selective, and in particular, translucent (plane translucent-aperture cavity antenna).

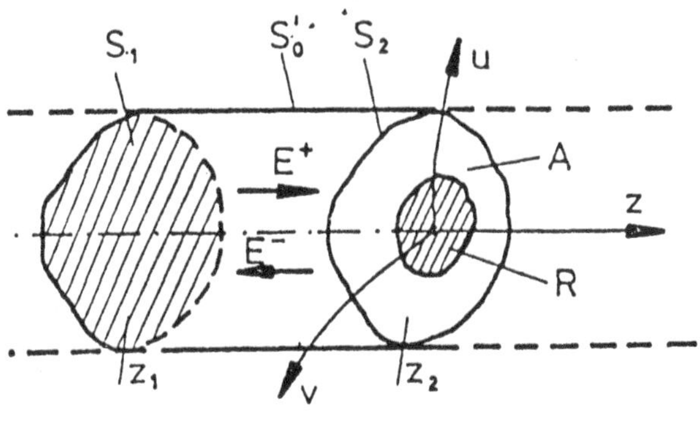

Figure 2.3 Cavity antenna as a piece of a cylindrical waveguide (from [1])

Let us assume that inside the waveguide there are two field modes [1,4–5]:

$$\left| \begin{aligned} \bar{E}_{\pm m} &= (\bar{e}_m \pm \bar{e}_{zm})\, e^{\mp j\beta_m z} \\ \bar{H}_{\pm m} &= (\pm \bar{h}_m + \bar{h}_{zm}) e^{\mp j\beta_m z} \end{aligned} \right. \tag{2.27}$$

$$\left| \begin{aligned} \bar{E}_{\pm n} &= (\bar{e}_n \pm \bar{e}_{zn}) e^{\mp j\beta_n z} \\ \bar{H}_{\pm n} &= (\pm \bar{h}_n + \bar{h}_{zn}) e^{\mp j\beta_n z} \end{aligned} \right. \tag{2.28}$$

where $\bar{E}_{\pm m,\pm n}(\tau, z)$ and $\bar{H}_{\pm m,\pm n}(\tau,z)$ are vector functions of the three coordinates: transverse $\tau \equiv u,v$ and longitudinal z; $\bar{e}_{m,n} \equiv \bar{e}_{m,n}(\tau)$, $\bar{e}_{zm,n} \equiv \bar{e}_{zm,n}(\tau)$. The superscript $(+)$ and $(-)$ indices in wave functions $e^{\pm j\beta_{m,n} z}$ correspond to the traveling waves in positive and negative z directions, respectively; $\beta_{m,n}$ is the phase constant of the mth or nth field mode.

Figure 2.4 illustrates an infinitely long cylindrical waveguide in which current sources $\bar{J}_0$ and $\bar{M}_0$, and aperture source S_0 are located in a region between z_1 and z_2. The field excited by these sources may be expressed as an infinite series in the orthogonal waveguide modes as follows:
For $z > z_2$:

$$\begin{aligned} \bar{E}_2 &= \bar{E}_2^+ + \bar{E}_2^- = \sum_m C_{2m}\bar{E}_m + \sum_m C_{2,-m}\bar{E}_{-m} \\ \bar{H}_2 &= \bar{H}_2^+ + \bar{H}_2^- = \sum_m C_{2m}\bar{H}_m + \sum_m C_{2,-m}\bar{H}_{-m} \end{aligned} \tag{2.29}$$

For $z < z_1$:

$$\bar{E}_1 = \bar{E}_1^+ + \bar{E}_1^- = \sum_m C_{1m}\bar{H}_m + \sum_m C_{1,-m}\bar{H}_{-m}$$
$$\bar{H}_1 = \bar{H}_1^+ + \bar{H}_1^- = \sum_m C_{1m}\bar{H}_m + \sum_m C_{1,-m}\bar{H}_{-m} \tag{2.30}$$

where $C_{1,m}$, $C_{1,-m}$, $C_{2,m}$, $C_{2,-m}$ are modal amplitude coefficients. The reflection coefficients Γ_{1m} and Γ_{2m} are defined

$$\Gamma_{1m} = \frac{C_{1m}}{C_{1,-m}} \qquad \text{at } z_1 \text{ plane} \tag{2.31}$$

$$\Gamma_{2m} = \frac{C_{2,-m}}{C_{2m}} \qquad \text{at } z_2 \text{ plane} \tag{2.32}$$

Using the Lorentz reciprocity lemma and mode orthonormality conditions, we discover

$$C_{1,-m} = \frac{\Gamma_{2m}C_m + C_{-m}}{\Gamma_{1m}\Gamma_{2m} - 1} \tag{2.33}$$

$$C_{2m} = \frac{\Gamma_{1m}C_{-m} + C_m}{\Gamma_{1m}\Gamma_{2m} - 1} \tag{2.34}$$

where

$$C_{\mp m} = \frac{Z_{wm}}{2\int_{S_i} (\bar{e}_m)^2 \mathrm{d}S} \left[\iint_{S_0} [\bar{E}_{\tau 0}, \bar{n}^0](\pm\bar{h}_m + \bar{h}_{zm})e^{\mp j\beta m^z}\mathrm{d}S \right.$$
$$\left. + \int_V \bar{J}_0(\bar{e}_m \pm \bar{e}_{zm})e^{\mp j\beta m^z}\mathrm{d}V - \int_V \bar{M}_0(\pm\bar{h}_m + \bar{h}_{zm})e^{\mp j\beta m^z}\mathrm{d}V \right] \tag{2.35}$$

In (2.35), Z_{wm} is the mth mode waveguide impedance and S_i is the waveguide cross sectional area. Thus, the expansion amplitude coefficients for the electromagnetic field in the cylindrical waveguide-type cavity antenna are determined in terms of the known primary sources and reflection coefficients.

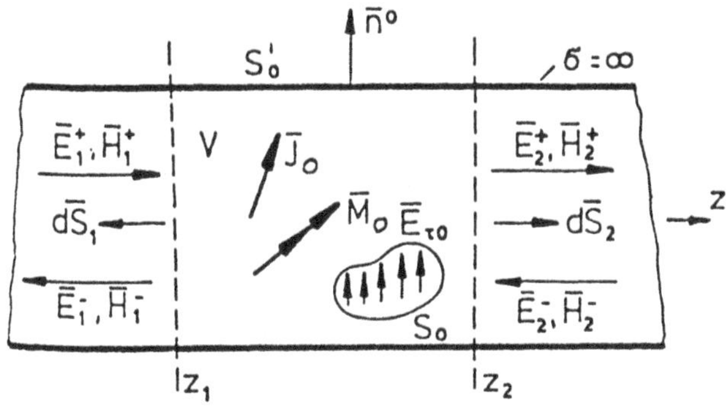

Figure 2.4 General excitation model of an infinite waveguide, with reflections at z_1 and z_2 planes

2.2.3 Moment Method Application to Cavity-Backed Aperture

Figure 2.5(a) represents the general problem of aperture coupling between two regions, (a) and (b). Each region is bounded by an electric (metal) wall S_E, so that region (a) represents the closed cavity and region (b) is the outer semi-infinite space. If the primary excitation (impressed) sources $\bar{J}_0$ and $\bar{M}_0$ are in the closed region, then the configuration shown actually illustrates a cavity antenna with a single radiation aperture.

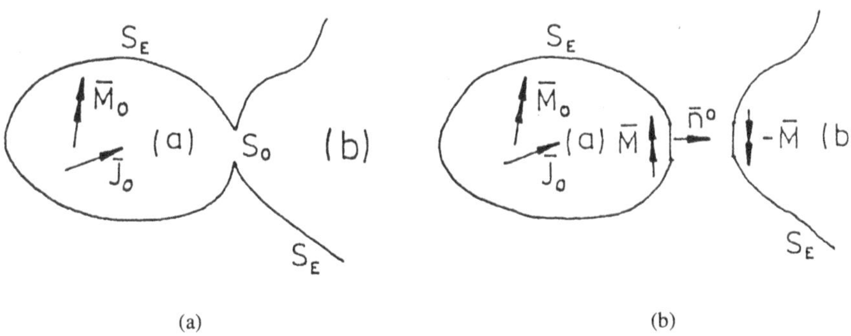

Figure 2.5 Cavity-backed radiation aperture (from [6]): (a) General problem formulation; (b) Original problem divided into two equivalent problems

For a small aperture S_0, the electromagnetic coupling is analyzed using Bethe's diffraction theory [7]. A general formulation for aperture problems is given by Harrington and Mautz [6]. Their approach is based on an application of the equivalence principle, and uses the generalized network formulation in terms of the moment method [8]. Later, this theory was completed and applied by Cheng and Liang [8a,9] to many aperture problems where the coupled regions are dissimilar.

We briefly summarize here the basic procedure for obtaining a generalized network representation of cavity-backed aperture problems. In region (a), the field is excited by the primary sources $\bar{J}_0$ and $\bar{M}_0$, plus the equivalent magnetic current $\bar{M}$

$$\bar{M} = [\bar{n}^0, \bar{E}_\tau] \tag{2.36}$$

over the aperture region, with the aperture covered by an electric conductor; $\bar{n}_0$ is a unit normal vector and $\bar{E}_\tau$ is a tangential component of the primary electric field within the aperture.

In region (b), the field is produced by the equivalent magnetic current $-\bar{M}$ over the aperture, with the aperture covered by an electric conductor. In order to ensure the continuity of the tangential component of the magnetic field across the aperture we required that

$$\bar{H}^a_\tau(\bar{M}) + \bar{H}^b_\tau(\bar{M}) = -\bar{H}^0_\tau \tag{2.37}$$

The magnetic field is given by the following integral equation:

$$\bar{H}^v_\tau(\bar{M}) = -\int_{S_0} \bar{\bar{G}}^V\left(\frac{\rho}{\rho'}\right) \cdot \bar{M}(\bar{\rho}')\mathrm{d}S' \tag{2.38}$$

where (for a plane aperture) $\bar{\rho} = x\bar{x}^0 + y\bar{y}^0$ is a radius vector of the field point and $\bar{\rho}' = \bar{x}'x^0 + \bar{y}'y^0$ is the radius vector of the source point. $\bar{\bar{G}}^V(\rho/\rho')$ is the Green tensor function for the region (v) and the tensor multiplication is marked by the symbol o.

After putting (2.38) into (2.37), the following first-order surface integral equation for the unknown magnetic current $\bar{M}(\bar{\rho}')$ is obtained:

$$\int_{S_0}\left[\bar{\bar{G}}^a\left(\frac{\rho}{\rho'}\right) + \bar{\bar{G}}^b\left(\frac{\rho}{\rho'}\right)\right] \cdot \bar{M}(\bar{\rho}')\mathrm{d}S' = \bar{H}^0_\tau(\bar{\rho}) \tag{2.39}$$

This integral equation has an exact solution only in some special cases and so a computer procedure, known as a moment method, is used for its approximate solution. A set of basic functions for the magnetic current $\bar{M}[\bar{M}_n(\bar{\rho}),\ n = 1, 2, \ldots, N]$ at the aperture S_0 is defined. Let

$$\bar{M} = \sum_{n=1}^{N} V_n \bar{M}_n \tag{2.40}$$

where the coefficients V_n are to be determined. After putting (2.40) into (2.39), a set of N linear equations is obtained:

$$\sum_{n=1}^{N} V_n \bar{H}_\tau^a(\bar{M}_n) + \sum_{n=1}^{N} V_n \bar{H}_\tau^b(\bar{M}_n) = -\bar{H}_\tau^0 \tag{2.41}$$

where the places of the sum and integral are interchanged.

Next, a symmetric scalar product of two vectors $\bar{P}$ and $\bar{Q}$ is defined:

$$\langle \bar{P},\bar{Q} \rangle = \int_{S_0} \bar{P}\bar{Q}\mathrm{d}S \tag{2.42}$$

as are a set of testing functions $[\bar{W}_m(\bar{\rho}),\ m = 1, 2, \ldots, N]$, which in the general case, are different from the basic functions. If $\bar{W}_m = \bar{M}_n$, the solution of the above set of equations follows the method of Galerkin. Taking the symmetric product of (2.41) with each testing function $\bar{W}_m$, and using the linearity of the symmetric product, the following set of linear algebraic equations is reached:

$$\sum_{n=1}^{N} V_n \langle \bar{W}_m,\bar{H}_\tau^a(\bar{M}_n) \rangle + \sum_{n=1}^{N} V_n \langle \bar{W}_m,\bar{H}_\tau^b(\bar{M}_n) \rangle = -\langle \bar{W}_m,\bar{H}_\tau^0 \rangle, \quad (m = 1, 2, \ldots, N) \tag{2.43}$$

Solving this set of linear equations determines the coefficients V_n and the magnetic equivalent current by (2.40). Once $\bar{M}$ is found, the fields and field-related parameters may be computed.

The above solution can be put into matrix notation as follows:

$$[Y^a + Y^b][V] = [I^0] \tag{2.44}$$

where the following admittance matrices are introduced

$$[Y_{mn}^a] = [-\langle \bar{W}_m, \bar{H}_\tau^a(\bar{M}_n) \rangle]_{N \times N} \tag{2.45}$$

$$[Y_{mn}^b] = [-\langle \bar{W}_m, \bar{H}_\tau^b(\bar{M}_n) \rangle]_{N \times N} \tag{2.46}$$

where $[Y_{mn}^a]$ and $[Y_{mn}^b]$ are square $(N \times N)$ aperture admittance matrices for the regions (a) and (b), respectively.

The elements of these matrices have the following network meaning: if $\bar{W}_m = \bar{M}_n$, the elements Y_{mn}^ν $(\nu = a,b)$ represent mutual $(m \neq n)$ or self- $(m = n)$ admittances of the aperture for any unit voltage ($V = 1$). $[I_m^0] = [\langle \bar{W}_m, \bar{H}_\tau^0 \rangle]_{N\times 1}$ and $[V_n]_{N\times 1}$ are $N \times 1$-dimensional matrices, known as *current vector* and *voltage vector*, respectively. Equation (2.44) can be interpreted in terms of generalized networks as two networks ($[Y^a]$ and $[Y^b]$) in parallel with the current source $[I^0]$ as shown in Figure 2.6. The resulting voltage vector

$$[V] = [Y^a + Y^b]^{-1}[I^0]$$

is then the vector of coefficients which give $\bar{M}$ according to (2.40).

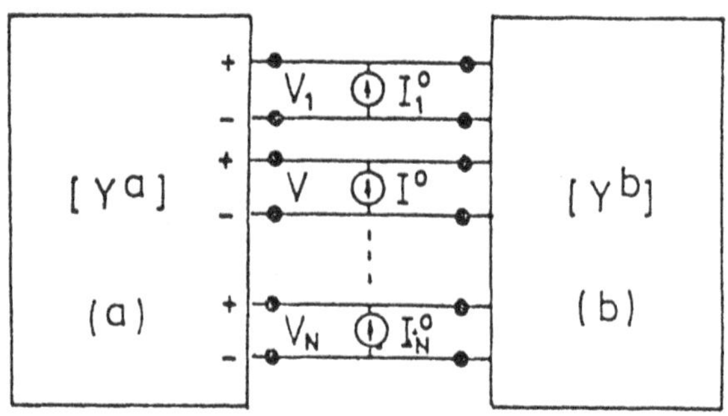

Figure 2.6 Generalized network interpretation of cavity-backed aperture problem (from [6])

The computation of $[Y^a]$ involves only region (a) and computation of $[Y^b]$ involves only region (b). Thus, the problem is divided into two parts, each of which may be formulated independently. Once $[Y]$ is computed for one region, it may be combined with $[Y]$ for the other region. For example, the same aperture admittance matrix for radiation into half-space would be useful for plane-wave excitation of the aperture, waveguide excitation, and cavity excitation.

2.3 CAVITY-BACKED SLOT ANTENNAS

2.3.1 Narrow Rectangular Slot in Large Ground Plane

A very important antenna element, not so much in its own right, but because of its derivatives, is the narrow rectangular slot in a large ground plane. On the basis of the duality principle, it is usually considered as a magnetic dipole with a surface current given by:

$$I_M = \int_V \bar{M}\bar{H}_m^* \mathrm{d}V = \int_{S_0} \bar{M}_S \bar{H}_{m\tau}^* \mathrm{d}S \tag{2.47}$$

where S_0 is the slot area; $\bar{M}_S$ is the surface magnetic current density, dual to the electric current density $\bar{J}_S(\bar{J}_S = [\bar{n}^0, \bar{H}], \bar{M}_S = -[\bar{n}^0, \bar{E}])$, where $\bar{n}^0$ = unit outer normal and $\bar{E}, \bar{H}$ are the electric and magnetic field intensities, on the dipole surface, respectively.

If w represents the slot width, the surface magnetic current I_M is

$$I_M = M_s \cdot 2w = -2E_{\tau 0}w = -2U \tag{2.48}$$

Thus, $U = E_{\tau 0}w$ or I_M represents voltage.

For a ξ-oriented half-wavelength slot (Figure 2.7), the voltage distribution is given by [10]:

$$U(\xi) = -j\frac{\cos \beta\xi}{2Y_s \sin \beta l}\int_{-0.5l}^{+0.5l} H_{\tau 0}(\xi) \cos \beta\xi d\xi \tag{2.49}$$

where

Y_S = the wave slot admittance;
$H_{\tau 0}$ = tangential component of the excitation magnetic intensity;
β = $\dfrac{2\pi}{\lambda}$ = free phase constant; and
l = slot length.

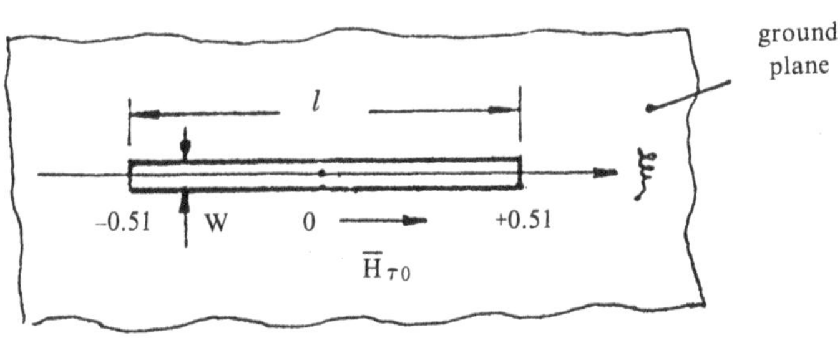

Figure 2.7 Narrow rectangular slot in a large ground plane

The admittance of the idealized, non-radiating slot is a real quantity (i.e., conductance G_S), and is found by:

$$Y_S = G_S = \frac{\ln\left(\dfrac{4l}{w} - 1\right)}{120\pi} = \frac{g_s}{120\pi} \tag{2.50}$$

where

$$g_s = \frac{1}{\pi} \ln\left(\frac{4l}{w} - 1\right) \quad (2.51)$$

Considering the approximate radiation effect, the real phase constant β in the sine function $\sin \beta l$ can be replaced by a complex propagation constant γ_s [10]

$$\bar{\gamma}_S^2 = \beta^2 - jn\eta \quad (2.52)$$

where

$$n = \frac{1}{2\pi g_s} = \frac{1}{2\left(\ln \frac{4l}{w} - 1\right)} \quad (2.53)$$

Because $n\eta/\beta^2 \ll 1$, $\bar{\gamma}$ is written approximately as

$$\bar{\gamma}_S \approx \beta - j\frac{n\eta}{2\beta} \quad (2.54)$$

or, after setting $\eta = \eta' + j\eta''$,

$$\bar{\gamma}_S = \left(\beta + \frac{n\eta''}{2\beta}\right) - j\frac{n\eta''}{2\beta} = \bar{\beta}_S - j\bar{\alpha}_S \quad (2.55)$$

Here $\bar{\beta}_S = \beta + \Delta\beta$ is a phase constant of the wave in the slot and $\Delta\beta$ is a phase correction due to wavelength shortening. The term $\bar{\alpha}_S$, given by

$$\bar{\alpha}_S = \frac{n\eta'}{2\beta} \approx -\frac{1.22}{\pi}\beta n \quad (2.56)$$

is an attenuation constant, which gives an account of the slot radiation phenomenon.

For the half-wavelength slot $(1/\lambda) = 0.5$), $\sin(\bar{\gamma}l) \approx jn(\eta' l/2\beta)$. Then, for the basic resonance (and small deviations from the resonant frequency), the half-wavelength narrow slot has a cosine voltage distribution given by:

$$U(\xi) = U_0 \cos \beta\xi \quad (2.57)$$

with an amplitude given by:

$$U_0 = -\frac{\omega\mu_0}{2\beta g_s \bar{\alpha}_s l} \int_{-0.5l}^{+0.5l} H_{\tau 0}(\xi) \cos \beta\xi d\xi \tag{2.58}$$

where $H_{\tau 0}$ is a tangential magnetic field intensity, which excites the slot from one side of the ground plane. When the slot is excited by a uniform plane wave, $H_{\tau 0}(\xi) = H_{\tau 0}$ = const. and

$$U_0 = U_{max} = -\frac{\omega\mu_0}{\beta^2 g_s \bar{\alpha}_s} \sin\left(\beta \frac{l}{2}\right) \tag{2.59}$$

If there are excitation fields on the two ground plane sides, the magnetic intensity in the slot is

$$H_{\tau 0} = \frac{1}{2}(H_\tau^{(1)} - H_\tau^{(2)}) \tag{2.60}$$

where $H_\tau^{(1)}$ and $H_\tau^{(2)}$ are tangential components in the vicinity of the slot in region (1) and region (2), respectively.

The slot shown in Figure 2.7 can be excited by a two-wire line in all ways except electromagnetically. According to the duality principle, a center-fed slot may be viewed as a magnetic dipole which is complementary to the center-fed electric dipole. The voltage distribution along the two-wire fed slot is also given by the cosine formula (2.57), where U_0 is a slot input voltage. One of the most important results in antenna theory is the relationship between the electric dipole and slot input impedances, which is

$$Z_{in}\,(\text{dipole}) \cdot Z_{in}\,(\text{slot}) = (120\pi)^2/4 \tag{2.61}$$

if the dipole and slot are located in the air space.

The same equation is valid for their radiation resistance. Because R_{rad}^{dipole} of the infinitely thin dipole is 73 ohms, using (2.61), R_{rad}^{slot} is calculated at 486 ohms.

There are few practical applications of slots cut in a large ground plane which radiate into both half-spaces. However, if the slot is backed at one side by a resonant cavity, its radiation and impedance properties are strongly affected.

2.3.2 Cavity-Backed Rectangular Slot

A coaxially-fed cavity-backed slot in a large ground plane is illustrated in Figure 2.8. If the cavity dimensions are large enough, the electric field (or voltage) distribution in the slot is approximately the same as before backing by a cavity. With the

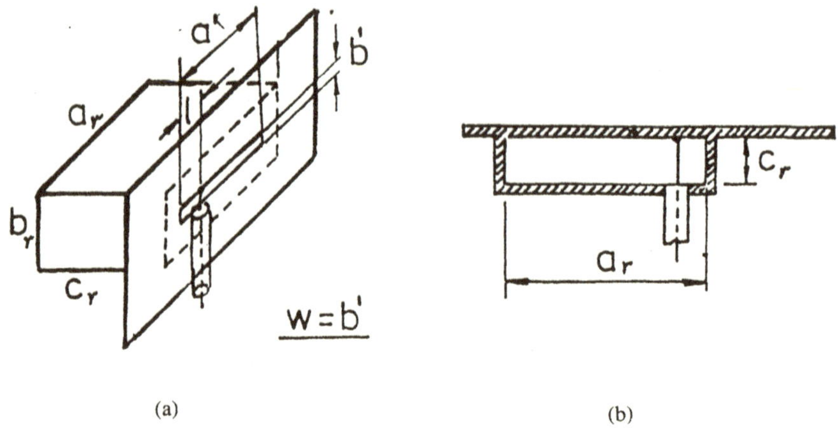

Figure 2.8 Cavity-backed coaxially-fed slot: (a) with a cable situated on the ground plane; (b) with a cable stretched through the back cavity wall

current distribution in the radiating face of the ground plane the same as though the cavity were not present, the admittance of the slot is

$$Y = Y_{\text{slot}} + Y_{\text{cavity}} = 2Z_{\text{dipole}}/(120\pi)^2 - j(a/b)(\beta/2\omega\mu_0) \cot \beta C_r \quad (2.62)$$

We see, then, that the slot conductance is half of what it was before backing, and the cavity input admittance is imaginary.

If the cavity dimensions and the slot are properly adjusted, the susceptible part of Y_{slot} can cancel Y_{cavity}, so that the input impedance becomes a resistive load of $(120\pi)^2/2R_{\text{dipole}}$. For a one-wavelength center-fed slot, the input resistance is on the order of 50 ohms (the characteristic impedance of the standard coaxial cable) for a properly adjusted slot length. The input impedance of the tuned (in resonance) half-wavelength slot, however, is in the range of 1000 ohms, and so some kind of impedance transformation is necessary. For example, using a T-bar transformer, an input VSWR under 1.5 over 30% frequency bandwidth can be achieved. The excitation of a cavity-backed slot directly in its ground plane has some advantages: the slot is tuned to resonance for a small cavity depth; and transformation problems can be overcome simply by moving the feeder connection points aside from the slot center.

A rather complete electromagnetic model of an antenna array of cavity-backed slot elements is made in [11–13]. An integral equation in relation to the unknown electric field distribution in a single cavity-backed slot is worked out and solved. A sketch of this antenna element is shown in Figure 2.9.

From the computer solution of the integral equation mentioned previously, valuable design data are obtained. Figure 2.10 illustrates the relationship between

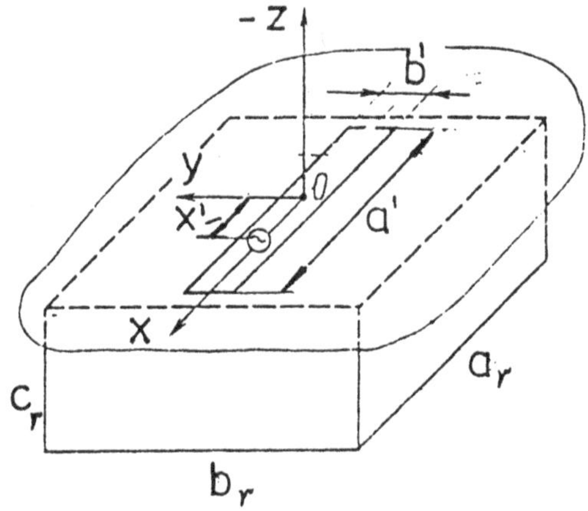

Figure 2.9 Cavity-backed slot geometry (from [11])

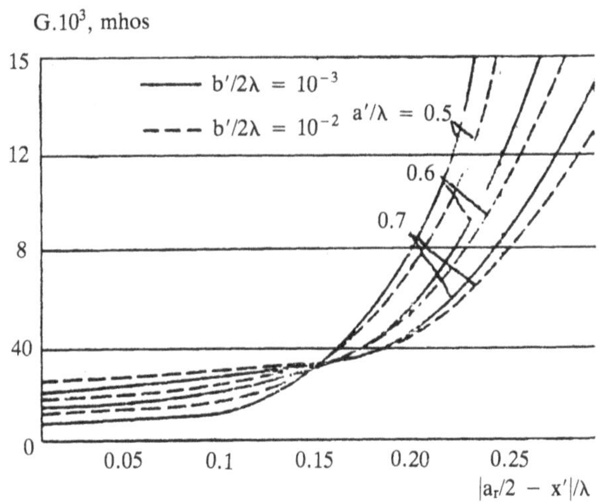

Figure 2.10 Input antenna conductance as a function of excitation point position (from [13])

the input antenna conductance G and the excitation point position for different slot widths and lengths. The resonant condition $B_{slot} + B_{cavity} = 0$ is kept.

Figure 2.11 shows a family of graphs for the relation between the cavity width and depth. Figure 2.12 gives VSWR, as a function of frequency, for three different slot widths.

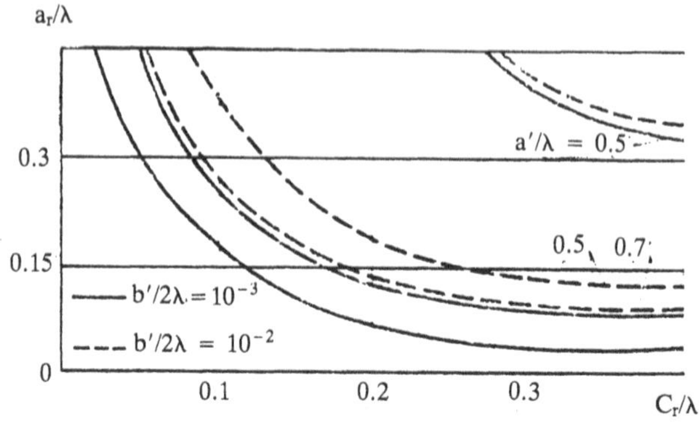

Figure 2.11 Relationship between cavity width and depth at resonance (from [13])

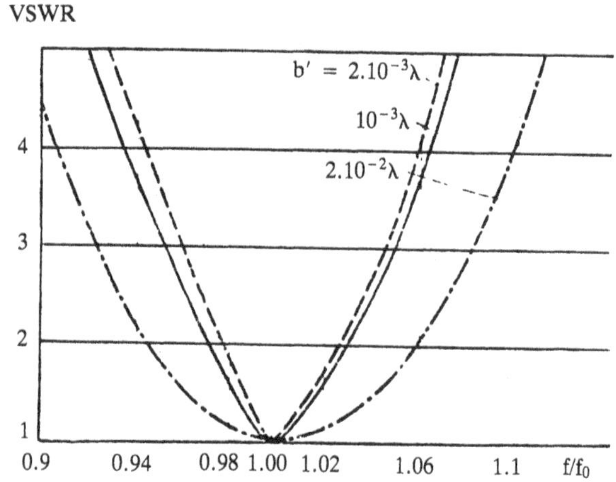

Figure 2.12 VSWR, as a function of frequency, for three different slot widths (from [13])

Exemplary practical designs of cavity-backed slots are shown in Figure 2.13: a rectangular cavity-backed slot, fed by a microcoaxial cable; and a crossed-slot antenna with circular polarization, fed by a system of microstrip lines.

The cavity-slot antenna, fed by a two-wire cable directly at the slot ground plane, is more suitable for the lower microwave frequency bands. At higher frequencies another cavity-slot design is a better choice (Figure 2.14). The slot is excited indirectly, by a cavity field. The cavity itself is coaxially-fed either from a probe, as shown here, or from a small loop. The use of dielectric or ferrite loading in the cavity permits the reduction of cavity volume and aperture size, at the expense of bandwidth and efficiency.

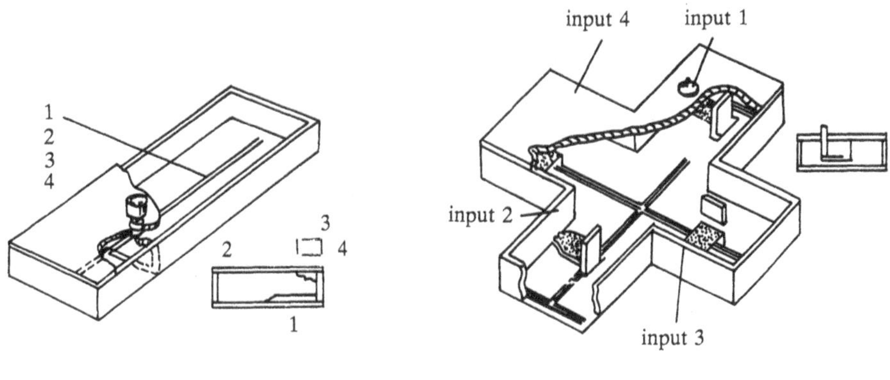

(a) (b)

Figure 2.13 Practical designs of cavity-backed slot antennas (from [13]). (a) rectangular cavity-backed slot (1-rectangular slot, 2-cavity, 3-high-frequency joint, 4-microcoaxial cable); (b) crossed-slot antenna (1-crossed slots, 2-crossed cavity, 3-microstrip line, 4-short circuit, 5-coaxial transition)

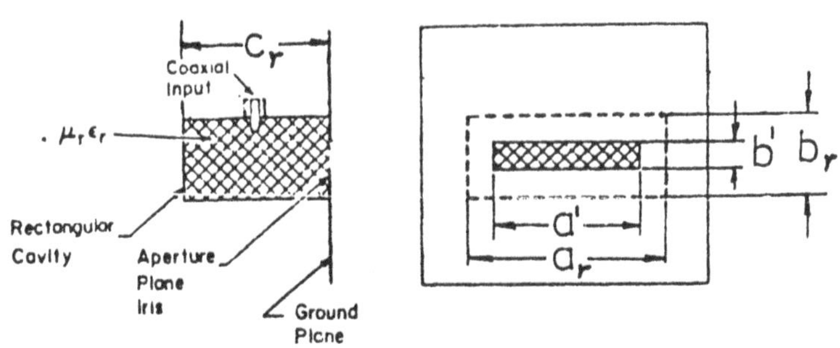

Figure 2.14 A loaded rectangular cavity slot antenna (from [14])

Adams [14] has performed an analysis of the loaded rectangular cavity slot, using variational methods in conjunction with simplified equivalent circuit theory. Actually, we have rigorously analyzed a semi-infinite waveguide radiator, and thus obtained a variational expression for the aperture admittance. Then, using an equivalent circuit and perturbational techniques, we have taken an account of the probe feed configuration, the finite size of the cavity, and the effect of the aperture plane iris.

The theoretical approach described here has been applied for a calculation of the basic parameters of a particular cavity slot antenna construction. It consists of a finite loaded cavity, backed by a short circuit and fed by a coaxial probe. The cavity is radiating out through one end of the cavity (i.e., $a_r = a'$, $b_r = b'$) (Figure 2.14). Experiments with a number of antenna versions are very interesting. Three of these antennas, loaded with air, ferrite powder, and solid ferrite, respectively, are shown from left to right in Figure 2.15(a). Figure 2.15(b) shows a closeup of the antenna model loaded by solid ferrite.

(a)

(b)

Figure 2.15 Loaded cavity antenna models (from [14])

Experimental measurements of resonant frequency, efficiency, bandwidth, and radiation pattern have been performed and compared with theoretical predictions. Table 2.1 summarizes the basic antenna characteristics, showing the size and weight reduction accomplished through the use of material loading. In addition, the characteristics of a solid dielectric loaded antenna (not shown in Figure 2.15) are given in the table. The dielectric and ferrite powder antenna models have equal dimensions. The loading materials have, at 20° C and 300 MHz, the following characteristics: ferrite powder—$\varepsilon_r = 8.3$, $\mu_r = 2.2$, and magnetic $Q = 30$; solid ferrite—$\mu_r = 6.63$, $\varepsilon_r = 12.6$, and magnetic $Q = 30$; and solid dielectric—$\varepsilon_r = 10$ and electric $Q > 1000$.

Table 2.1 (from [14])

Loading	*Air*	*Ferrite Powder*	*Ferrite Solid*	*Dielectric*
Size (inches)	30 by 7-1/2 by 10	12 by 3 by 4	5 by 2 by 1-1/2	12 by 3 by 5
Volume (cubic inches)	2250	144	15	180
Bandwidth				
at VSWR = 3		22 MHz	19 MHz	10 MHz
at VSWR = 6		50 MHz	34 MHz	18 MHz
Efficiency	90 per cent	65 per cent	30 per cent	85 per cent
Directivity	5.8 dB	5 dB	5 dB	5 dB
Weight (pounds)	25-3/4	16-3/4	3.6	14.5
Frequency (MHz)	300	320	352	316

From Table 2.1 it is evident that cavity loading really gives rise to antenna miniaturization. The linear dimensions are reduced significantly by factors of from 2.1 to 6.1. This size reduction is accomplished at the expense of bandwidth and efficiency, which is tolerable in many cases. On the other hand, the operating frequency of a ferrite-loaded slot antenna can be shifted through the application of magnetic bias. By applying a magnetic field parallel to the *b* dimension, it is possible to change the resonant frequency from 243 to 320 MHz.

Practical design and optimization results for a dielectric-filled, cavity-backed slot antenna have been published by Lagerlof [15]. The antenna configuration is the same as that of Figure 2.14. The disturbance of the field in the cavity and slot due to the excitation probe is supposed to be negligible. Assuming a narrow slot, an approximate electric field distribution in the slot is chosen, as follows:

$$\bar{E}(x,y,0) = \bar{x}^* E_0 f(x) \cos\left(\pi \frac{y}{a'}\right) \tag{2.63}$$

where $f(x)$ is the transverse field variation, given by a quasi-static approximation.

The slot self-admittance Y_s can be defined through the complex power S radiated by the slot into the cavity and into the outer space:

$$S = \frac{1}{2}|V|^* = \frac{1}{2}|V|^2 Y_s^* \tag{2.64}$$

where V is the maximum complex value of the voltage across the slot at its center. Thus, the slot admittance may be resolved into two parts:

$$Y_s = Y_{si} + Y_{so} \tag{2.65}$$

where Y_{si} is the admittance looking into the cavity and Y_{so} is the admittance looking at outer space.

In Lagerlof's paper Y_{si} is given by the following expression:

$$\begin{aligned} Y_{si} = {} & \frac{32\pi^2 a'^2}{k_o a_r b_r b'^2} \sqrt{\frac{\varepsilon_0}{\mu_0}} \sum_{m=0}^{\infty} \frac{k^2 - k_{my}^2}{[\pi^2 - k_{my}^2 a'^2]^2} \cdot \cos^2(k_{my} a'/2) \\ & \cdot \left\{ \frac{b^2}{4} \left(\frac{\coth(jk_{mOz} c_r)}{k_{mOz}} + j\frac{b}{\pi} \ln \frac{1}{\sin\left(\frac{\pi}{2}\frac{b'}{b_r}\right)} \right) \right. \\ & \left. + 2 \sum_{n=1}^{\infty} \frac{\sin^2(k_{nx} b'/2)}{k_{nx}^2} \left(\frac{\coth(k_{mnz} c_r)}{k_{mnz}} - \frac{j}{k_{nx}} \right) \right\} \end{aligned} \tag{2.66}$$

where the propagation constants are

$$k_{mnz} = \pm\sqrt{k^2 - k_{my}^2 - k_{nx}^2} \tag{2.67}$$

the constant is positive when k_{mnz} is real, and negative when k_{mnz} is imaginary;

$$k = \frac{2\pi}{\lambda}\sqrt{\varepsilon_r} = k_0\sqrt{\varepsilon_r}; \quad k_{nx} = 2n\frac{\pi}{b_r}; \quad \text{and} \quad k_{my} = (2m+1)\frac{\pi}{a_r}$$

The real part of Y_{so}, called a radiation conductance, is calculated by

$$G_{so} = \frac{16}{3\pi}\sqrt{\frac{\varepsilon_0}{\mu_0}}\left(\frac{a'}{\lambda}\right)^2\left[1 - 0.37392\left(\frac{a'}{\lambda}\right)^2 + 0.12967\left(\frac{a'}{\lambda}\right)^4 - 0.03390\left(\frac{a'}{\lambda}\right)^6\right] \tag{2.68}$$

The above expression is valid for very narrow slots (i.e., $b' \to 0$).

The resonance frequency of the cavity-backed slot is found from the equation $B_s = 0$, and the quality factor Q at resonance frequency is computed by using the formula:

$$Q = \frac{\omega_0}{2G_{so}} \frac{\partial B_s}{\partial \omega}\bigg|_{\omega = \omega_0} \tag{2.69}$$

The Q factor and resonance frequency are strongly dependent on the dimensions of the slot and the cavity. By varying two dimensions of the cavity, a minimum Q factor can be obtained, while the resonance frequency is kept constant by adjusting the third dimension. For the minimum Q factor, the cavity antenna has a maximum bandwidth.

In Figure 2.16, the curves illustrate the relationships between Q-factor and cavity dimensions, for $\varepsilon_r = 2.5$. At a specific value of the cavity width and depth, the Q factor has a minimum. The curves in Figure 2.17 are useful for optimum design. Cavity dimensions are shown in terms of slot lengths to achieve a prescribed resonant frequency. Furthermore, these cavity dimensions give the smallest possible Q factor for the given slot dimensions and the relative dielectric constant of the cavity filling at that resonance frequency.

The standard cavity-backed slot antenna is inherently a narrowband device. Some slot antenna designs, which have much wider frequency bandwidth, are briefly described below. Figure 2.18 shows two varieties of the so-called *tee-fed cavity-backed antennas* with a slot size equal to the cavity cross section of the cavity. In part (a) of this figure, a round T-bar coaxial-to-waveguide transition, comprised of a tapered probe P connected to the center point of a stub S, and shorted at each end with the side cavity walls, is shown. The cavity-backed T-bar fed slot in Figure 2.18(b) is excited coaxially at the top wall, and the center conductor is flared to a flat T-bar, which extends to touch the side walls. An experimentally-optimized design has an input bandwidth of 2.3:1 for VSWR = 2.0 [16,17].

In 1973, Mayes [18] patented a monopole-slot antenna, which uses a broadband, unidirectional design. It consists of two resonant dual-type elements (monopole and slot), fed simultaneously from the same transmission line. The slot, backed

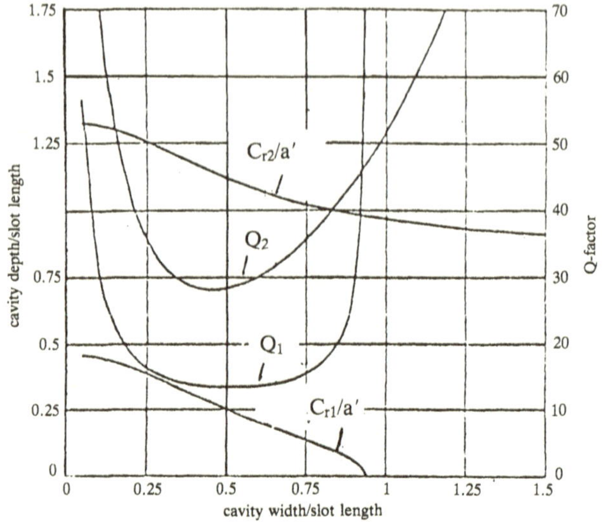

Figure 2.16 Cavity depth required to keep resonant frequency constant when cavity width is varied, plus corresponding Q factor (ε_r = 2.5, a'/λ = 0.45, b'/λ = 0.05, a_r/a' = 1.2) (from [15])

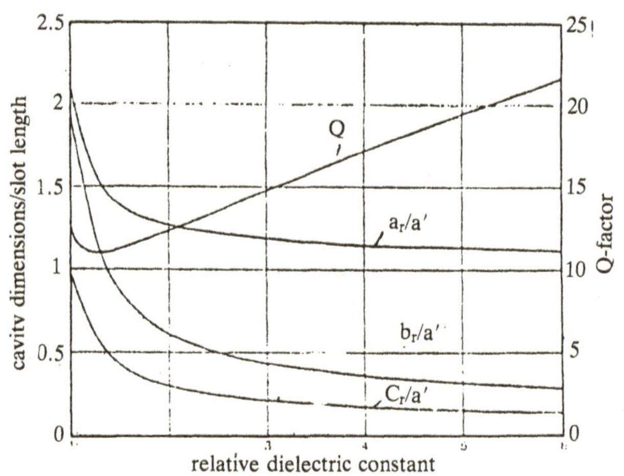

Figure 2.17 Cavity dimensions in slot lengths at optimal Q factors, as functions of the relative dielectric constant (a'/λ = 0.45, b'/a' = 0.05) (from [15])

by a cavity, is fed via a microstrip line, which is normal to the slot. One port is terminated in a resistor approximately equal in value to the image impedance. A sketch of the monopole plus cavity-backed slot antenna is shown in Figure 2.19. This antenna has reasonably constant impedance, and a radiation pattern over 10:1

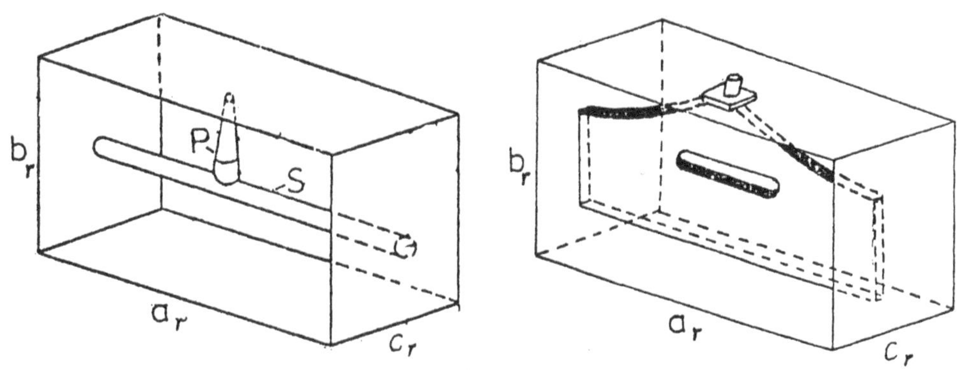

(a) (b)

Figure 2.18 Tee-fed cavity-backed slot antennas: (a) with round T-bar configuration; (b) with flat T-bar configuration

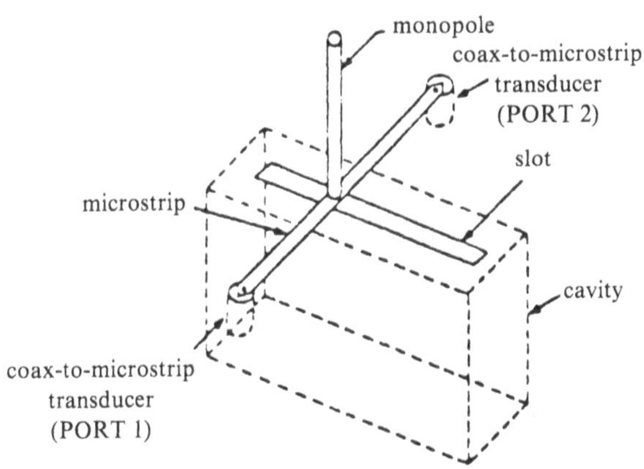

Figure 2.19 Monopole-slot antennas—monopole plus cavity-backed slot, fed from the same line (from [18])

frequency band, at the expense of efficiency at the lower frequencies. In the horizontal plane, the pattern is unidirectional cardioid; its maximum direction may be switched 180°, depending on which of the two ports is excited [19].

Clavin, *et al.* [20] have proposed antenna elements consisting of a cavity or two wire-fed slots and two symmetrically-placed, parasitic monopoles (Figure 2.19). The same structure has been applied to the cavity-backed cross-slot antenna, to obtain circular polarization over a wider angular range and smaller coupling between the elements in a circularly-polarized array (see Figure 2.20) [21].

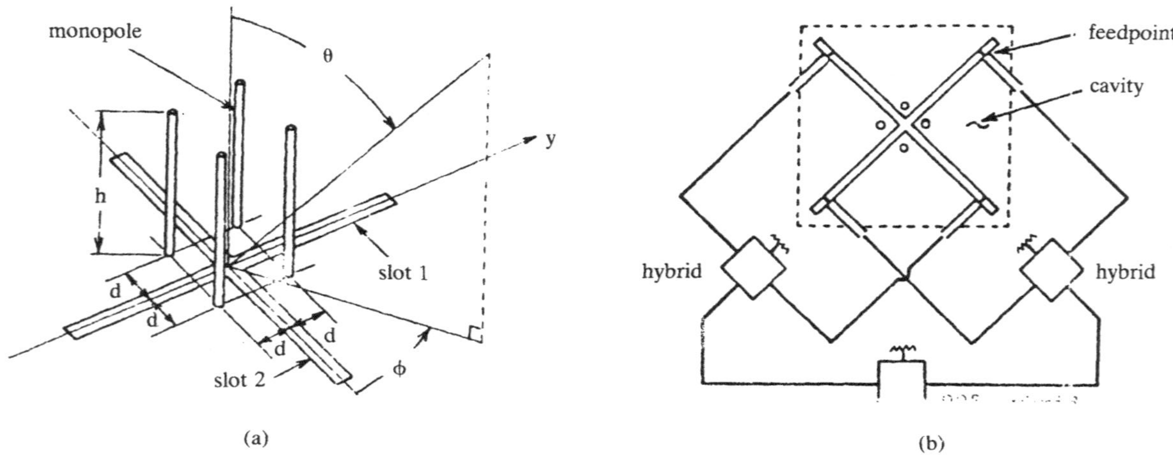

Figure 2.20 Cavity-backed crossed-slot antenna with coupled monopoles (from [21]): (a) Antenna configuration; (b) Feed network for circular polarization

So far, we have considered a rectangular cavity-backed slot in a radiation mode. Using the equivalence principle and generalized network formulation in terms of the moment method (described in Section 2.2.3), Liang and Cheng [9,22] have analyzed the electromagnetic coupling of an incident plane wave through a slot-aperture backed by lossy rectangular cavity (Figure 2.21).

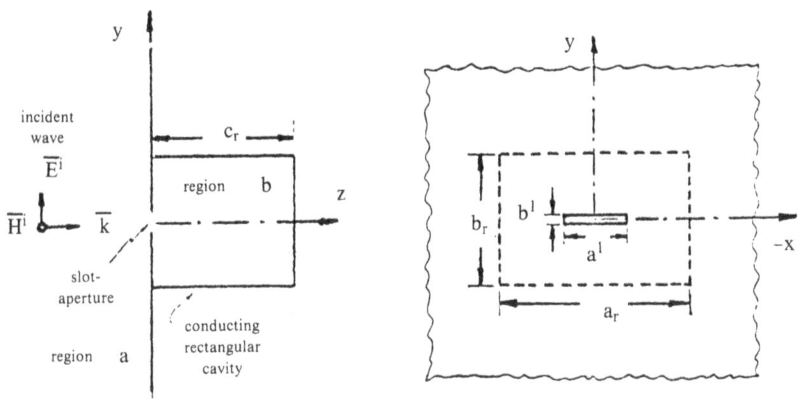

Figure 2.21 Plane wave incident normally on slot-aperture backed by cavity (from [22])

Initially, the analysis is started from the single-mode representation of the equivalent magnetic current M. According to (2.36),

$$\bar{M} = [\bar{z}^0, \bar{E}^i] = V_0 \bar{M}_0 \tag{2.75}$$

over the aperture region. In this special case, the matrix equation (2.44) is written

$$(Y^a + Y^b)V_0 = I \tag{2.76}$$

where Y^a and Y^b are generalized admittances at the aperture in region a and region b, respectively, and I is the generalized current source. An equivalent network for the cavity-backed slot-aperture is shown in Figure 2.22(a), in which the total admittance seen by the current source is

$$Y_0 = Y^a + jB + A_{01}^2 Y_{in} = G^a + j(B^a + B) + A_{01}^2 Y_{in} \tag{2.77}$$

For convenience, we let $B' = B^a + B$ (i.e., B' is the sum of the generalized half-space susceptance of region a and the transformed susceptance due to the non-propagating high-order modes in region b). Depending on aperture size, B' can be

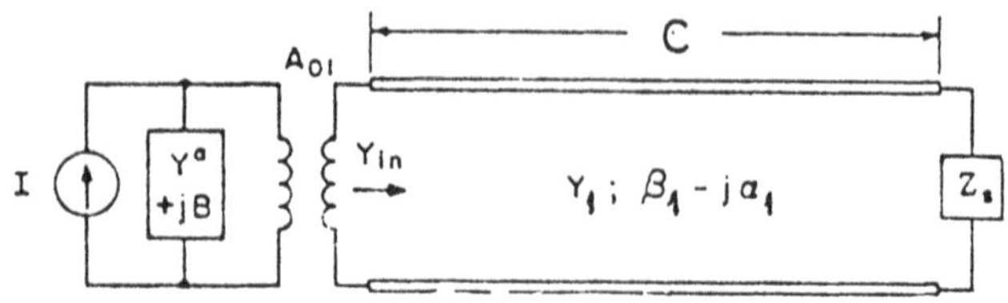

Figure 2.22 (a) Generalized network with terminated lossy transmission line (from [22])

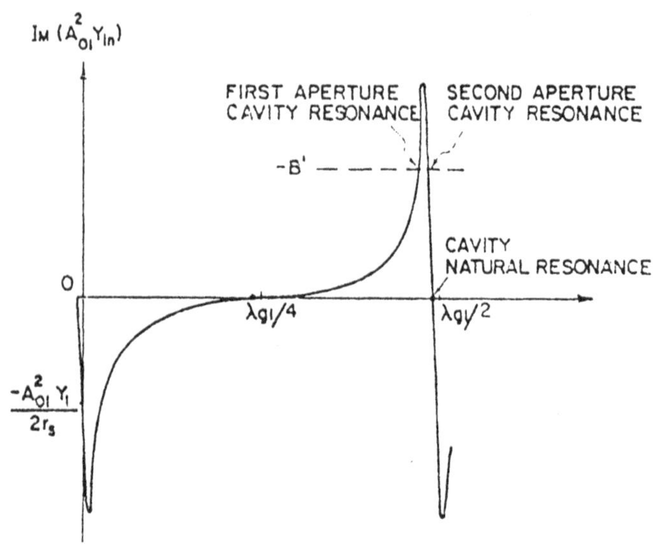

Figure 2.22 (b) Double aperture cavity resonances [22]

either a positive or a negative quantity. Then, (2.77) can be rewritten as:

$$Y_0 = [G^a + Re(A_{01}^2 Y_{in})] + j[Im(A_{01}^2 Y_{in}) + B'] \tag{2.78}$$

In Figure 2.22(b), $A_{01} = n_2/n_1$ is a turns-ratio of the idealized transformer;

$$Z_s = (1 + j)R_s = (1 + j)\sqrt{\frac{\pi f \mu}{\sigma}}$$

represents the surface impedance of the end wall (σ = conductivity of the cavity material); Y_1 = real characteristic admittance of the dominant mode; $\gamma_1 = \beta_1 - j\alpha_1(\beta_1 = 2\pi/\lambda_{g1}$ = dominant-mode phase constant and α_1 = dominant-mode attenuation constant, due to power loss in cavity walls).

In (2.78), G^a is the radiation conductance and Y_{in} is given by:

$$Y_{in} = Y_1\left[\frac{1 + jz_s \tan(\beta_1 c_r - j\alpha_1 c_r)}{z_s + j \tan(\beta_1 c_r - j\alpha_1 c_r)}\right] \tag{2.79}$$

where

$$z_{S'} = Z_s Y_1 = (1 + j)R_s Y_1 = (1 + j)r_s$$

is the normalized terminating impedance and $r_s = R_s Y_1$. The aperture cavity resonances are found from $Im(Y_0) = 0$, or

$$Im(A_0^2 Y_{in}) = -B' \tag{2.80}$$

A parameter t is introduced such that

$$t = \frac{A_{01}^2 Y_1}{2r_c|B'|} \tag{2.81}$$

$$r_c = r_s + \alpha_1 c$$

Further analysis shows that apart from the cavity's natural resonance, two types of aperture-cavity resonance may occur, provided a parameter $t \geq 1$ exists. The first type of aperture-cavity resonance is parallel, and leads to high aperture and cavity fields (Figure 2.23(a)). The second type of aperture-cavity resonance is *series,* which differs little from the natural resonance, both in resonance frequency and cavity field amplitude (Figure 2.23(b)). At natural resonance, maximum power penetration takes place at $t = 0.5$. Thus, a practical condition for avoiding the parallel-type aperture-cavity resonance, as well as restraining the power penetration, is to make $t < 0.5$. The high-order mode contributions are relatively unimportant, except in the immediate vicinity of the slot-aperture.

2.3.3 Cavity-Backed Slot Arrays

Designs for microwave antenna applications involving slots usually unify the feeding and radiating structures by placing the slots on one of the walls of a rectangular waveguide. These resonant waveguide-fed slot arrays may be interpreted as a special type of cavity antenna. Because they are described well in the literature, we will not cover them. Likewise, we will not cover arrays consisting of individual cavity-backed slots.

In certain cases, groups of slots or even whole slot arrays can be backed and illuminated by a single cavity resonator. As far as we know, this idea originated in 1963 [23]. Its main point is the utilization of the plane field configuration inside a

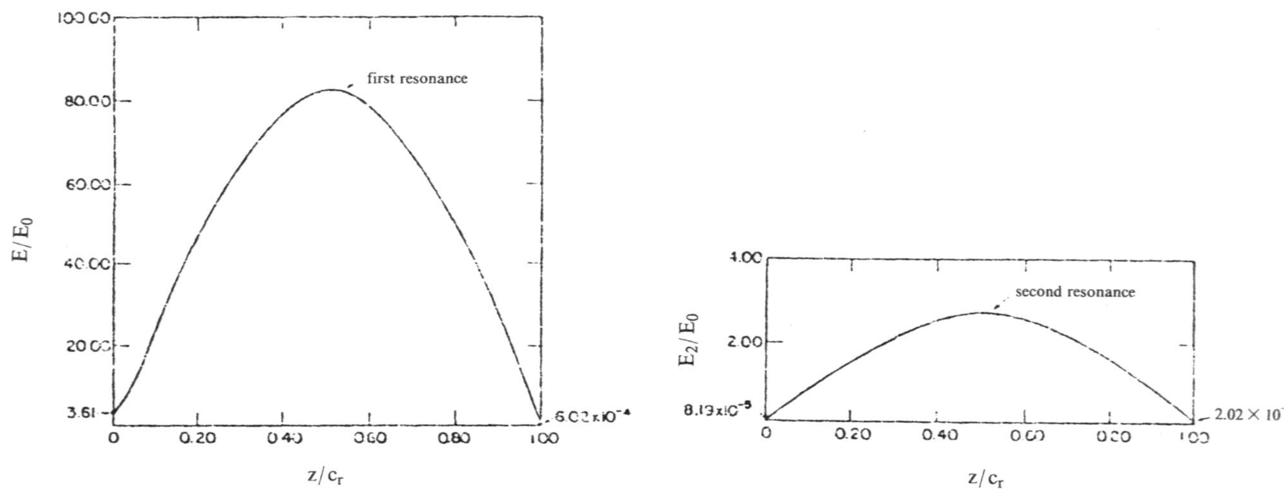

(a)

(b)

Figure 2.23 Dominant-mode electric field distribution inside a copper cavity at aperture-cavity resonances [22]: (a) At first aperture-cavity resonance, and (b) at second aperture-cavity resonance single-mode cavity to directly excite the metal plate comprising an array of small openings (holes, slots, cut-off guides, *et cetera*). The electromagnetic coupling (through this array) with the outer space should be sufficiently small to give the cavity resonator a proper Q factor for one selection among the different modes of the resonator. The paper cited above will be examined in detail in Chapter 3.

A broadside array consisting of narrow resonant slots backed by a single rectangular cavity has been proposed and examined by Paoloni [24]. This array is illustrated in Figure 2.24. The cavity, filled with a low-loss dielectric, supports the $TE_{0\eta\zeta}$ mode. The upper *y-z* wall can be divided into $\eta \cdot \zeta$ imaginary cells, defined by the node points of the electric field given by:

$$\bar{E} = E_0 \sin(\eta\pi y/b_r) \sin(\zeta\pi z/c_r)\bar{x}^0 \tag{2.82}$$

Each slot is positioned within its cell a distance y_s from the edge. The simplest array is one for which the cell width equals the slot length (i.e., $c_r/\zeta = \lambda_0/2$ and $b_r/\eta = \lambda/\sqrt{\varepsilon_r}$), where ε_r is the relative dielectric constant of the cavity filling and λ_0 is the free space wavelength. In this manner, a *broadside array* is produced, in which the slot elements are driven by the standing waves within the cavity.

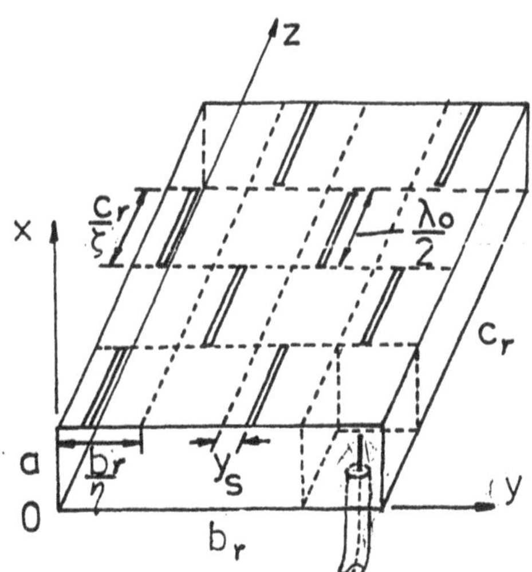

Figure 2.24 Cavity-backed slot array—slot locations on upper cavity wall (from [24])

Because the numbers η and ζ are chosen to be large, the cavity is highly overmoded, and it is necessary to ensure the presence of only the $TE_{0\eta\zeta}$ mode, to provide the correct excitation of the array. This is achieved by the inclusion of a filter, consisting of metallic pins which short circuit the top and bottom walls at all the node points. Thus, resonances with mode numbers not equal to $(0,\eta,\zeta)$ or their harmonics cannot exist within the cavity, and the filter will not affect the resonance frequency of the $TE_{0\eta\zeta}$ mode.

By neglecting the mutual coupling between the radiating slots, internal cavity losses and field disturbance, the following approximate expression for the quality factor is found:

$$Q \approx Q_r = 2\pi\eta_0 G \frac{\eta\zeta}{m_s}\left(\frac{a_r}{\lambda_0}\right)\frac{\varepsilon_r}{(\varepsilon_r - 1)^{3/2}}\sec^2(\eta y_g/b_r) \tag{2.83}$$

We suppose that the cavity is fed by a coaxial-waveguide transition (i.e., an electric probe that enters the cavity enclosure from the lower large wall and is placed in the center of any one cell) as shown in Figure 2.24.

At the TE_{0np} mode resonance frequency, $\omega = \omega_{0\eta\zeta}$ (the input impedance) should have only a resistive component, given by:

$$R = \frac{Q}{\eta\zeta}\frac{W_0}{2\pi^3}\frac{(\varepsilon_r - 1)^{1/2}}{\varepsilon_r^2}\left(\frac{\lambda_0}{a_r}\right)\sec^4(\pi\sqrt{\varepsilon_r}\, g/\lambda_0) \tag{2.84}$$

The input reactive component is due to the harmonic modes only. It is small because, for the working frequency, the harmonic modes are much smaller in amplitude comparing with TE_{0np} mode.

In the above formula G is cavity-backed slot conductance, which equals 1.03×10^{-3} mho if the slot length is approximately half a free-space wavelength; $W_0 = 120\pi$ ohms = free-space characteristic impedance and g = probe length.

The input resistance is inversely proportional to the number of slots, and depends upon the probe length and the position of each slot within its cell. An antenna of given geometry, therefore, can be matched to the feeder by selection of the coupling probe length and the slot location. An experimental antenna has been designed for the TE mode with $\xi = 0$, $\eta = 9$ and $\zeta = 11$, at 10 GHz resonant frequency. The cavity, loaded with polyethelene of $\varepsilon_r = 2.3$, has dimensions as follows: $a_r \times b_r \times c_r$ = 8 mm × 135 mm × 145 mm. Each slot is 15 mm long and 1 mm wide.

Two arrays have been studied experimentally. The first one, shown in Figure 2.24 contains 18 slots. The coupling probe is positioned in the center cell and is connected to the 50-ohm coaxial cable. The optimum probe length is 6 ± 0.1 mm. At the operating frequency (9.84 GHz), the VSWR is 1.05 and remains below 2.0 for a frequency deviation of 30 MHz on either side of resonance.

The second antenna model contains 32 similarly-arranged elements and is placed on an identical cavity. The measured gain of this antenna is 18 dB, and its radiation patterns are shown in Figure 2.25 (Figure 2.25(a)—*H*-plane radiation pattern and Figure 2.25(b)—*E*-plane radiation pattern).

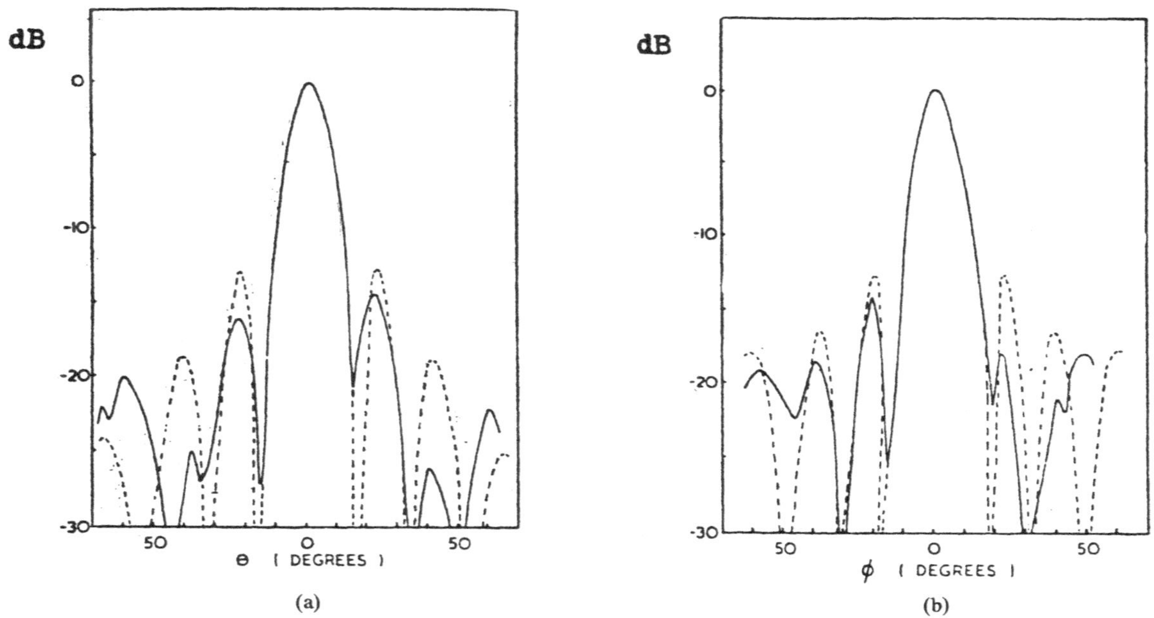

Figure 2.25 Radiation patterns of 32-element slot array backed by dielectric-loaded cavity (from [24]) (solid line—measured, dashed line—calculated): (a) *H*-plane radiation pattern; (b) *E*-plane radiation pattern

Another design of a cavity-backed slot antenna array is studied both theoretically and experimentally in [1,25]. Its geometry is shown in Figure 2.26. In the general case, the array, consisting of $k = k_x \times k_y$ equal slots, each 21 in length and 2Δ in width, is illuminated by a rectangular cavity of dimensions $a_r \times b_r \times c_r$. The cavity is excited from a TE_{10} mode rectangular waveguide by means of end-wall slot, $2l_0$ in length and $2\Delta_0$ in width. The array plane is $z = c_r$; d_x and d_y are the element lattice spacings. For the sake of convenience, the slots are numbered as follows: the exciting slot at the $z = 0$ plane with $j = 0$, and the radiating slots at the $z = c_r$ plane with $j = 1, 2, 3, \ldots, k$. The coordinates of the jth slot center are (x_j, y_j).

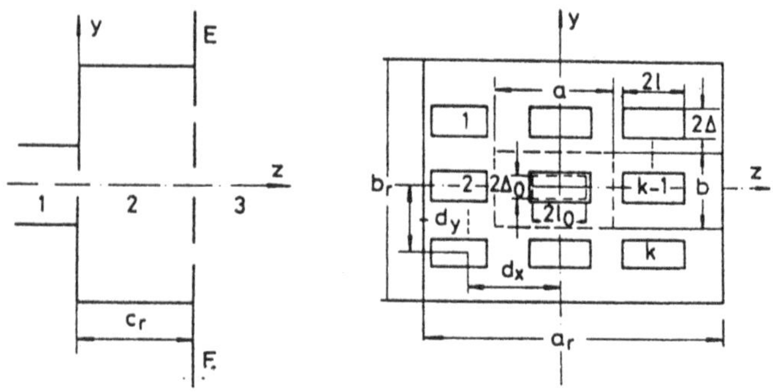

Figure 2.26 General antenna geometry (from [25])

For the antenna analysis, the generalized network formulation in terms of the moment method is used further, as explained in Section 2.2.3. The equivalence principle allows the use of the concept of an equivalent magnetic current

$$M_j = \pm[z^0, E_{\tau j}] \tag{2.85}$$

over each slot j, with a slot aperture covered by an electric conductor. In this case, the plus and minus signs have a bearing on the left- and right-hand-side regions, respectively, with regard to the slot. In this way, the general electromagnetic problem can be reduced to three equivalent partial network problems, one for each of the three independent regions 1, 2, and 3. To ensure continuity of the magnetic field tangential components across the slot apertures it is required that:

$$H_\tau^{(1)}(M_0) + H_\tau^{(2)}(M_0) + \sum_{j=1}^{k} H_\tau^{(2)}(M_j) = -H_\tau^p \quad (j = 0) \tag{2.86}$$

$$H_\tau^{(2)}(M_0) + \sum_{j=1}^{k} H_\tau^{(2)}(M_j) + \sum_{j=1}^{k} H_\tau^{(3)}(M_j) = 0 \quad (j = 1, 2, \ldots k) \tag{2.87}$$

where $\boldsymbol{H}_\tau^\nu(\boldsymbol{M}_j)$ is a tangential magnetic field intensity in the region ν, excited by the magnetic current $\boldsymbol{M}_j$, where $\nu = 1, 2, 3$, and $\boldsymbol{H}_\tau^p$ is a magnetic field intensity of the primary source, which in this case is a normalized TE_{10} rectangular waveguide mode ($|\boldsymbol{E}_\tau^p| = 1$).

In (2.86) and (2.87), the superposition principle is used to consider the mutual coupling between the slots. Thus, we obtain a system of $k + 1$ coupled vector operator equations for determining the equivalent magnetic currents. The moment method is then used to make an approximate solution of the above system. For this purpose, we expand $\boldsymbol{M}_j$ in terms of a set of expansion functions $\boldsymbol{M}_{jn}$:

$$\boldsymbol{M}_j = \sum_{n=1}^{N} V_{jn}\boldsymbol{M}_{jn} \tag{2.88}$$

where the coefficients V_{jn} are to be determined. We now consider narrow slots ($2\Delta \ll \lambda$, $2\Delta_0 \ll \lambda$), and for this case it is suitable in a first approximation to choose only one sine-type expansion function for each of the slots.

Thus, for the exciting slot ($j = 0$)

$$\boldsymbol{M}_{01} = \boldsymbol{x}^0 \frac{\sin[\beta(l_0 - |x - x_0|)]}{2\Delta_0 \sin \beta l_0} \tag{2.89}$$

and for the radiating slot ($j = 1, 2, 3, \ldots k$)

$$\boldsymbol{M}_j = \boldsymbol{x}^0 \frac{\sin[\beta(l - |x - x_j|)]}{2\Delta \sin \beta l} \tag{2.90}$$

where $\beta = 2\pi/\lambda$ is a free-space phase constant; and $\boldsymbol{x}^0$ = unit vector in the x-direction.

Following the moment method procedure, an algebraic set of $k + 1$ unknown quantities V_j ($j = 1, 2, 3, \ldots k$) is obtained:

$$(Y_{00}^{(1)} + Y_{00}^{(2)})V_0 + \sum_{j=1}^{k} Y_{0j}^{(2)}V_j = I_0^P \quad (i = 0) \tag{2.91}$$

$$Y_{i0}^{(2)}V_0 + \sum_{j=1}^{k} (Y_{ij}^{(2)} + Y_{ij}^{(3)})V_j = 0 \quad (i = 0, 1, 2, \ldots k) \tag{2.92}$$

where

$$Y_{00}^{(1)} = -\langle \boldsymbol{M}_{01}, \boldsymbol{H}_\tau^{(1)}(\boldsymbol{M}_{i1})\rangle_0$$

$$
\begin{aligned}
Y_{00}^{(2)} &= -\langle \boldsymbol{M}_{01}, \boldsymbol{H}_{\tau}^{(2)}(\boldsymbol{M}_{01})\rangle_0 \\
Y_{0j}^{(2)} &= -\langle \boldsymbol{M}_{01}, \boldsymbol{H}_{\tau}^{(1)}(\boldsymbol{M}_{j1})\rangle_0 \\
Y_{i0}^{(2)} &= -\langle \boldsymbol{M}_{i1}, \boldsymbol{H}_{\tau}^{(2)}(\boldsymbol{M}_{01})\rangle_i \\
Y_{ij}^{(2)} &= -\langle \boldsymbol{M}_{i1}, \boldsymbol{H}_{\tau}^{(2)}(\boldsymbol{M}_{j1})\rangle_i \\
Y_{ij}^{(3)} &= -\langle \boldsymbol{M}_{i1}, \boldsymbol{H}_{\tau}^{(3)}(\boldsymbol{M}_{j1})\rangle_i \quad (i, j = 1, 2, \ldots k)
\end{aligned}
$$

are so-called *generalized admittances* at the slot apertures and

$$
I_0^P = \langle \boldsymbol{M}_{01}, \boldsymbol{H}_{\tau}^P \rangle
$$

is the generalized current source.

To solve the above set of linear algebraic equations, we make a specification of its coefficients for each electromagnetic region.

Region 1: Rectangular Waveguide

The generalized admittance of region 1 can be expressed as follows:

$$
Y_{00}^{(1)} = -\int_{S_0} \boldsymbol{M}_{01} \boldsymbol{H}_{\tau}^{(1)}(\boldsymbol{M}_{01}) \mathrm{d}S \tag{2.93}
$$

where $\boldsymbol{M}_{01}$ is given by (2.89) and $\boldsymbol{H}_{\tau}^{(1)}(\boldsymbol{M}_{01})$ is represented as a sum of all possible waveguide mode vectors:

$$
\boldsymbol{H}_{\tau}^{(1)}(\boldsymbol{M}_{01}) = \sum_{\nu=1}^{N_{\nu}^{(1)}} y_{\nu}^{(1)} R_{\nu 0}^{(1)} e^{\gamma_{\nu}(1) z} [z^0, e_{\nu}^{(1)}] \tag{2.94}
$$

Here

$\gamma_{\nu}^{(1)}$ are mode propagation constants

$y_{\nu}^{(1)}$ = mode admittances

$N_{\nu}^{(1)}$ = total number of modes, inclusive propagating ($\nu = 1$) and evanescent ($\nu \neq 1$) modes;

$R_{\nu 0}^{(1)}$ = excitation coefficients, which are given by:

$$
R_{\nu 0}^{(1)} = \int_{S_g} \boldsymbol{M}_{01} \boldsymbol{H}_{\tau}^{(1)}(\boldsymbol{M}_{01}) \mathrm{d}S \tag{2.95}
$$

The expressions for the rectangular waveguide mode vectors are given in many textbooks (i.e., [4]). Substituting $\boldsymbol{M}_{01}$ and $e_\nu^{(1)}$ in (2.95) and integrating, we obtain:

$$R_{\nu 0}^{(1)} = \begin{cases} R_{\nu 0}^{(1)h} = \dfrac{\xi}{a} A_{\xi n} P_{\xi 0} Q_{n0} & \text{for H modes} \\ \\ R_{\nu 0}^{(1)e} = -\dfrac{\eta}{b} A_{\xi \eta} P_{\xi 0} Q_{\eta 0} & \text{for E modes} \end{cases} \tag{2.96}$$

$$A_{\xi\eta} = \sqrt{\frac{ab x_\xi x_\eta}{(\eta a)^2 + (\xi b)^2}} \quad x_\eta = \begin{cases} 1 & \eta = 0 \\ 2 & \eta \neq 0 \end{cases} \tag{2.97}$$

$$P_{\xi 0} = \frac{2\sin\left[\dfrac{\xi\pi}{2}\left(1 + \dfrac{2x_0}{a}\right)\right]\left(\cos\dfrac{\xi\pi l_0}{a} - \cos\beta l_0\right)}{\sin\left[\beta l_0\left[1 - \left(\dfrac{\xi\pi}{\beta a}\right)^2\right]\right]} \tag{2.98}$$

$$Q_{\eta 0} = \frac{\cos\left[\dfrac{\eta\pi}{2}\left(1 + \dfrac{2y_0}{b}\right)\right]\sin\dfrac{\eta\pi\Delta_0}{b}}{\dfrac{\eta\pi\Delta_0}{b}} \tag{2.99}$$

The generalized current source is given by:

$$I_0^P = \int_{S_0} \boldsymbol{M}_{01} \boldsymbol{H}_\tau^P \mathrm{d}S \tag{2.100}$$

According to image theory, the magnetic field intensity on the conductor is doubled. After integrating over S_0 for an incident TE_{10} mode, we obtain:

$$I_0^P = 2R_{10}^{(1)h} y_1^{(1)h} \tag{2.101}$$

Region 2: Rectangular Cavity Resonator

In region 2 the following four types of generalized admittances are specified as:

$$Y_{00}^{(2)} = -\int_{S_0} \boldsymbol{M}_{01}\boldsymbol{H}_{\tau}^{(2)}(\boldsymbol{M}_{01})\mathrm{d}S \tag{2.102}$$

$$Y_{0j}^{(2)} = -\int_{S_0} \boldsymbol{M}_{01}\boldsymbol{H}_{\tau}^{(2)}(\boldsymbol{M}_{j1})\mathrm{d}S \tag{2.103}$$

$$Y_{i0}^{(2)} = -\int_{S_0} \boldsymbol{M}_{i1}\boldsymbol{H}_{\tau}^{(2)}(\boldsymbol{M}_{01})\mathrm{d}S \tag{2.104}$$

$$Y_{ij}^{(2)} = -\int_{S_0} \boldsymbol{M}_{i1}\boldsymbol{H}_{\tau}^{(2)}(\boldsymbol{M}_{j1})\mathrm{d}S \tag{2.105}$$

where $i, j = 1, 2, 3, \ldots k$. If $i \neq j$, Y_{ij} is the slot mutual admittance and if $i = j$, Y_{ij} is the slot self admittance.

The field excited in the resonator by an equivalent magnetic current $\boldsymbol{M}_{01}$ ($z = 0$) is

$$\boldsymbol{E}_{\tau}^{(2)}(\boldsymbol{M}_{01}) = \sum_{\nu=1}^{N_{\nu}^{(2)}} R_{\nu 0}^{(2)} \frac{\sinh[\gamma_{\nu}^{(2)}(c_r - z)]\boldsymbol{e}_{\nu}^{(2)}}{\sinh(\gamma_{\nu}^{(2)}c_r)} \tag{2.106}$$

$$\boldsymbol{H}_{\tau}^{(2)}(\boldsymbol{M}_{01}) = -\sum_{\nu=1}^{N_{\nu}^{(2)}} y_{\nu}^{(2)}R_{\nu 0}^{(2)} \frac{\cosh[\gamma_{\nu}^{(2)}(c_r - z)][\boldsymbol{z}^0,\boldsymbol{e}_{\nu}^{(2)}]}{\sinh(\gamma_{\nu}^{(2)}c_r)} \tag{2.107}$$

where $\gamma_{\nu}^{(2)}$, $y_{\nu}^{(2)}$, $\boldsymbol{e}_{\nu}^{(2)}$ and $N_{\nu}^{(2)}$ are referred to the multimode rectangular waveguide, with dimensions $a_r \times b_r$. The expressions for the excitation coefficients $R_{\nu}^{(2)}$ are analogous to those in (2.95)

$$R_{\nu 0}^{(2)} = \int_{S_0} \boldsymbol{M}_{01}[\boldsymbol{z}^0,\boldsymbol{e}_{\nu}^{(2)}]\mathrm{d}S \tag{2.108}$$

There are similar expressions for the field excited by the current $\boldsymbol{M}_{j1}$ ($j = 1, 2, 3, \ldots k$, at $z = c_r$ plane):

$$\boldsymbol{E}_{\tau}^{(2)}(\boldsymbol{M}_{j1}) = -\sum_{\nu=1}^{N_{\nu}^{(2)}} R_{\nu j}^{(2)} \frac{\cosh(\gamma_{\nu}^{(2)}z)}{\sinh(\gamma_{\nu}^{(2)}c_r)} \boldsymbol{e}_{\nu}^{(2)} \tag{2.109}$$

$$\boldsymbol{H}_{\tau}^{(2)}(\boldsymbol{M}_{j1}) = -\sum_{\nu=1}^{N_{\nu}^{(2)}} R_{\nu j}^{(2)} y_{\nu}^{(2)} \cdot \frac{\cosh(\gamma_{\nu}^{(2)}z)}{\sinh(\gamma_{\nu}^{(2)}c_r)} [\boldsymbol{z}^0,\boldsymbol{e}_{\nu}^{(2)}] \tag{2.110}$$

and for the coefficients $R_{\nu j}^{(2)}$ ($j = 1, 2, 3, \ldots k$)

$$R_{vj}^{(2)} = \int_{S_1} \boldsymbol{M}_{j1}[\boldsymbol{z}^0, \boldsymbol{e}_v^{(2)}] \mathrm{d}S \tag{2.111}$$

After integration, the formulas for $R_{vj}^{(2)}$ differ from those given by (2.96) in that a, b, x_0, y_0, l_0 and Δ_0 are replaced by a_r, b_r, x_j, y_j, and Δ, respectively.

Substituting (2.106), (2.107), (2.109) and (2.110) into (2.102)–(2.105) and bearing in mind (2.85), the following expressions for the admittances are obtained:

$$Y_{00}^{(2)} = \sum_{v=1}^{N_v^{(2)}} [R_{v0}^{(2)}]^2 y_v^{(2)} \coth(\gamma_v^{(2)} c_r) \tag{2.112}$$

$$Y_{0j}^{(2)} = -\sum_{v=1}^{N_v^{(2)}} R_{v0}^{(2)} R_{vj}^{(2)} y_v^{(2)} \frac{1}{\sinh(\gamma_v^{(2)} c_r)} = Y_{j0}^{(2)} \tag{2.113}$$

$$Y_{ij}^{(2)} = \sum_{v=1}^{N_v^{(2)}} R_{v0}^{(2)} R_{vj}^{(2)} y_v^{(2)} \coth(\gamma_v^{(2)} c_r) \tag{2.114}$$

Region 3: Free Half-Space

The admittances in region 3 are expressed by

$$Y_{ij}^{(3)} = -\int_{S_i} \boldsymbol{M}_{ij} \boldsymbol{H}_\tau^{(3)}(\boldsymbol{M}_{j1}) \mathrm{d}S \quad (i,j = 1, 2, \ldots k) \tag{2.115}$$

where the terms with $i \neq j$ correspond to the external mutual coupling. The field presentation in this region is more complicated, because it is free half-space and the mode spectrum is continuous. Instead of using (2.115), the calculations of $Y_{ij}^{(3)}$ are made after considering the slot array as a dual problem to the complementary strip dipole array, radiating into a full space

$$Y_{ij}^{(3)} = \frac{2Z_{ij}^d}{W_0^2} \tag{2.116}$$

where W_0 is a free-space wave impedance and Z_{ij}^d are the equivalent strip-dipole impedances. Some of the basic antenna characteristics are represented further in a computational form.

Radiation Pattern and Directivity

The antenna directivity D is defined as:

$$D = \frac{|F_{max}|^2}{\frac{1}{4\pi}\int_\theta \int_\phi |F(\theta,\phi)|^2 \sin\theta \, d\theta \, d\phi} = \frac{|F_{max}|^2}{\langle |F|^2 \rangle} \tag{2.117}$$

where $F(\theta,\phi)$ is the slot radiation pattern and $F(0,\phi) = F_{max}$. The integration in (2.117) is simplified, considering the radiation patterns in the H- and E-planes only (H-plane: $\phi = \pi/2$, E-plane: $\phi = 0$). The slot array radiation pattern is given by:

$$F(\theta,\phi) = F_1(\theta,\phi) \sum_{i=1}^{k} U_i(\theta,\phi) \cdot V_i \tag{2.118}$$

where

$$F_1(\theta,\phi) = \frac{\cos(\beta l \psi_x) - \cos(\beta l)}{\sqrt{(1 - \psi_x^2)}} \cdot \frac{\sin(\beta \Delta \psi_y)}{\psi_y} \tag{2.119}$$

is a slot element radiation pattern and

$$U_i(\theta,\phi) = e^{j\beta(x_i\psi_x + y_i\psi_y)} \tag{2.120}$$

is a phase factor ($\psi_x = \sin\theta \cos\phi$, $\psi_y = \sin\theta \sin\psi$).

Hence, for $\theta = 0$, $\psi_x = \psi_y = 0$ and

$$|F_{max}|^2 = \beta\Delta(1 - \cos\beta l)\left|\sum_{i=1}^{k} V_i\right|^2 \tag{2.121}$$

In the H-plane, $\psi_x = \sin\theta$, $\psi_y = 0$ and the following quadratic form for the average power is obtained:

$$\langle |F|^2 \rangle_H = \sum_{i=1}^{k}\sum_{j=1}^{k} S_{ij}^{CH} V_i V_j^* \tag{2.122}$$

where the coefficients $S_{ij}^{c(H)}$ are given by:

$$S_{ij}^{CH} = \frac{(2\beta\Delta)^2}{8}\int_0^\pi \left[\frac{\cos(\beta l \sin\theta) - \cos(\beta l)}{\cos\theta}\right]^2 \times \sin\theta e^{j\beta(x_i - x_j)\sin\theta} d\theta \tag{2.123}$$

In the E-plane $\psi_x = 0$, $\psi_y = \sin\theta_d$

$$\langle |F|^2 \rangle_E = \sum_{i=1}^{k} \sum_{j=1}^{k} S_{ij}^{CE} V_i V_j^* \tag{2.124}$$

where

$$S_{ij}^{CE} = 0.5[1 - \cos(\beta l)]^2 \times \int_0^{\pi} \frac{[\sin(\beta \Delta \sin\theta)]^2}{\sin\theta} e^{j\beta(y_i - y_j)\sin\theta} d\theta \tag{2.125}$$

After a numerical computation of D_H and D_E from (2.117), the average antenna directive is given by $D = D_H \cdot D_E$.

Transmission Coefficient

In general, the antenna transmission coefficient is defined as:

$$|T|^2 = \frac{P_{tr}}{P_{in}} \tag{2.126}$$

where the input power P_{in} is given by:

$$P_{in} = y_1^{(1)h} \tag{2.127}$$

and the transmitted (radiated) power is

$$P_{tr} = Re\left[\sum_{i=1}^{k} \sum_{j=1}^{k} Y_{ij}^{(3)} V_i V_j^*\right] \tag{2.128}$$

Here the loss power within the cavity is neglected.

Quality Factor

The quality factor definition is

$$Q = \frac{\omega_{res} W_{res}}{P_{tr}} = \frac{P_{res}}{P_{tr}} \tag{2.129}$$

where the transmitted power is given by (2.128) and the power, stored in the cavity at resonance P_{res} is

$$P_{res} = \sum_{i=1}^{k} \sum_{j=1}^{k} H_{ij}^{(2)} V_i V_j^* \tag{2.130}$$

The antenna dimensions are chosen so that only the presence of TE_{101} mode inside the cavity will be ensured. In this case, the inner surface of the radiation wall is streamed by electric currents which excite all slots in phase.

In present antenna design, the transverse dimensions a_r and b_r are much bigger than the corresponding waveguide dimensions a and b, and there are many unwanted modes with the same or close resonant frequencies. For $a_r = b_r = 2.5\lambda_{101}$ ($\lambda_{101} = \lambda_c^{TE_{101}}$ = cut-off wavelength of the waveguide, with $a = a_r$ and $b = b_r$). The relative resonant frequencies of the most probable higher modes TM can be computed by

$$\frac{f_{101}}{f_{\xi\eta 0}} = \frac{2}{\sqrt{\xi^2 + \eta^2}} \tag{2.131}$$

The calculated values of this ratio for ξ, η = 1, 2, 3, 4 and 5 are given in Table 2.2.

Table 2.2

η \ ξ	1	2	3	4	5
1	3.53	2.24	1.58	1.12	0.98
2	2.24	1.77	1.39	1.12	0.93
3	1.58	1.39	1.18	1.00	0.86
4	1.12	1.12	1.00	0.84	0.78
5	0.98	0.93	0.86	0.78	0.71

From the table we see that the most undesired modes are TM_{430} and TM_{340}, because their resonant frequencies are equal to the basic mode frequency (i.e., $\lambda_{430} = \lambda_{340} = \lambda_{101}$). But because of the orthogonality between the magnetic lines in the excitation slot and the TM_{430}-mode magnetic lines (see Figure 2.27(a)), this mode does not arise. The higher mode TM_{340} can easily be damped by a pin-type filter (see Figure 2.27(b)—filter metal pins are shown by cross-stitches).

We will now consider a cavity-backed slot antenna model comprising $k = 4 \times 4 = 16$ equal slots with $2l = 14$ mm and $2\Delta = 1.5$ mm, and a rectangular cavity resonator of dimensions 80 × 80 × 16 mm. The cavity is excited by a standard X-band waveguide through a slot of dimensions $2l_0 = 14$ mm and $2\Delta_0 = 3$ mm. The antenna frequency response is illustrated by the power transmission coefficient $|T|^2$ (Figure 2.28) and by the VSWR (Figure 2.29) as functions of the frequency, measured and calculated using the described theory. Figure 2.30 shows the radiation

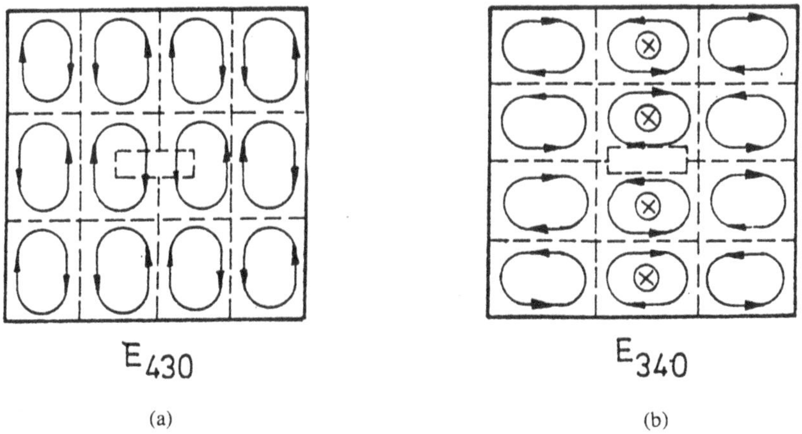

Figure 2.27 Higher modes *TM* magnetic field lines: (a) TM_{430} magnetic lines and (b) TM_{340} magnetic lines with pin-type filter

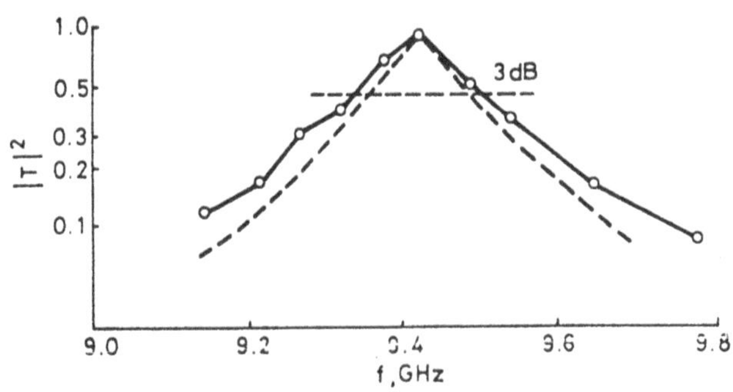

Figure 2.28 Transmission coefficient (from [25]):
——— measured
----- calculated

pattern in the *H*-plane, for the optimum frequency $f_0 = 9.41$ GHz. The calculations have been made with $N_\nu^{(1)} = 10$ waveguide modes in region 1 and with $N_\nu^{(2)} = 60$ waveguide modes in region 2.

Table 2.3 summarizes the main antenna parameters—calculated and measured ones, and shows that there is good agreement between theory and experiment.

The examined rectangular cavity-backed slot array with TE_{101} fundamental mode is a typical frequency selective (narrowband) miniature antenna. Compared, for example, with the optimum pyramidal horn having the same gain, this antenna is about five times shorter in length and more than two times smaller in volume.

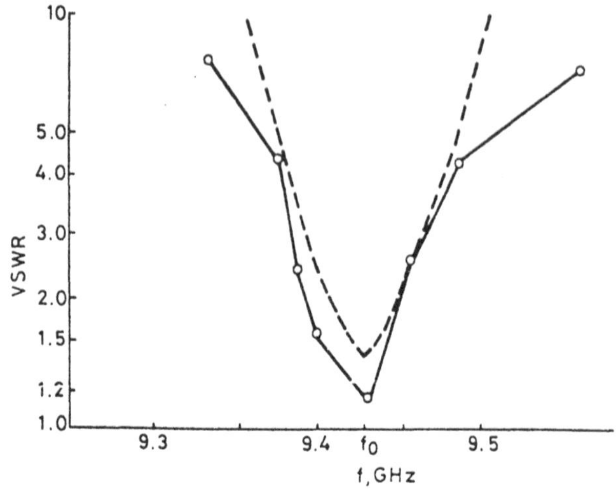

Figure 2.29 Voltage standing wave ratio (VSWR) (from [25]):
——— measured
----- calculated

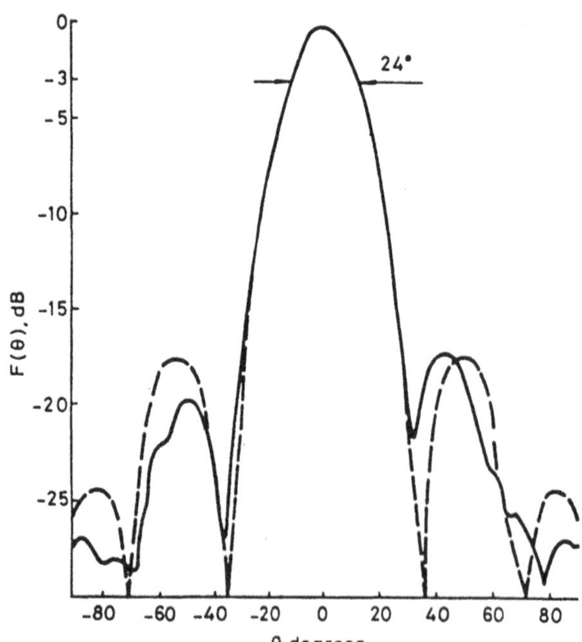

Figure 2.30 *H*-plane radiation pattern (from [25]):
——— measured
---- calculated

Table 2.3

Antenna parameters	*Calculated*	*Measured*
Radiation pattern		
(a) *H*-plane:		
3 dB Beamwidth	20°	22°
maximum sidelobe, dB	−19	−20
(b) *E*-plane:		
3 dB Beamwidth	24°	24°
maximum sidelobe, dB	−17.5	−17
Resonant frequency, GHz	9.41	9.43
Frequency bandwidth, %	1.45	1.7
Antenna *Q*-factor	69	60
VSWR for f_0	1.4	1.15
Directivity, dB	17.3	16.9

The slot array can be backed by a standing-wave cavity as we have shown here. However, an alternative design is the circular slot array illuminated by a travelling-wave ring-shaped cavity resonator [26–27] (see Figure 2.31). The antenna consists of: a TE_{10} mode rectangular waveguide, an excitation slot, radiation slots and a ring-shaped cavity resonator. The cavity dimensions are: r_2 = inner ring radius, r = outer ring radius, a_r = cavity depth.

For directional coupling through the excitation slot, a travelling wave rises into the ring-shaped cavity and the array radiation field is of circular polarization. When there is non-directional coupling, a standing wave originates in the cavity and the radiation field is of linear polarization.

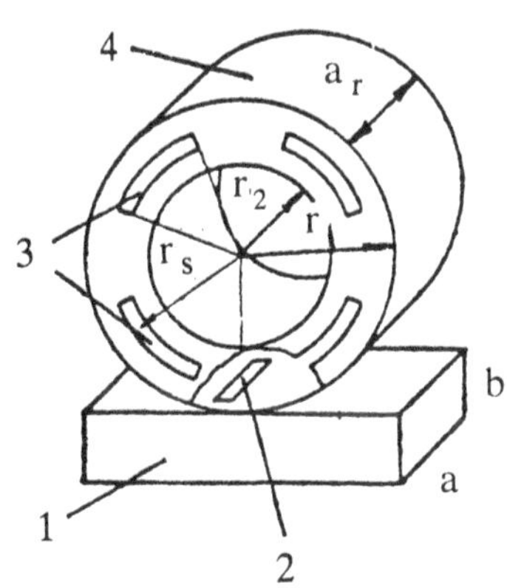

Figure 2.31 Ring-shaped cavity antenna geometry (from [25])

REFERENCES

1. Hristov, H.D., "Cavity Antennas," D. Sc. Thesis 1987, Central Institute for Scientific and Technical Information, Sofia, Bulgaria (in Bulgarian).
2. Vainshtein, L.A., *Open Resonators and Waveguides*, Sovjetskoe Radio Publishing, Moscow, 1966 (in Russian).
3. Lavrenko, J.E., "Field Presentation in a Cavity, Bounded by Electric and Magnetic Walls," *Proc. Higher Institute of Mech. and Elect. Engineering Conference*, Varna, Bulgaria, 1985 (in Russian).
4. Collin, R.E., *Field Theory of Guided Waves*, McGraw-Hill, New York, 1960.
5. Hong, M.H., D.P. Nyquist, and K.M. Chen, "Radiation Fields of Open-Cavity Radiators and Backfire Antenna," *IEEE Trans. Antennas and Propagation*, Vol. AP-18, No. 6, November 1980, pp. 813–815.
6. Harrington, R.F., and J.R. Mautz, "A Generalized Network Formulation for Aperture Problem," *IEEE Trans. Antennas and Propagation*, Vol. AP-24, No. 6, November 1976, pp. 870–873.
7. Beth, H.A., "Theory of Diffraction by Small Holes," *Physical Review*, Vol. 66, October 1944, pp. 163–182.
8. Harrington, R.F., *Field Computation by Moment Method*, Macmillan, New York, 1968.

8a. Chang, D.K., and C.H. Liang, "Optimum Directivity of Elliptic Loop Antennas," *Electron. Letter.*, Vol. 17, No. 17, October 1981, pp. 736–738.

9. Liang, C.H., and D.K. Cheng, "Generalized Network Representation for Small Aperture Coupling between Dissimilar Regions," *IEEE Trans. Antennas and Propagation*, Vol. AP-31, No. 1, January, 1983, pp. 177–182.
10. Feld, J.N., *Basic Theory of Slot Antennas*, Sovjetskoe Radio Publishing House, Moscow, 1948 (in Russian).
11. Grinev, A.J., and J.V. Kotov, "Scanning Characteristics of Antenna Structure Consisting of Cavity-Backed Slots, Covered by Dielectric Plate," *Proc. USSR Radiotechnika*, Vol. 21, No. 12, 1978, pp. 1822–1833 (in Russian).
12. Illinski, A.S., A.J. Grinev, and J.V. Kotov, "Investigation of Electromagnetic Characteristics of Cavity-Backed Radiators with Excitation Sources in the Slot Plane," *Radiotechnika Electronika*, Vol. XXIII, No. 5, 1978, pp. 922–930 (in Russian).
13. Voskresenski, D.I. (Editor), *Antennas and Microwave Devices (Design of Antenna Phased Arrays)*, Radio Svjaz Pub. House, Moscow, 1981.
14. Adams, A.T., "Flush Mounted Rectangular Cavity Slot Antennas—Theory and Design," *IEEE Trans. Antennas and Propagation*, Vol. AP-15, No. 3, May 1967, pp. 342–351.
15. Lagerlof, R.O.E., "Optimization of Cavities for Slot Antennas," *Microwave Journal*, Vol. 16, No. 10, October 1973, pp. 12C–12F.
16. Crews, M.R., and G.A. Thile, "On the Design of Shallow Depth T-Bar Fed Slot Antennas," *IEEE Trans. Antennas and Propagation*, Vol. AP-25, No. 6, November 1977, pp. 833–836.
17. Andersen, J.B., "Low- and Medium-gain Microwave Antennas," *The Handbook of Antenna Design* (Editors: Rudge A.W., K. Milne, A.D. Olver, and P. Knight), Vol. 1, Chapter 7, Peter Peregrinus, New York, 1982.
18. Mayes, P.E., "Small, Broadband, Unidirectional Antenna," US Patent No. 3 710340, 9th January 1973.
19. Mayes, P.E., "The Monopole Slot: A Small Broadband Unidirectional Antenna," *IEEE Trans. Antennas and Propagation*, Vol. AP-20, No. 4, July 1972, pp. 489–493.
20. Clavin, A., D.A. Huebner, and F.J. Kilburg, "An Improved Element for Use in Array Antennas," *IEEE Trans. Antennas and Propagation*, Vol. AP-22, No. 4, July 1974, pp. 521–526.

21. Numazaki, T., M. Ono, and M. Mizusawa, "Cavity-Backed Crossed-Slot Antenna with Coupled Monopoles," *Proc. Int. Conf. on Antennas and Propagation,* Sendai, Japan, August 1978, pp. 167–170.
22. Liang, C.H., and D.K. Cheng, "Electromagnetic Fields Coupled into a Cavity with a Slot-Aperture under Resonant Conditions," *IEEE Trans. Antennas and Propagation,* Vol. AP-30, No. 4, July 1982, pp. 664–672.
23. Jacobsen, S., E. Andersen, and M. Groundlund, "An Antenna Illuminated by a Cavity Resonator," *Proc. IEEE,* Vol. 51, No. 11, November 1963, pp. 1431–1435.
24. Paoloni, F.J., "A Cavity-Backed Resonant Slot Array—Theory and Measurements," *IEEE Trans. Antennas and Propagation,* Vol. AP-28, No. 2, March 1980, pp. 259–263.
25. Savov, S.V., and H.D. Hristov, "Cavity-Backed Slot Array Analysis," *Proc. IEE,* Vol. 134, Pt. H, No. 3, June 1987, pp. 280–284.
26. Shubarin, J.V., and T.N. Anischenko, "Waveguide-Slot Radiator, USSR Patent No. 154 300, 1963.
27. Gorobetz, N.N., and M.V. Nesterenko, "Slot Array on Ring Cavity Antenna with Travelling Wave," *Proc. USSR's Higher Schools-Radioelektronika,* Vol. XXV, No. 2, 1982, pp. 83–85, (in Russian).

Chapter 3
Translucent-Aperture Cavity Antennas

3.1 PLANE TRANSLUCENT-APERTURE ANTENNAS

3.1.1 Some Translucent-Aperture Plates

We now consider a two-layer plate infinite in extent, and a normally-incident electromagnetic plane wave (Figure 3.1). The media (1) and (4) are infinite half-spaces, with wave impedances Z_{w1} and Z_{w4}, respectively. The layers (2) and (3) have depths w_2 and w_3, and wave impedances Z_{w2} and Z_{w3}.
We define the reflection coefficient Γ as:

$$\Gamma = \frac{E_1^-}{E_1^+} \tag{3.1}$$

and the transmission coefficient T as:

$$T = \frac{E_4^-}{E_1^+} \tag{3.2}$$

where E_1^+ and E_1^- are the complex amplitudes of the incident and reflected waves at the input boundary plane $A - A'$; and E_4^- = complex amplitude of the outgoing wave at the boundary plane $C - C'$.

The reflection and transmission coefficients characterize the degree of plate transparency. Using the well known electromagnetic signal flow graph method, an oriented signal graph corresponding to the two-layer structure is shown in Figure 3.2. In accordance with Maysson's formula for complicated signal graphs, the following expressions for the reflection and transmission coefficients are obtained [1]:

$$\Gamma = S_{11} = \frac{T_1(1 - L_1^{(1)} - L_2^{(1)} - L_3^{(1)} + L_1^{(2)}) + T_2(1 - L_2^{(1)}) + T_3}{1 - L_1^{(1)} - L_2^{(1)} - L_3^{(1)} + L_1^{(2)}} \tag{3.3}$$

$$T = S_{41} = \frac{T_{41}}{1 - L_1^{(1)} - L_2^{(1)} - L_3^{(1)} + L_1^{(2)}} \tag{3.4}$$

where

$$L_1^{(1)} = \Gamma_{22}\Gamma'_{22}e^{-2\gamma_2 W_2}; \quad L_2^{(1)} = \Gamma_{33}\Gamma'_{33}e^{-2\gamma_3 W_3}$$

$$L_1^{(2)} = L_1^{(1)}L_2^{(1)}; \quad L_3^{(1)} = \Gamma_{22}\Gamma'_{33}T_{32}T_{23}e^{-2\gamma_2 W_2}e^{-2\gamma_3 W_3}$$

$$T_2 = T_{21}T_{12}\Gamma'_{22}e^{-2\gamma_2 W_2}; \quad \Gamma_{22} = -\Gamma_{11}; \quad T_1 = \Gamma_{11}$$

$$T_3 = T_{21}T_{32}T_{23}T_{12}\Gamma'_{33}e^{-2\gamma_2 W_2}e^{-2\gamma_3 W_3}$$

$$T_{41} = T_{21}T_{32}T_{43}e^{-\gamma_2 W_2}e^{-\gamma_3 W_3}; \quad \Gamma_{33} = -\Gamma'_{22}$$

$$\Gamma_{11} = \frac{Z_{W2} - Z_{W1}}{Z_{W2} + Z_{W1}}; \quad \Gamma'_{22} = \frac{Z_{W3} - Z_{W2}}{Z_{W3} + Z_{W2}}$$

$$\Gamma'_{33} = \frac{Z_{W4} - Z_{W3}}{Z_{W4} + Z_{W3}}; \quad T_{21} = \frac{2Z_{W2}}{Z_{W2} + Z_{W1}}$$

$$T_{12} = \frac{2Z_{W1}}{Z_{W2} + Z_{W1}}; \quad T_{32} = \frac{2Z_{W3}}{Z_{W2} + Z_{W3}}$$

$$T_{23} = \frac{2Z_{W2}}{Z_{W2} + Z_{W1}}; \quad T_{43} = \frac{2Z_{W4}}{Z_{W4} + Z_{W3}}$$

and γ = propagation constant.

Metalized Dielectric Plate

In practice, a very convenient method for producing a translucent wall is to use a metalized dielectric plate. Let $Z_{w1} = Z_{w4} = 120\pi$ ohms (i.e., (1) and (4) are air half-spaces). Assume that layer (2) is a thin metal film and layer (3) is a low-loss dielectric plate.

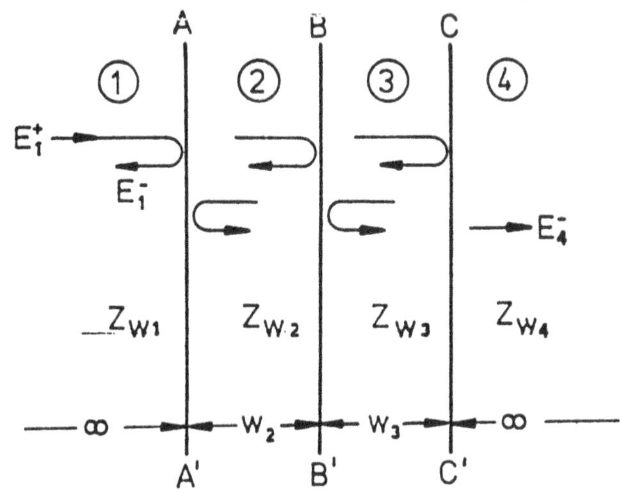

Figure 3.1 Transmission through and reflection from a two-layer plate

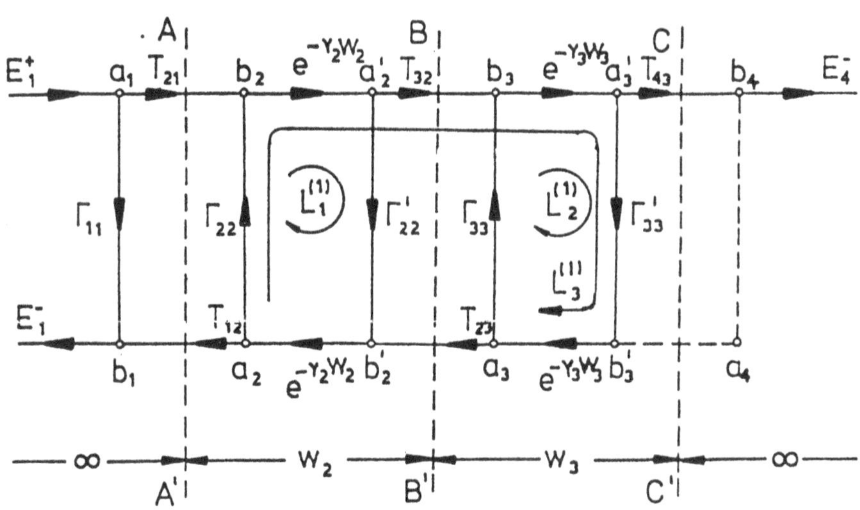

Figure 3.2 Two-layer plate signal flow graph—general case (from [1])

In this case, the wave impedance Z_{w3} is equal to the surface metal impedance Z_s^σ

$$Z_S^E = Z_S^\sigma = R_S^\sigma + jX_S^\sigma \cong (1 + j)\frac{\delta_S \omega \mu}{2} \tag{3.5}$$

where

$$\delta_S = \frac{1}{\sqrt{\pi f \mu \sigma}} \tag{3.6}$$

is the skin depth.

If the metal film depth is commensurable with or less than the skin depth δ_s, its shielding effect diminishes (i.e., the metal film becomes translucent or semitransparent). From (3.3) and (3.4), the reflection and transmission coefficients of the metalized dielectric plate can be computed.

One-Layer Dielectric Plate

This is a special case of a semi-transparent plate, for which $w_3 = 0$, $w_2 = w$, $L_2^{(1)} = L_1^{(2)} = 0$, $T_3 = T_{41} = 0$. Thus, from (3.3) and (3.4), we find

$$\Gamma = \frac{\Gamma_{11}(1 - \Gamma_{11}^2 e^{-\gamma w}) - T_{12}T_{21}\Gamma_{11}e^{-2\gamma w}}{1 - \Gamma_{11}^2 e^{-2\gamma w}} \tag{3.7}$$

$$T = \frac{T_{12}T_{21}e^{-\gamma w}}{1 - \Gamma_{11}^2 e^{-\gamma w}} \tag{3.8}$$

where

$$\Gamma_{11} = \frac{Z_w - 120\pi}{Z_w + 120\pi}; \quad T_{12} = \frac{2Z_w}{Z_w + 120\pi}$$

$$T_{21} = \frac{240\pi}{Z_w + 120\pi}; \quad Z_w \approx \sqrt{\frac{\mu}{\varepsilon}}\left(1 + j\frac{1}{2}\tan\delta\right)$$

$$\gamma = \alpha + j\beta \approx \pi f\sqrt{\mu\varepsilon}\left[\tan\delta + j^2\left(1 + \frac{1}{8}\tan^2\delta\right)\right]$$

Perforated Metal Plate

Figure 3.3 shows a translucent perforated thin metal plate, as a square array of round holes. If the holes are small enough, an equivalent surface impedance can

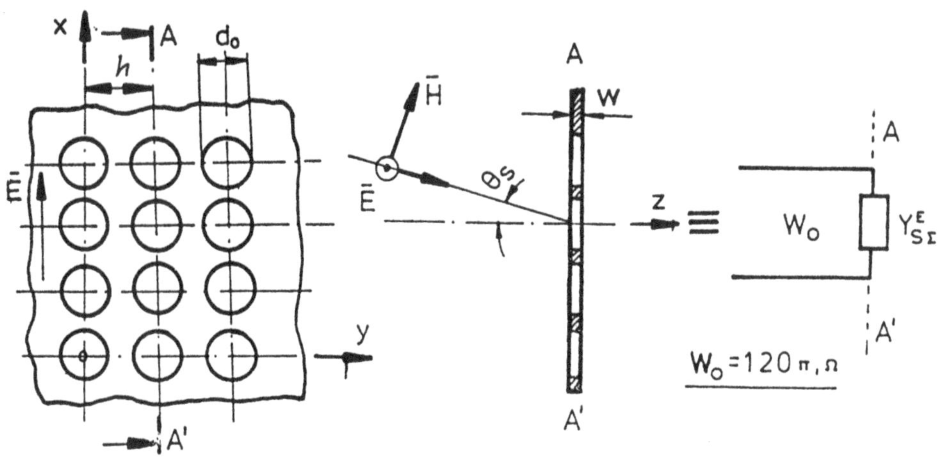

(a) (b)

Figure 3.3 Translucent perforated metal plate: (a) perforated metal plate geometry; (b) transmission line equivalency (from [2,3])

be introduced. θ_s is the incidence angle of a plane electromagnetic wave. As shown in Figure 3.3(b), the translucent plate can be represented as an equivalent transmission line loaded by a complex admittance $Y^E_{S\Sigma} = 1/Z^E_{S\Sigma}$. Its normalized value is given as: [2–3]

$$\tilde{Y}^E_{S\Sigma} = \frac{120\pi}{Z^E_{S\Sigma}} = 1 - j\tilde{B}^E_{S\Sigma} \tag{3.9}$$

where

$$\tilde{B}^E_{S\Sigma} \approx \frac{3\lambda h^2}{\pi d_0^3 \cos\theta_s}\left[1 - 0.24\left(\frac{d_0}{h}\right)^3\right]\left[1 - \left(\frac{\lambda_{c0}}{\lambda}\right)^2\right] \tag{3.10}$$

and

$$\cos\theta_s = \frac{\lambda}{\lambda_g} \tag{3.11}$$

In addition, λ_{co} is the cut-off wavelength for the TE_{11} mode in a circular waveguide of the same diameter d_0 as the holes; and λ_g = wavelength in the waveguide, shunted by the perforated plate.

The second multiplier in (3.10) accounts for the mutual coupling between the holes, and the third one corrects the value of the equivalent susceptance for larger holes ($d_0 \leq (\lambda/\pi)$).

Grating of Thin Parallel Wires

The translucent wall can also be formed by a grating of thin parallel wires or strips (Figure 3.4), characterized by a polarization anisotropy. For a parallel polarization of the incident plane wave ($\bar{E} = E_y \bar{y}^0$), the reflection coefficient of a dense grating ($h \ll \lambda$), consisting of round thin wires ($(d_0/2) \ll \lambda$), can be computed by:

$$\Gamma \approx -\frac{1}{1 + j\dfrac{2h}{\lambda}\cos\theta_s \ln\dfrac{h}{\pi d_0}} \tag{3.12}$$

For the same grating, when the polarization is perpendicular ($\bar{H} = H_y \bar{y}^0$), the reflection coefficient is very small ($|\Gamma| < 0.1$). If the grating is positioned in a dielectric plate with $w \gg h$, then $h\sqrt{\varepsilon_r} < 0.8\lambda$, and the reflection coefficient for a parallel polarization and $\theta_s = 0$ is given by the following approximate formula:

$$\Gamma \approx \frac{(v^2 - u^2)T_0 e^{j\beta w}}{v^2 e^{j\beta_E h} - 2\Gamma_0 uv - u^2 e^{j2(\phi_0 - \beta_E h)}} - 1 \tag{3.13}$$

where

$$u = 1 - \frac{1}{\sqrt{\varepsilon_r}}; \quad v = 1 + \frac{1}{\sqrt{\varepsilon_r}}$$

$$\beta_c = \beta\sqrt{\varepsilon_r}; \quad \phi_0 = \text{arc cos } |T_0|$$

Here, Γ_0 and T_0 are the reflection and transmission coefficients, respectively, for the same grating placed in air space, for $\theta_s = 0$.

3.1.2 Closed Cavity Antennas with Translucent Plane Apertures

We have already emphasized that research on closed cavity antennas with translucent apertures was started 25 years ago. At that time, the Microwave Laboratory at the Danish Academy of Sciences set forth the idea of the possibility of making an aperture illumination, without phase error, by means of a light coupling between a

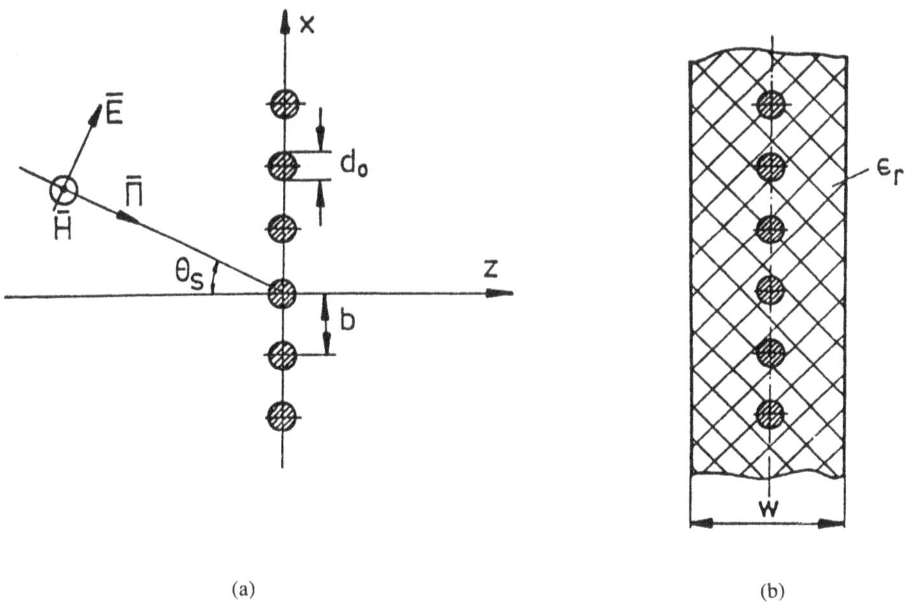

Figure 3.4 Grating of thin wires as a translucent wall (a) grating in air space; (b) grating in dielectric plate

cavity resonator and an antenna aperture of the same cross section and size as the resonator.

A sketch of the first such antenna is shown in Figure 3.5. Its dimensions in the x and y directions are large compared with the wavelength, whereas the z dimension is about $\lambda_g/2$. Utilizing the basic TE_{101} mode, there is, inside the resonator, a current on the aperture plate with a constant phase that induces a plane-phase aperture field on the outside. Because the resonator's internal losses are negligible compared with the radiation power, the unloaded cavity is represented approximately by a loss-free transmission line, short-circuited on both ends, with a length $c_r = \lambda_g/2$, and a characteristic impedance $Z_c = Z_0\lambda_g/\lambda$, where $Z_0 = W_0 = 120\pi$ ohms and the propagation constant is $\gamma = j\beta = j(2\pi/\lambda_g)$.

Figure 3.6(a) shows the equivalent diagram for the cavity which couples the generator to the load. Here Z_G is the characteristic impedance of the feeding waveguide, jX_1 represents the reactance of the coupling iris in an infinitely thin plate, and $Z_L = Z_0 = 120\pi$ ohms is the load represented by an infinitely long transmission line of characteristic impedance equal to Z_L. The terms L and C represent the resonator for small frequency variations $\Delta\omega$, around the resonance frequency ω_0:

$$L = \frac{Z_\gamma}{2} \cdot \frac{\lambda_g^2}{\lambda^2} \cdot \frac{\pi}{\omega_0} \tag{3.14}$$

$$C = \frac{1}{\omega_0^2 L} \tag{3.15}$$

When $|X_1| \ll Z_G$, and $|X_2| \ll Z_L$, the diagram of Figure 3.6(a) can be made equivalent to the diagram of Figure 3.6(b), where $R_1 = X_1^2/Z_G$ and $R_2 = X_2^2/Z_L$. When $R_1 = R_2$, and the sum of all the reactances becomes zero, a maximum output will dissipate in Z_L:

$$\frac{X_1^2}{Z_G} = \frac{X_2^2}{Z_L} \tag{3.16}$$

$$X_1 + X_2 + \omega_r L - \frac{1}{\omega_r C} = 0 \tag{3.16a}$$

The relation between the resonance frequency ω_r of the total system and the resonance frequency ω_0 of the unloaded resonator is, for $|X_1|$, $|X_2| \ll \omega_0 L$, $1/\omega_0 C$ given by the following:

$$\omega_0 = \omega_r\left(1 + \frac{X_1 + X_2}{2\omega_r L}\right) \tag{3.17}$$

The loaded Q-factor of the total system is

$$Q_L = \frac{\omega_r L + X_1 + X_2}{R_1 + R_2} \tag{3.18}$$

A metal perforated plate has been chosen for the aperture plate shown in Figure 3.3, which is a two-dimensional array of circular holes. To avoid beam splitting in the radiation pattern, the distance between the holes should be

$$h \leq \frac{\lambda}{1 + |\sin \theta_s|} \tag{3.19}$$

The aperture plate is equivalent to jX_2 and the waveguide diaphragm is equivalent to jX_1.

A sketch of the antenna model studied with a square perforated aperture of dimension $a_r = 100$ mm and surrounding frame of $a_r' = 130$ mm is shown in Figure 3.7. The cavity has dimensions $a_r \times b_r \times c_r = 100 \times 100 \times 15$ mm^3 for a resonant frequency $f_0 = 9.8$ GHz. The excitation rectangular waveguide with an inductive diaphragm is oriented such that its polarization vector $\bar{E}$ is directed along the cavity

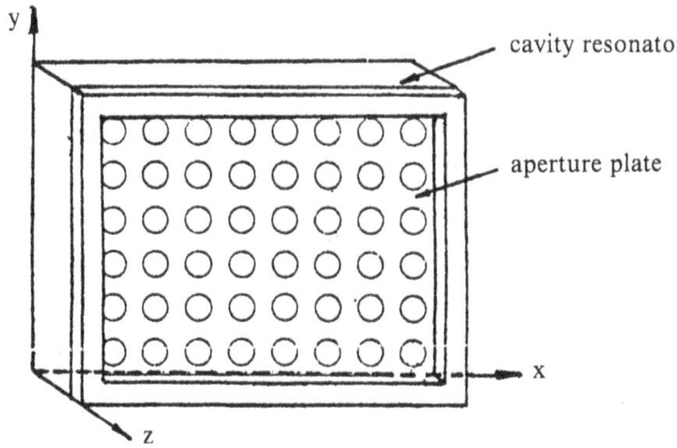

Figure 3.5 Sketch of a rectangular resonator as a translucent aperture antenna (from [2])

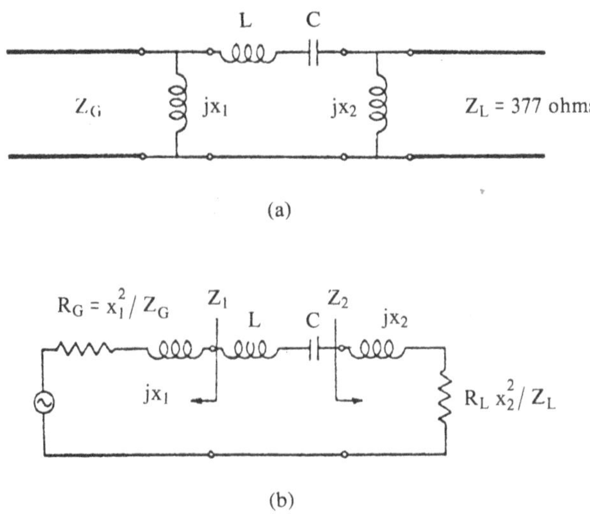

Figure 3.6 Equivalent circuit diagram for one mode cavity antenna (from [2]): (a) equivalent diagram for the cavity coupling the generator to the load; (b) final circuit diagram

diagonal and the cavity modes TE_{011} and TE_{101} are equally excited and radiated. The diagonally-illuminated square aperture antenna should have very small pattern sidelobes.

The aperture plate is a brass plate of a thickness of $t = 1.5$ mm. The hole diameter is made $d_0 = 7.6$ mm with a distance $b = 10$ mm between the centers of the holes. The measured and calculated radiation patterns are shown in Figure 3.8. In Table 3.1 the experimental results are compared with the theoretical ones.

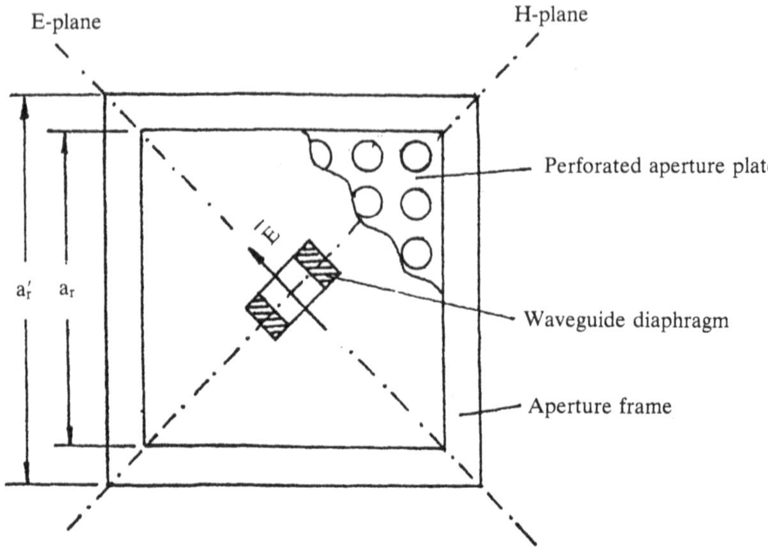

Figure 3.7 Diagonally-fed cavity antenna with semitransparent perforated aperture plate

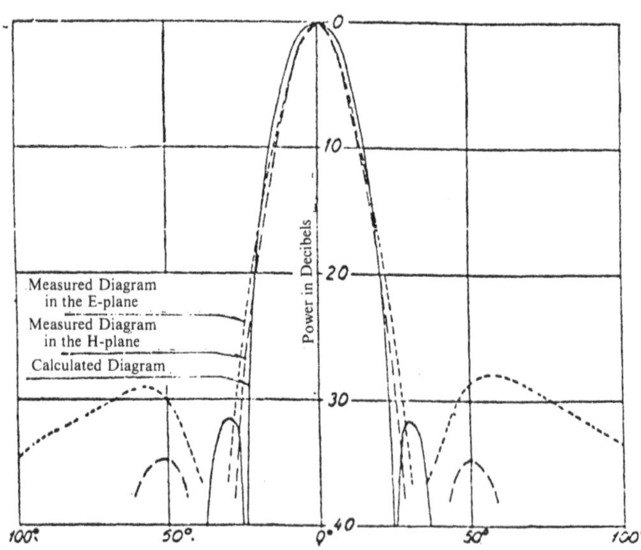

Figure 3.8 Measured *H*-plane and *E*-plane radiation patterns drawn together with calculated pattern (from [2])

Table 3.1 (from [2]).

Parameter	Calculated	Measured	
Loaded Q, Q_L	70	90	
Transmission coefficient T	-13.47 dB	-14.57 dB	
Gain G	20.3 dB	20.0 dB	
		E-plane	H-plane
3 dB beamwidth	17.9°	15.0°	14.0°
10 dB beamwidth	30.9°	28.3°	26.7°
Angular position 1 minimum	24.8°	34.2°	33.2°
Angular position 2 maximum	29.4°	57.3°	50.0°
Level of first sidelobe	31.5 dB	28.5 dB	34.7 dB

By considering the antenna radiation pattern, we will see that the main lobe has very good axial symmetry, and the sidelobes are as low as about -30 dB. Using the excitation theory of generalized cavity antennas described in Chapter 2, let us perform an electromagnetic analysis of the same antenna configuration, fed by means of a waveguide end-wall circular hole of diameter $2a_0'$ (Figure 3.9). According to Bethe's theory, for small aperture holes ($\beta a_0 \ll 1$), the perforated plate can be replaced by an equivalent array of radiating magnetic dipoles. The polarization magnetic moment of the arbitrary νth hole is

$$\bar{p}_{m_\nu} = -\mu_0 \alpha_m \bar{H}_{x\nu} \tag{3.20}$$

The electric field intensity in the νth hole is given by:

$$\bar{E}_\nu = j\omega p'_{m\nu} \bar{y}_0 \tag{3.21}$$

where

$$p'_{m\nu} = \frac{p_{m\nu}}{S_0(a_0)} = \frac{p_{m\nu}}{\pi a_0^2} \tag{3.22}$$

and

$$\alpha_m = \alpha_{mx} = \frac{4}{3} a_0^2 \tag{3.23}$$

where α_m is the magnetic polarization susceptibility for a circular hole. The normalized components of the cavity field for the TE_{101} operating mode are

$$\tilde{E}_y = \frac{2}{\sqrt{\varepsilon_0 a_r b_r c_r}} \sin\left(\frac{\pi}{a_r} x\right) \sin\left(\frac{\pi}{c_r} z\right) \tag{3.24}$$

$$\tilde{H}_x = -\frac{j}{\omega\mu_0} \cdot \frac{2\pi}{c_r \sqrt{\varepsilon_0 a_r b_r c_r}} \sin\left(\frac{\pi}{a_r} x\right) \cos\left(\frac{\pi}{c_r} z\right) \tag{3.25}$$

$$\tilde{H}_z = -\frac{j}{\omega\mu_0} \cdot \frac{2\pi}{c_r \sqrt{\varepsilon_0 a_r b_r c_r}} \sin\left(\frac{\pi}{a_r} x\right) \cos\left(\frac{\pi}{c_r} z\right) \tag{3.26}$$

After putting $H_{x\nu} = H_{x\nu}(x_\nu, z = c_r/2)$ (from (3.25)), p'_m (from (3.23)) and α_m (from (3.24)) into (3.20), the final expression for $E_\nu(x_\nu)$ is obtained. Thus, $E_\nu(x_\nu)$ is proportional to $\sin((\pi/a_r)x_\nu)$. In the y-direction, the amplitude distribution is constant (i.e., $E_y(y_\nu) = \text{const}$). The total far field of the antenna is calculated as a sum of the radiation fields of all array elements:

$$E(\theta,\phi) = \sum_\nu E_\nu(\theta,\phi) \tag{3.27}$$

From (2.18), the radiation quality factor is given by:

$$Q_{101}^{\Sigma} = \frac{N_{101}\omega_{101}}{\sum_{\nu=1}^{M} \int_{S_0} Re Z_{S\Sigma\nu} |\bar{H}_{\tau 101\nu}|^2 dS} \tag{3.28}$$

where

M = $N_x N_y$ is the number of holes;

S_0 = hole area;

$Z_{s\Sigma\nu}$ = equivalent surface impedance (radiation impedance) of the hole, which is given by the following approximate expression:

$$Z_{S\Sigma\nu} \approx 80\pi \frac{a_0}{\lambda}\left[\left(\frac{ka_0}{9\pi}\right)^2 + j\right] \tag{3.29}$$

For a square array, $N_x = N_y = N$ and $M = N^2$.

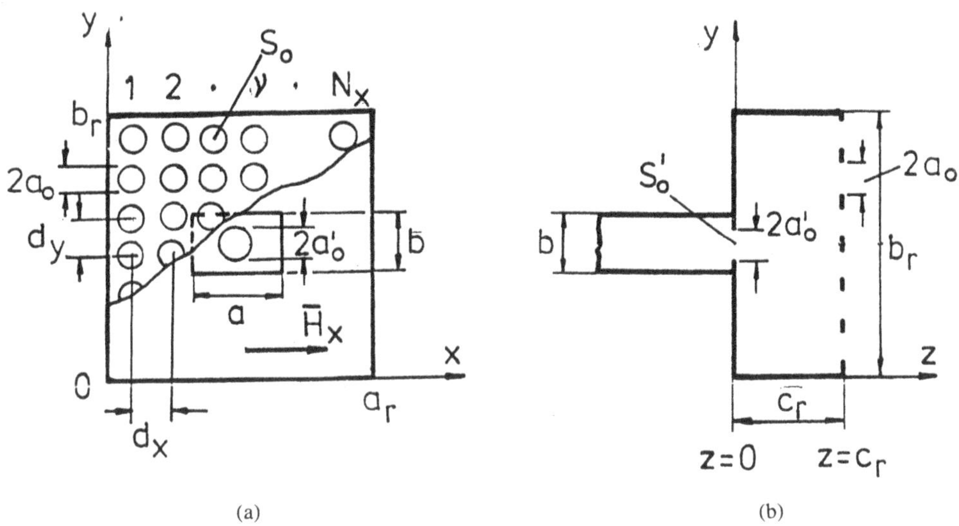

Figure 3.9 Cavity antenna with perforated plate aperture, fed by means of end-wall circular hole (from [1]): (a) antenna front view; (b) antenna cross section

The normalized magnetic transverse component $\tilde{H}_{\tau 101} = \tilde{H}_{X101}$ (for $N_{101} = 1$) is taken from (3.25). If the mutual coupling between the radiating holes is neglected, we can assume that $ReZ_{S\Sigma}(M) =$ const. Under these conditions, (3.28) takes the following form:

$$Q_{101}^{\Sigma} = \frac{9\omega_{101}^3 \varepsilon_0 \mu_0^2 a_r b_r c_r^3}{160\pi^2 a_0^2 N(\beta a_0)^4 \sum_{\nu=1}^{N} \sin^2\left(\frac{\pi}{a_r} d_x n_x\right)} \tag{3.30}$$

The normalized input impedance, at $z = 0$, is given by:

$$\tilde{Y}_a = -j\frac{3ab\lambda_{10}}{8\pi a'^3_0} + \frac{2\pi Q_{A101}\lambda_{10}[1 - j[1 + F_{101}(\omega)]]}{\omega_{101}^2 \mu_0 \varepsilon_0 a_r b_r c_r^3[1 + j[1 + F_{101}(\omega)]]} \tag{3.31}$$

where

$$F_{101}(\omega) = Q_{A101}(\omega/\omega_{101} - \omega_{101}/\omega)$$

The ideal matching condition is

$$\tilde{Y}_a = \tilde{G}_a + j\tilde{B}_a = 1 \tag{3.32}$$

i.e.,

$$\tilde{G}_a = 1 \tag{3.33}$$

and

$$B_a = 0 \tag{3.34}$$

From (3.33), we find a quadratic equation for matching frequency ω:

$$\omega^2 \cdot \frac{1}{\omega_{101}} - \frac{\sqrt{\dfrac{2\pi Q_{A101} ab\lambda_{10}}{\omega_{101}^2 \mu_0 \varepsilon_0 a_r b_r c_r^3} - 1} - 1}{Q_{A101}} \cdot \omega - \omega_{101} = 0 \tag{3.35}$$

and from (3.34), we derive a formula for the matching radius a_0:

$$a_0' = \sqrt{\frac{-3ab\lambda_{10}}{8\pi\left[1 + Q_{A101}\left(\dfrac{\omega}{\omega_{101}} - \dfrac{\omega_{101}}{\omega}\right)\right]}} \tag{3.36}$$

In the above expressions λ_{10} and Q_{A101} are defined as:

$$\lambda_{10} = \frac{\lambda}{\sqrt{1 - \left(\dfrac{\lambda}{2a}\right)^2}} \tag{3.37}$$

$$\frac{1}{Q_{A101}} = \frac{1}{Q_{101}^{\sigma}} + \frac{1}{Q_{101}^{\Sigma}} + \frac{1}{Q_{101}^{W}} \tag{3.38}$$

With Q_{101}^{σ}, Q_{101}^{Σ} and Q_{101}^{W} the unloaded Q factor of the cavity, the radiation quality factor, and the waveguide-cavity coupling Q factor, are designated, respectively.

However, the following relationship is also valid:

$$\frac{1}{Q_{101}^{W}} = \frac{1}{Q_{101}^{\sigma}} + \frac{1}{Q_{101}^{\Sigma}} \tag{3.39}$$

Thus, if the matching condition is fulfilled, we obtain:

$$Q_{A101} = \frac{1}{2} Q_{101}^{W} \tag{3.40}$$

The cavity antenna efficiency is given by:

$$\eta_A = \frac{P_\Sigma}{P_W} = 1 - \frac{Q_{A101}}{Q_{101}^{\sigma}} \tag{3.41}$$

Figure 3.10 shows computed graphs of the cavity antenna quality factor Q_{A101} and the efficiency η_A as functions of a_0, with the input matching condition maintained.

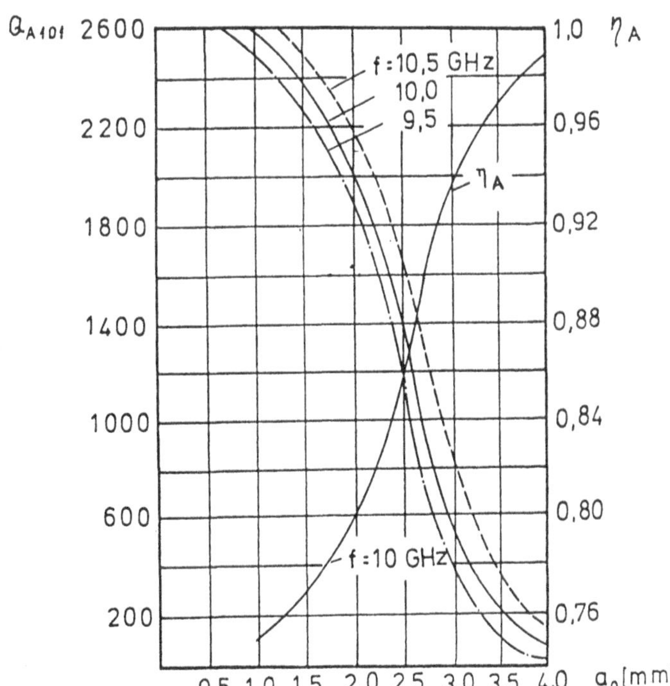

Figure 3.10 Antenna quality factor and efficiency in terms of radiating hole radius (from [1])

We see from the above figure that for a small transparency ($a_0 < 3$ mm), $\eta_A < 0.9$ and $Q_{A101} > 400$. Thus, for the X-band it is necessary that the hole radius a_0 be about 3.5 mm to 4.0 mm (0.12λ to 0.13λ). In this case, the antenna efficiency and the bandwidth will have satisfactory values. When $a_0 > 5$ mm, the electromagnetic coupling with the outer space increases too much, and the cavity field and radiation patterns become distorted in form (i.e., the antenna directivity goes down). Figure 3.11 shows the calculated relationship between a_0' and a_0, for which the input matching condition is fulfilled.

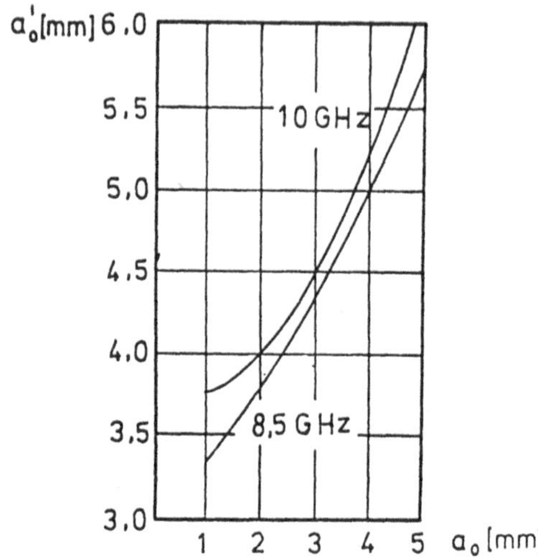

Figure 3.11 Relationship between radiating hole radius a_0 and excitation hole radius a_0', for cavity dimensions $a_r = b_r = 80$ mm, $c_r = c_r(a_r, f, a_0)$, ($f = 8.5$ GHz and $f_2 = 10$ GHz) (from [1])

In antenna technology, it is well known that better axial pattern symmetry and lower sidelobes are obtained from circular antenna shapes. Many circular cavity antennas with translucent apertures have been examined [1]. As a theoretical example of a circular cavity antenna, we consider the configuration shown in Figure 3.12. This antenna is seen as a piece of a circular waveguide, short-circuited at $z_1 = -c_r/2$, and having a semi-transparent perforated plate at $z_2 = +c_r/2$. In the cross-section ($z = 0$), a symmetrical thin wire radiator (electric dipole) is placed, with length 21 and a current density vector defined as:

$$\bar{J}(x,y,z) = \bar{y}_0 \frac{I_0}{\sin \beta_0 l} \delta(x)\delta(z) \sin \beta_0(l - |y|) \quad (-l \le y \le l) \tag{3.42}$$

where $\delta(x)$ and $\delta(z)$ are Dirac delta functions.

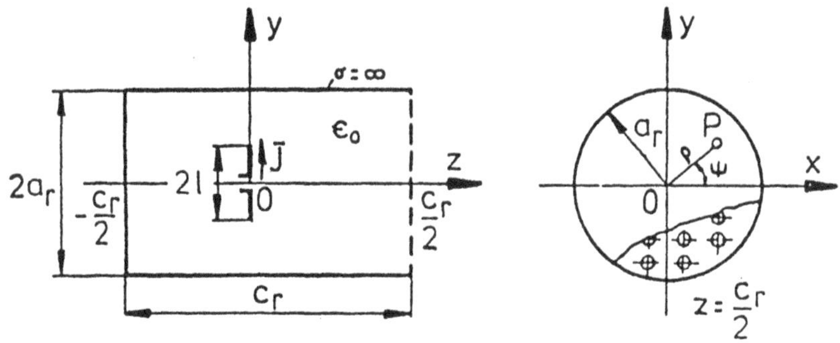

Figure 3.12 Circular closed-cavity antenna with perforated aperture wall and dipole feed (from [1])

Using the excitation theory presented in Chapter 2, the antenna has been studied for the cavity mode TE_{111}. The reflection coefficients at the basic cavity walls for this particular case are: $\Gamma_{1,11}$ at $z = -c_r/2$ and $\Gamma_{2,11}$ at $z = +c_r/2$. When $c_r = 0.5\lambda$, the reflection coefficients, transferred to the $z = 0$ plane, are given by:

$$\Gamma'_{1,11} = +1 \tag{3.43}$$

$$\Gamma'_{2,11} = j\frac{\tilde{B}_{2,11}}{2 + j\tilde{B}_{2,11}} \tag{3.44}$$

where $\tilde{B}_{2,11}$ is the normalized susceptance of the perforated aperture plate, given by (3.10). The input resistance of the tuned resonant feed dipole is found as [1,4]:

$$R_A^{(11)} = \frac{1}{|I_0|^2} \cdot \frac{|C_{2,11}|^2}{Z_{11}^H}(1 - |\Gamma'_{2,11}|^2)I''_{11} \tag{3.45}$$

where

$$C_{2,11} = -\frac{I_0 Z_{11}^H}{2\sin[(2\pi/\lambda)l]} \cdot \frac{I'_1}{I''_{11}} \cdot \frac{2}{1 + \Gamma'_{2,11}}$$

$$I'_{11} = \frac{2}{\beta_{C11}} \int_0^l \frac{1}{\rho} J_1(\beta_{C11}\rho)\sin\beta(l - \rho)\mathrm{d}\rho$$

$$I''_{11} = \pi \int_0^{a_r} \left[\frac{1}{\beta_{C11}^2} + \frac{1}{\rho} J_1^2(\beta_{C11}\rho) + \rho J_1'^2(\beta_{C11}\rho)\right]\mathrm{d}\rho$$

$$\beta_{C11} = \frac{P'_{1,11}}{a_r} = \frac{1,841}{a_r}; \quad \beta = \frac{2\pi}{\lambda}$$

and where Z^H_{11} is the TE_{11} mode waveguide impedance; J_1 is a Bessel function of first kind, first order.

The perforated aperture plate, for a slightly perturbed inner field, is represented as an array of radiating holes, illuminated by a plane wave. The electric field intensity of this non-uniform plane wave at the arbitrary hole center is

$$\bar{E}(\rho,\psi) = -j\omega\mu_0 \frac{4a_0}{3\pi} [\bar{H},\bar{z}^0] = -jE_0[\bar{h},\bar{z}^0] \tag{3.46}$$

or, in a normalized form:

$$|\tilde{E}(\rho,\psi)| = \left|\frac{E(\rho,\psi)}{E_0}\right| = \sqrt{h_x^2(\rho,\psi) + h_y^2(\rho,\psi)} \tag{3.47}$$

where

$$h_x = h_\rho \cos\psi - h_\psi \sin\psi \tag{3.48}$$

$$h_y = h_\rho \sin\psi + h_\psi \cos\psi \tag{3.49}$$

For the TE mode, the normalized components h_ρ and h_ψ are expressed as:

$$h_\rho(\rho,\psi) = \frac{1}{2Z^H_{11}} \left[J_0\left(\frac{1,841}{a_r}\rho\right) - J_2\left(\frac{1,841}{a_r}\rho\right)\right] \tag{3.50}$$

$$h_\psi(\rho,\psi) = \frac{1}{Z^H_{11}\beta_{C11}} \cdot \frac{J_1\left(\frac{1,841}{a_r}\rho\right)}{\rho} \sin\psi \tag{3.51}$$

where J_0 and J_2 are Bessel functions of zero and second order, respectively. The electric dipole is not an appropriate feed element for circular cavity antennas at higher microwave frequency bands. Instead, a cable-to-cavity transition in the form of a probe, or waveguide-to-cavity transition in the form of an iris are most popular in practice. Figure 3.13 shows sketches of several round closed-cavity antenna designs studied recently [1,5].

The basic circular cavity antenna is of length $c_r \approx 0.5\lambda$ and diameter d_r, with W_g = rectangular waveguide, and I = excitation iris. Its aperture wall is a perforated

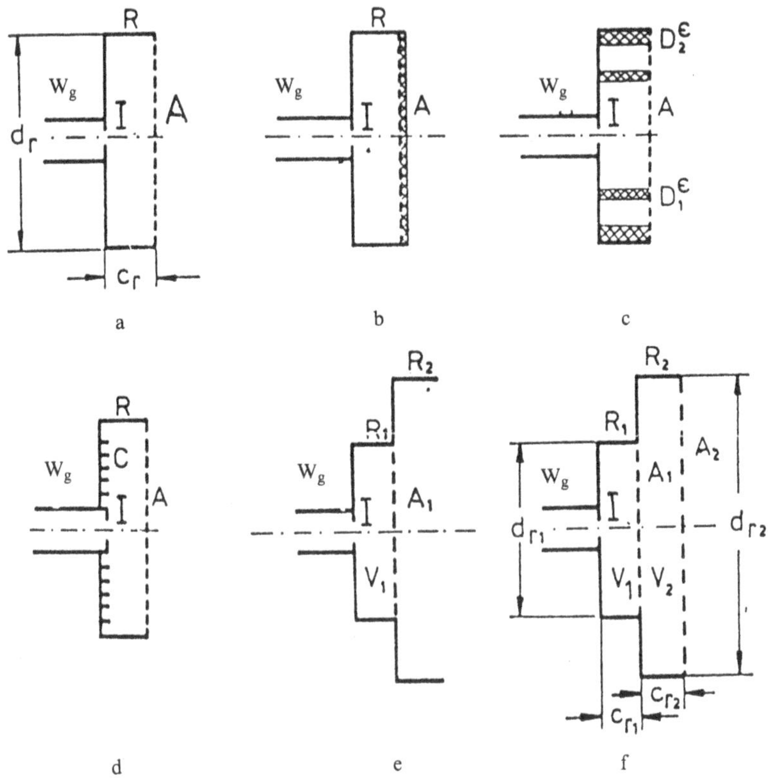

Figure 3.13 Round closed-cavity antenna versions: (a) with perforated aperture plate; (b) with printed translucent aperture plate; (c) with dielectric rings $D^{\varepsilon}_{1,2}$; (d) with corrugated back wall C; (e) with additional metal rim R_2; and (f) two-cavity antenna.

metal plate in the form of a square lattice having small holes 7 mm in diameter, spaced 10 mm apart. This basic configuration is shown in Figure 3.13(a). The semi-transparent aperture can be manufactured by a printed technology (Figure 3.13(b)).

In order to make some suitable changes in the amplitude-phase aperture distribution, dielectric rings $D^{\varepsilon}_{1,2}$ (Figure 3.13(c)) or a corrugated back wall C (Figure 3.13(d)) can be included in the cavity structure. Table 3.2 summarizes the main characteristics of single closed-cavity antenna models (numbers 1–3) which differ only in diameter. For comparison, the characteristics of an optimum conical horn with its aperture diameter equal to that of the number 1 antenna model (i.e., $d_c = d_{r1} = 90$ mm) are also given.

The table shows that having the same aperture area and almost equal gain G and efficiency e_A, the number 1 cavity antenna is six times shorter in length and two times smaller in volume than the optimum conical horn. This comparison, however,

Table 3.2

Antennas	3 dB beamwidth		Maximum sidelobe levels, dB		G	L.10^{-1}	V.10^{-3}	$e_A = \frac{G}{D_0}$
	H-plane	E-plane	H-plane	E-plane	[dB]	[mm]	[mm^3]	
Optimum conical horn (apert. dia. d_c = 90 mm)	25°	23°	−28	−20	17.5	9	191	0.63
Circular cavity antenna No 1 (d_{r1} = 90 mm)	25°	26°	−27	−26.5	17.2	1.5	95	0.59
Circular cavity antenna No 2 (d_{r2} = 120 mm)	24°	22°	−20	−18	18.1	1.5	169	0.41
Circular cavity antenna No 3 (d_{r3} = 170 mm)	16°	22°	−19	−20	19.0	1.5	340	0.25

is only valid for a very narrow frequency band (about 1.0 to 1.5% relative to the resonant frequency); and increasing the single-cavity antenna diameter causes an increase in the gain. This effect occurs only to a certain extent however. If $d_r >$ 4–5λ, the antenna efficiency and bandwidth decrease rapidly; therefore, we conclude that there is an optimum diameter of about 3.0–3.5λ for the single-cavity antenna configuration.

Figure 3.14 shows the *H*-plane theoretical (dashed line) and experimental (solid line) radiation patterns of the antenna model marked by number 1 in Table 3.2, for an optimum resonant frequency of 8.88 GHz. A promising idea for reducing the inherent limitation of the single-cavity antenna with respect to its gain, efficiency and bandwidth, was proposed recently [1,5], and materialized in an original design called the *two-cavity antenna* (Figure 3.13(f)). It consists of a waveguide feed W_g plus end-wall iris I, passage cavity V_1 and radiation cavity V_2. The latter is excited by a translucent wall A_1, through which a distributed electromagnetic coupling between V_1 and V_2 takes place. A double-cavity antenna model of dimensions d_{r1} = 90 mm, d_{r2} = 200 mm and $c_r = c_{r1} + c_{r2}$ = 30 mm has been examined experimentally at *X*-band. The radiation aperture plate A_2 and the aperture-exciter A_1 can be in the same form of the square hole array described previously. In Table 3.3, the performance of a two-cavity antenna is compared with that of an optimum conical horn; both have equal radiation aperture areas. The antenna efficiency in Tables 3.2 and 3.3 is defined as a ratio of the power gain G, and the normal aperture directivity $D_0 = 4\pi \times a/\lambda^2$, where a = aperture area.

From Table 3.3, we conclude that for almost the same gain and efficiency, the two-cavity antenna is fifteen times shorter in length and about eight times smaller

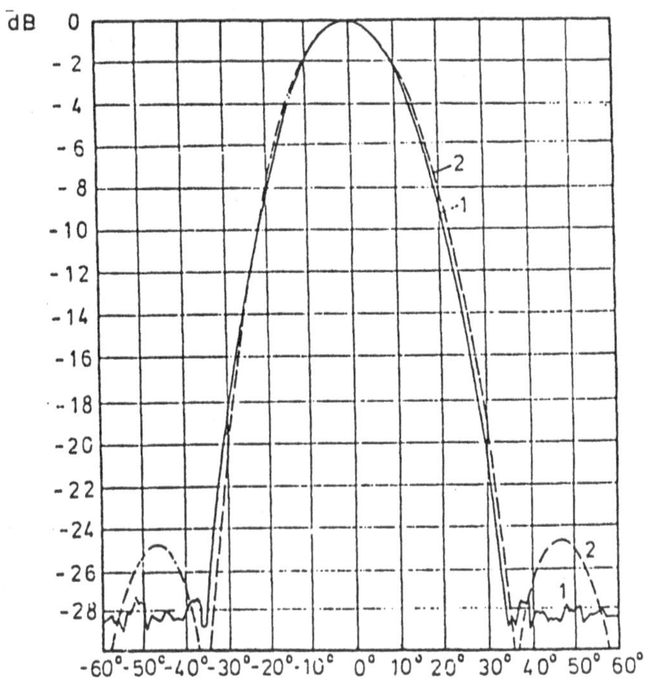

Figure 3.14 *H*-plane experimental (number 1 = solid line) and calculated (number 2 = dashed line) radiation patterns of circular single-cavity antenna 90 mm in diameter for f = 8.88 GHz (from [1])

Table 3.3

Antennas	*Gain* G[dB]	*Length* $L \cdot 10^{-1}$ [mm]	*Volume* $V \cdot 10^{-3}$ [mm^3]	*Efficiency* $e_A = \frac{G}{D_0}$
Two-cavity antenna	22.5	3	567	0.40
Optimum conical horn	23.8	45	4712	0.54

in volume than an optimum conical horn. After removing the radiating semi-transparent plate A_2, the antenna is transformed into a single-cavity antenna with a radiation aperture A_1 and peripheral flange-rim reflector (Figure 3.13(e)). This antenna construction also deserves attention. It has a gain of 19.5 dB and low sidelobes; −27 dB in *H*-plane, and −23 dB in *E*-plane. If we compare this antenna with one which has multisection band filters, a multicavity antenna filter can be realized. Figure 3.15 shows a model configuration, with three cavities having close resonant frequencies: $f_1 \approx f_2 \approx f_3$. In this way, we achieve frequency band broadening (i.e., $(\Delta f)_3 > (\Delta f)_1$). Therefore, the idea of multicavity antennas is also fruitful: a gain higher than the single-cavity antenna gain is achieved, and the antenna frequency bandwidth can be widened.

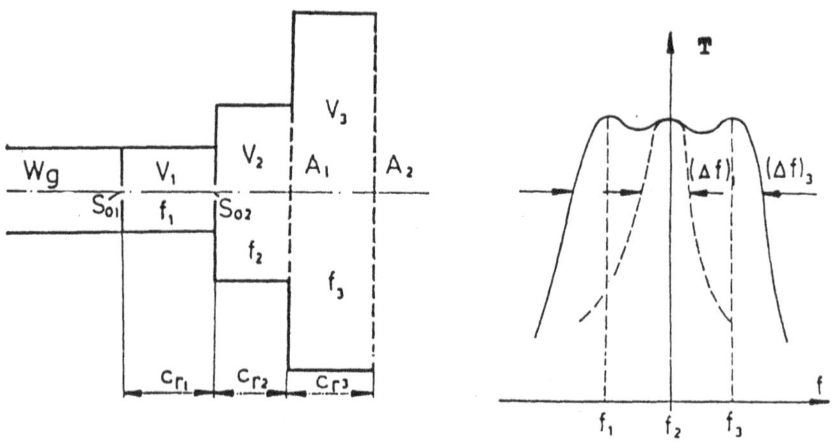

Figure 3.15 Three-cavity antenna and its transfer characteristic (transmission coefficient T in terms of frequency) (from [1])

3.1.3 Open Cavity Antennas with Translucent Plane Apertures

The so-called *partially reflecting sheet arrays* ([6,7]) or *image element arrays* are perhaps the first representatives of open cavity antennas. In general, the image element antenna consists of a radiating source between two planes, one a total reflector and the other a partially reflecting (translucent) surface.

A brief description of this antenna radiation mechanism is given in [8]. Figure 3.16 illustrates an image element configuration in which the total reflecting plane is a metal plate M, S is a primary source (i.e., wire dipole) and P is a partially-reflecting dielectric plate. The source is positioned between the two plates at distances x from M and y from P.

The parallel plates M and P cause multiple reflections (or images) S_1', S_1'', S_2' and S_2'', and due to this electromagnetic phenomenon, a substantial increase in the antenna gain is possible. The source radiation and multiple source reflections (images) result in energy radiation by rays 1, 2, 3, ... n. The total electric far-field intensity is expressed as a geometric progression, the sum of which is given by [8]:

$$E = \frac{TE_0(1 - A)}{1 + \Gamma AB} \tag{3.52}$$

where

Γ = reflection coefficient;
T = transmission coefficient;
A = $\exp(-j(2\pi/\lambda) \cdot 2x \cdot \cos \phi)$
B = $\exp(-2(2\pi/\lambda) \cdot 2y \cdot \cos \phi)$

If we assume that a broadside array is to be used ($\phi = 0$) and set $x = 0.25\lambda$ and $y = 0.5\lambda$, then (3.52) is reduced to:

$$E = \frac{2TE_0}{1 - \Gamma} \tag{3.53}$$

Because $T = 1 + \Gamma$, (3.53) may be expressed as:

$$E = \frac{2E_0(1 + \Gamma)}{1 - \Gamma} \tag{3.54}$$

Thus, (3.54) requires that E has a real, positive value, so that the antenna gain can be realized. For example, in the theoretical (lossless) case, when $\Gamma = 0.6$, $E = 8E_0$ (i.e., gain of 18.06 dB is obtained). Figure 3.17 shows an image element antenna which has an iris (slot) for the source of radiation. Slot S is located in the plane of reflector M. In this case,

$$E = \frac{TE_0}{1 + \Gamma B} \tag{3.55}$$

where

$$B = \exp(-j(2\pi/\lambda) \cdot x' \cdot \cos \theta)$$

If we set the spacing $x' = 0.25$, and substituting $(1 + \Gamma)$ for T, (3.55) reduces to:

$$E = \frac{E_0(1 + \Gamma)}{1 - \Gamma} \tag{3.56}$$

resulting in gain for positive values of Γ. Again, as for the previous antenna configuration, if $\Gamma = 0.6$, $E = 4.125E_0$, corresponding to a gain of 12.31 dB. If a plurality of source elements is combined in the structure, so that their radiation fields are added, an antenna system having much greater gain with fewer source elements for a given array aperture than conventional broadside antenna arrays is realized. Figure 3.18 shows an exploded view of an antenna system for flying objects (missiles, planes, *et cetera*), comprising a circular image element antenna fed by a square array of four slots.

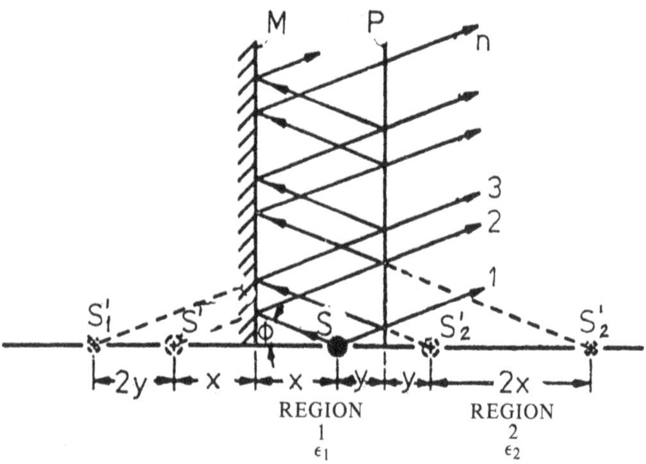

Figure 3.16 Image element antenna with primary source between reflector plane and partially reflecting plane (from [8])

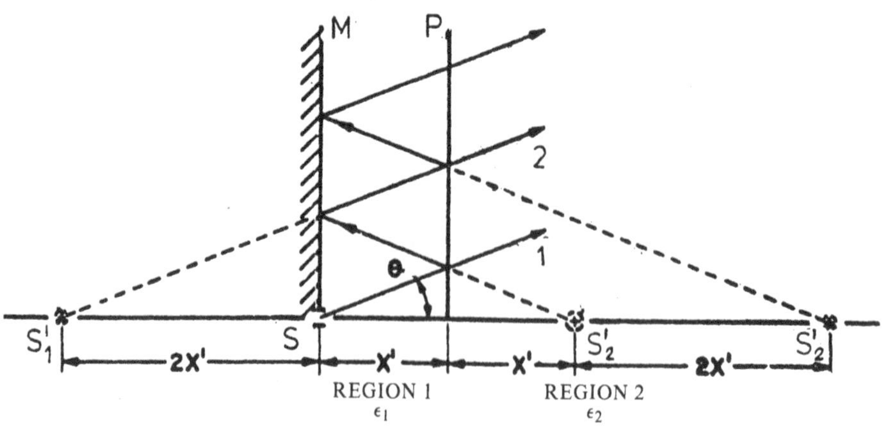

Figure 3.17 Image element antenna with primary source located in the plane of reflection (from [8])

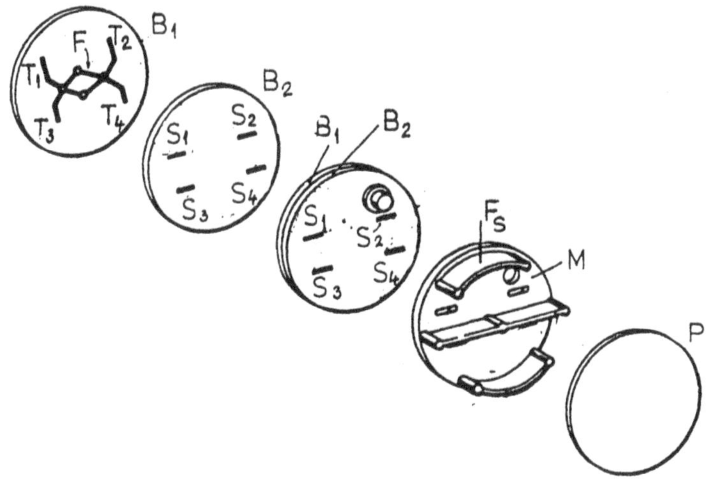

Figure 3.18 Exploded view of an antenna system for flying objects (from [8])

Feed network F is fabricated using conventional stripline techniques on a teflon (PTFE) fiberglas board B_1. Slot source board B_2, also of copper-clad fiberglas, includes the four slot sources S_1–S_4 and is mated to feed network board B with the alignment of output terminals $T_1 - T_4$ of feed network F, which are in line with the slot sources. Board B_2 is then aligned and fixed to reflector M. Dielectric disk plate P is attached to reflector board M, thus completing the open-cavity antenna. The dielectric plate is spaced from reflector M appropriately (as discussed earlier in regard to the antenna in Figure 3.17), by spacer flanges F_s. Thus, this kind of open-cavity antenna comprises four individual image element antennas. By using stripline comparators in the feed network, sum channel signal information, and two difference channels, signal information is developed for guiding the flying object.

A modification of the two-cavity antenna, is made by removing the metal rim R_2 [1]. In this way, the closed radiation cavity V_2 is transformed into an open radiation cavity. The perforated translucent plate is spaced by a dielectric (or absorbing) spacer rim R_d (Figure 3.19(a)). As a result, the new antenna configuration comprises a closed-type passage cavity V and an open-type radiation cavity V_2. Figure 3.19(b) shows a picture of the antenna in its disassembled form. The perforated aperture disk, labeled 1, is made on a fiberglas board using printed circuit technology. The dielectric rim spacer is labeled 2, and the passage perforated disk is labeled 3. The antenna is fed by a rectangular waveguide, labeled 4, through a capacitive iris, labeled 5. For X-band, the examined antenna model has the following dimensions: $c_{r1} = 15$ mm, $c_{r2} = 30$ mm, $d_{r1} = 120$ mm, and $d_{r2} = 200$ mm.

Figure 3.20 shows the E-plane (left) and H-plane (right) radiation patterns for the optimum frequency = 10090 MHz (solid line) and for a higher frequency = 10150 MHz (dashed line). We conclude that the antenna pattern is characterized by

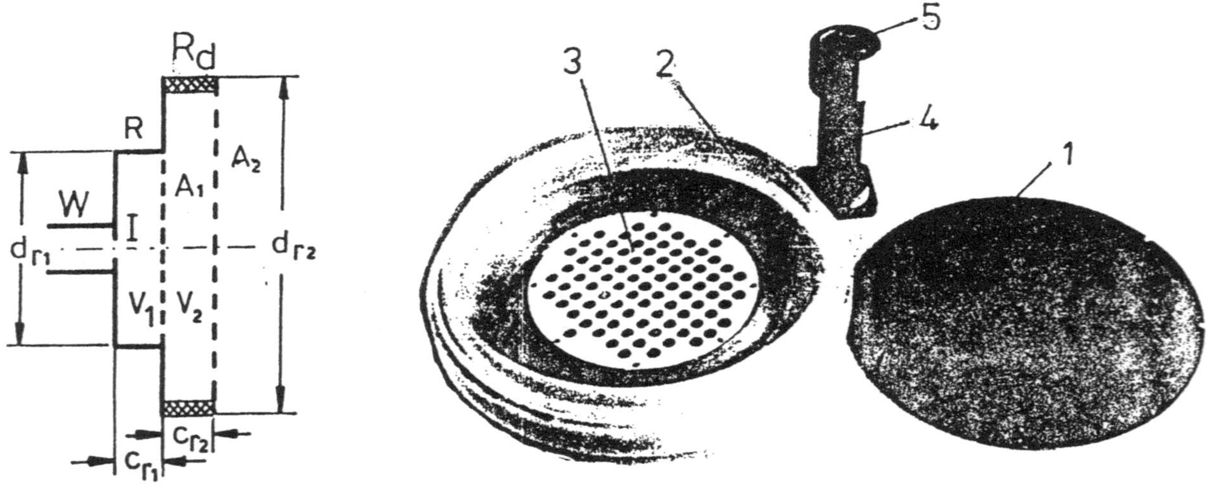

(a) (b)

Figure 3.19 Two-cavity antenna with open radiation cavity: (a) antenna geometry and (b) picture of disassembled antenna

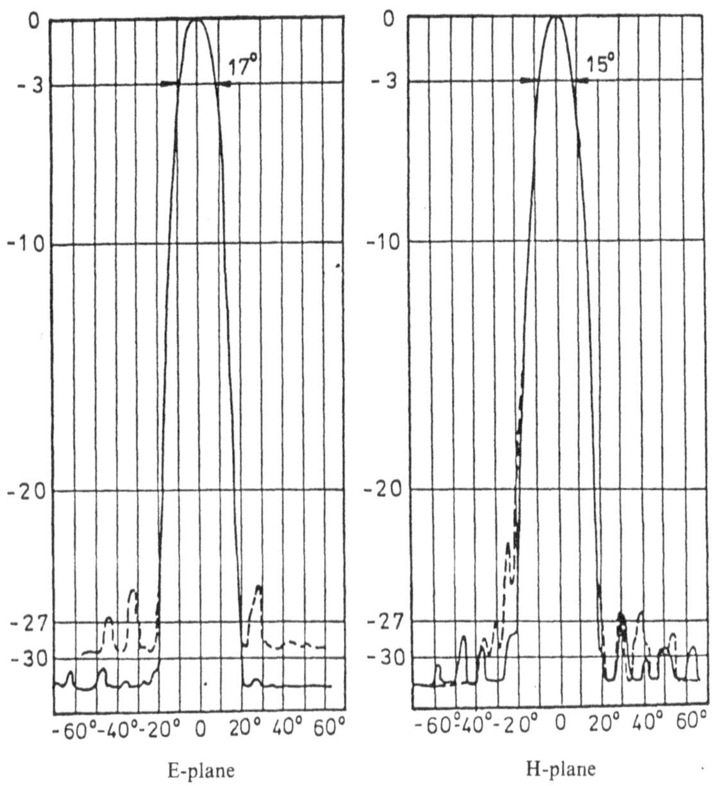

Figure 3.20 Antenna radiation patterns: f = 10090 MHz (solid line) and f = 10150 MHz (dashed line)

good axial symmetry and very low sidelobes for the aperture size. Klass [9] has developed a plane-spherical configuration of the open-cavity antenna (Figure 3.21(a)) to provide a Gaussian-shaped microwave beam for helicopter model stationkeeping. It consists of a horn-fed spherical metal reflector with microwave energy at its center, over which a metal perforated aperture disk 228 mm in diameter is suspended. A sheet of low-loss plastic with a thickness of 6.3 mm is placed on top of the perforated plate.

To suppress energy spillage and to decrease the radiation pattern sidelobes, a metal baffle plate and a ring of absorbing material around the resonator are added. At X-band, the overall experimental antenna efficiency is about 80–85%. The rectangular open beam waveguide has been used by Brauer [10–11] to form a closed metal wall cavity (Figure 3.22). Then, by cutting small openings along the $z = 0$ plane wall, a cavity antenna with a Gaussian-type radiation pattern in the $x0z$ plane is made.

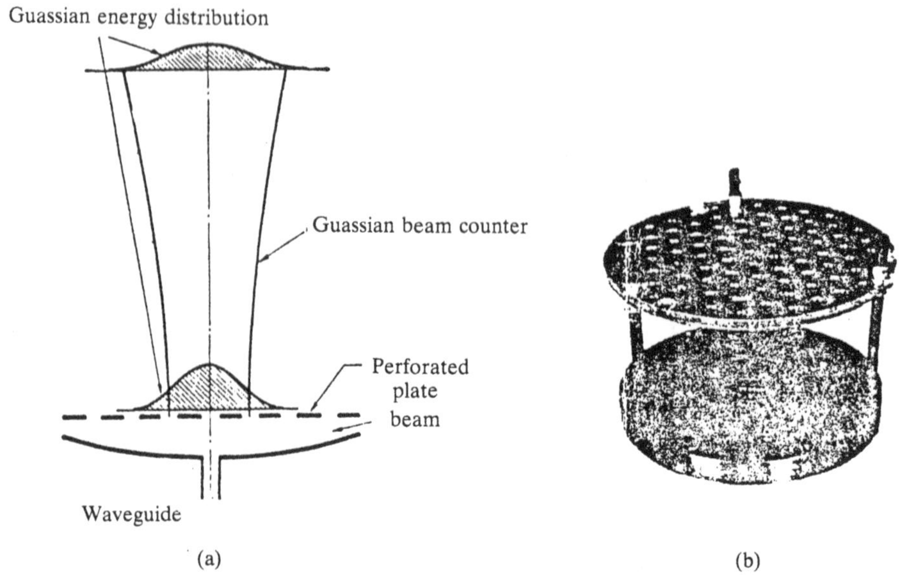

Figure 3.21 Circular beam-waveguide cavity antenna with Gaussian beam: (a) antenna sketch; (b) antenna picture (from [9])

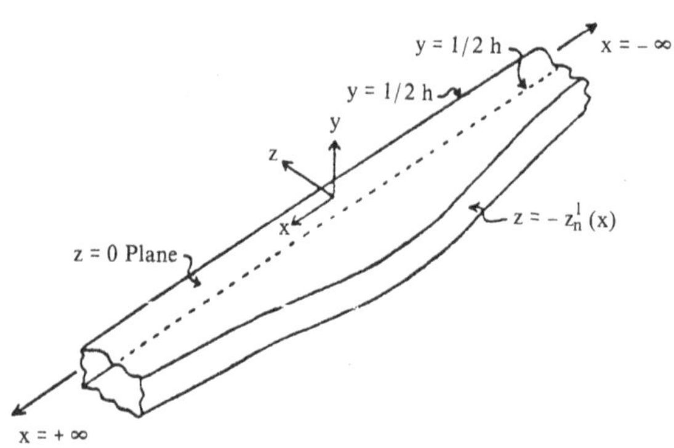

Figure 3.22 Rectangular beam waveguide cavity (from [11])

The x dependence of nth mode is given by:

$$g_n(x) = He_n(2u_0x) \exp(-u_0^2x^2) \tag{3.57}$$

where

He_n = the Hermite polynomial of degree n;
u_0 = parameter determining field variation along the x direction; and

$\exp(-u_0^2x^2)$ = Gaussian function. Because the Gaussian factor gives small $g_n(x)$ at $|u_0x| \gtrsim 3$, the cavity can be made of finite length L in the x direction. We assume, therefore, that the cavity extends from $x = -0.5L$ to $x = +0.5L$. So for finite length L, the cavity is not completely enclosed by conductors, but has two small openings to free space at $x = +0.5L$, and $-0.5L$. If L is large enough, these openings will not allow leakage of fields. If openings are cut in the $z = 0$ wall of the cavity to serve as a line source of radiation on the x axis, then the near-field distribution is given by (3.57).

At the resonant frequency of the $n = 0$ mode, the line source distribution is Gaussian, which should give a Gaussian radiation pattern. At the resonant frequency of the $n = 1$ mode, the distribution function is the asymmetric function:

$$g_1(x) = 2u_0x \exp(-2u_0^2x^2) \tag{3.58}$$

The Poynting vector of the radiated field produced by the line source is derived in terms of the coordinate system of Figure 3.23(a), and given by:

$$\bar{S} = \left(\frac{1}{\eta}\right) \bar{u}_R(\sin^2\phi + \cos^2\theta\cos^2\phi)\frac{I_y^2\pi}{\lambda^2R^2u_0^2} \cdot He_n^2\left(\frac{k\sin\theta\cos\phi}{u_0}\right)\exp\left(\frac{-k^2\sin^2\theta\cos^2\phi}{2u_0^2}\right) \tag{3.59a}$$

The following assumptions are made to derive (3.59a): the holes in the $z = 0$ wall open into the conducting ground plane $z = 0$ and $L_s = \infty$. The radiation patterns are calculated for $n = 0$ mode and $n = 1$ mode. (a) For symmetrical $n = 0$ mode, from (3.59a), the radiation pattern is essentially Gaussian and given by:

$$\bar{S} = \left(\frac{\bar{u}_R}{\eta}\right)(\sin^2\phi + \cos^2\theta\cos^2\phi)\frac{I_y^2\pi}{\lambda^2R^2u_0^2} \cdot \exp\left(\frac{-k^2\sin^2\theta\cos^2\phi}{2u_0^2}\right) \tag{3.59b}$$

Figure 3.23(b) shows the experimental and calculated radiation patterns of the antenna for $n = 0$ mode. For the asymmetrical $n = 1$ mode, the radiation pattern is a derivative of the Gaussian function, and is given by:

$$\bar{S} = \left(\frac{\bar{u}_R}{\eta}\right)(\sin^2\phi + \cos^2\theta\cos^2\phi)\frac{I_y^2\pi}{\lambda^2R^2u_0^4}k^2\sin^2\theta\cos^2\phi$$
$$\cdot\exp\left(\frac{-k^2\sin^2\theta\cos^2\phi}{2u_0^2}\right) \quad (3.60)$$

Figure 3.23(c) shows the experimental and theoretical radiation patterns of the antenna for the $n = 1$ mode. In (3.59a) and (3.60) $I_y = \omega\mu_0 D^3\alpha_n/3a\eta$, and $\eta = 120\pi$ ohms. The radiation pattern produced by the examined beam of the waveguide cavity antenna is of interest in monopulse radar tracking. The equation for the curved reflector surfaces $z_n'(x)$ of the cavity is given by:

$$z_n'(x) \cong \frac{\pi/k}{1 - \left(n + \frac{1}{2}\right)(\omega_0/kz_0) + \omega_0^2x^2/2z_0^2} \quad (3.61)$$

where

w_0 = $2u_0^2z_0^2/k$ is a dimensionless parameter;
$2z_0$ = spacing between beam waveguide phase transformers (typically hundreds of wavelengths)
k = $2\pi/\lambda$—free-space phase constant.

To verify the theory, Brauer [11] designed the antenna model to have the resonant frequency of its $n = 0$ mode ($f_{n=0}$) at about 9 GHz. The ratio w_0/z_0 has been chosen as $(5/100\lambda_0) = (1/20\lambda_0) = (1/3.33\text{ m})$, resulting in frequency spacing between resonant modes $f = f_{n+1} - f_n = 72$ MHz. The total antenna length $L = 760$ mm. The aperture plate has a width $w = 19$ mm. $z_0'(0) = 16.7$ mm and z_0' (380 mm) $= 14.3$ mm (i.e., $z_n'(x)$ decreases very gradually with increasing (x^2)).

The coaxial-to-waveguide junction feeds a piece of rigid 50-ohm coaxial cable force-fitted into a threaded sleeve. The center conductor of the cable extends 3 mm beyond the end of the sleeve, and acts as the probe. A hole is drilled and tapped in the resonator at the position ($x = 0$, $z = -8.4$ mm) to receive the probe. For the $n = 1$ mode, the field distribution is zero at $x = 0$ and, in this case, two probes have been used, one at each maximum in the field distribution (at $z = -8.4$ mm, $x = +60$ mm, and $x = -60$ mm). The probes have been excited 180° out of phase by means of a hybrid tee.

In Table 3.4, the experimental and theoretical values of quality factor and total gain for the $n = 0$ mode are listed in terms of the hole diameter and the spacing between two adjacent holes. We find from experiments that even for $d_0 = 4.44$ mm and $b = 5.1$ mm, the radiation pattern differs from the theoretical Gaussian function.

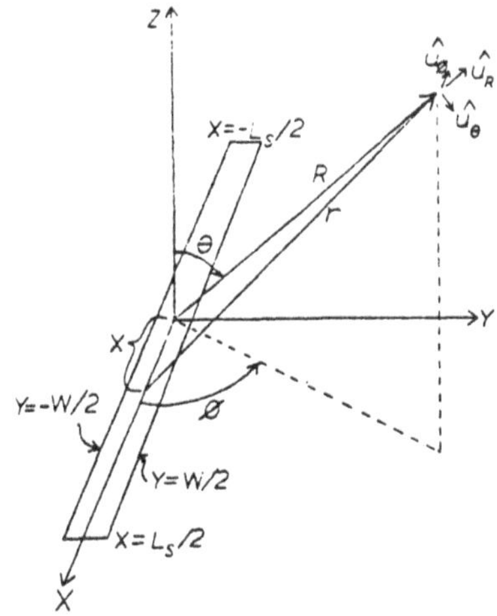

Figure 3.23(a) Coordinate system of the antenna [11]

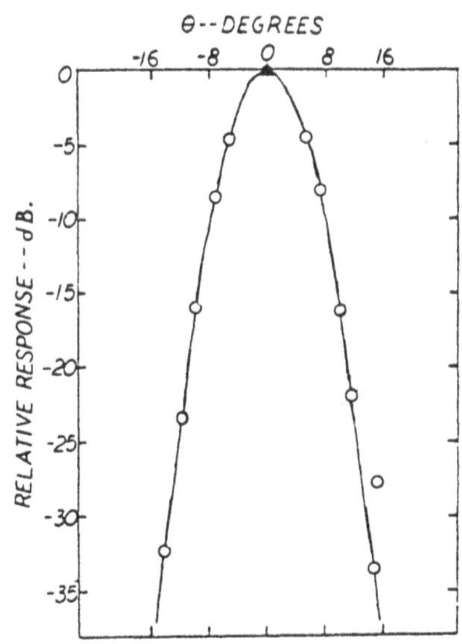

Figure 3.23(b) Radiation pattern of the antenna for n = 0 mode (from [11]); experimental (d_0 = 3.8 mm) = O-O-O, theoretical = ———

The proposed beam-waveguide cavity antenna, although very narrow-banded, is attractive in several ways. It is easy to fabricate and low in cost. In addition, the

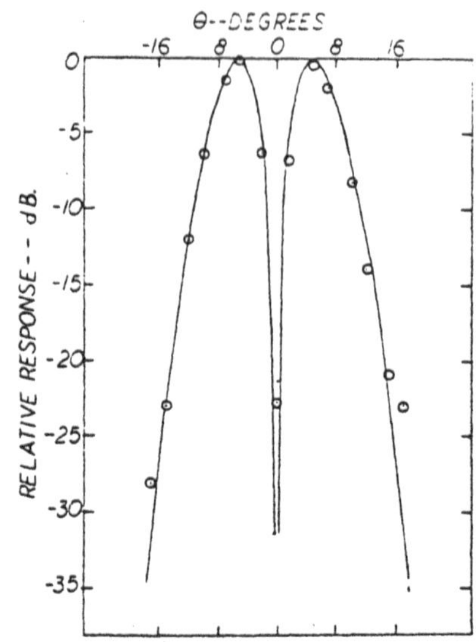

Figure 3.23(c) Radiation patterns of the antenna of n = 1 mode (from [11]); experimental (d_0 = 3.8 mm) = O-O-O, theoretical = ———

Table 3.4 (from [11])

Hole diameter (mm)	*Center-to-center hole spacing* (mm)	*Resonant Frequency* (GHz)	Q_{00_T} *Theory*	Q_{00_T} *Experiment*	G_T *Theory* (dB)	G_T *Experiment* (dB)
2.5	3.2	8.940	2130	1400	11.6	8.6
3.8	4.44	8.892	820	840	12.6	12.0
4.44	5.1	8.878	480	620	13.0	13.4

radiation patterns produced by the antenna are promising for use in monopulse radar and tracking radio systems.

3.2 SHAPED TRANSLUCENT-APERTURE CAVITY ANTENNAS

3.2.1 Antenna Shape Synthesis Method

There are currently few examples in which the antenna configuration is predicted directly by a rigorous electromagnetic solution. That sort of case is the creation of a new cavity antenna by Katzenelenbaum *et al.* [12–14], which we will call a shaped translucent-aperture cavity antenna. This antenna is essentially a cavity resonator with a volume V^+ enclosed between two appropriately-shaped surfaces (S_1 and S_2, Figure 3.24). The surface S_1 is a total reflector and S_2 is an electromagnetically-translucent surface. The outer infinite space is marked by V^- and $\bar{N}$ is a normal vector from V^+ to V^-.

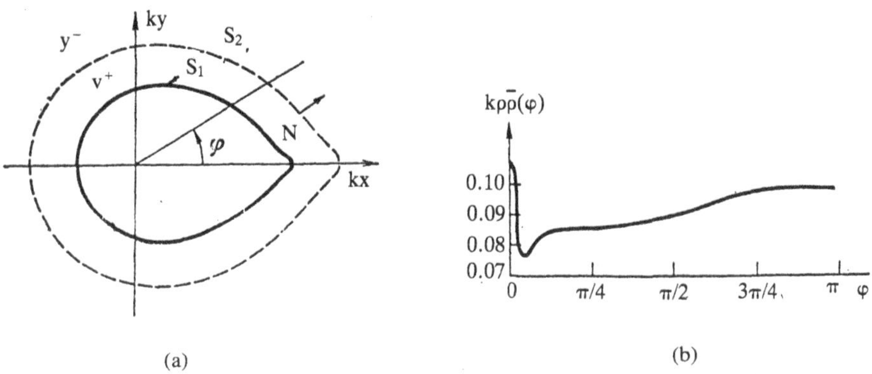

Figure 3.24 Shaped translucent-aperture cavity antenna (from [12]): (a) two-dimensional antenna geometry; (b) transparency distribution on the outer contour

The antenna synthesis method considered later [12,14] provides a possibility of finding the shapes of the two cavity surfaces and the transparency distribution on the outer surface S_2. The well known method of characteristic modes is widespread for a solution of endoresonant diffraction problems. According to this method, the field excited by an excitation function J in a closed cavity is represented as an expansion of accessory functions called *eigenfunctions* or *characteristic modes* of the examined cavity.

The nonuniform wave equation for the scalar function U is written as:

$$\Delta U + k^2 U = J \tag{3.62}$$

Here U is defined in the closed region V^+ and satisfies the following condition on the boundary surface S:

$$U|_S = 0 \tag{3.63}$$

The excited cavity field function U is expanded as

$$U = \sum_n A_n u_n \tag{3.64}$$

where u_n is the nth characteristic mode, and A_n is its amplitude coefficient.

In constructing solutions in the form (3.64), the excitation function J is not taken into consideration. The characteristic mode u_n satisfies the uniform wave equation as follows:

$$\Delta u_n + k_n^2 u_n = 0 \tag{3.65}$$

and the boundary condition

$$u_n|_S = 0 \tag{3.66}$$

In the non-uniform problem, k is a given value ($k = 2\pi/\lambda = 2\pi f/c$) and in the uniform problem k_n is a characteristic value ($k_n = 2\pi f/c$), for the air-filled cavity. The characteristic value is a real number expressed by its characteristic mode:

$$k_n^2 = \frac{\int |\nabla u_n|^2 \mathrm{d}V}{\int |u_n^2| \mathrm{d}V} \tag{3.67}$$

The amplitude coefficients A_n are given by:

$$A_n = \frac{1}{k^2 - k_n^2} \frac{\int J u_n \mathrm{d}V}{\int u_n^2 \mathrm{d}V} \tag{3.68}$$

The method of characteristic modes is convenient when the solution of (3.62) is given by a one-term expression:

$$U \sim \frac{C_n}{k^2 - k_n^2} u_n \tag{3.69}$$

which is valid for the cavity with a high quality factor. The excitation currents are included in the coefficient C_n. Because the wave number k is proportional to the frequency f, this method is called the k method. It cannot be carried directly over to open or half-open cavities (i.e., for solving outer radiation problems), because the cavity mode functions increase infinitely when the observation point is taken to infinity.

The k method was used in Chapter 2 to obtain an approximate excitation solution for a generalized cavity antenna model. It represents a special case of a more general diffraction method, described in [14] and called a *generalized method of the characteristic modes*. This generalized method may be used for all kinds of resonators, open and closed, and in general for various diffraction problems concerning finite bodies. According to its basic idea, solutions of the uniform problem are used as characteristic functions, in which a characteristic value in general is not k (or frequency), but some other material parameter (i.e., dielectric constant, in which case the method is called ε *method*). What kind of electromagnetic material parameter is chosen as a characteristic value depends on the diffraction problem type. In all varieties of this generalized method, the characteristic functions correspond to real frequencies.

The shaped translucent-aperture cavity antenna can be synthesized using a special variety of the generalized method named ζ method, after the aperture electromagnetic transparency function marked with ζ. The radiating cavity is synthesized in such a way that one of its characteristic fields in the outer space is equal to some field u^s, which for a two-dimensional case at infinity, has the form:

$$u^s(r,\phi) \simeq \Phi(\phi)\frac{e^{-j\kappa r}}{\sqrt{kr}} \quad (kr \to \infty) \tag{3.70}$$

where $\Phi(\phi)$ is given radiation function (pattern). On the translucent surface S_2, the boundary conditions are

$$u^+ = u^- \tag{3.71}$$

$$u = \rho\tilde{\rho}(s)v \tag{3.72}$$

$$v = \frac{\partial u^-}{\partial N} - \frac{\partial u^+}{\partial N} \tag{3.73}$$

In (3.72), $\tilde{\rho}(s)$ is a function characterizing the transparency distribution, $v = v(s)$ is called a current on the surface s_2 and ρ is a numerical parameter, defined by the degree of transparency. The total field $\boldsymbol{u}$ in the excitation problem satisfies (3.62), the boundary conditions of (3.71)–(3.73) on S_2 and the Sommerfeld radiation condition.

The characteristic mode, u_n, corresponding to the field u^s (3.70) also satisfies (3.62), and the analogous boundary conditions on the surface s_2:

$$u_n^+ = u_n^- \tag{3.74}$$

$$u_n = \rho_n\tilde{\rho}(s)v \tag{3.75}$$

$$v_{n_2} = \frac{\partial u_n^-}{\partial N} - \frac{\partial u_n^+}{\partial N} \tag{3.76}$$

where $\tilde{\rho}(s)$ is the same function as in (3.72) and ρ_n = the characteristic value of the corresponding uniform problem. The radiation condition for u_n is similar to (2.02)

$$u_n(r,\phi) \simeq \Phi(\phi)\frac{e^{-j\kappa r}}{\sqrt{kr}} \quad r \to \infty \tag{3.77}$$

In accordance with the ρ method, the total excitation field in the high quality factor cavity antenna with a translucent aperture has the following form [12]:

$$u \simeq U^0 + A_n u_n = \frac{D_n}{\frac{1}{\rho} - \frac{1}{\rho}} u_n \tag{3.78}$$

where

$$D_n = \frac{\int_s u_n \frac{\partial U^0}{\partial N} ds}{\int_s (u_n^2/\tilde{\rho}(s))ds} \tag{3.79}$$

where U^0 is the inside field of the same primary sources, when S_2 is also a total reflecting (metalized) surface. When the sources are inside the cavity $U^0 \equiv 0$ out of the cavity.

If the amplitude coefficient A_n(ρ method) from (3.78–3.79) is compared with the amplitude coefficient A_n(k method) from (3.68), we see that they have similar resonant factors—$1/(k^2 - k_n^2)$ and $1/(1/\rho_n - 1/\rho)$, respectively. The resonant factor $1/(1/\rho_n - 1/\rho)$ as a function of k is a frequency resonant characteristic. The characteristic value is a fast-changing complex function of the frequency. The resonant frequency is the frequency for which $|1/\rho_n - 1/\rho|$ is minimum, and for this frequency, marked by k_*, $Re(1/\rho_n) = 1/\rho$.

The synthesis problem of the high quality factor cavity according to (3.70) has no unique solution. One possible way for the antenna synthesis is described below. The characteristic field mode u_n^- is restored in the whole volume out of the cavity according to the condition $u_n^- = u^s$, where u^s is a solution of the uniform wave equation:

$$u^s = \sqrt{\frac{\pi}{2}}\, e^{-j\pi/4} \sum_m \alpha_m e^{-j\pi \frac{m}{2}} H_m^{(2)}(kr) \cos m\phi \tag{3.80}$$

and α_m is a Fourier coefficient in the given radiation pattern. A closed surface on which u^s has a constant phase is chosen as an antenna surface. For that purpose, one of the surfaces will be found using the following expression:

$$\bar{u} = Re\, u^s + \cot \psi_2\, Im\, u^s \tag{3.81}$$

where ψ_2 is the phase on S_2 and it is proved that

$$\bar{u}|_{s_2} = 0 \tag{3.82}$$

Inside the cavity, the next solution is constructed as follows:

$$u_n^+ = \frac{\cot \psi_2 + j}{\cot \psi_1 - \cot \psi_2} (Re\, u^s + \cot \psi_1\, Im\, u^s) \tag{3.83}$$

The field (3.83) (in the general case) is not a solution because of its irregularity for $r = 0$. We could show that (3.83) has no irregularities if it satisfies the condition $\phi^*(\rho) = je^{+j2\psi_1}\phi(\rho + \pi)$, where ψ_1 is the phase on the inner surface.

If the above condition is not fulfilled, then the cavity must consist of two closed surfaces and the inner surface has to be metalized (total reflector) as is shown in Figure 3.24(a). Thus, the field (3.83) is equal to zero on the metal surface S_1

$$Re\, u^s + \cot \psi_1\, Im\, u^s = 0|_{S_1} \tag{3.84}$$

This is how the antenna configuration is obtained as a result of the theory.

To be more specific, one possible synthesis solution of the two-dimensional cavity antenna can be performed as follows:

a) Given u_n^- and u_n^+, the current distribution ν_2 on the aperture contour is found:

$$\nu_{n2} = \frac{-\cot \psi_1 + j}{\cot \psi_2 + \cot \psi_1} (\partial \bar{u}/\partial N) \tag{3.85}$$

b) For $k = k_*$, the characteristic value ρ_n is calculated as follows:

$$\rho_n(k_*) = \frac{\cot \psi_2 - \cot \psi_1}{\cot \psi_1 \cot \psi_2 + 1 - j(\cot \psi_2 - \cot \psi_1)}; \tag{3.86}$$

c) If $u^s = |u^s|e^{j\psi}$, the transparency distribution function $\tilde{\rho}(s)$ is

$$\tilde{\rho}(s) = \frac{(\cot^2 \psi_2 + 1) Im\, u^s}{\dfrac{\partial \bar{u}}{\partial N}} = \frac{1}{\dfrac{\partial \Psi}{\partial N}(s)} \tag{3.87}$$

$$\rho = \frac{1}{Re(1/\rho_n(k_*))} = \frac{\cot \psi_2 - \cot \psi_1}{\cot \psi_1 \cot \psi_2 + 1} \tag{3.88}$$

d) Current distribution on the inner surface S_1 is given by:

$$v_{n1} = (\cot \psi_2 + j) \left. \frac{\partial Im\, u^s}{\partial N} \right| \tag{3.89}$$

The surfaces S_2 and S_1 are equiphase ones, but ψ_2 is not equal to ψ_1. According to (3.86), the characteristic value ρ_n is in the order of $1/\cot \psi_2$, while $Im\rho_2$ - is in the order of $1/\cot^2 \psi_2$.
The antenna quality factor is calculated by:

$$Q \approx \frac{p\pi}{2} \cot^2 (\psi_2 - \psi_1), \quad p \neq 0 \tag{3.90}$$

3.2.2 Some Examples of Shaped Translucent-Aperture Cavity Antennas

The antenna configuration in Figure 3.24(a) has been synthesized for a given radiation pattern $\Phi(\phi) = \sin^8(\phi/2)$, $|\cot \psi_1| = \infty$ and $|\cot \psi_2| = 8$. The contours of the surface S_2 and S_1 are calculated from (3.82) and (3.85), respectively. Figure 3.24(b) shows the dimensionless distribution quantity $k\rho\tilde{\rho}(\phi)$, proportional to the transparency distribution function $\tilde{\rho}(\phi)$ on the aperture surface S_2. The experiment (Figure 3.25(a,b)) shows that the cavity antenna described by Katzenelenbaum creates (in the outer region V^-) very small reactive fields, and the radiation pattern is practically the same when the distance between the radiating horn and receiving test antenna is changed from 0.35 m to 2.5 m. The cavity antenna we have examined has a very wide radiation pattern which in theory is sidelobeless. Its x-y cross section has a maximum dimension of about 2.5λ. The corresponding linear antenna with a constant amplitude and phase distribution has a 3 dB beamwidth $= 23°$ and high sidelobes of about -7 dB.

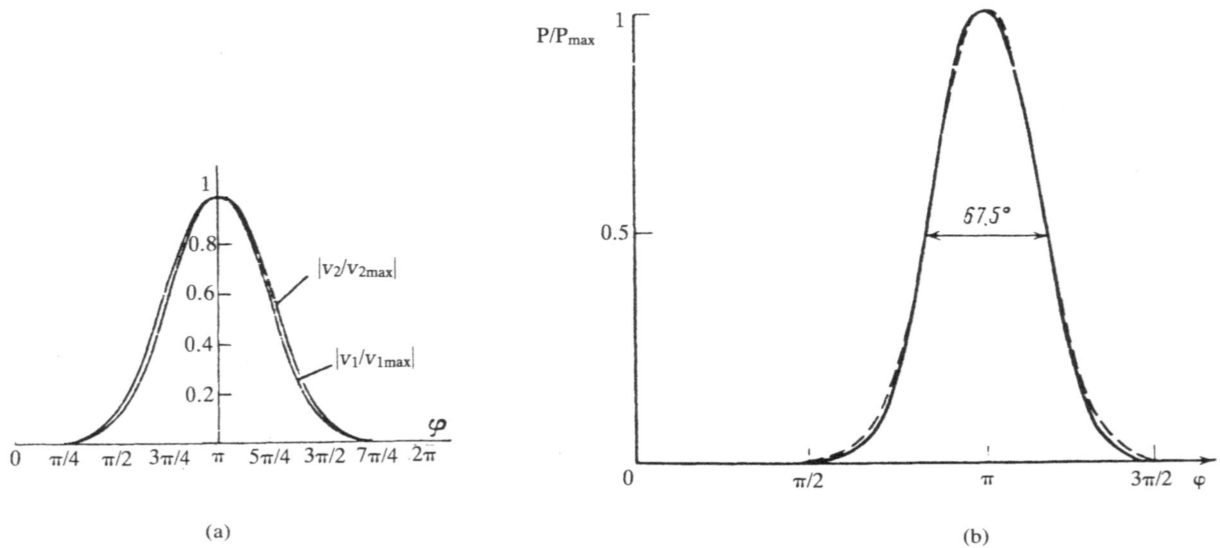

Figure 3.25 Polar antenna diagrams: (a) current distributions on the shaped surfaces S_1 and S_2 (from [12]); (b) *H*-plane radiation patterns (measured-solid line and calculated-dashed line) (from [13])

REFERENCES

1. Hristov, H.D., "Cavity Antennas," D.Sc. Thesis, 1987. Available from Central Inst. for Scientific and Technical Information (Tz.I.N.T.I.), Sofia (in Bulgarian).
2. Jacobsen, S., E. Andersen, and M. Groundlund, "An Antenna Illuminated by a Cavity Resonator," *Proc. IEEE*, November 1963, pp. 1431–1435.
3. Collin, R.E., *Field Theory of Guided Waves*, McGraw-Hill, New York, 1960.
4. Hong, M.H., D.P. Nyquist, and K.M. Chen, "Radiation Fields of Open-Cavity Radiators and a Backfire Antenna," *IEEE Trans. Antennas and Propagation*, Vol. AP-18, November 1980, pp. 813–815.
5. Hristov, H.D., "Resonant Antennas with Cavity-Backed Semi-Transparent Apertures," *Fifth Int. Conf. on Antennas and Propagation*, York, UK, March 30–April 2, 1987, pp. 75–78.
6. Vainshtein, L.A., *Open Resonators and Waveguides*, Sovjetskoe Radio Publishing, Moscow, 1966 (in Russian).
7. Voskresenski, D.I. (Editor), *Antennas and Microwave Devices (Design of Antenna Phased Arrays)*, Radio i Svjaz Publ., Moscow, 1981, (in Russian).
8. Belee, E.C., R.C. Breithaupt, D.L. Godwin, and S.H. Walker, "Image Element Antenna Array for a Monopulse Tracking System for a Missile," US Patent No 3 990 078, November 2, 1976.
9. Klass, P.J., "Helicopter Stabilized by Microwave Beam," *Aviation Week and Space Technology*, April 22, 1968, pp. 85–89.
10. Brauer, J.R. "A Rectangular Waveguide Resonator and Antenna," Ph.D. Dissertation, University of Wisconsin, Madison, June 1969.
11. Brauer, J.R., "A Rectangular Beam Waveguide Resonator and Antenna," *IEEE Trans. Antennas and Propagation*, Vol. AP-20, No 5, September 1972, pp. 595–601.
12. Katzenelenbaum, B.Z., E.N. Korshunova, L.I. Pangonis, and A.N. Sivov, "Resonant Antenna Synthesis," *Radiotechnika i Electronika*, USSR's Academy of Sciences Periodical, Vol. XXVIII, No 5, 1983, pp. 854–863 (in Russian).
13. Katzenelenbaum, B.Z., E.N. Korshunova, L.I. Pangonis, A.N. Sivov, "Experimental Investigation of Resonant Antenna Synthesized at a Given Radiation Pattern," *Radiotechnika i Electronika*, Vol. XXVIII, No 7, 1983, p. 1245 (in Russian).
14. Katzenelenbaum, B.Z., E.N. Korshunova, L.I. Pangonis, A.N. Sivov, and A.D. Shatrov, "On the Resonant Antenna Synthesis," *Radiotechnika i Electronika*, Vol. XXIX, No 11, 1984, pp. 2116–2123 (in Russian).

Chapter 4
Cavity-Backed, Crossed-Dipole-Fed, Helical and Spiral Antennas

Cup dipole cavity-backed radiator antennas have been used in a number of communication applications at VHF, UHF and microwave frequencies. Since the late 1970s, this type of radiator has become prominent in a number of satellite applications due to the following advantages: a high-gain element, a lightweight antenna element, compactness and ease of manufacture, including low construction cost, good axial ratio for circular polarization (CP) applications, easy to adopt dual-polarization characteristics, easy to match, low sidelobes and backlobes, easy to design for high-power applications, easy to adopt to environmental protection requirements, ease of mechanical design.

The cup dipole antenna is described in Section 4.1. The experimental and theoretical considerations of this antenna are discussed in Section 4.2. Cavity-backed helical and spiral antennas are discussed in Section 4.3. The subject of multipactor breakdown has gained considerable interest in recent years because of its occurrence in components designed for operation in space. Multipactor breakdown is discussed in Section 4.4 for cup dipole antennas, cavity-backed helical antennas, and cavity-backed spiral antennas as a result of increasing interest in high-power applications in space. Applications as single-element or multielement (array) antennas are discussed in Section 4.5. It is important to note that these antennas comprise a special case of SBF antennas, but there will be no discussion of the SBF antenna in this chapter.

4.1 CUP DIPOLE ANTENNA DESCRIPTION

The basic concept of the cavity-backed crossed-dipole radiator antenna for *circular polarization* (CP) is the use of crossed dipoles. These dipoles are mounted in a cylinder of circular, square or other polygonal cross section. Crossed dipoles, which are the most commonly used sources of CP, generate two orthogonal signals. Two of the most efficient techniques used to feed the input coaxial cable of the cup dipole antenna (to produce CP without resorting to power splitting and phasing network) are single coaxial input and dual coaxial input.

4.1.1 Single Coaxial Input Cup Dipole Antennas

In this technique [1], the CP is produced by two orthogonal dipoles driven in parallel by a common or single coaxial line. The lengths of the dipoles are chosen so that the magnitude of their input impedances is equal, and the angle of the input impedances differ by 90°. When these two conditions are met, the resulting radiated electromagnetic wave (in a normal direction) will be circularly polarized.

Figure 4.1(a) shows two orthogonal cylindrical dipoles fed in parallel and a representative equivalent circuit. The mutual interaction between two dipoles should be negligible, as long as the radii of one dipole are placed at the zero potential plane of the other dipole. In the figure, hx, hy, ax, and ay are related by the expression [2]:

$$\frac{hx}{ax} = \frac{hy}{ay} = 75.2 \tag{4.1}$$

Figure 4.1(b) shows the calculated input impedance of a single dipole (as a function of its length) for lengths in the neighborhood of a half-wavelength. The crossed-dipole system in a cup, described by Silberberg [3], is shown in Figure 4.2. The operation of the dipole can be interpreted from various points of view. The simplest picture is that the radiating system is energized by a voltage impressed across the slot. The origin of the voltage becomes evident upon consideration of mode relationships in the slotted region. In the absence of the short-circuiting post, the *TEM* mode and possibly higher modes would be generated in the open-ended termination or, in the case of a wide slot, generated by the slot itself. All these modes, however, would be symmetrical with respect to the plane containing the axes of the slots, and would not create an impressed field across the slot; under these conditions, the dipoles are not excited. With the insertion of the post, modes are generated that are symmetrical with respect to the plane determined by the axis of the post and the axis of the inner conductor. These modes, when superimposed on the preceding set, must give rise to a field in which the tangential electric field is zero over the surface of the post. The modes are generated by the post and propagate through the slotted region. The superposition of the *TEM* and TE_{11} modes in the slotted region of the slot-fed dipole is also shown in Figure 4.2. The resultant configuration gives a field that is zero along the post, and increases with angle to a maximum value directly

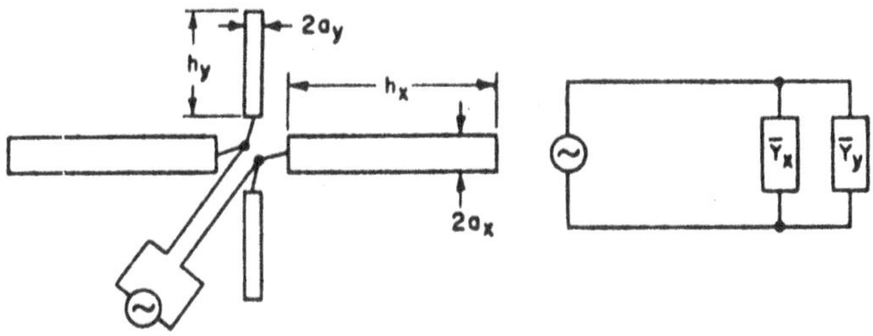

Figure 4.1(a) Arrangement of dipoles and equivalent circuit (from [1])

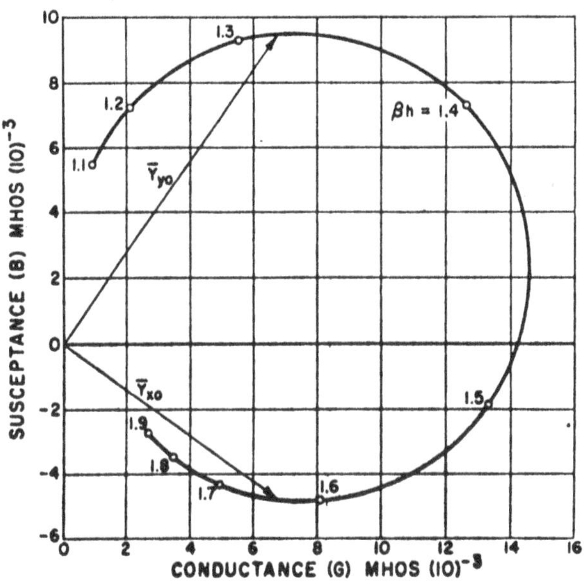

Figure 4.1(b) Dipole input impedance (from [1])

opposite the post. We can readily see that this impresses a voltage across the slot, resulting in excitation of the dipoles.

Figure 4.3(a) and (b) shows the sense of polarization, either *right-hand circular polarization* (RHCP) or *left-hand circular polarization* (LHCP), by the physical orientation of the dipole with respect to the shorting post. The results of experiments by Katulski [4] involving the polarization ratio and the standing wave ratio are presented in Figures 4.4(a) and (b), respectively, for a set of crossed-dipole lengths $((l_{xi},l_{yi})$, where i = 1, 2, 3, 4: therefore, we have (82,46), (84,48), (86,49.5), (88,51)),

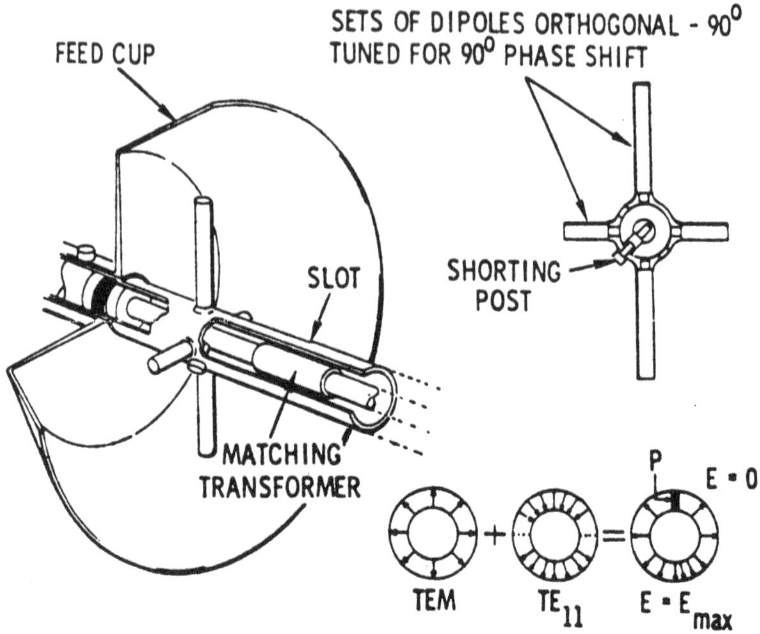

Figure 4.2 Cup dipole system described by Silberberg (from [3])

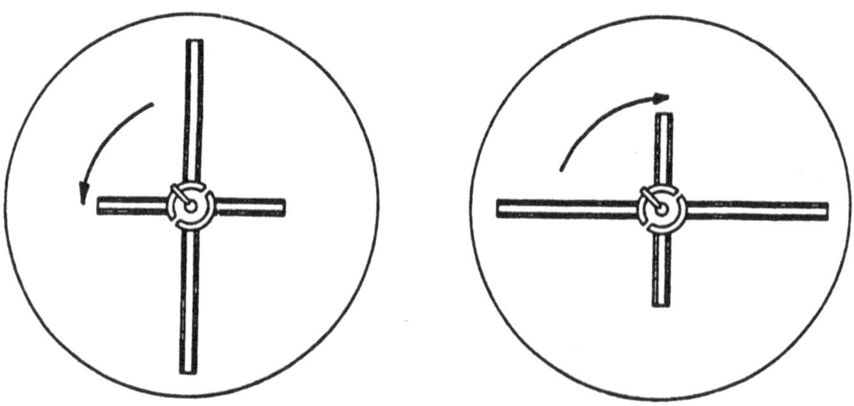

(a) (b)

Figure 4.3(a) Left-hand circular polarization
(b) Right-hand circular polarization

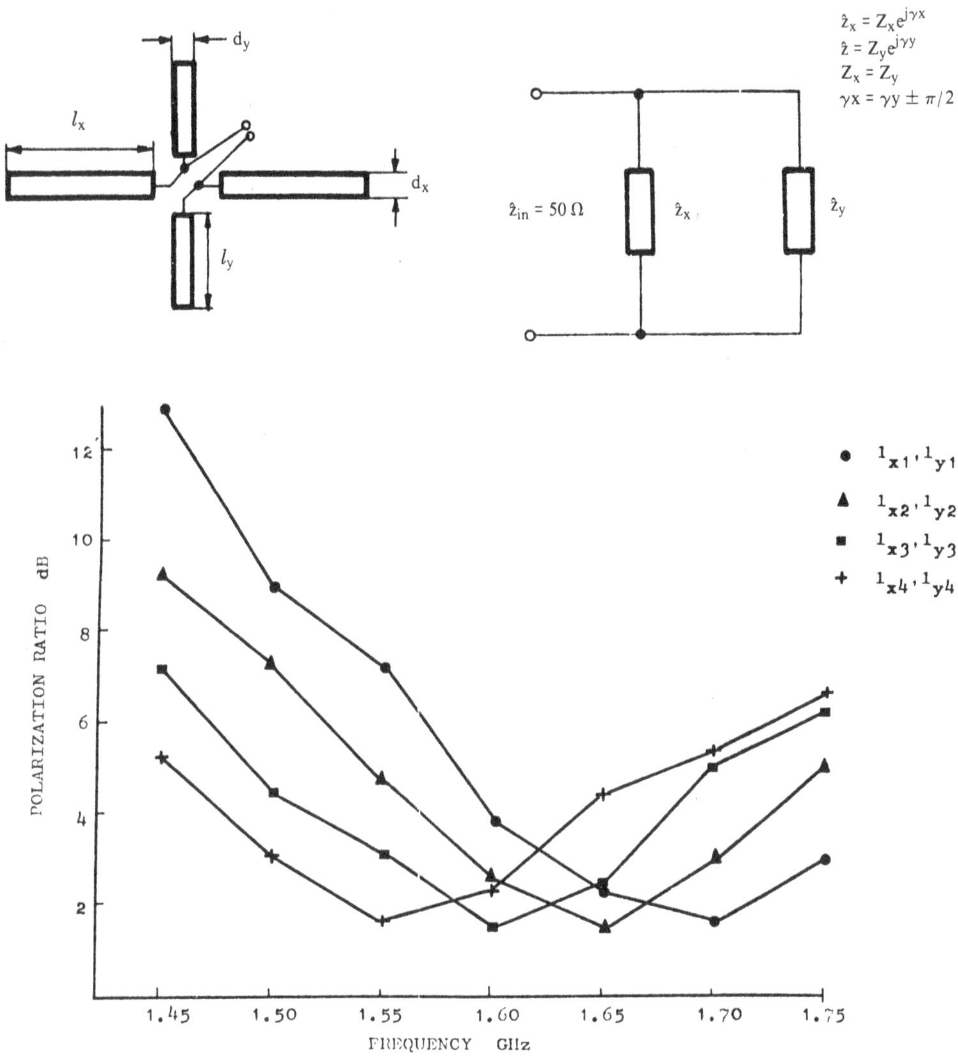

Figure 4.4(a) Polarization ratios for different lengths of crossed-dipoles *versus* frequency (from [4])

as a function of frequency. The dimensions shown are in millimeters.

The frequency at which the value of the polarization ratio is minimal increases when the dipole length decreases. The variation of the short dipole length essentially influences the polarization ratio (Figure 4.4(c)). This property can be used for the

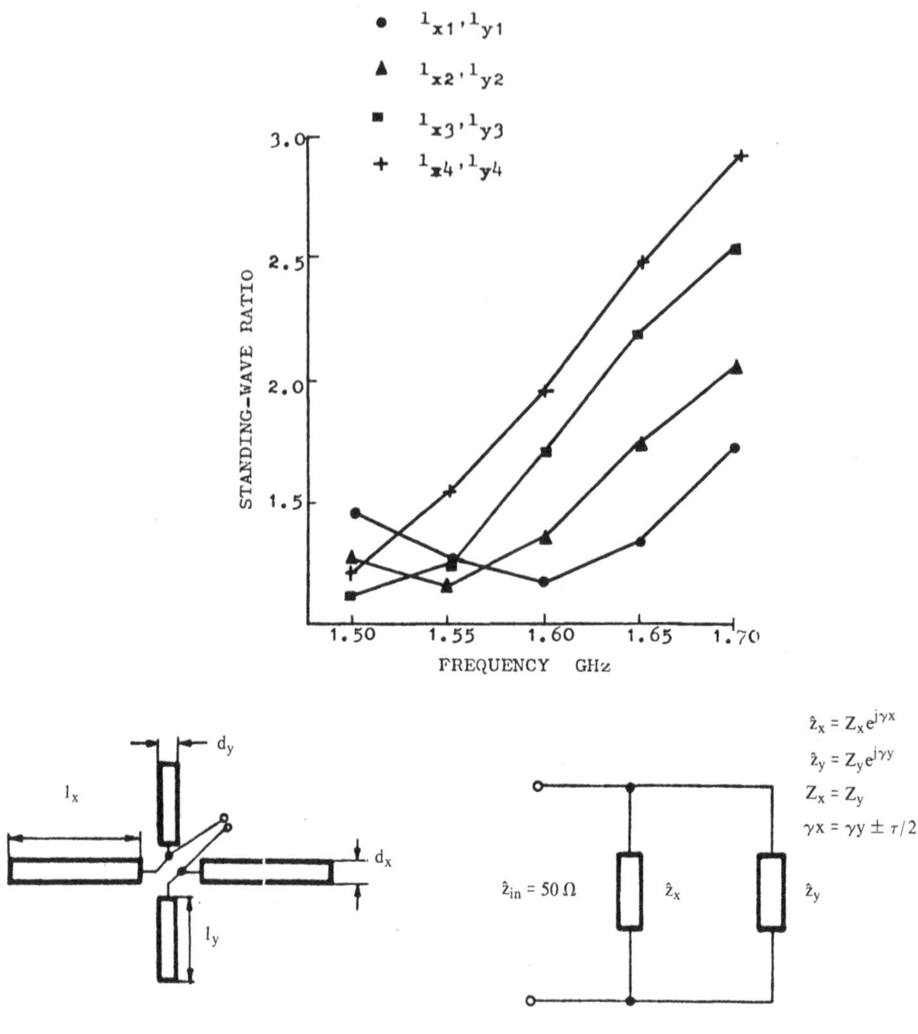

Figure 4.4(b) Standing wave ratios for different lengths of crossed-dipoles *versus* frequency (from [4])

optimization of the crossed dipoles. We can generalize, on the basis of the experimental results produced by Katulski, that the input impedance and axial ratio of crossed dipoles in a cup are functions of the dipole lengths as well as cup geometry.

4.1.2 Dual Coaxial Input Cup Dipole Antenna

For this type of arrangement, the two inputs are fed with a power divider and 90° phasing loop to provide right-hand or left-hand circular polarization. Figure 4.5(a)

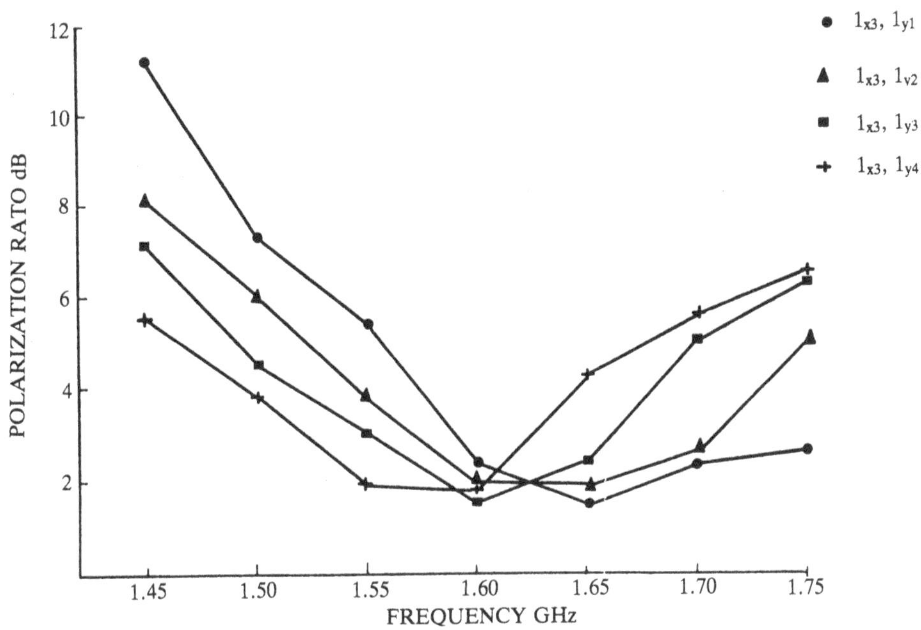

Figure 4.4(c) Polarization ratios of different crossed-dipoles (with short dipole lengths fixed and long dipole lengths varied) *versus* frequency (from [4])

Figure 4.5(a) Photograph of a typical cavity-backed crossed-dipole antenna (from [7])

shows a photograph of a typical cavity-backed crossed-dipole antenna. In the antenna, the crossed-dipole feed network is made up of a hybrid and a compensated balun [5,6,7] for each dipole.

The dual-polarized feed element has been designed in the frequency range of 3.6 to 4.3 GHz. The cup is approximately 7.6 cm in diameter and 3.8 cm deep. The two dipoles excite the cylindrical cup-type cavity in orthogonal TE_{11} modes. These orthogonal dipoles are driven by a 3 dB hybrid to produce CP from the element. In order to produce a low axial ratio, the antenna radiation pattern must be rotationally symmetric (equal *E*-plane and *H*-plane beamwidths), and the element phase center must be unique. The antenna provides an axial ratio less than 0.3 dB on axis over the frequency band 3.6 to 4.3 GHz. Figure 4.5(b) shows a typical axial ratio pattern at 3.95 GHz. The figure shows that the axial ratio is less than 0.5 dB over ±36 degrees. The aperture efficiency of the antenna is higher than 99 percent. The design of the antenna incorporates a balun and a coaxial transformer section that matches the dipole impedance to a 50 ohm line with an input VSWR less than 1.06:1 over the frequency range.

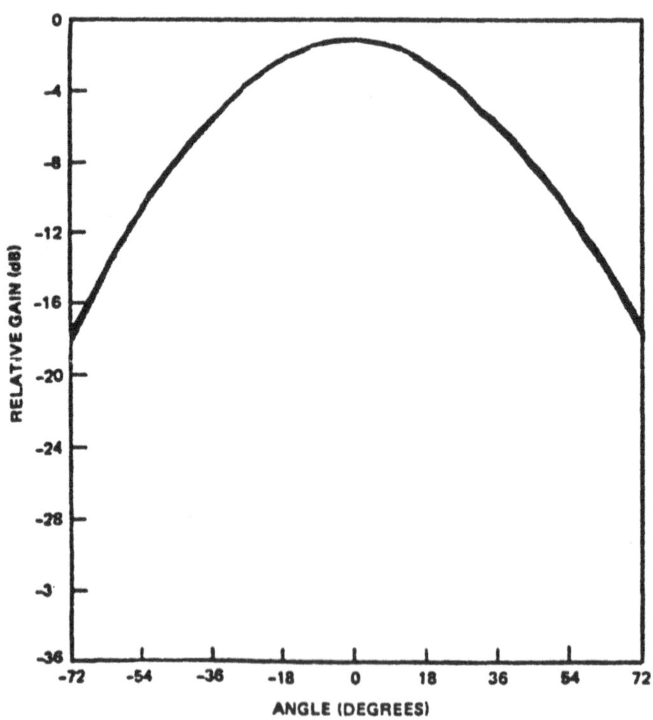

Figure 4.5(b) Axial ratio pattern of a cup-dipole antenna at 3.95 GHz (from [7])

Figure 4.6(a) and (b) shows the coaxial balun structure and its equivalent circuit. In the figure, the input coaxial line (Z_a) has its outer conductor attached to one terminal of the balanced output line, and its inner conductor is connected (in series) to the inner conductor of the parallel compensating line (Z_b). The outer conductors of the two parallel coaxial lines form a shorted section of two-wire transmission line (Z_{ab}), which interacts with the open-circuited (Z_b) line to produce a compensated impedance. The open-circuited line and two-wire line are both designed to be one quarter-wavelength long at band centered frequency (f_0). For broadest bandwidth, it is desirable to make Z_{ab} as large as possible relative to the balanced load impedance $R(Z_c)$.

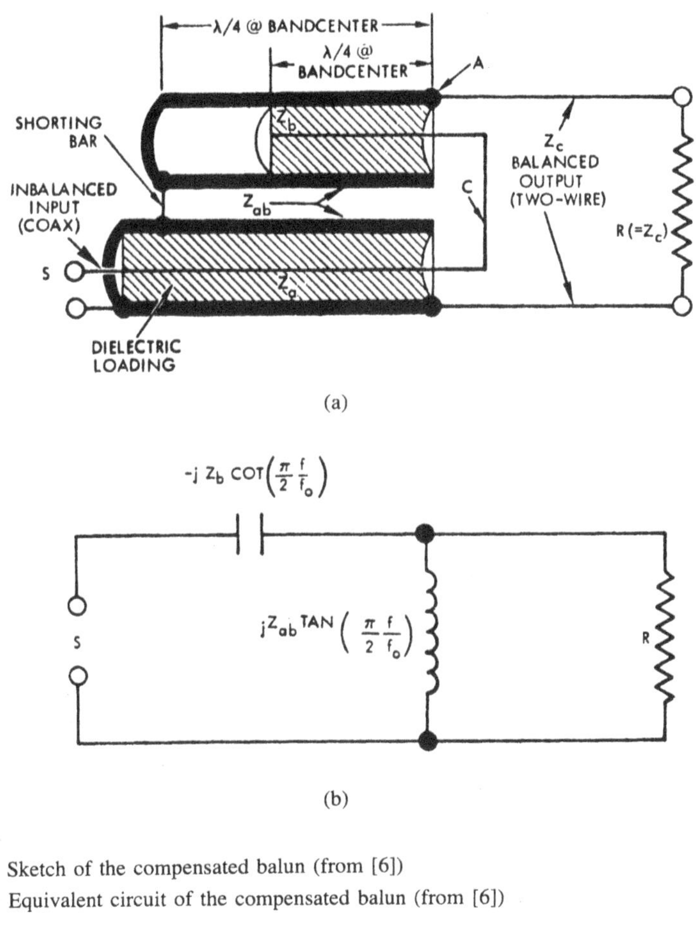

Figure 4.6(a) Sketch of the compensated balun (from [6])
(b) Equivalent circuit of the compensated balun (from [6])

A typical Smith chart impedance plot of a compensated balun with optimum design is shown in Figure 4.7. The design is optimized by adjusting the position of the loop of balun impedance symmetrically about the generator impedance ($S = Z_a$). Figures 4.6(a) and 4.7 show that the dipole arms are excited with a 50 ohm coaxial transmission line (Z_a) with one arm screwed to the outer conductor of Z_a. The center conductor of Z_a is connected to that of Z_b by a thin strip. The other arm is similarly screwed to another cylindrical conductor of the same diameter as the outer conductor of the input coaxial line (Z_a). This cylindrical conductor and the outer conductor of the coaxial form a two-wire transmission line section shorted at one end by the cup base, which functions as a balanced-to-unbalanced balun impedance matching transformer.

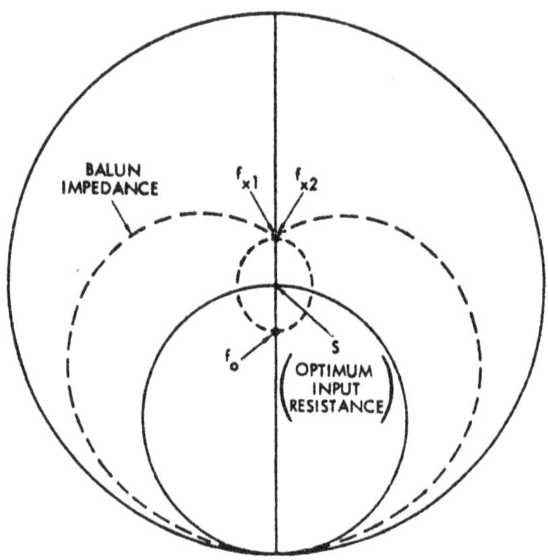

Figure 4.7 A typical Smith chart impedance plot of a compensated balun with optimum design (from [6])

Figure 4.8 shows a summary of Oltman's analysis for optimally-matched compensated baluns [6]. These curves relate the balun design parameters (Z_a,Z_{ab},Z_b,R) to the maximum *voltage standing wave ratio* (VSWR) and normalized frequency (f_{x1}/f_0) or bandwidth ratio (f_{x2}/f_{x1}). The relationship between optimum impedance (α) and maximum VSWR can be given by:

$$\text{VSWR}_{(max)} = \frac{2}{1 + \alpha} \tag{4.2}$$

Oltman [6] has described two experimental versions of the compensated balun design (Figure 4.9(a)). In one, two identical 4-to-1 bandwidth baluns were arranged back-to-back, so that the input and output are both in 50-ohm coaxial line. This arrangement was used because a well-matched load in two-wire line was not available. Oltman used the balun component as follows: $Z_a = Z_c = 50$ ohms, $Z_b = 12.5$ ohms, and $Z_{ab} = 200$ ohms. The measured and theoretical VSWR and insertion loss are shown in Figure 4.9(b).

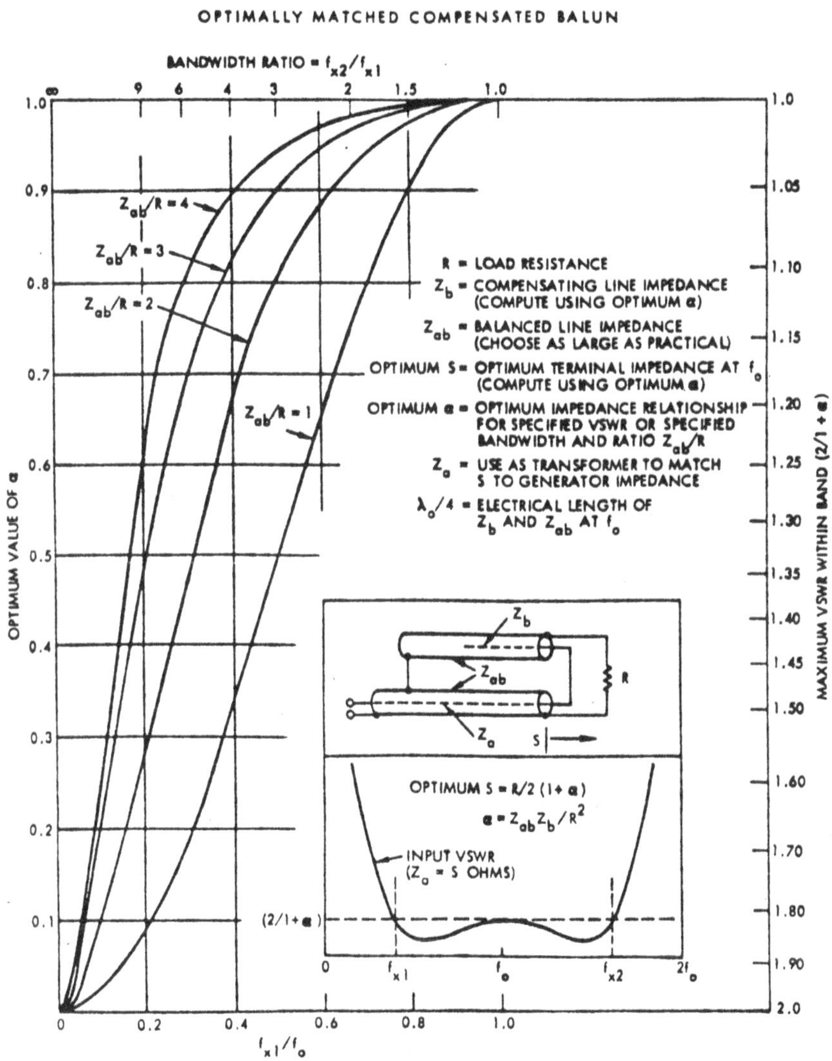

Figure 4.8 Oltman's design curves for an optimally-matched balun (from [6])

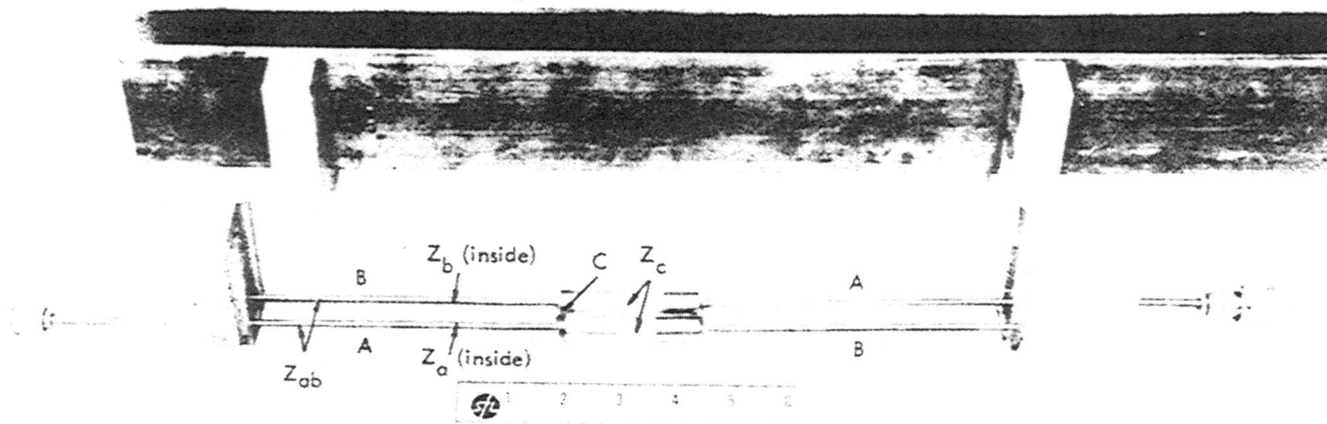

Figure 4.9(a) Experimental arrangement of back-to-back compensated baluns (from [6])

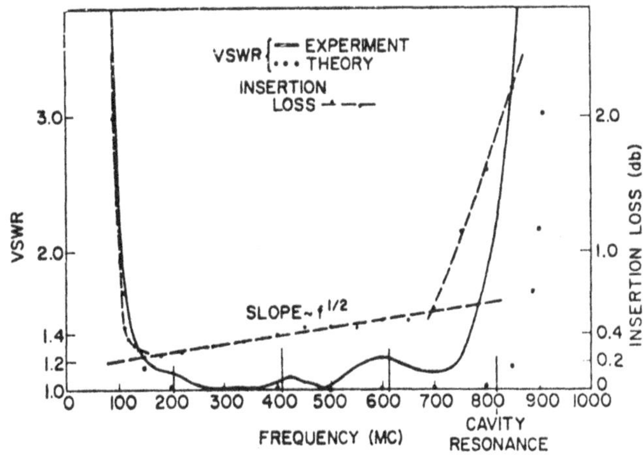

Figure 4.9(b) *VSWR* and insertion loss of two back-to-back compensated baluns (from [6])

4.2 EXPERIMENTAL AND THEORETICAL CONSIDERATIONS OF CROSSED-DIPOLE-FED CAVITY ANTENNAS

The radiation fields of an open-cavity radiator have been studied theoretically and experimentally by Hong, Nyquist, and Chen [8]. In their study, the antenna consisted of a simple open-ended circular cavity with a primary radiator, which included a dipole, a dipole array, a terminated transmission line, and a loop, placed at an appropriate location inside the cavity.

Figure 4.10 shows geometries of the open-ended circular cavity radiators with four kinds of exciters. These cavities have aperture diameters $2A(2\lambda_0)$ and variable lengths L. The position of the dipole radiator with respect to the shorted end and the aperture is shown in the figure. The fields inside the cavity are the Fourier sum of waveguide modes. The diffraction and coupling effects at the open end of the cavity are neglected. In the theoretical analysis for an open-cavity radiator, the waveguide excitation theory is employed to calculate the normal modes excited by the exciter in the cavity. Subsequently, the radiation field is calculated from the aperture fields. The total radiation field (E_q^r) produced by the $E_{\theta q}$ and $H_{\phi q}$ of the aperture fields of the qth mode is obtained by

$$E_q^r = \hat{\theta} E_{\theta q}^r + \hat{\phi} E_{\phi q}^r \tag{4.3}$$

where

$$E^r_{\theta q} = \frac{jk_0}{4\pi} \frac{\exp(-jk_0 r)}{r} C_q\left(1 + \frac{\zeta_0}{Z_q}\cos\theta\right)$$

$$\cdot(\exp(-j\beta_q l_2) + \Gamma_{2q}\exp(j\beta_q l_2))F_{\theta q} \tag{4.4}$$

and

$$E^r_{\phi q} = \frac{jk_0}{4\pi} \frac{\exp(-jk_0 r)}{r} C_q\left(\frac{\zeta_0}{Z_q} + \cos\theta\right)$$

$$\cdot(\exp(-j\beta_q l_2) + \Gamma_{2q}\exp(j\beta_q l_2))F_{\phi q} \tag{4.5}$$

with

$$B_q = \frac{\Gamma_{2q}M_q + N_q}{\Gamma_{1q}\Gamma_{2q} - 1}, \quad C_q = \frac{\Gamma_{1q}N_q + M_q}{\Gamma_{1q}\Gamma_{2q} - 1}$$

$$M_q = \frac{\int_V \boldsymbol{E}_q - \cdot \boldsymbol{J}_a \mathrm{d}V}{2\int_{CS} (\boldsymbol{e}_q \times \boldsymbol{h}_q)\cdot \hat{z}\mathrm{d}S} \quad \text{and} \quad N_q = \frac{\int_V \boldsymbol{E}_q + \cdot \boldsymbol{J}_a \mathrm{d}V}{2\int_{CS} (\boldsymbol{e}_q \times \boldsymbol{h}_q)\cdot \hat{z}\mathrm{d}S}$$

where $\int_V \mathrm{d}V$ is the volume integral enclosing the current source $\boldsymbol{J}_a$ and $\int_{CS}\mathrm{d}S$ is the surface integral over the *cross section* (*CS*) of the cavity.

$$F_{\theta q} = j^{n-1} 2\pi \sin n\phi \boldsymbol{J}_n(\beta_0 A \sin\theta) \cdot \begin{cases} \dfrac{n}{k_{cq}\beta_0 \sin\theta} \boldsymbol{J}_n(k_{cq}A), & \text{for } TE \text{ mode} \\[2ex] \dfrac{-A\beta_0 \sin\theta}{k^2_{cq} - \beta_0 \sin\theta)^2} \boldsymbol{J}'_n(k_{cq}A), & \text{for } TM \text{ mode} \end{cases} \tag{4.6}$$

$$F_{\phi q} = \begin{cases} j^{n-1} \dfrac{2\pi A k_{cq} \cos n\phi}{k^2_{cq} - (\beta_0 \sin\theta)^2} \boldsymbol{J}_n(k_{cq}A)\boldsymbol{J}'_n(\beta_0 A \sin\theta), & \text{for } TE \text{ mode} \\[2ex] 0, & \text{for } TM \text{ mode} \end{cases} \tag{4.7}$$

$$\Gamma_{1q} = -\exp[-j2\beta_q l_1] \tag{4.8}$$

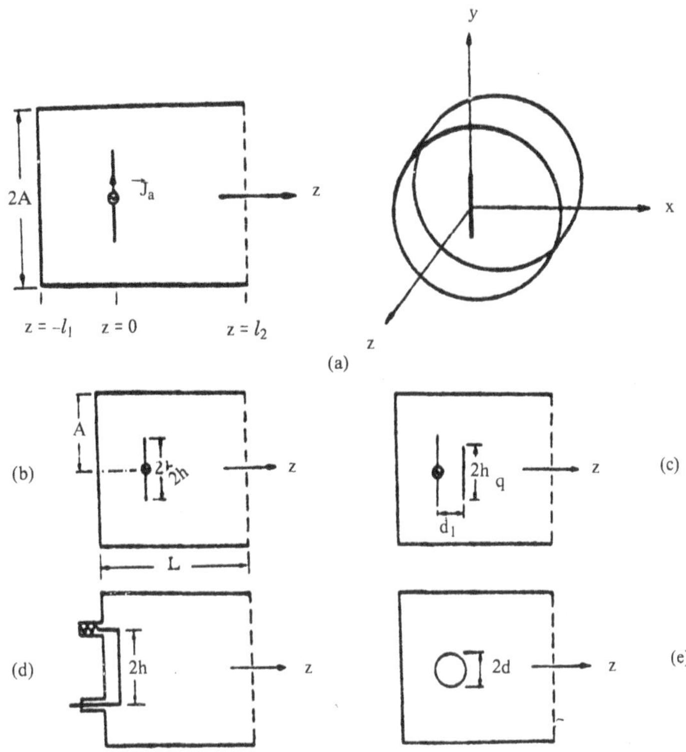

Figure 4.10 Open-cavity radiator: (a) configuration; (b) radiator excited by dipole; (c) radiator excited by dipole array; (d) radiator excited by transmission line; (e) radiator excited by circular loop (from [8])

$$\Gamma_{2q} = \frac{\zeta_0 - Z_q}{\zeta_0 + Z_q} \exp\,[-j2\beta_q l_2] \tag{4.9}$$

where ζ_0 and Z_q are the wave impedances of the free space and of the qth waveguide mode, respectively. The eigenvalue of the qth *TE* or *TM* waveguide mode is k_{cq}.

The theoretical and experimental radiation patterns of an open-cavity antenna with a half-wave dipole source placed at $0.25\lambda_0$ from the shorted end of the cavity of a length of $1.2\lambda_0$ are shown in Figure 4.11(a) by a dotted line and a solid line, respectively. Figure 4.11(b) shows the patterns of a parasitic element with a length of $0.44\lambda_0$ placed $0.25\lambda_0$ in front of the dipole. These results (Figure 4.11(a) and (b)) show that with the parasitic reflector, the *H*-plane radiation pattern has no sidelobe and the radiator behaves much like a short backfire antenna (described in Chapter 5).

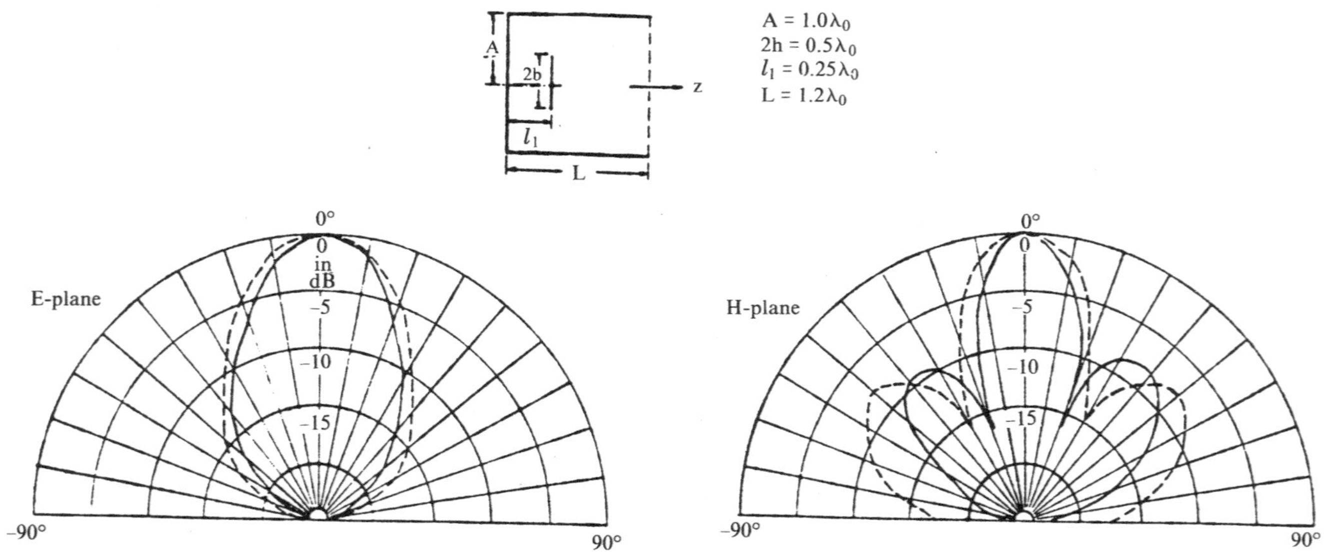

Figure 4.11(a) Radiation patterns of an open-cavity radiator with dipole exciter ($h = 0.25\lambda_0$, $l_1 = 0.25\lambda_0$, $L = 1.2\lambda_0$) (from [8]). Experimental results—solid line; theoretical results—dashed line

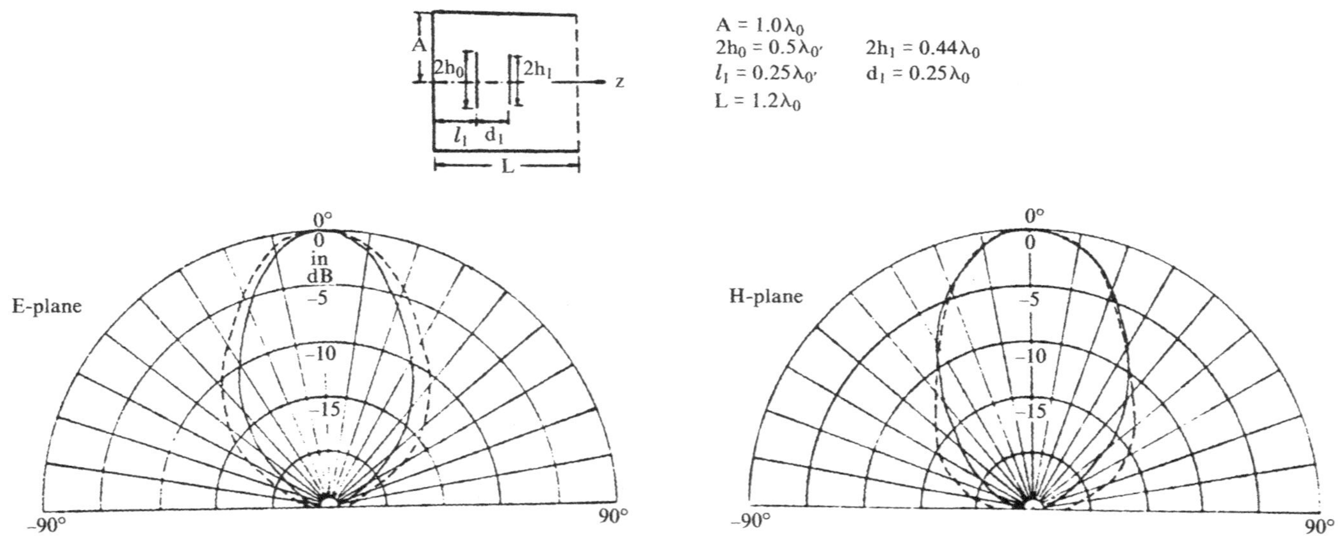

Figure 4.11(b) Radiation patterns of an open-cavity radiator with dipole exciter ($h = 0.25\lambda_0$, $h = 0.22\lambda_0$, $l = 0.25\lambda_0$, $d = 0.25\lambda_0$, $L = 1.2\lambda_0$) (from [8]). Experimental results—solid line; theoretical results—dashed line

Figure 4.11(c) shows the radiation patterns of the radiator with a terminal transmission line ($0.5\lambda_0$) source placed at $0.045\lambda_0$ from the shorted end of the cavity of length of $1.2\lambda_0$. This kind of radiator has a broadband characteristic. Figure 4.11(d) shows the radiation patterns of the antenna with a circular loop exciter placed $0.25\lambda_0$ from the shorted end of the cavity of a length of λ_0 and the radius of the loop at $0.6\lambda_0$. King and Wong [9] studied the characteristics of a balun-fed open-sleeve dipole antenna in the 225 to 400 MHz band. A cylindrical dipole radiator is called a *sleeve antenna* when the exciter of the coaxial transmission line is used as a radiating element. The main advantage of the sleeve antenna over a thin conventional dipole antenna is that it operates in wider bandwidth. Figure 4.12(a) shows the dipole reflector. A coaxial line is used to feed the dipole, and a balun is added to the antenna structure. The dipole and sleeve are constructed of circular cylindrical elements, with a 0.358 cm (0.141 inch) diameter copper-clad semirigid coaxial cable as the feed line. The balanced line of the balun is also a length of semirigid cable, but without the center conductor and with the short circuit of this line coincident with the reflector surface. King and Wong made the angle $\theta = 90°$ at the feed point for all dipoles except for the 2.858 cm (1.125 inch) diameter dipole, which was inadvertently made 45°. The dipoles are screwed into the feed terminals, and the sleeves are supported by Styrofoam. A parametric study, shown in Figure 4.12(b–d), was made of the VSWR response as a function of dipole and sleeve diameter, sleeve length, and sleeve-to-dipole spacing. The VSWR response became flatter and the required dipole length shortened as the diameter increased. To show the wideband characteristics of the open-sleeve antenna, the VSWR response of an unsleeved cylindrical dipole with a diameter of 2.858 cm, driven through the same balun, is also included in Figure 4.12(b). The measured coupling between two adjacent sleeve dipoles has been found to be less than −20.6 dB in the frequency range of 225 to 400 MHz. In Figure 4.12(c), the curve is typical of the expected change as the spacing between the dipole and the sleeve varies. Sleeve spacing has relatively little effect on VSWR response for diameters of less than 0.9525 cm. Figure 4.12(c) and (d) shows how critical the sleeve dimensions are, and illustrates how the VSWR response at the upper end of the band can be modified at the expense of the midband region. In addition, the sleeves and the dipole need not be coplanar, as shown in Figure 4.12(a). The sleeves can be placed in any parallel position with respect to the dipole, provided the proper sleeve-dipole spacing is maintained. Furthermore, an open-sleeve dipole can be operated over a bandwidth of 1.8:1, as compared with an operating bandwidth of approximately 1.25:1 for an unsleeved dipole with the same diameter.

The dimensions and the radiation patterns for a sleeve mounted in front of a flat reflector are shown in Figure 4.13. A cosine voltage pattern is also plotted as a dotted line. The measured and calculated (using pattern integration) gain values are shown in Table 4.1 [9]. Figure 4.14 shows the crossed-dipole arrangement to achieve circular polarization. The sleeves, which are arranged in tic-tac-toe configuration, can be supported by four metallic posts attached directly to the reflector. The VSWR

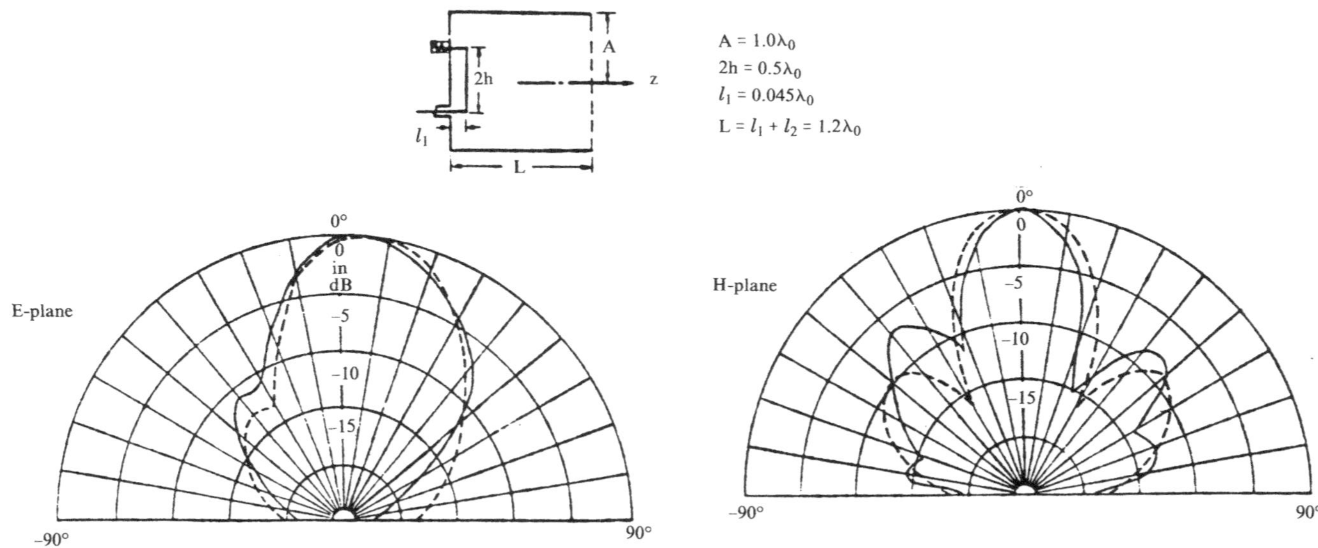

Figure 4.11(c) Radiation patterns of an open-cavity radiator with transmission line exciter ($2h = 0.5\lambda_0$, $l_1 = 0.045\lambda_0$, $L = 1.2\lambda_0$) (from [8]). Experimental results—solid line; theoretical results—dashed line

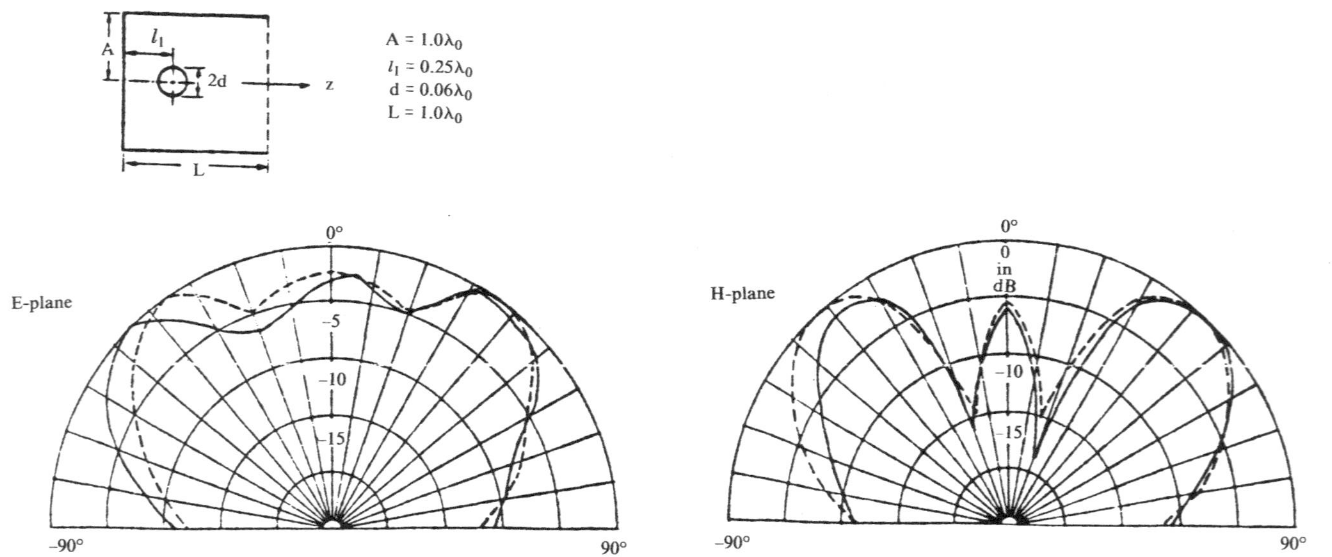

Figure 4.11(d) Radiation patterns of an open-cavity radiator with circular loop exciter placed in a longitudinal plane ($d = 0.06\lambda_0$, $L = \lambda_0$) (from [8]). Experimental results—solid line; theoretical results—dashed line

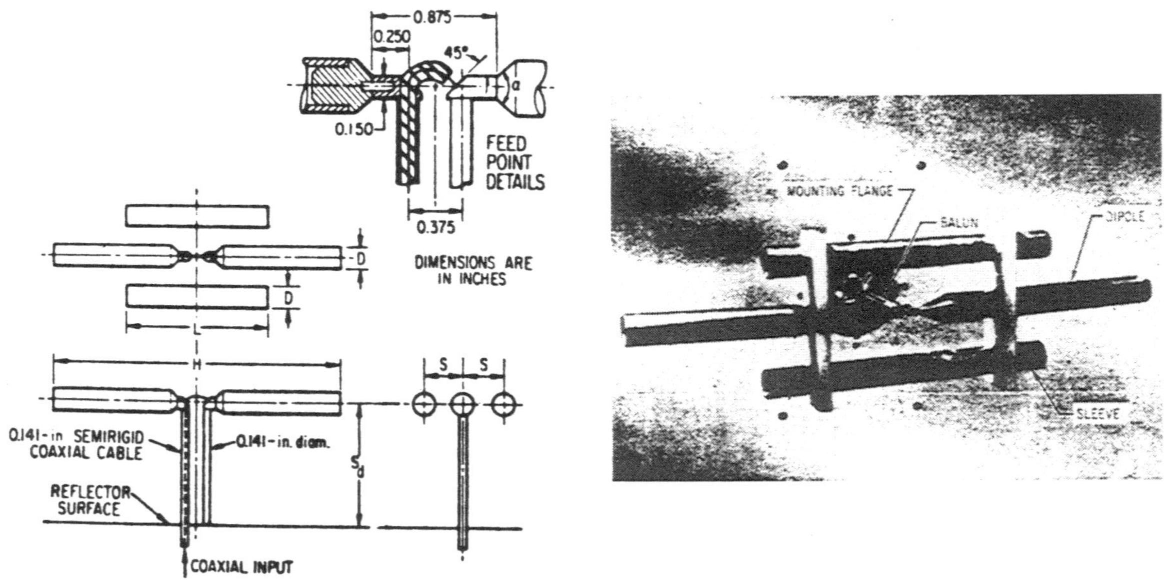

Figure 4.12(a) Open-sleeve dipole reflector assembly (from [9])

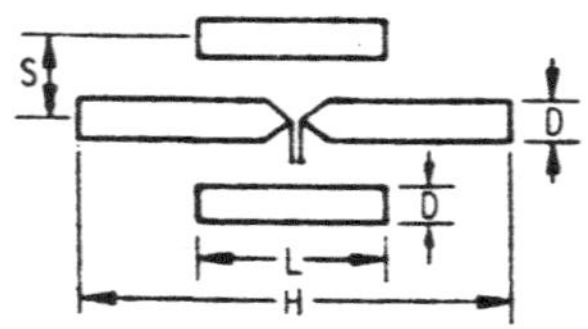

D	H	L	S
3/8	22.2	13.5	1.29
1/2	21.7	13.0	1.29
3/4	21.7	12.2	1.29
1-1/8	20.2	11.38	2.00
1-3/8	19.63	10.88	2.75

DIMENSIONS ARE IN INCHES

DIPOLE-TO-REFLECTOR SPACING = 8-5/8 in.

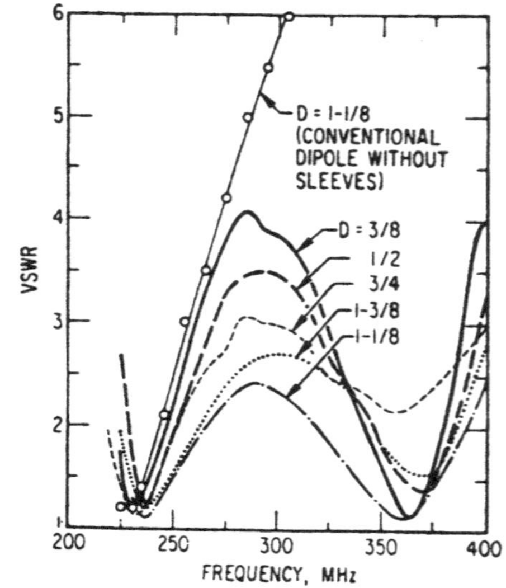

Figure 4.12(b) VSWR response of open-sleeve dipoles for various dipole and sleeve diameters (from [9])

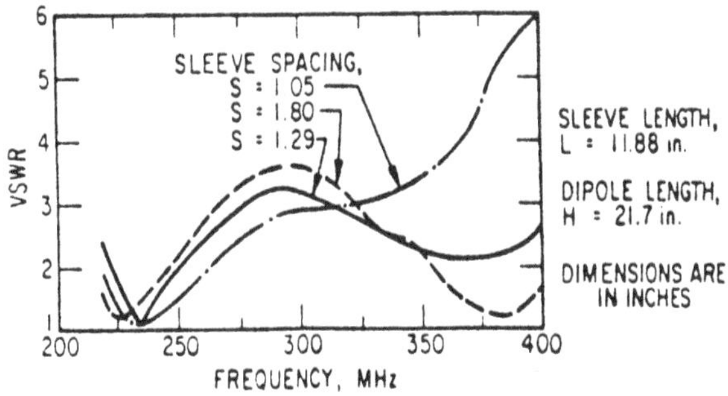

Figure 4.12(c) VSWR for 1.905 cm (0.75 inch) diameter open-sleeve dipole with sleeve spacing as parameter (from [9])

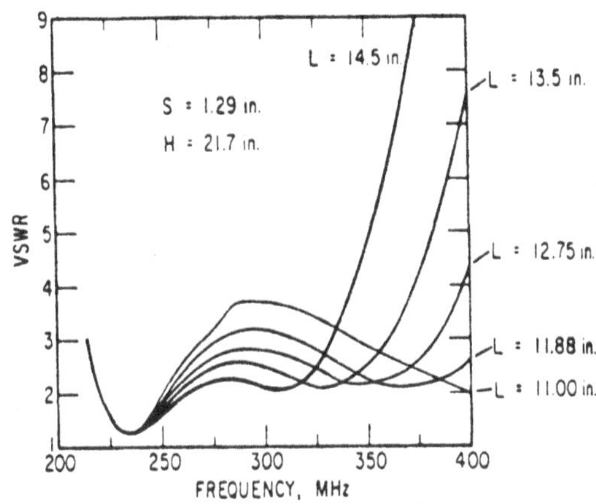

Figure 4.12(d) VSWR for 1.905 cm diameter open-sleeve dipole with sleeve length as parameter (from [9])

response is relatively insensitive to the support method of the sleeves. For the tic-tac-toe sleeve configuration, the isolation between the two orthogonal dipoles is greater than 25 dB throughout the 225 to 400 MHz band. Figure 4.15 is a photograph of an open-sleeve crossed-dipole backed by a shallow circular cavity. This type of antenna gives broadband characteristics as explained earlier. The measured VSWR, axial ratio, half-power beamwidths, and gain are shown in Figure 4.16. Typical measured radiation patterns for the double-sleeve dipole arrangement are shown in Figure 4-17(a–c) at 450 MHz, 600 MHz and 800 MHz, respectively.

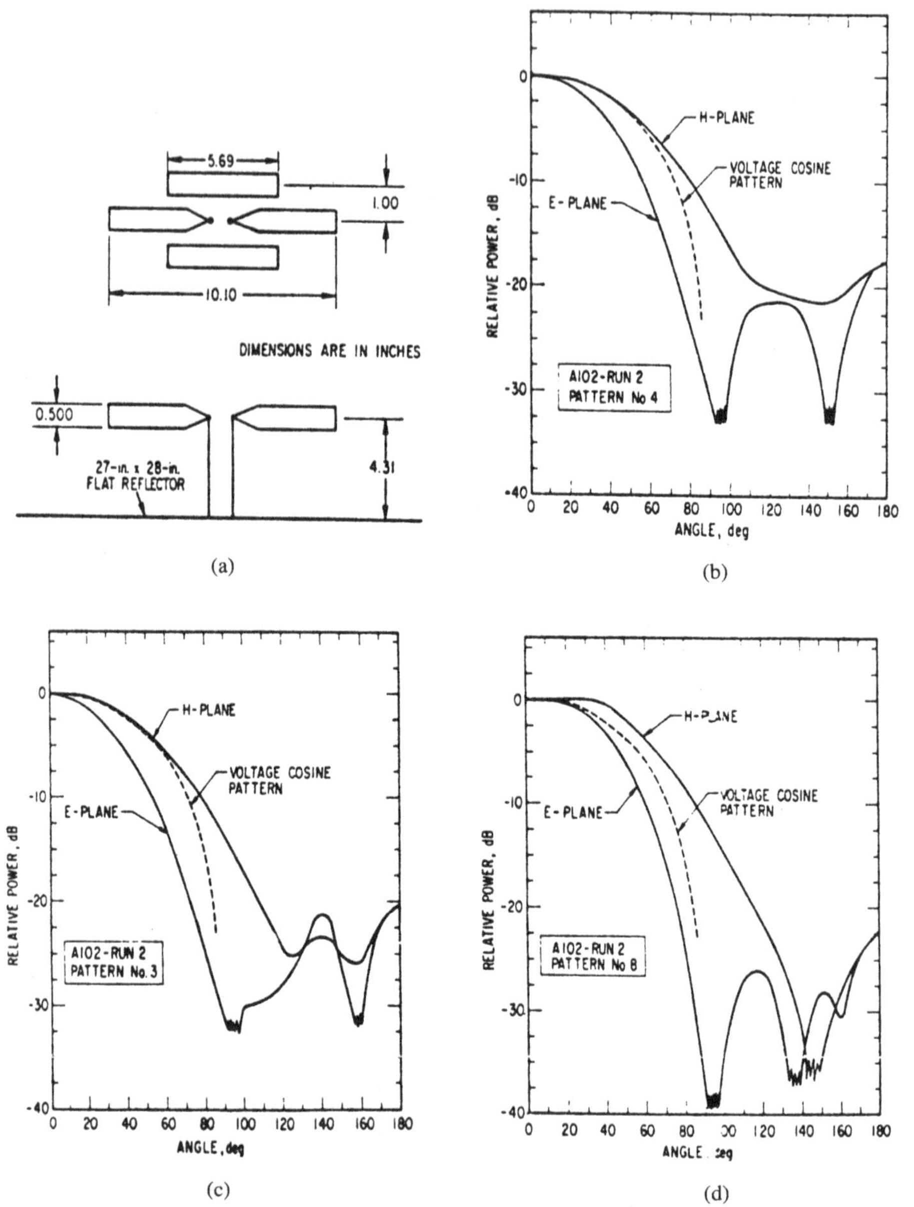

Figure 4.13 *E*-plane and *H*-plane patterns of an open-sleeve dipole reflector assembly (from [9]): (a) dimensions; (b) 450 MHz; (c) 600 MHz; (d) 800 MHz

Table 4.1 Gain of Open-Sleeve Dipole in Front of Reflector

	Gain in dB	
Frequency in MHz	*Calculated*	*Measured*
450	8.62	9.1
600	9.03	9.1
800	7.24	7.0

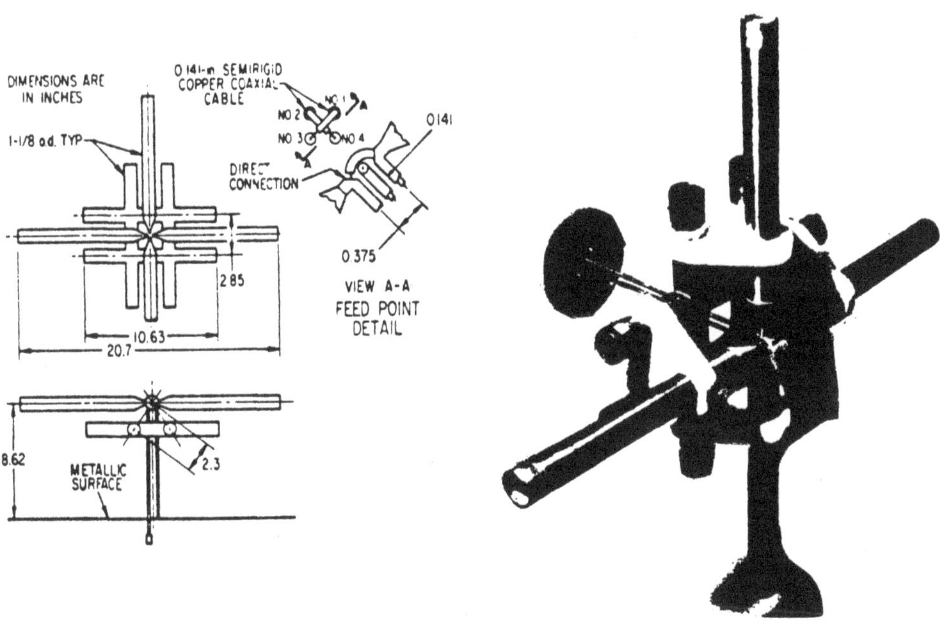

Figure 4.14 Crossed-dipole assembly (from [9])

A triple-sleeve dipole antenna, consisting of a top sleeve added to the double-sleeve arrangement, has even better VSWR response than that of the double-sleeve configuration at the high end of the frequency band (as shown in Figure 4.17). Selection of dimensions is based on a compromise between VSWR and radiation pattern behavior in the frequency band. The radiation patterns of the triple-sleeve dipole

Figure 4.15 Photograph of an experimental antenna (from [10])

antenna are similar to those of Figure 4.16. A cavity-backed open-sleeve dipole antenna gives moderate gain over a 1.8:1 bandwidth. The antenna can be operated in either a linearly- or circularly-polarized mode. Broadband VSWR response and radiation patterns with relatively good front-to-back ratio are available.

In 1974, Ehrenspeck [11] described the cup dipole antenna as a medium-sized center-fed reflector antenna consisting of a reflecting surface (*A*), a peripheral rim (*B*) and a feed system (*F*), located on the reflector, as shown in Figure 4.18. The feed system (*F*), located on the reflector central axis at a distance d_F between a quarter-wavelength and a half-wavelength from *A*, must efficiently illuminate reflector *A* for high-directive gain achievement. This condition is satisfied when a single dipole illuminates a reflector area of about two square wavelengths in cup dipole antennas. When the reflector area is greater than two square wavelengths, a feed system with effective apertures larger than that of a dipole is required. For improved illumination of a reflector with an area ranging from 2 to 6 square wavelengths, the single-dipole feed is replaced by a dipole with a reflector disk in front of it. This constitutes a short backfire antenna (SBFA), which is discussed in Chapter 5. In Figure 4.18, a linear dipole provides linear polarization. Crossed dipoles or any other radiators that provide the desired polarization may also be used. With crossed dipoles as the feed system, measurements indicate isolation of approximately 30 dB between the horizontally- and vertically-polarized field components.

Figure 4.18(b) shows that the rim *B* of the experimental antenna can be moved to the right any distance d_B, until it reaches its outer position (dashed lines), where $d_B = W_B$. The plane of the rim edge, which also constitutes the aperture plane of

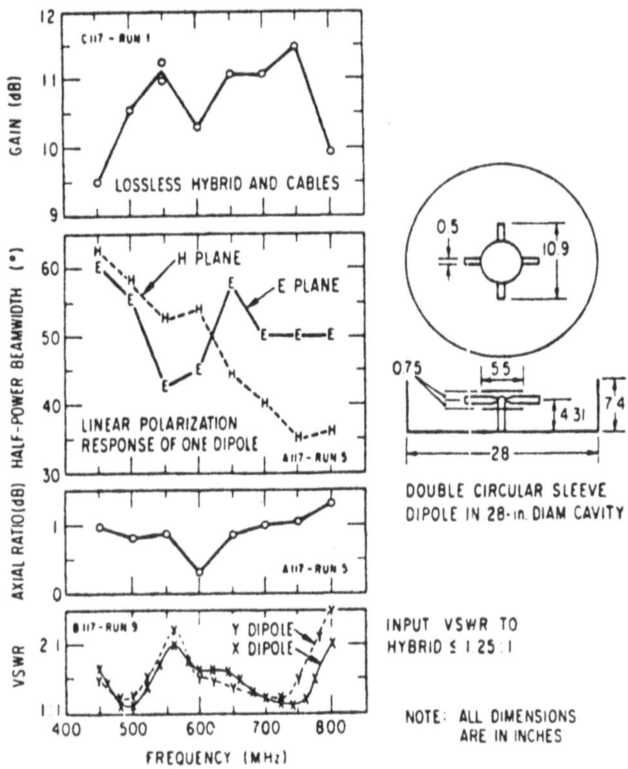

Figure 4.16 Measured electrical characteristics of a cavity-backed double-sleeve dipole antenna (from [10])

the antenna, is denoted by the dotted line P extending parallel to and spaced a distance d_B in front of A. The cup feed can be visualized as composed of two sources which, when combined in proper amplitude and phase, yield a resulting radiation pattern that is optimized. These sources are the crossed dipoles and the parasitically excited cup. Two modes of radiation exist: due to the crossed dipoles; and due to the open aperture of the parasitically excited cup. If the cup depth is longer than a few waveguide lengths, the cup dipole is shortened so that the dipoles are exposed and radiate fields which add to that of the open-ended waveguide. If the cup is rimless, the dipoles in front of the cup base reflecting surface radiate freely in space. Hence, by varying the cup depth between these two extremes, we can optimize the resulting radiation pattern of the cup antenna.

These results [11] suggest analyzing the antenna as a combination of two radiators: the feed in the center of the reflecting surface A and edge E of the rim in front of A. The rim edge is excited by radiation; therefore, its edge E acts as a secondary radiator when the dipole is excited. Highest directivity is obtained when

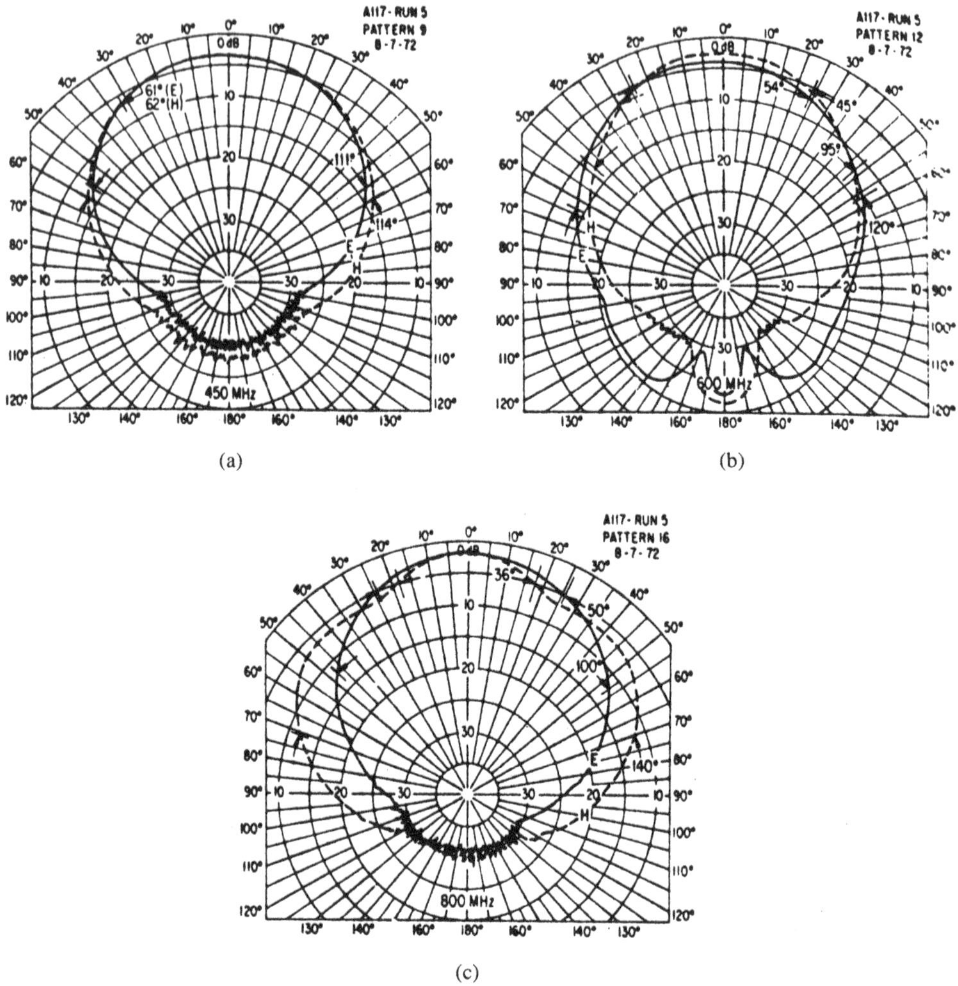

Figure 4.17(a–c) Measured radiation patterns of a cavity-backed double-sleeve dipole antenna (from [10])

far-field contributions of both radiators are in phase, as shown in Figure 4.19. Ehrenspeck has given the following three conditions for producing the highest directivity:

1. The diameter of the edge radiator E is such that $d_{A,opt} \simeq (n + 0.35)\lambda_0$, where $n = 1, 2, 3, \ldots$ This condition is valid for cup dipole radiators.
2. The rim width d_B of the rim is adjusted so that the far fields of the feed and edge radiators coincide in phase for the direction normal to the plane of

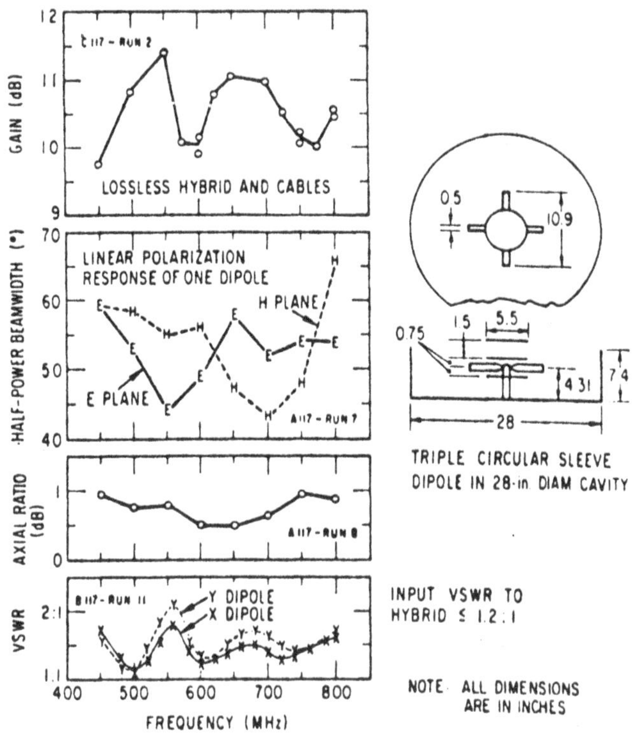

Figure 4.17(d) Measured electrical characteristics of a cavity-backed triple-sleeve dipole antenna (from [10])

reflector *A*.

3. The radiation coupling between feed *F* and edge radiator *E* is strong enough to contribute appreciably to the far field of the two-source combination.

For a spiral or helical waveguide, the radiation coupling and the current distribution on the rim edge may be different; therefore, the above conditions should be modified.

Figure 4.20(a) and (b) shows the antenna configuration and the field distribution across the aperture measured in both the *E*-plane and *H*-plane of the feed radiator, respectively. The radiating aperture of the antenna extends far outside the physical dimensions of the structure, and power has decreased by 20 dB from the antenna axis at distances of approximately $0.4\lambda_0$ from the edge of the rim at points h on the x-axis in the E-plane, and approximately $0.9\lambda_0$ at points v in the H-plane. The phase distribution across the aperture is also required, in order to calculate the increase in directive gain of the antenna. However, the field distribution curves indicate that there is certainly an increase in the directive gain of the antenna.

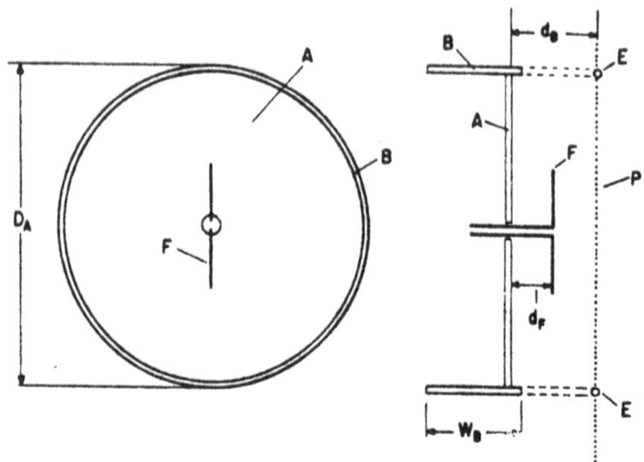

Figure 4.18 Sketch of a cup dipole antenna (from [11]): (a) front view, (b) cross section

Ehrenspeck [11] used a theory similar to that given by Hong, *et al.* [8] to analyze the cup dipole antenna shown in Figure 4.18. The large discrepancies between Ehrenspeck's results and those of Hong, Nyquist and Chen [8], however, indicate that the theory of the latter does not seem to apply to the antennas described by Ehrenspeck. Most likely, this is for the following reasons:

1. For a favorable excitation of propagating cavity modes, Hong, *et al.* assumed much higher cavity walls than Ehrenspeck did (Figure 4.18).
2. Hong, *et al.* decided to neglect the diffraction and coupling effects at the radiating cavity end. According to Ehrenspeck's results, these effects have a most decisive influence on the directive gain.
3. In contrast to the procedure followed in Ehrenspeck's work, Hong *et al.* apparently did not investigate the influence of the height of the cavity wall on pattern performance, nor did they direct their analysis toward a directive-gain optimization of their antennas.

In 1972, the Harris Broadcasting Product Division of the U.S. initiated a survey of potential design approaches to set objectives for a VHF CP antenna. The main requirements for the antenna were given as follows: satisfactory gain, multiplexing suitability, good axial ratio over a broadband of frequencies (equal *E*-plane and *H*-plane patterns), an element independent of mounting structure and adjacent elements, good impedance matching over a broadband of frequencies, minimum susceptibility to environment (including ice and snow) of impedance matching of the antenna, the vertical or horizontal inputs independent for emergency use, suitability for horizontal polarization transmission during transition period, grounded elements for lighting protection, omnidirectional pattern with three or four elements around supporting structure, minimum weight and windloading, and the capability to move easily in any direction.

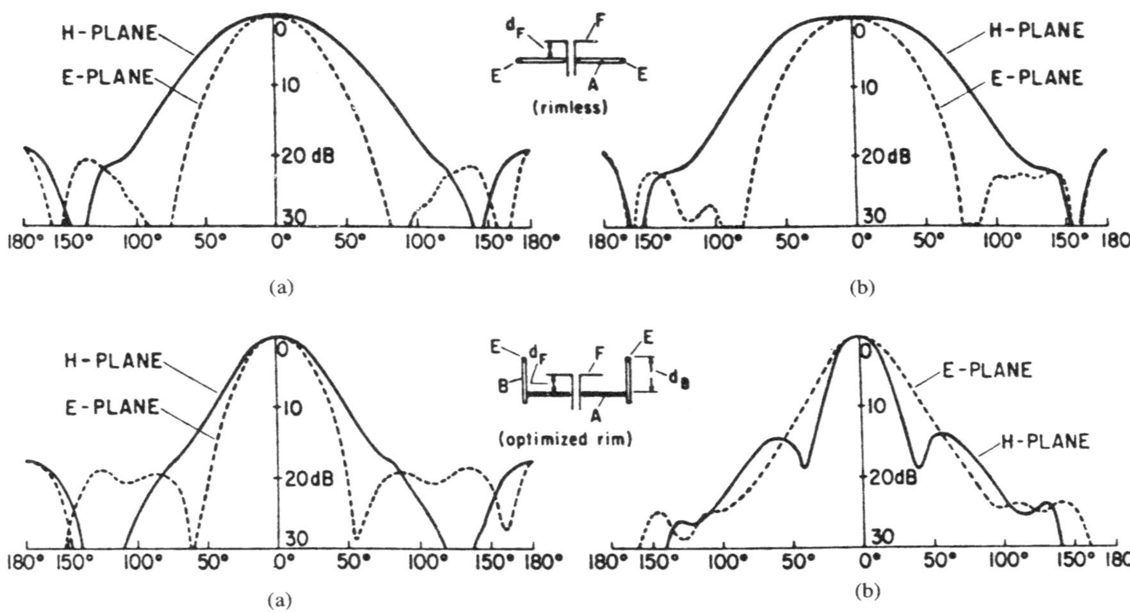

Figure 4.19 *E*-plane and *H*-plane radiation patterns before and after optimum directive gain adjustment (from [11]): (a) $D_A = 1.35\lambda_0$, (b) $D_A = 1.75\lambda_0$

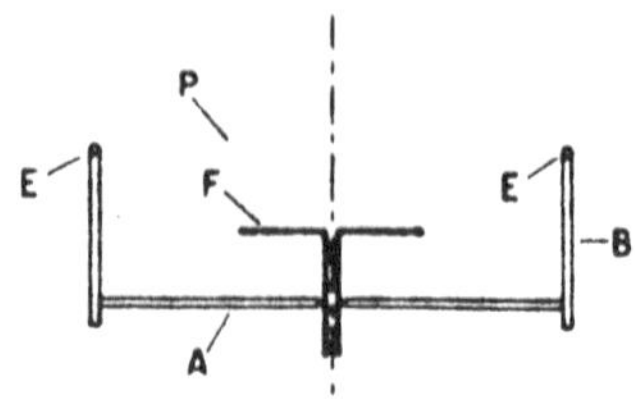

(a)

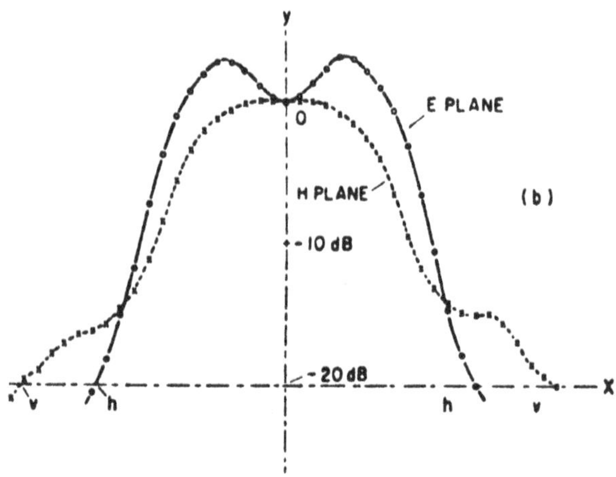

(b)

Figure 4.20(a) Antenna configuration (from [11])
(b) Amplitude field distribution at the aperture of the antenna (Figure 4.20(a)) (from [11])

During that time, a new requirement was added for a circularly-polarized antenna, and a program was initiated to develop a unit employing crossed dipoles on a printed circuit board located on the open face of the cavity. The cavity-backed CP antenna became a very attractive candidate for application to TV transmitting service, due to electrical and mechanical properties. The Harris Corporation initiated a program to develop a high-power cavity-backed coaxial crossed-dipole antenna.

Fisk and Donovan [12] have discussed the design and experimental results of the cavity-backed antenna in detail. A typical square cavity-backed crossed-dipole receiving antenna is shown in Figure 4.21. The square shape is physically better suited to the printed circuit dipoles and the mounting requirements for the receiving antenna. Due to round corners, the aperture forms an octagonal shape, which is similar to a circular backing cavity shape. For broadcasting applications, a circular

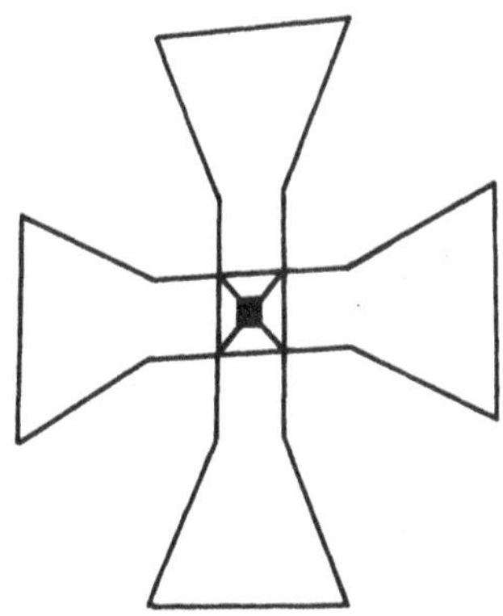

Figure 4.21 A typical microstrip cavity-backed receiving antenna (from [12])

cavity shape was chosen, with a grid construction to minimize windload and physical size, as shown in Figure 4.22(a). Figure 4.22(b) shows a wideband flat dipole, which is used to excite the cavity and to handle the required power level. The shape of the flat dipole offers advantages in mounting and in control of its impedance characteristics. Each dipole is mechanically supported and fed from a special balun, as shown in Figure 4.22(c). The balun serves the usual function of covering the coaxial input to the balanced circuit, while also providing a dc ground for the dipole elements. The cavity has separate inputs to the horizontal and vertical elements. In this way, the cavity can be fed for either horizontal, vertical, or circularly-polarized radiation. A detailed sketch and equivalent circuit of a compensated balun is shown in Figure 4.6(a) and (b).

Duncan, Hamada and Ingerson [12a] have developed a dual-polarization multiple-beam antenna for frequency reuse satellites that require amplitude isolation and polarization orthogonality between regional coverage beams. They have used an offset paraboloidal reflector fed by dual-polarized cup-dipole elements, as shown in Figure 4.22(d). The offset paraboloid is defined by the focal length F, the projected aperture diameter D, angles θ_1 and θ_2 that define the extremities of the reflector in the XZ plane, the angle of feed offset θ_0, and the coordinate X_c, which locates the center of the aperture relative to the focal axis of the reflector. The diameter of the cup-dipole feed is 7.6 cm, as discussed earlier in this chapter. Each element can simultaneously radiate both RHCP and LHCP. Duncan, *et al.* [12a] have used 20 of the cup-dipole elements in an array feed. The cup-dipole elements are spaced about 7.6 cm apart in an equilateral triangle lattice of contiguous apertures. When the element spacings are on the order of one wavelength, mutual coupling can seriously degrade the axial ratio of the individual elements. This happens because mu-

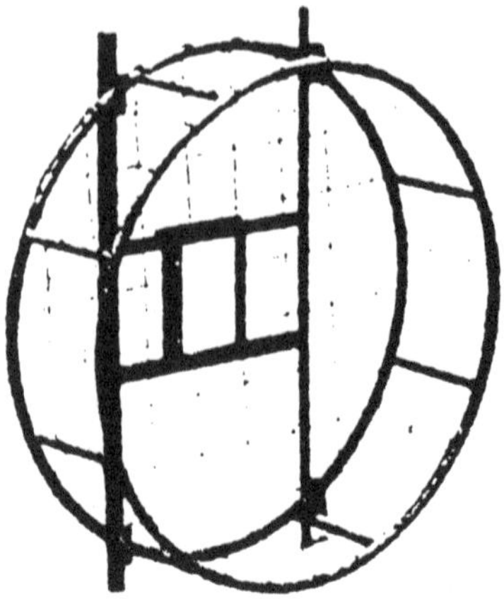

(a)

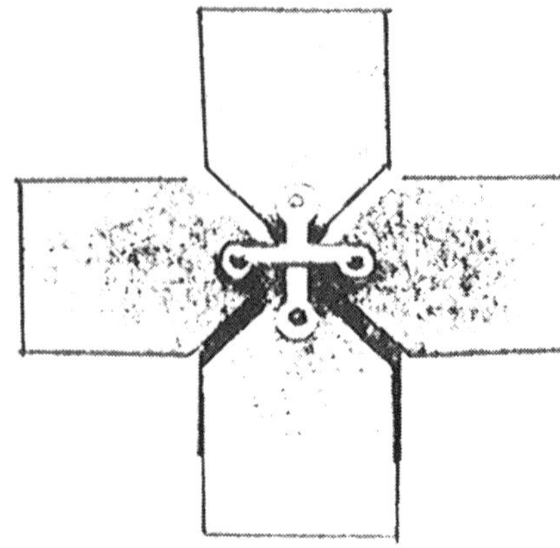

(b)

Figure 4.22(a) Grid wire cavity-backed radiator (from [12])
(b) Flat dipole arrangement (from [12])

Figure 4.22(c) Mechanical arrangement of dipole and balun (from [12])

tual coupling is different for the two orthogonal polarization components of a circularly-polarized element. We have found that mutual coupling is a function of element spacing, array lattice, wavelength and types of elements. Mutual coupling between cup-dipole elements perturbs the excitation coefficients and axial ratios of the array radiators. Circular aperture horns are better than square aperture horns in a circularly-polarized array. A photograph of a similar single-element cup-dipole antenna is shown

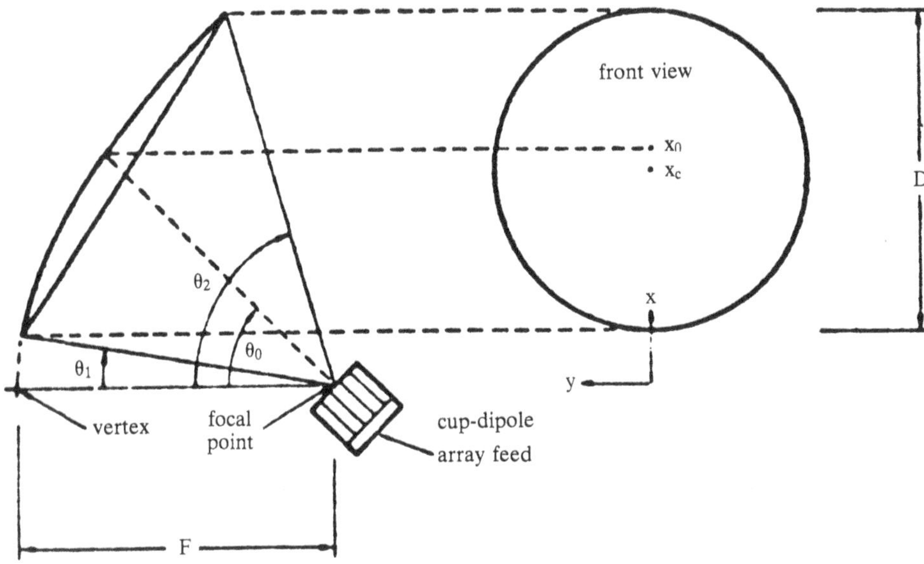

Figure 4.22(d) Offset paraboloidal reflector antenna illuminated by a multi-element array feed (from [12a])

in Figure 4.5. Each cup-dipole element is connected to a 90° hybrid. A 20-way stripline power divider board is connected to the hybrids by semi-rigid coaxial cable, in order to provide the required power distribution. Because the elements are dual polarized, two identical power divider boards were fabricated, one board was connected to the LHCP ports, and the other board to the RHCP ports. They have used 25 dummy elements out of a 45 element array cup-dipole. These dummy loads are terminated in matched loads in order to evaluate coupling effects.

Figure 4.22(e) shows the calculated far-field radiation pattern from near-field measurement data. Figure 4.22(f) shows the far-field measured radiation pattern of the cup-dipole array in an anechoic chamber, using a rotating linear source antenna. These measurements were performed at 3.83 GHz.

Near-field probing of feed aperture verified that mutual coupling had indeed perturbed the axial ratios of the array cup-dipole elements. It was found that the effect, due to mutual coupling on the axial ratio, can be minimized by small adjustments in the amplitude and phase of the cup-dipole elements. Following these adjustments, the cup-dipole array was mounted in the offset reflector configuration and measured. LHCP and RHCP contour plots are shown in Figure 4.22(g and h) at 3.88 GHz. Experimental results have substantiated the basis of the antenna design, which incorporates low sidelobe beam synthesis with a dual circularly-polarized cup-dipole feed element, to achieve orthogonally regional coverage beams. The cup-

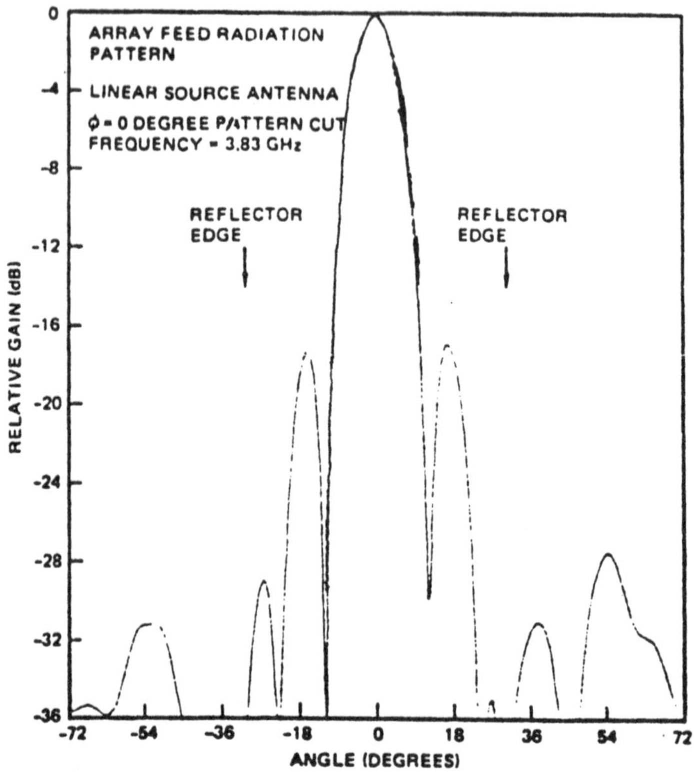

Figure 4.22(e) Array feed radiation pattern calculated from near-field cup-dipole measurements (from [12a])

dipole element produces a low axial ratio, and it can be adjusted to compensate for mutual coupling in the array environment.

Kumar [13] has developed software to calculate the *E*-plane and *H*-plane patterns of a cup dipole antenna at 8 GHz. The cup dipole model is shown in Figure 4.23(a). Figure 4.23(b) shows the *E*-plane and *H*-plane patterns. There is good agreement between the experimental and the theoretical radiation patterns.

Experimental verification conducted by many professionals [14, 15, 16] has shown that the axial ratio performance of a cylindrical cup dipole antenna with a diameter larger than about a wavelength deteriorates in terms of frequency bandwidth. In other words, the *E*-plane and *H*-plane beamwidths are unequal when the cylindrical cup diameter is larger than a wavelength. Flaring the cup into a conical shape [17] is a technique for controlling horn pattern symmetry over the frequency band. A conical cup with small flare angle significantly improves the axial ratio and

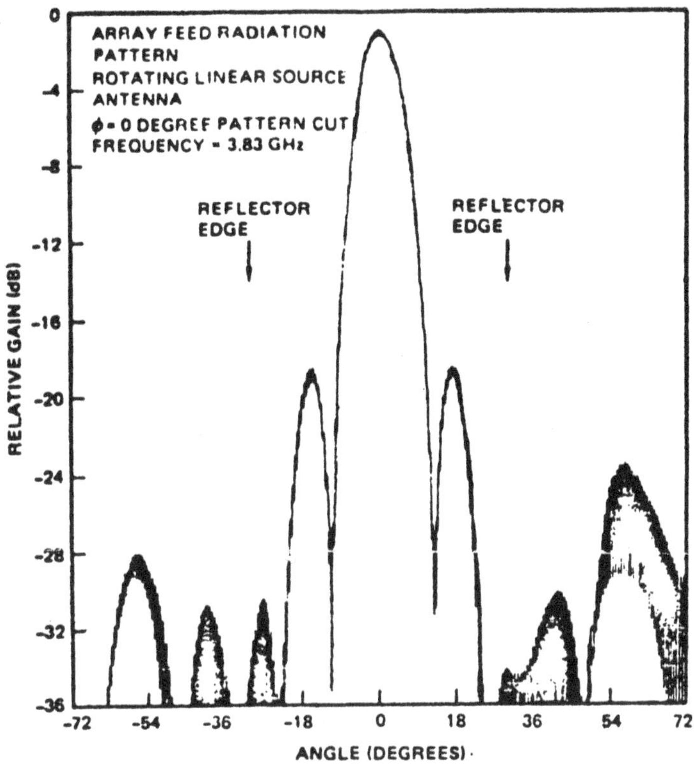

Figure 4.22(f) Measured far-field pattern of the cup-dipole array (from [12a])

return loss of the antenna. However, this is accomplished at the expense of antenna gain, due to a broadened beam. The flare angle of the cone is usually determined by a trade-off between the gain and axial ratio requirements.

Wong [18] described the design, development, and testing of an *X*-band corrugated horn antenna fed by a crossed-dipole, which was optimized to meet the earth coverage gain requirement of 16.5 dBi, with a small differential between the peak gain and the edge of earth gain (±8.7 degrees off from boresight), while keeping a compact mechanical design. The corrugated horn shown in Fig. 4.23(c) has a simple feed, consisting of a crossed dipole of unequal length. The diameter of the horn is 17.7 cm, and the overall length is only 19.3 cm, including the height of the waveguide transition. RHCP is provided by the unequal length crossed-dipole, which is excited by a split balun. When the dipole arms are adjusted to provide the necessary phase quadrature, a coverage axial ratio of less than 1 dB over 3% operating frequency bandwidth is obtained. The impedance of the element is matched by correctly

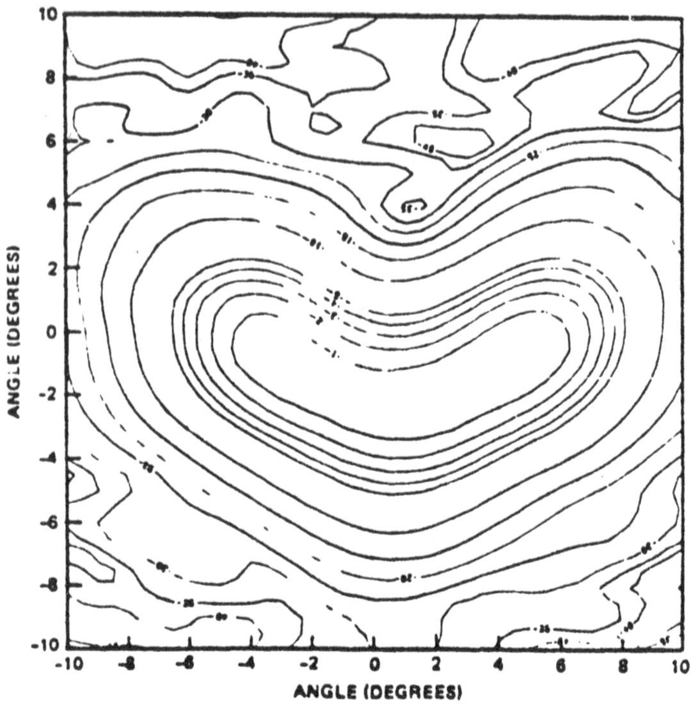

Figure 4.22(g) Measured LHCP (co-polarization) contour plot of the antenna at 3.88 GHz (from [12a])

sizing the dipole dimension and location in the cavity. The dipole arm lengths are the dimensions most critical for obtaining good axial ratio and impedance match at the input port. An optimized dipole geometry is shown in Fig. 4.23(d). After matching, the antenna VSWR has been less than 1.2 to 1.0 over 3% operating frequency bandwidth. At center frequency, the VSWR of the antenna is better than 1.1 to 1.0.

The dipoles of unequal length are machined copper parts attached to a 0.358 cm diameter coaxial cable. The coaxial cable is connected to a machined copper support base. Attached to the far end of the coaxial cable is a machined copper probe, which extends into the waveguide mount assembly. The probe is soldered to the coaxial center conductor using an SN96 high-temperature structure solder. The waveguide mount assembly is a dip-brazed aluminum WR112 waveguide assembly. A standard WR112 waveguide flange has been incorporated for input interfacing. All three components are assembled with screws.

Table 4.2 summarizes the performance of the antenna. A 0.5 dB gain improvement has been obtained by operating the antenna with a very simple teflon lens consisting of a stack of three 1.905 cm thick teflon disks. The disks are 2.54 cm,

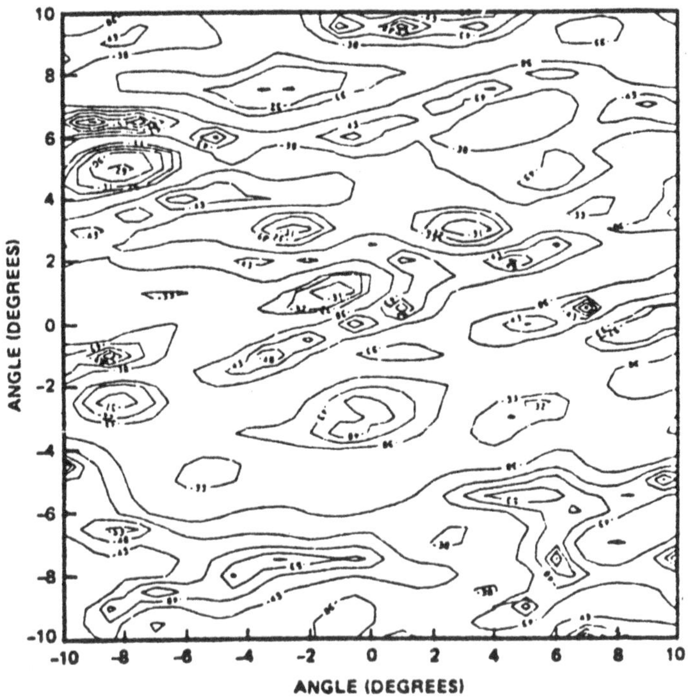

Figure 4.22(h) Measured RHCP (cross-polarization) contour plot of the antenna at 3.88 GHz (from [12a])

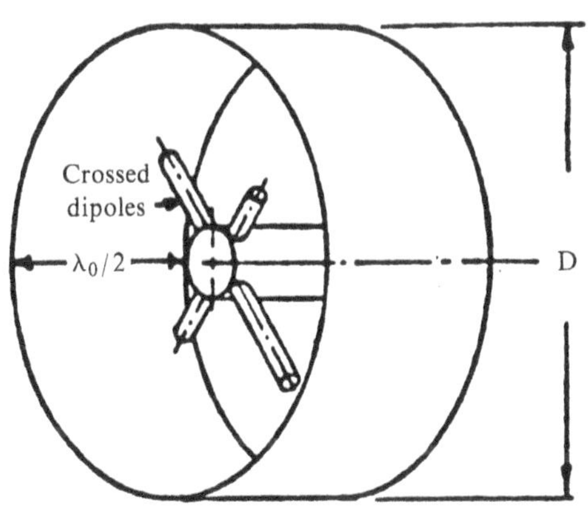

Figure 4.23(a) Geometry of a cup dipole antenna (from [13])

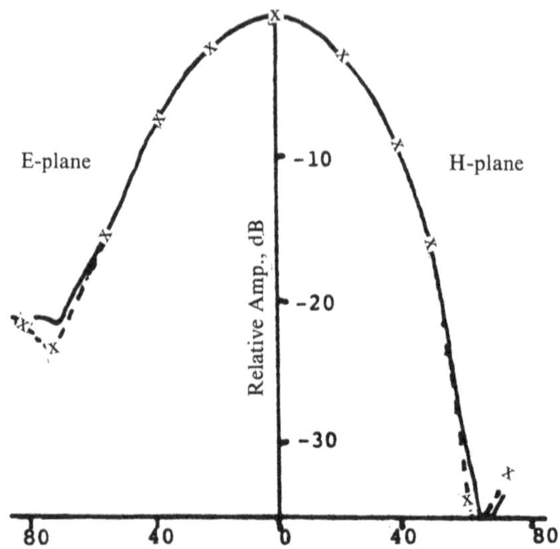

Figure 4.23(b) *E*-plane and *H*-plane patterns of a cup dipole antenna at 8 GHz (from [13]). Theoretical———-----; Experimental—x----x

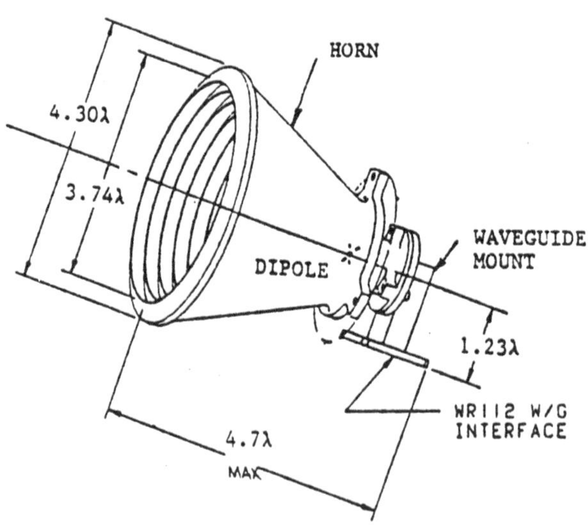

Figure 4.23(c) Sketch of corrugated horn antenna (from [18])

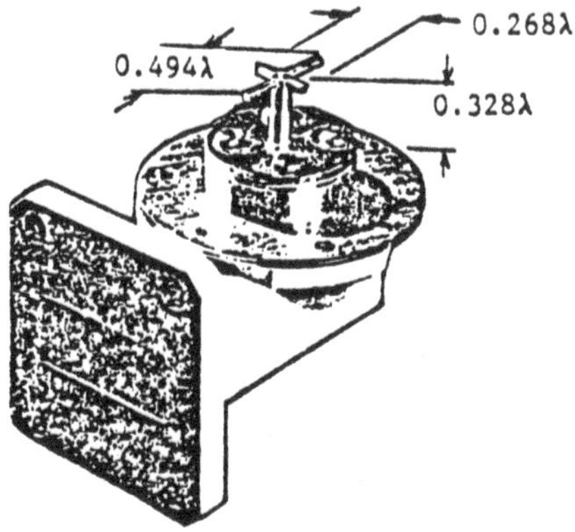

Figure 4.23(d) Sketch of the crossed-dipole and waveguide transition (from [18])

Table 4.2 Performance of an X-Band Corrugated Horn (from [18])

X-Band Corrugated Horn	
Frequency:	7900 to 8100 MHz
Gain	
Peak:	18.0 dBi
Earth Coverage:	16.5 dBi
Polarization:	RHCP
Axial Ratio:	1.0 dB
Horn Depth:	19.3 cm (7.6 in)
Including transition weight:	0.5 kb (1.1 lb)

5.08 cm and 7.62 cm in diameter and are placed in front of the horn aperture. The radiation pattern performance with and without the lens is shown in Figure 4.23(e).

Recently, Kumar [19,50] has developed two CP square horns at 6.433 GHz, which are shown in Figure 4.24(a) and (b). The cup depth and dipole lengths are optimized to obtain the axial ratio on axis of better than 0.4 dB. The crossed-dipoles are placed diagonally, to increase the gap between the tip of the dipole and wall of the horn. This gap is an important factor to avoid multipactor breakdown due to high power in the feed.

Figure 4.24(c) and (d) show the measured axial ratio pattern of the square feed shown in Figure 4.24(a) and (b), respectively. Over the frequency band (6.433 ± 0.250 GHz) the maximum value of the axial ratio is less than 0.4 dB; this increases

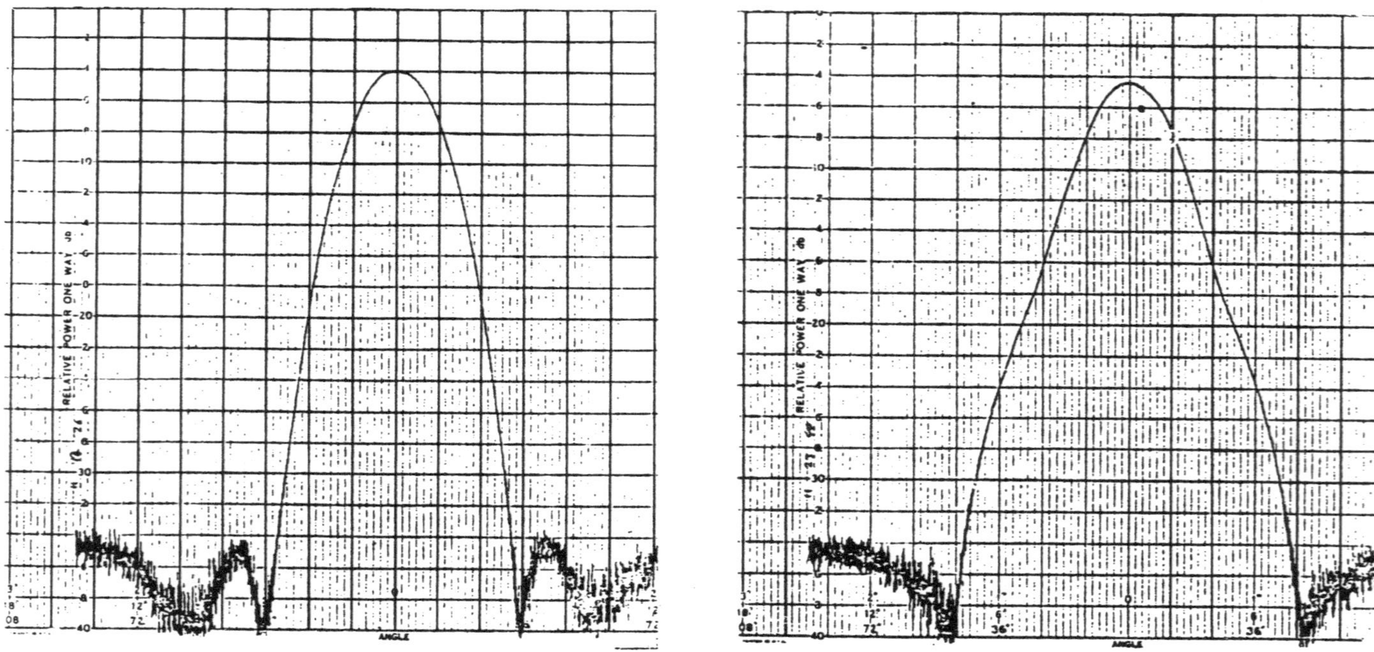

With simple teflon lens

Without lens

Figure 4.23(e) Radiation pattern of the antenna (from [18])

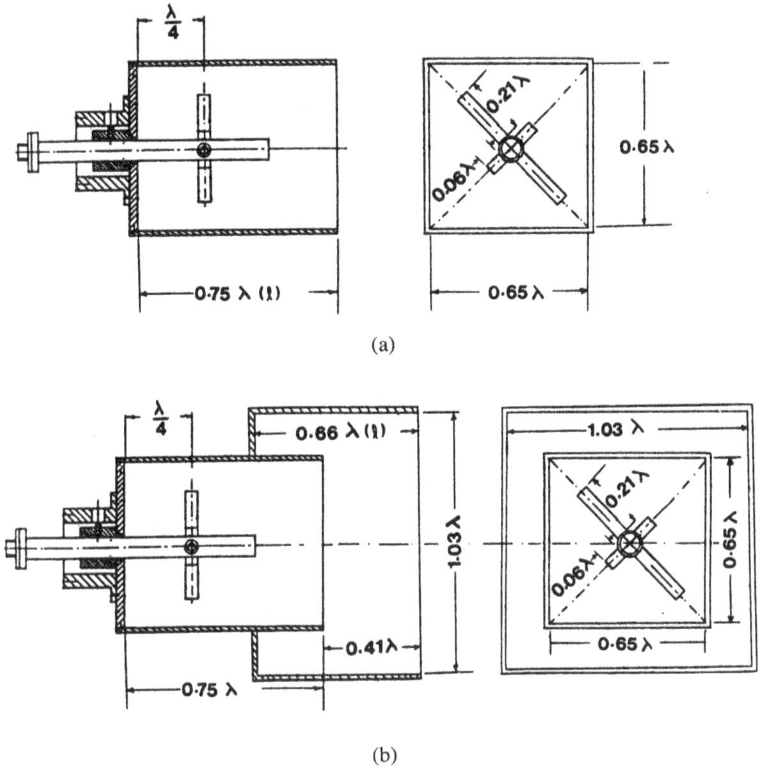

Figure 4.24(a) Crossed-dipole-fed square horn designed at 6.433 GHz (from [19,50])
(b) Crossed-dipole-fed dual-band square horn (from [19,50])

to 3 dB off-axis (−36 degrees to +36 degrees). The aperture efficiency of these antennas is about 95%. A return loss of better than −18 dB can be achieved by using a step transformer on the center conductor of the coaxial balun. Three different sizes of cup dipole antenna have been studied by Kumar [20,20a] at *L*-band: 1.3 λ_0 diameter circular cup (Figure 4.24(e)); 0.65 λ_0 diameter circular cup (Figure 4.24(f)); and 0.65 λ_0 square cup (Figure 4.24(g)). Each antenna assembly comprises a center conductor, a center conductor support, a matching transformer, crossed-dipoles, an outer conductor, a connector, and a cup. For optimum axial ratio, the length of dipoles and cup depth are optimized experimentally; these dimensions are presented in Table 4.3. Figure 4.24(h–j) shows the axial ratio pattern for optimized 1.3 λ_0 diameter circular, 0.65 λ_0 diameter circular and 0.65 λ_0 square aperture cup antennas. The aperture efficiency of the circular cup antenna is about 99%, while that of the square cup antenna has a poorer aperture efficiency (about 95%). The worst-case return loss of the 1.3 λ_0 cup is better than −25 dB in the frequency range of 1.5 to

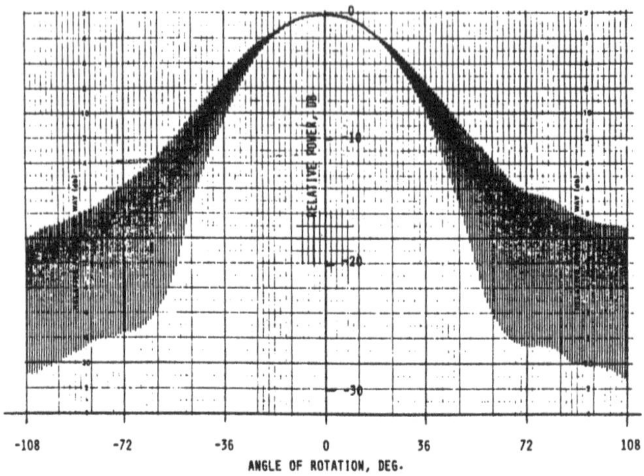

(c)

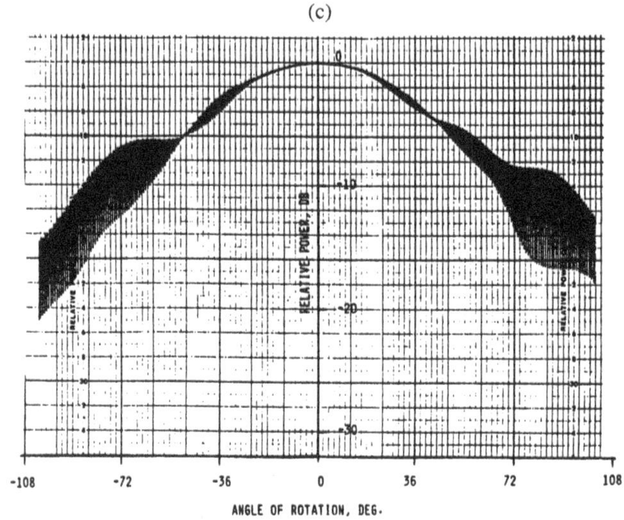

(d)

Figure 4.24(c) Measured axial ratio pattern (Feed shown in Figure 4.24(a)) (from [19,50])
(d) Measured axial ratio pattern (Feed shown in Figure 4.24(b)) (from [19,50])

1.6 GHz. The return loss of the 0.65 λ_0 circular (or square) aperture cup is better than -17 dB in this frequency range. The return loss of each can be improved to -20 dB or better if bandwidth is sacrificed. The worst-case axial ratio is better than 0.9 dB for the 0.65 λ_0 circular (or square) cup in the frequency range of 1.5 to 1.6 GHz. The axial ratio of the 1.3 λ_0 circular cup antenna deteriorates sharply with

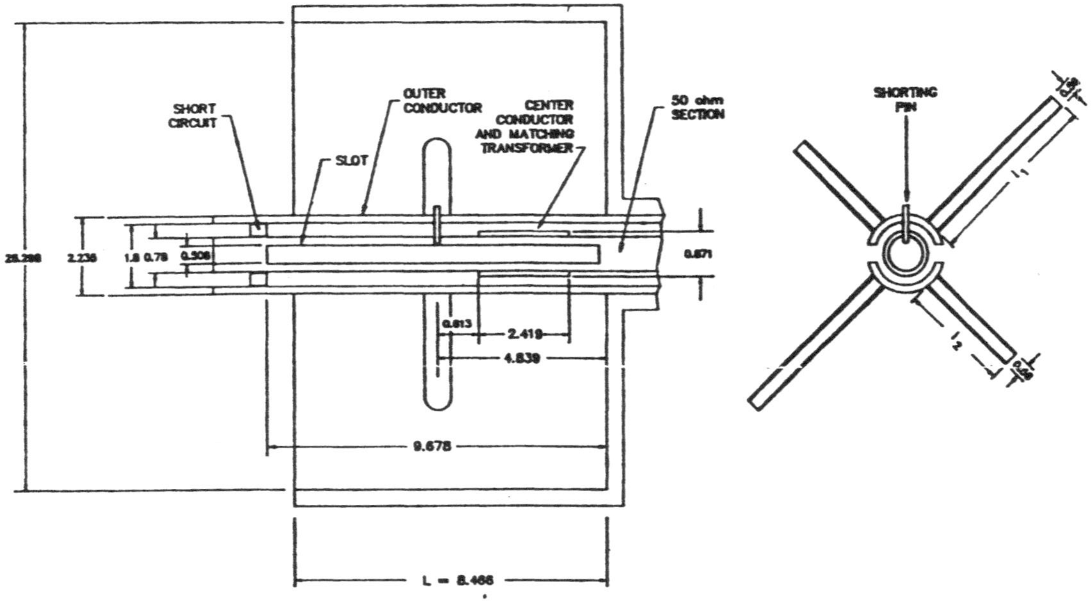

Figure 4.24(e) 1.3 λ_0 diameter circular crossed-dipole-fed cup (from [20,20a])

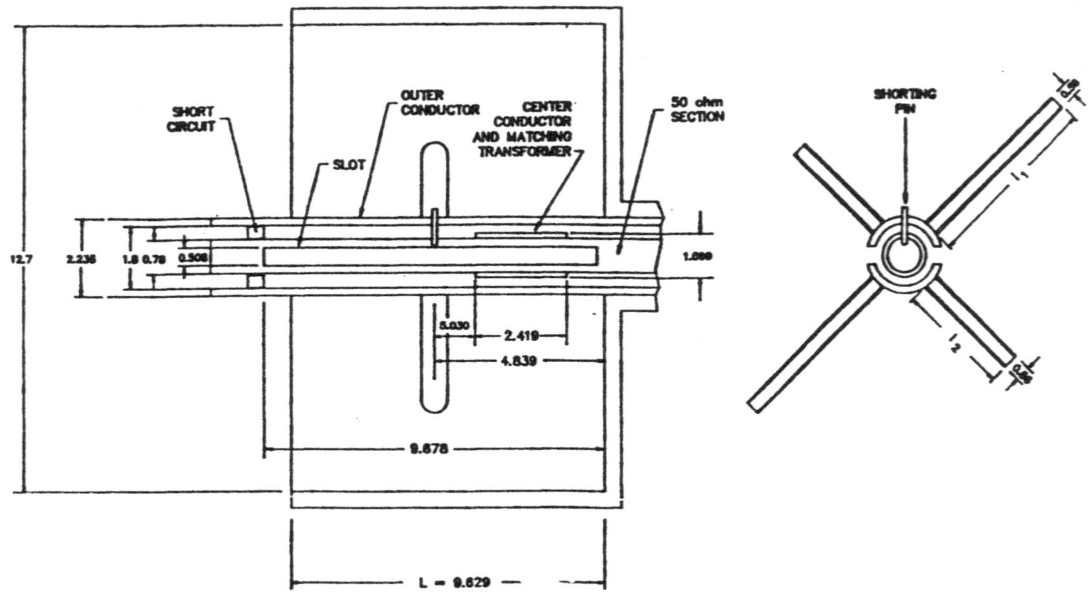

Figure 4.24(f) 0.65 λ_0 diameter circular crossed-dipole-fed cup (from [20,20a])

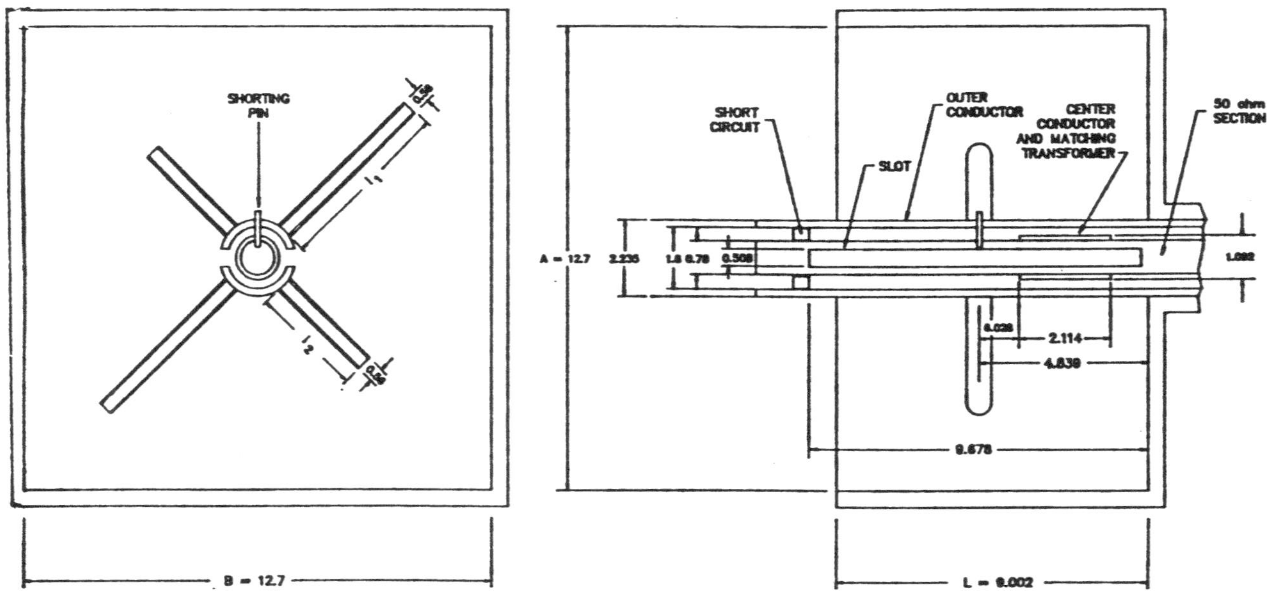

Figure 4.24(g) 0.65 λ_0 square crossed-dipole-fed cup (from [20,20a])

Table 4.3 Optimized Cup Dipole Antenna Dimension at 1.55 GHz with Axial Ratio on Axis < 0.3 dB

Cup Aperture in Wavelength	*Cup Depth in cm*	*Dipole Length in cm*
1.3 (Circular)	8.466	l_1 = 4.263
		l_2 = 3.045
0.65 (Circular)	9.629	l_1 = 4.120
		l_2 = 1.926
0.65 (Square)	9.002	l_1 = 4.383
		l_2 = 2.212

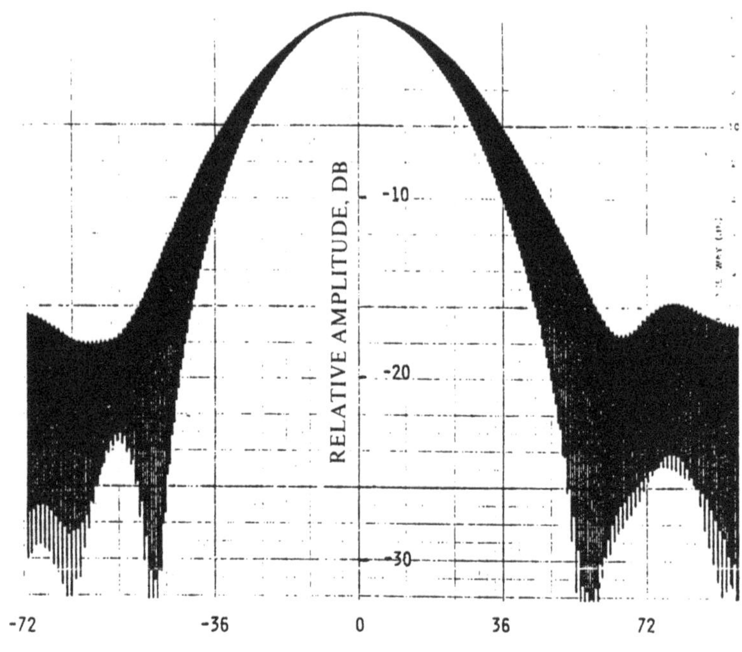

Figure 4.24(h) Axial ratio pattern of the 1.3 λ_0 diameter circular cup at 1.55 GHz (from [20,20a])

and reaches about 3.5 dB in the aforementioned frequency range. On the basis of these results, we conclude that the axial ratio deteriorates faster for the larger cup. These results, together with the return loss data, suggest that a compromise should be made on the aperture size of the cup to get both a good axial ratio and acceptable

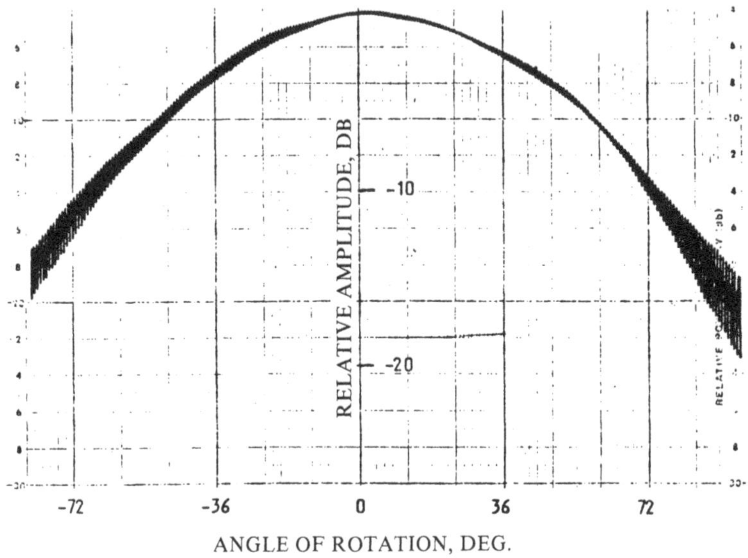

Figure 4.24(i) Axial ratio pattern of the 0.65 λ_0 diameter circular cup at 1.55 GHz (from [20,20a])

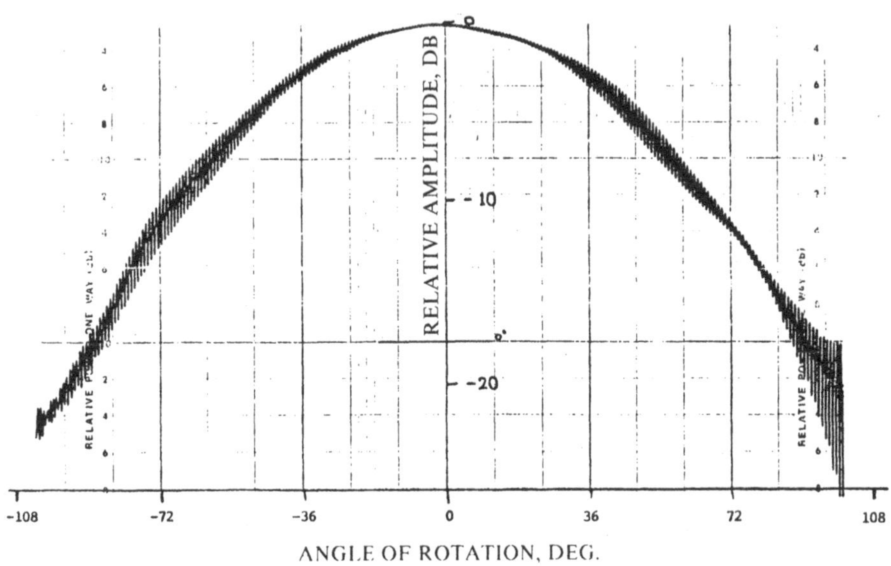

Figure 4.24(j) Axial ratio pattern of the 0.65 λ_0 square cup at 1.55 GHz (from [20,20a])

frequency, and reaches about 3.5 dB in the aforementioned frequency range. On the basis of these results, we conclude that the axial ratio deteriorates faster for the larger cup. These results, together with the return loss data, suggest that a compromise should be made on the aperture size of the cup to get both a good axial ratio and acceptable return loss. These radiators can be used either as feed elements or arrays for future M-Sat or other satellite programs.

4.3 CAVITY-BACKED HELICAL AND SPIRAL ANTENNAS

4.3.1 Cavity-Backed Helical Antennas

In recent years, the need for broadband elliptically-polarized antennas has become important in aerospace industries. Helical antennas fulfill this requirement, and in addition, are mechanically simple and relatively small in wavelength. Many types of helices have been studied by various authors [21–25]. We will discuss only cavity-type helical antennas in this section. The backfire type of cavity antenna will be discussed in the Chapter 5.

Bystrom and Berntsen [26] have discussed cavity-mounted cylindrical helical antennas. Figure 4.25 shows three types of cavity-mounted helical antennas. Figure 4.26 shows a typical two-turn cavity-mounted helical antenna. Figure 4.27(a) shows the effect of the number of helix turns on the half-power beamwidths, with the helices mounted in square cavities. The half-power beamwidths are independent of the number of helix turns used. The VSWR *versus* normalized frequency, the average half-power beamwidth *versus* normalized frequency, and the axial ratio *versus* normalized frequency are shown in Figure 4.27(b). As expected, the half-power beamwidths tend to be narrower for larger cavity sizes. The measured and estimated beamwidths [27] are approximately the same for both square and round cavity shapes. Axial ratios are generally higher than those for the helix mounted on a ground plane, but this effect could be reduced by the proper choice of other parameters.

Figure 4.27(a) shows a typical two-turn cavity-mounted helical antenna on a dielectric rod. The size of the aperture is 0.6 λ_0 square. Figure 4.27(b–d) show the VSWR *versus* normalized frequency, axial ratio *versus* normalized frequency and half-power beamwidth *versus* normalized frequency, respectively.

While pattern beamwidth and axial ratio bandwidths depend on exact performance requirements, the antenna shown in Figure 4.27(e) is considered to have a useful bandwidth of 1.7:1, with half-power beamwidths between 80° and 50°, on-axis axial ratios below 4 dB, and negligible pattern tilt (Figure 4.27f–h).

The geometry of a typical helical design [26a] is shown in Figure 4.27(i). The antenna operates over a 15.5 cm diameter cup ground plane. The diameter of the helix is 5.9 cm, the helix length is 53 cm, and the pitch angle is 14°. In terms of antenna dimensions, the calculated antenna performance is half-power beamwidth = 32.7°; first nulls beamwidth = 74.4°.

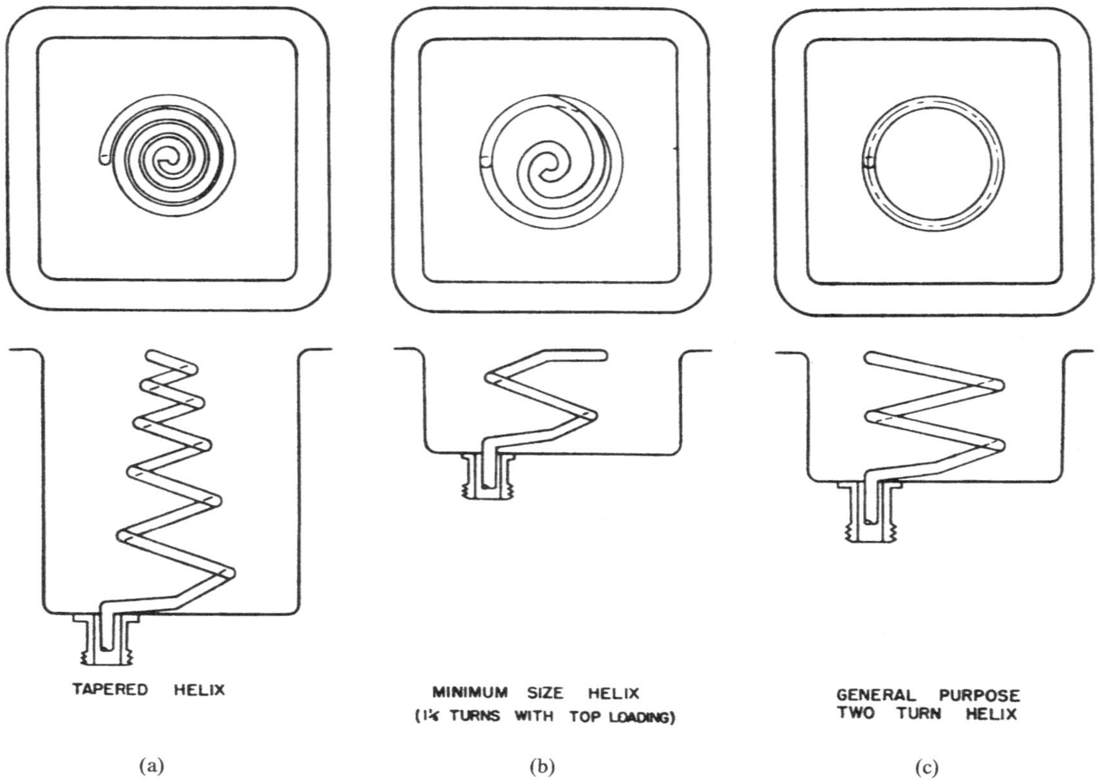

Figure 4.25 Three types of cavity-mounted helical antennas (from [26])

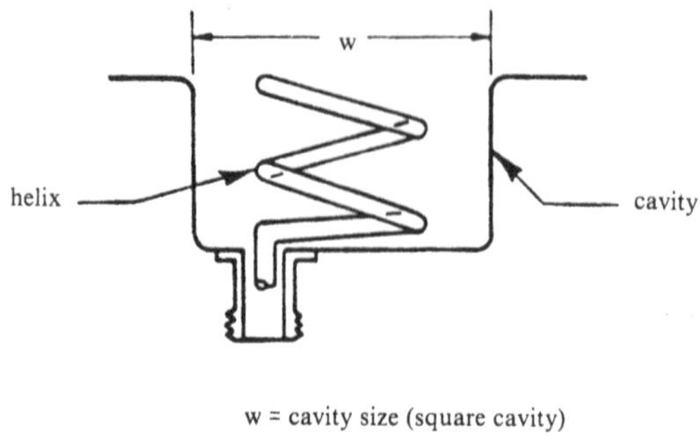

Figure 4.26 Two-turn helical antenna mounted in a cavity (from [26])

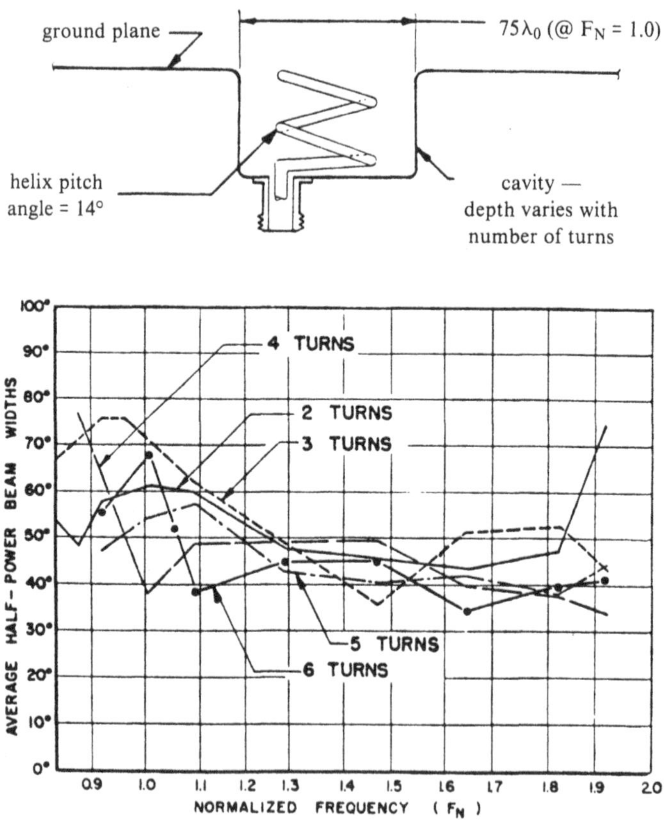

Figure 4.27(a) The effect of number of helix turns on half-power beamwidths (from [26])

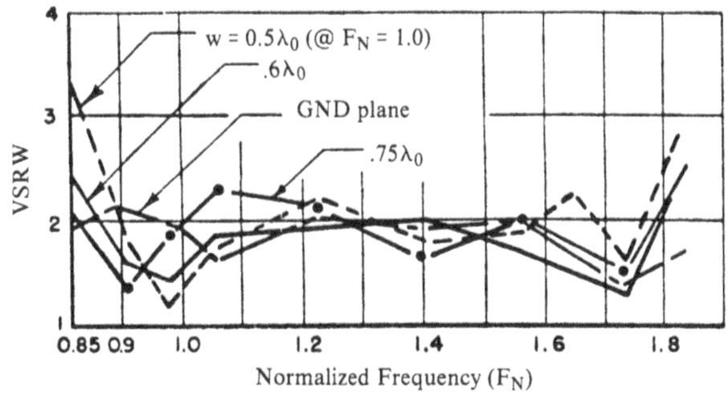

(b)

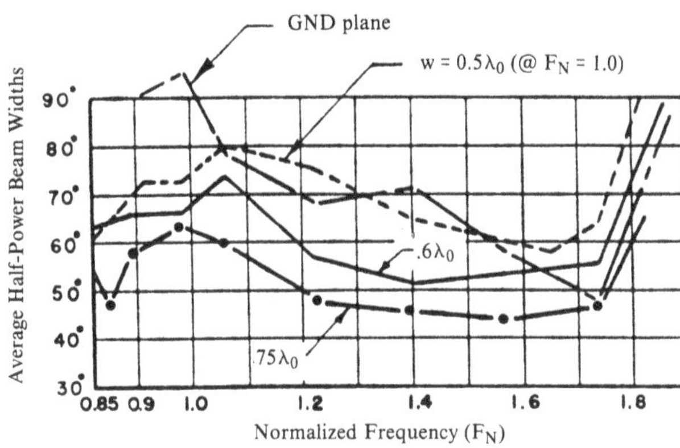

(c)

Figure 4.27(b) VSWR *versus* normalized frequency of a two-turn helical antenna mounted inside a cavity (from [26])

(c) Axial ratio *versus* normalized frequency of a two-turn helical antenna mounted inside a cavity (from [26])

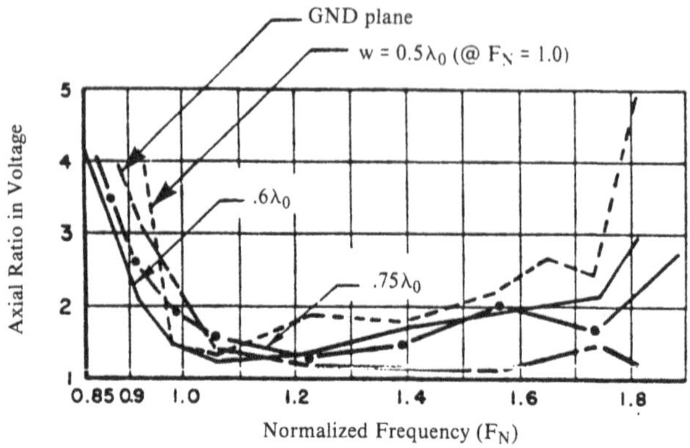

(d)

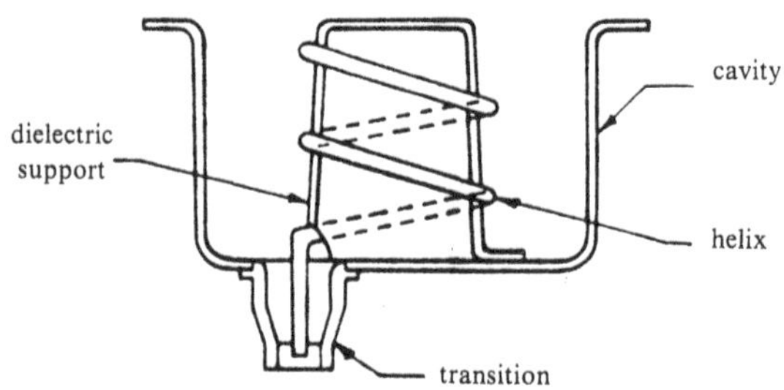

(f)

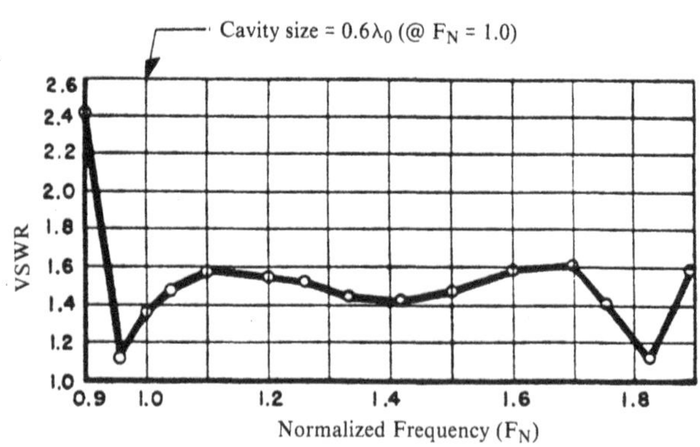

(e)

Figure 4.27(d) Half-power beamwidth *versus* normalized frequency of a two-turn, helical antenna in a cavity (from [26])

(e) Two-turn helix on a dielectric support inside a cavity (from [26])

(f) VSWR *versus* normalized frequency of a dielectrically-supported helical antenna inside a cavity (from [26])

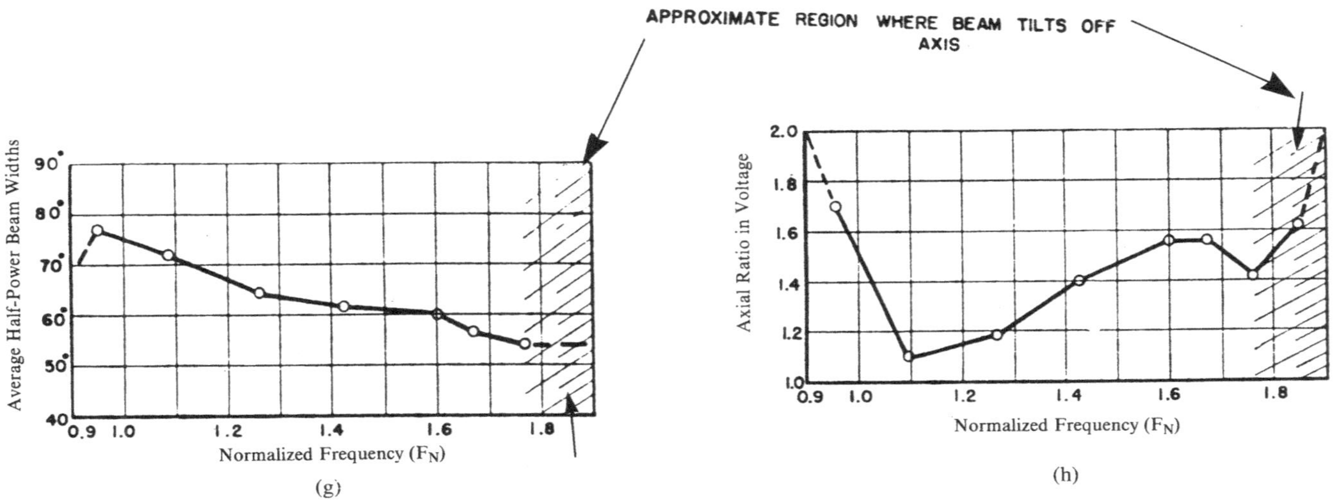

Figure 4.27(g) Average half-power beamwidths of a helical antenna supported by a dielectric tube and mounted in a square cavity (from [26])

(h) Axial ratio of a helical antenna supported by a dielectric tube and mounted in a square cavity (from [26])

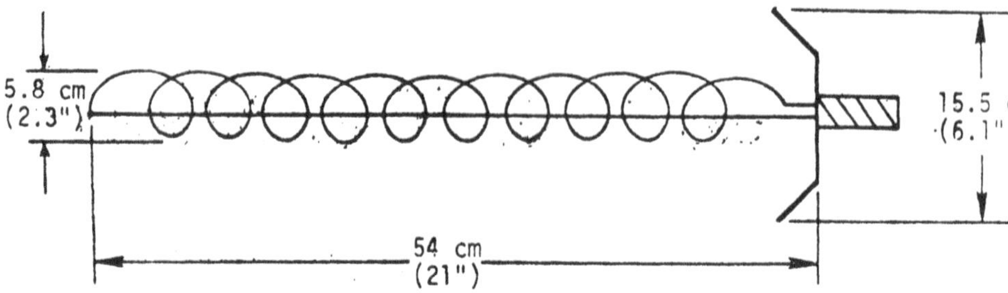

Figure 4.27(i) Typical cavity-backed helical antenna geometry (from [26a])

Wong and King have published several papers [27–29] on cavity-backed helical antennas. Their experimental measurements were taken in UHF and *s*-band. In their work, the helices are constructed by winding 4.76 mm diameter copper tubing around a styrofoam cylindrical form. The helix diameter is defined as the center-to-center distance of the copper wire. The helix is backed by a 261.62 mm diameter × 127 mm high circular cavity, rather than a conventional ground plane, to reduce back radiation, as shown in Figure 4.28. A 119.38 mm long linear-taper microstrip transformer, constructed from a printed circuit board and placed on the bottom of the cavity, was used to match the helix impedance from 140 ohm to a 50 ohm coaxial input [23]. Kraus [30] showed that the antenna can be matched by bounding a thin metal strip between the feed point and the beginning of the helix. Kumar [31] also showed that the presence in the cavity of a thin metal strip improves the matching of cavity-backed helical antennas. Figure 4.29 shows a metal strip on the feed point of a cavity-backed antenna. The return loss is better than −25 dB for more than 15% of the bandwidth.

Gain and radiation pattern measurements were made on a variety of helix configurations [24, 25, 27, 28, 29], including fixed length helices with variable pitch angle and diameter, and variable length helices with a constant diameter and constant pitch angle.

Agrawal and Wong [32] have reported on the effect of the cavity on the helix. They chose the cavity diameter to be 15.24 cm in order to cutoff the TE_{01} mode which, if present, will produce off-axis radiation. Figure 4.30(a) shows the helix with tapered termination and feed cavity. The main function of the feed cavity is to launch a slow wave on the end-fire structure, so that a feed cavity with good efficiency will convert most of the energy into this mode [33]. There are two main sources to produce the final radiation pattern of this type of antenna: directly by the feed, or by the end-fire helix. These two field components are added or subtracted to produce the final radiation pattern, either constructively or destructively, depending on the design. Agrawal and Wong [32] found that a 12.7 cm cavity length can

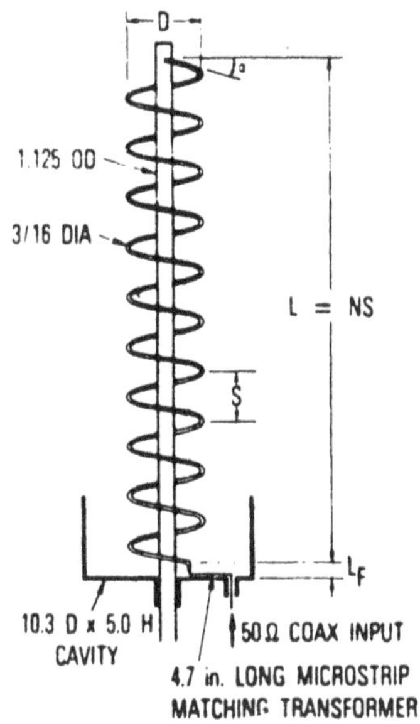

Figure 4.28 Cavity-backed helical antenna (from [27–29])

produce the highest gain. Figure 4.30(b) shows the measured pattern of the antenna at 2.287 GHz. The feed cavity also plays an important role in controlling the side-lobes of the antenna.

The mutual coupling between helices has been studied by many researchers [34–36]. The aperture efficiency of these elements is very high; therefore, high mutual coupling has been found between them. The mutual coupling can be reduced by increasing the gap between elements, and by using a feed cavity.

Figure 4.30(c and d) shows measured patterns of the element in Figure 4.30(a), and those of an identical element with an undersized cavity. Although the cavity produced the same edge gain from the isolated element, the performance in cluster was degraded, due to beam narrowing. One reason for the reduced coupling in the large cavity is that the mode near the feed end, which is known to travel with free-space velocity [33], is not allowed to radiate. Such modes produce strong off-axis radiation and cause large coupling, as discussed by Yen [35].

Shiokawa, Karasawa and Yokoi [36] have studied the characteristics of a helical array antenna composed of four helical antennas. It may be significant and interesting to consider the effect of mutual coupling on the properties of the array antenna, but here we will discuss this effect on the properties of the helical antenna. Figure 4.31(a–c) shows the geometrical configuration of a helical array antenna of

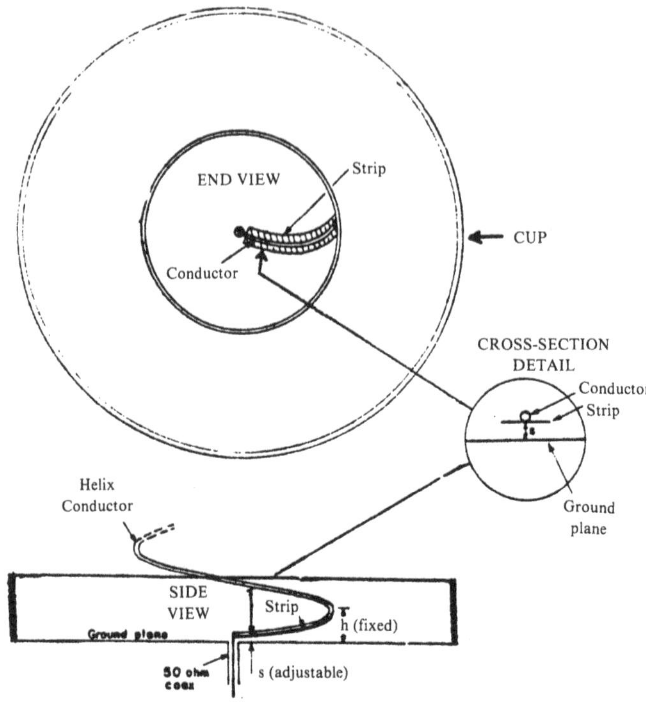

Figure 4.29 Matching of a cavity-backed antenna by a metal strip (from [30–31])

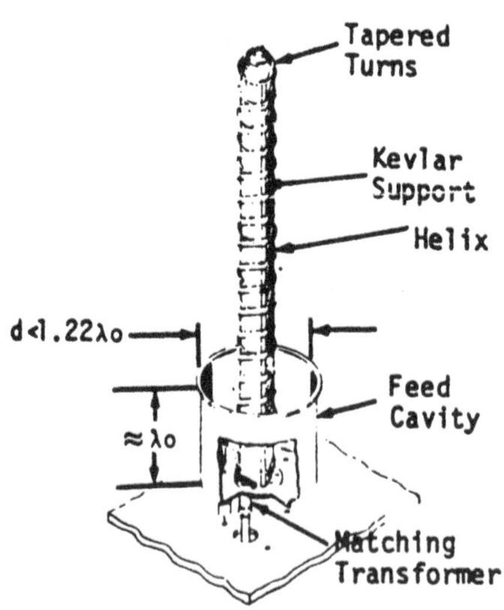

Figure 4.30(a) Cavity-backed helical antenna (from [32])

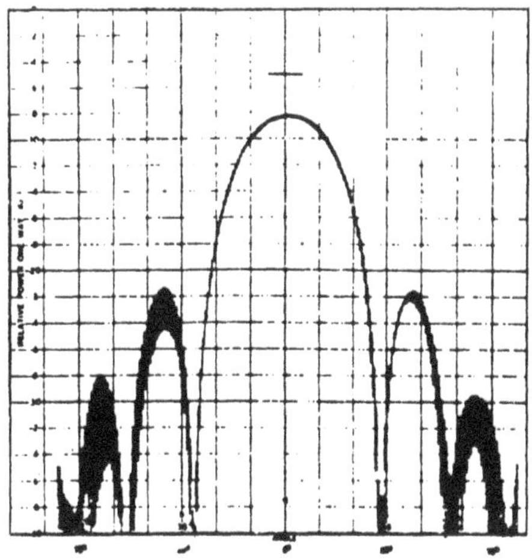

(b)

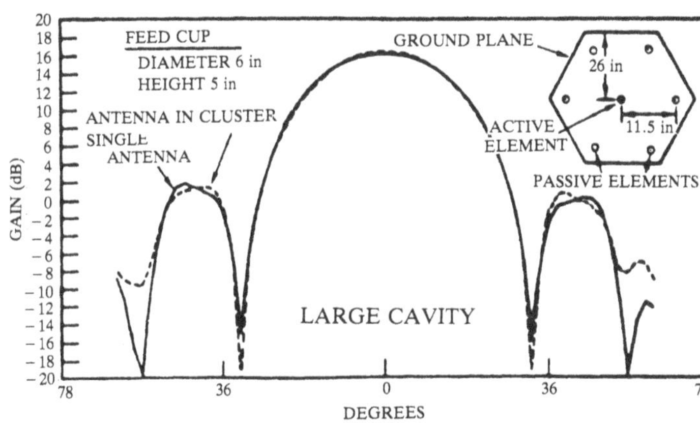

(c)

Figure 4.30(b) Measured pattern of the cavity-backed helical antenna of Figure 4.30(a) at 2.287 GHz (from [32])

(c) Measured patterns of the large cavity-backed element placed alone and in cluster (from [32])

square arrangement whose elements are identically oriented. In the figure

F_n = feed point of nth element;
S_n = length measured along the nth element from the feed point;
δ = element spacing; and
D = reflector diameter.

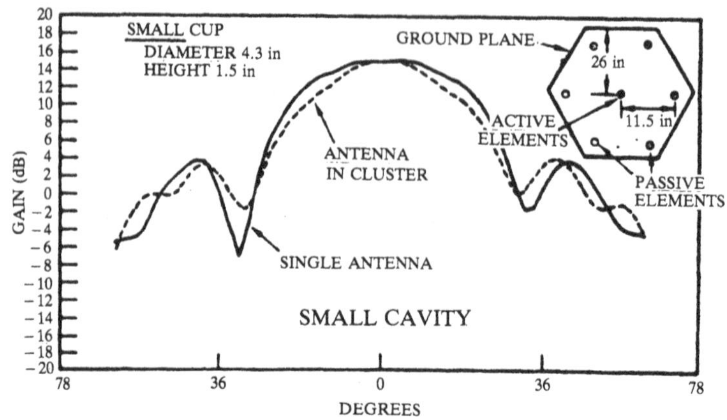

Figure 4.30(d) Measured patterns of the small cavity-backed element placed alone and in cluster (from [32])

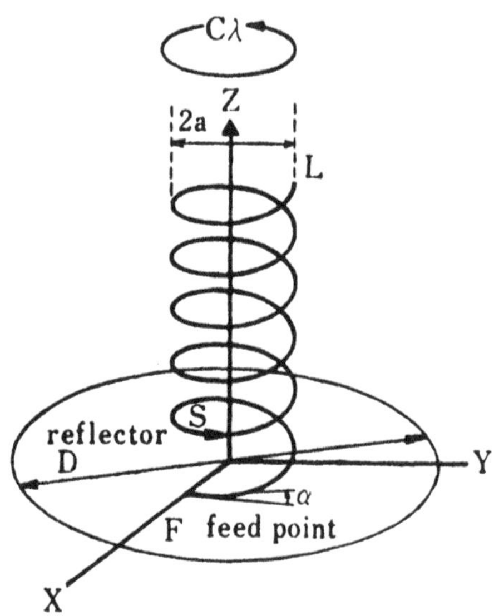

Figure 4.31(a) Single element helical antenna (from [36])

Figure 4.31(d) shows the effect of reflector diameter on the gain and axial ratio of a single-element helical antenna [36]. The large reflector diameter *D* (Figure 4.31(c)) is assumed to be infinite on the basis of the current distribution [38]. The practical helical array antenna is reduced to the balanced helical array antenna shown in Figure 4.31(a–c). The current distribution in region *A* and region *B* has equal value in

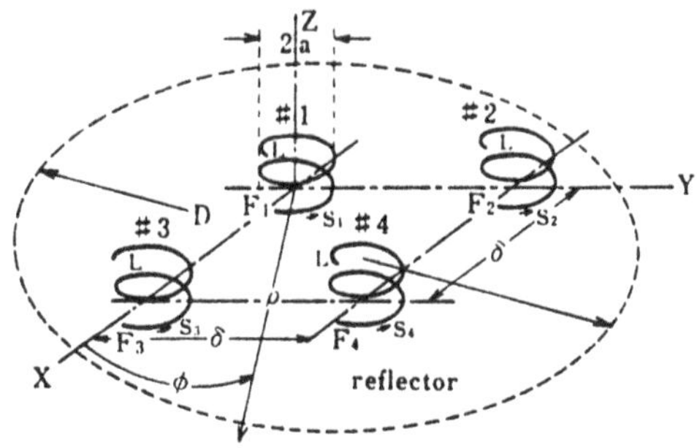

(b)

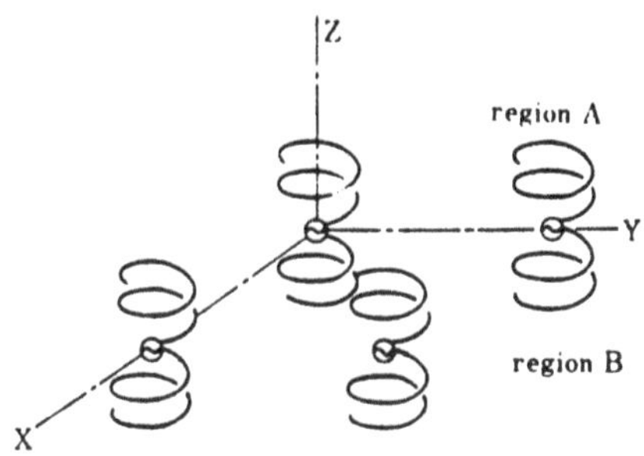

(c)

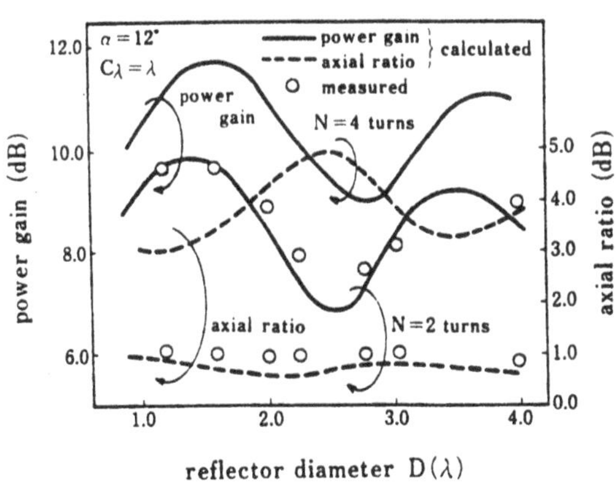

(d)

Figure 4.31(b) and (c) Co-ordinates for a four-element helical array antenna (from [36])

(d) Power gain and axial ratio of a single-element helical antenna (from [36]) magnitude, but opposite in phase. Thus, the Hallen type integral equation [24] is given by:

$$\int_0^L \sum_{n=1}^{4} I_n(S_n) \prod_{n,m} (S_m, S_n)\, dS_n = B_m \cos kS_m^{-jV_m}/2Z_0 \sin kS_m \qquad (4.10)$$

where $I_n\ (S_n)$ is the current distribution on nth element and B_m is the constant of integration, which is to be determined by the condition that each current vanishes at the open end of helix. V_m and Z_0 are the voltage applied at the mth element, and at the impedance of free space, respectively. $\Pi_{n,m}\ (S_m, S_n)$ is written as:

$$\prod_{n,m} (S_m, S_n) = \prod_{n,m}^{1} (S_m, S_n) + \prod_{n,m}^{2} (S_m, -S_n)$$

where

$$\prod_{n,m}^{l} (S_m, S_n) = G_l\ (S_m, S_n)(\hat{S}_m \cdot \hat{S}_n)_l$$

$$- \int_0^{S_m} \left\{ \frac{\partial G_l\ (\xi_m, S_n)}{\partial \xi_m} (\hat{\xi}_m \cdot \hat{S}_n)_l + \frac{\partial G_l\ (\xi_m, S_n)}{\partial S_n} \right.$$

$$\left. + G_l\ (\xi_m, S_n) \frac{\partial(\hat{\xi}_m \cdot \hat{S}_n)_l}{\partial\ \xi_m} \right\} \cos k\ (S_m - \xi_m)\ d\xi_m$$

in which

$$(\hat{\xi}_m \cdot \hat{S}_n)_l = (-1)^{l-1} \cos^2 \alpha \cos\left(\frac{\xi_m - S_n}{\alpha} \cos \alpha\right) + \sin^2 \alpha$$

$$\frac{\partial}{\partial \xi_m} (\hat{\xi}_m \cdot \hat{S}_n)_l = (-1)^l \frac{\cos^3 \alpha}{\alpha} \sin\left(\frac{\xi_m - S_n}{\alpha} \cos \alpha\right)$$

$$G_1 = e^{-jkrl/4\pi r_1}, \quad G_2 = e^{-jkr2/4\pi r_2}$$

and r_1, r_2 are given by:

$$r_1^2 = 2\,\alpha^2\left\{1 - \cos\left(\frac{S_m - S_n}{\alpha}\cos\alpha\right)\right\} + 2\delta^2 + (S_m - S_n)^2\sin^2\alpha$$
$$+ 2\alpha k_1\left\{\cos\left(\frac{S_m}{\alpha}\cos\alpha\right) - \cos\left(\frac{S_n}{\alpha}\cos\alpha\right)\right\}$$
$$+ 2\alpha k_2\left\{\sin\left(\frac{S_m}{\alpha}\cos\alpha\right) - \sin\left(\frac{S_n}{\alpha}\cos\alpha\right)\right\}$$
$$r_2^2 = 2\alpha^2\left\{1 - \cos\left(\frac{S_m + S_n}{\alpha}\cos\alpha\right)\right\} + 2\delta^2 + (S_n - S_m)^2\sin^2\alpha$$
$$+ 2\alpha k_1\left\{\cos\left(\frac{S_m}{\alpha}\cos\alpha\right) - \cos\left(\frac{S_n}{\alpha}\cos\alpha\right)\right\}$$
$$+ 2\alpha k_2\left\{\sin\left(\frac{S_m}{\alpha}\cos\alpha\right) + \sin\left(\frac{S_n}{\alpha}\cos\alpha\right)\right\}$$

where k_1, k_2 and δ are given as follows:

	k_1	k_2	δ		k_1	k_2	δ		k_1	k_2	δ		k_1	k_2	δ
$\Pi_{1,1}$	0	0	0	$\Pi_{2,1}$	0	$-\delta$	δ	$\Pi_{3,1}$	$-\delta$	0	δ	$\Pi_{4,1}$	$-\delta$	$-\delta$	δ
$\Pi_{1,2}$	0	δ	δ	$\Pi_{2,2}$	0	0	0	$\Pi_{3,2}$	$-\delta$	$-\delta$	δ	$\Pi_{4,2}$	$-\delta$	0	δ
$\Pi_{1,3}$	δ	0	δ	$\Pi_{2,3}$	δ	δ	δ	$\Pi_{3,3}$	0	0	0	$\Pi_{4,3}$	0	$-\delta$	δ
$\Pi_{1,4}$	δ	δ	δ	$\Pi_{2,4}$	δ	0	δ	$\Pi_{3,4}$	0	δ	δ	$\Pi_{4,4}$	0	0	0

Figure 4.32 shows the calculated current distributions of a four-element helical array antenna obtained from (4.10) by using the point matching method. In Figure 4.32(a), the current distribution, where $\alpha = 12°$, $C_\lambda = \lambda_0$, $N = 2$ turns, and δ is fixed at 0.7 λ_0, is presented in the case in which element one in Figure 4.31(b and c) is driven and elements two to four are parasitic. We see that unlike a single helical antenna, a standing wave exists near the open end of element one. This is because the reflected wave, which attenuates rapidly as it travels from the open end of helix, rises on element one, because of mutual coupling.

In Figure 4.32(b), all elements are driven with equal phase voltage. The shapes of the current distributions on each element are slightly different from each other when $\delta = 0.7\ \lambda_0$. Figure 4.32(c) shows that the current distribution on each element is nearly equal to that of a single helical antenna ($\delta = 0$) when δ is more than λ_0 for $N = 2$ or 3 turns, and 1.2 λ_0 for $N = 4$ turns. Therefore, the effect of mutual

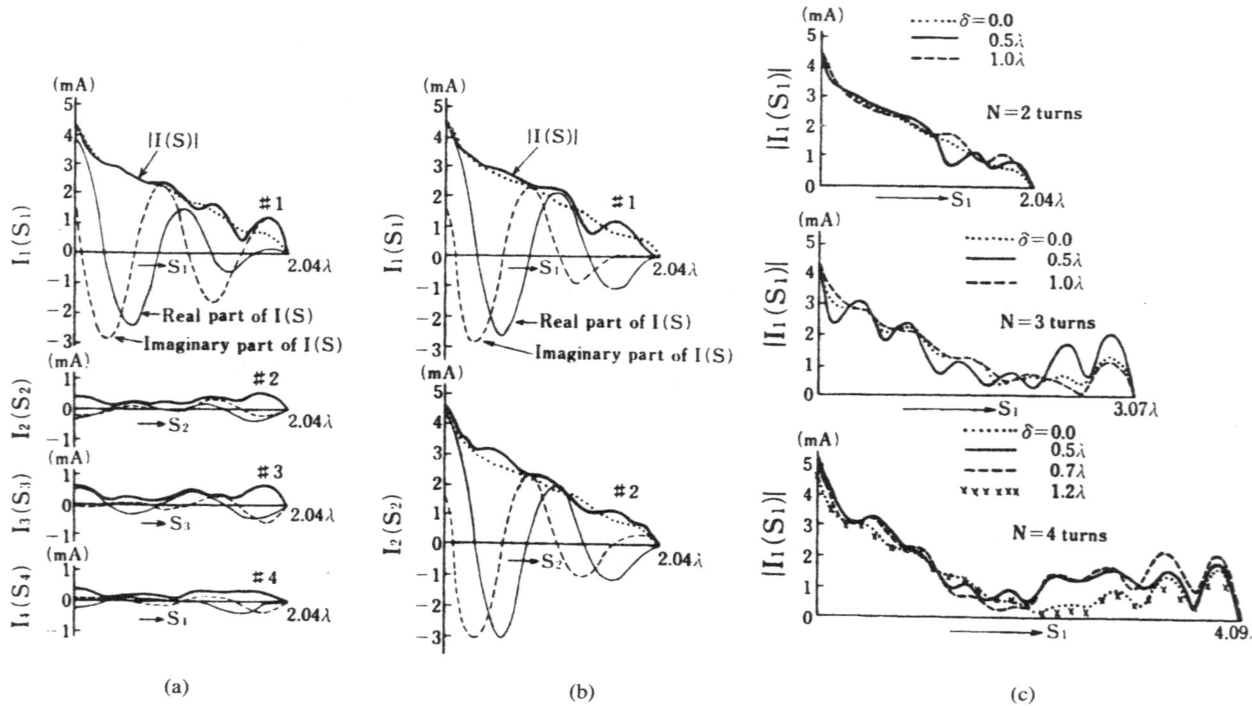

Figure 4.32 Current distribution of a four-element helical array antenna where $\alpha = 12°$ and $C_\lambda = \lambda_0$ (from [36])

coupling is significant as long as the element spacing is approximately 1.2 λ_0 in the case of a short helical antenna.

The current density in ρ and ϕ components on the finite reflector is

$$J_\rho = -\frac{1}{2\pi}\int_0^L \sum_{n=1}^{4} I_n\,(S_n)\left(\frac{\partial G_n}{\partial Z}\cos\alpha \sin\left(\phi - \frac{S_n}{\alpha}\cos\alpha\right) - \frac{\partial G_n}{\partial \rho}\sin\alpha\right) dS_n|_{Z=0} \tag{4.11}$$

$$J_\phi = \frac{\cos\alpha}{2\pi}\int_0^L \sum_{n=1}^{4} I_n\,(S_n)\left(\frac{1}{\rho}\frac{\partial G_n}{\partial \phi}\tan\alpha - \frac{\partial G_n}{\partial Z}\cos\left(\phi - \frac{S_n}{\alpha}\cos\alpha\right)\right) dS_n|_{Z=0} \tag{4.12}$$

$$G_n = e^{-jkR_n}R_n \tag{4.13}$$

where

$$\begin{aligned} R_n^2 = &\left\{\rho\cos\phi - \delta_1 - \alpha\cos\left(\frac{S_n}{\alpha}\cos\alpha\right)\right\}^2 \\ &+ \left\{\rho\sin\phi - \delta_2 - \alpha\sin\left(\frac{S_n}{\alpha}\cos\alpha\right)\right\}^2 + (S_n\sin\alpha)^2 \end{aligned} \tag{4.14}$$

$$\begin{cases} n = 1, & \delta_1 = \delta_2 = 0 \\ n = 2, & \delta_1 = 0 \;\; \delta_2 = \delta \\ n = 3, & \delta_1 = \delta \;\; \delta_2 = 0 \\ n = 4, & \delta_1 = \delta_2 = \delta \end{cases} \tag{4.15}$$

The theoretical values of power gain and axial ratio *versus* element spacing δ are shown in Figure 4.33(a and b) in comparison with the case where mutual coupling is disregarded, as well as the experimental values for $\alpha = 12°$, $C_\lambda = \lambda_0$, $D = 1.7\ \lambda_0$, and $N = 2$ or 4 turns. The mutual coupling primarily causes the degradation of axial ratio in an array antenna composed of short helical antennas.

Figure 4.34 shows a four-element cavity-backed helical antenna array. The cavities control the mutual coupling of the array. Figure 4.35 shows the experimental values of power gain and axial ratio *versus* rim length l when $\alpha = 12°$, $C_\lambda = \lambda_0$, $D = 1.7\ \lambda_0$, $N = 2$ turns, δ is fixed at 0.7 λ_0, and rim diameter is fixed at 0.7 λ_0. The best values of power gain and axial ratio are obtained experimentally at $l = 0.25\ \lambda_0$. There is an improvement of about 0.4 dB in gain and about 3.5 dB in axial ratio over the case where $l = 0$.

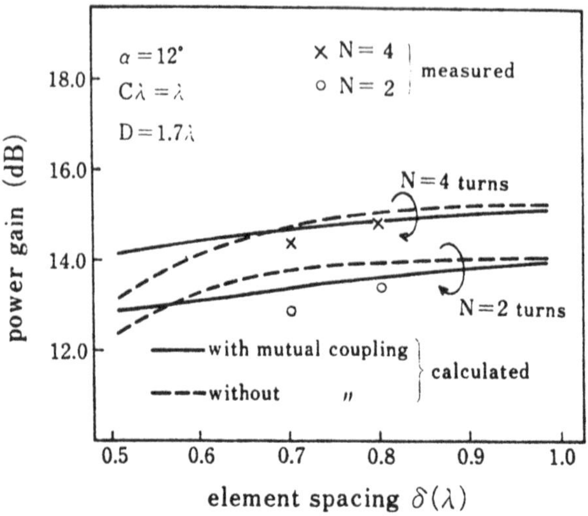

(a)

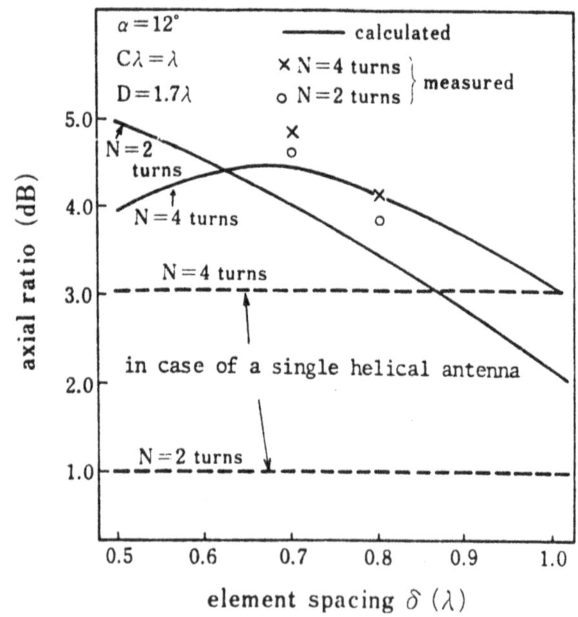

(b)

Figure 4.33(a) Power gain *versus* element spacing of a four-element helical array antenna (from [36])
(b) Axial ratio *versus* element spacing of a four-element helical array (from [36])

Figure 4.34 Four-element helical array (from [36])

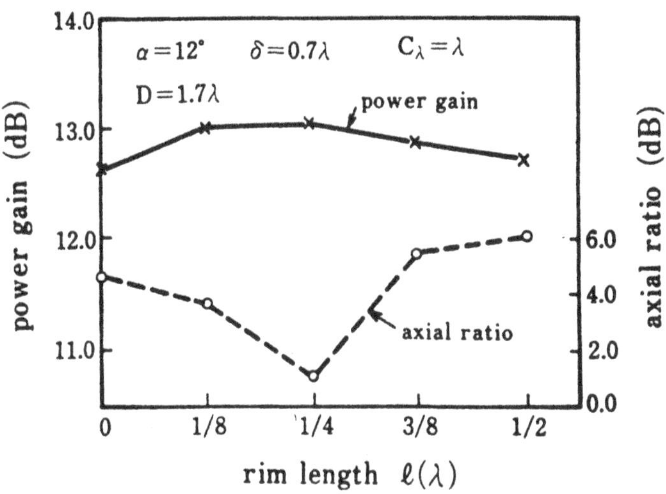

Figure 4.35 Measured power gain and axial ratio of a four-element helical array antenna (from [36])

4.3.2 Cavity-Backed Spiral Antennas

Planar wideband antennas in the microwave and the lower millimeter-wave bands frequently use spirals of either the equiangular or the Archimedean type. Both spirals exhibit operating characteristics that are independent of frequency over very broad bands [37]. The spiral antenna is small in size and fabricated as a two-start spiral etched on copper-clad substrate. When the spiral arms are fed in antiphase at the center, the spiral radiates energy in bidirectional beams perpendicular to its plane.

Figure 4.36 shows both the Archimedean and the quiangular spirals. These are generated by the equations $(r_0 + a\theta)$ and $(r = r_0 \exp(a\theta))$ respectively, where r is the radius of the spiral, r_0 is the radius at the start, a is the growth rate, and θ is its angular position. When the spiral arms are fed in antiphase at the center, radiation takes place from a region on wavelength in circumference. Figure 4.37 shows the radiation region of an Archimedean spiral. If current is launched in antiphase at the center of the spiral, then at A and A_1, the currents will still be in antiphase, because both have progressed an equal amount along their respective arms. If at a given frequency the path length from A to B is a half wavelength, then at A_1 and B the currents will be in phase. Under these conditions, both spiral arms radiate energy from a circular region one wavelength in circumference. When the wavelength is large compared to the circumference of the inner turns of the spiral, the currents in the adjacent arms are all essentially in antiphase, and the radiation from both arms cancel, as they do in a twin-wire transmission line. Thus, the inner turns of the spiral can be looked upon as a twin-wire transmission line which feeds the one-wavelength circumference radiating region. From this reasoning, we see that the frequency band of the isolated spiral is limited only by its physical dimensions. The energy not radiated at the one-wavelength circumference will continue along the spiral as if it were a twin-wire transmission line, and be partially radiated each time the circumference becomes an odd number of wavelengths long. Nakano, *et al.* [38] have described this type of higher-order mode of radiation.

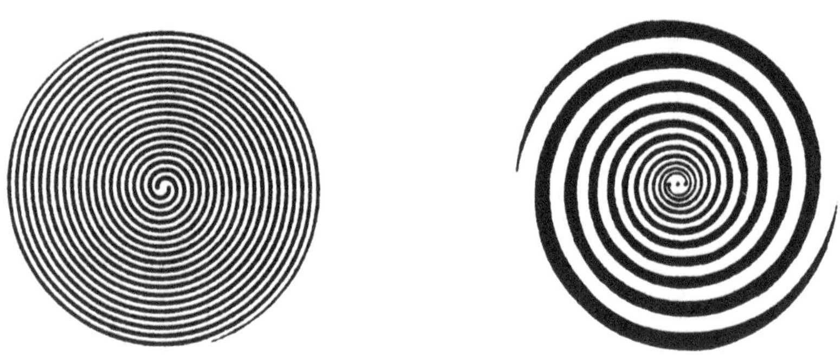

Figure 4.36 Spiral configurations

These antennas are inherently circularly polarized, and their radiation patterns bidirectional when a planar configuration is used. Spiral antennas are usually made unidirectional over moderately large bands through the use of back cavities. A back cavity, with a rear wall spaced approximately a quarter-wavelength from the plane of the spiral, will yield directional behavior roughly analagous to that of a dipole a quarter-wavelength away from the gound plane (see Section 4.2.1). The cavity spacing, because it is frequency-dependent, limits operation to about a 3:1 or 4:1 range of frequencies [38–40].

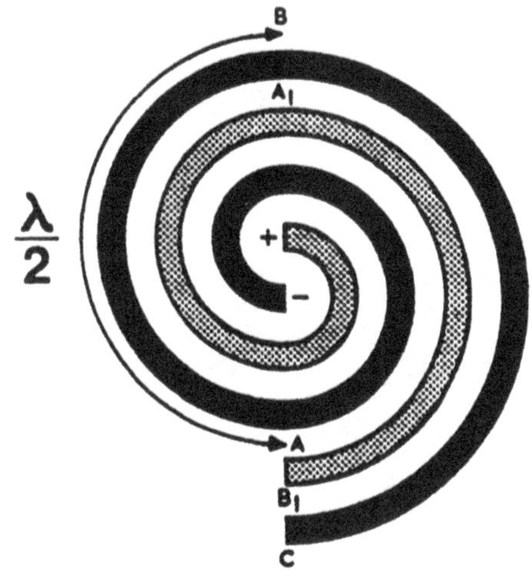

Figure 4.37 Radiation region of an Archimedean spiral

In many designs, a transmission line structure is brought to the antenna's center feed point along a path normal to the plane of the spiral, with the spiral frequently entering by a small hole through the cavity. The dual-arm spiral requires balanced excitation, and if an unbalanced input is used, some type of balun is required. The most popular balun transformers are the matched balun [41] and, where beam squint is not important, the taper transformer [42]. The bandwidth limitations of these transformers have been discussed by Morgan, [43] who has shown that a virtual squint-free performance is achieved over a 9:1 frequency band using a Marchand balun.

Saul [44] has described a cavity-backed equiangular spiral, which is shown in Figure 4.38. One of the spiral arms is visible in the photo; its companion arm is located on the opposite side of the dielectric, directly behind the tapered microstrip feed line. This method of fabrication allows photoetching of the entire structure, with the microstrip line already connected to its proper arm. In theory, both arms should be in the same plane. With the dielectric sufficiently thin, however, this condition is closely approximated, and satisfactory operation is obtained. The *Naval Ocean Systems Center* (NOSC) in San Diego, designed the spiral antennas operating in the 20 to 40 GHz range using ten mil thick Duroid 5880. The VSWR of the antenna (Figure 4.38) is better than 2:1 over the 26.5 to 40 GHz band, including a scissor-type transition effect from the waveguide to the microstrip [44]. Recently, five mil thick dielectric Duroid 5880 has been used for antennas operating at 94 GHz.

Morgan [45] developed an 18 to 40 GHz spiral antenna at Marconi Defence Systems in the U.K. The antenna output is double-ridged waveguide WRD180C24. The antenna has an aperture diameter of 12 mm, which is increased to 16 mm with

Figure 4.38 Cavity-backed equiangular spiral (from [44])

the addition of an integral radome. Its overall length, including the radome, is 48 mm. Figure 4.39 shows the spiral antenna, including choke, radome, and backing cavity. There is squint in the radiation pattern (Figure 4.40) without a quarter-wavelength choke.

Figure 4.40(b and c) shows the return loss and the gain of the antenna. Typical measured radiation patterns at 18, 26.5, 34, 38 and 40 GHz for horizontally- and vertically-polarized signals are shown in Figure 4.40(d–h). The squint and axial ratio are determined from the above radiation patterns and plotted in Figure 4.40(i–j). The radiation patterns of three similar spiral antennas are plotted at 36 GHz in Figure 4.40(k) to assess the reliability of design and manufacturing by tracking performance.

4.4 MULTIPACTOR BREAKDOWN IN THE ANTENNA

4.4.1 Introduction

Multipactor breakdown is a secondary electron resonance breakdown. It became a serious failure mechanism with the introduction of medium- and high-power UHF

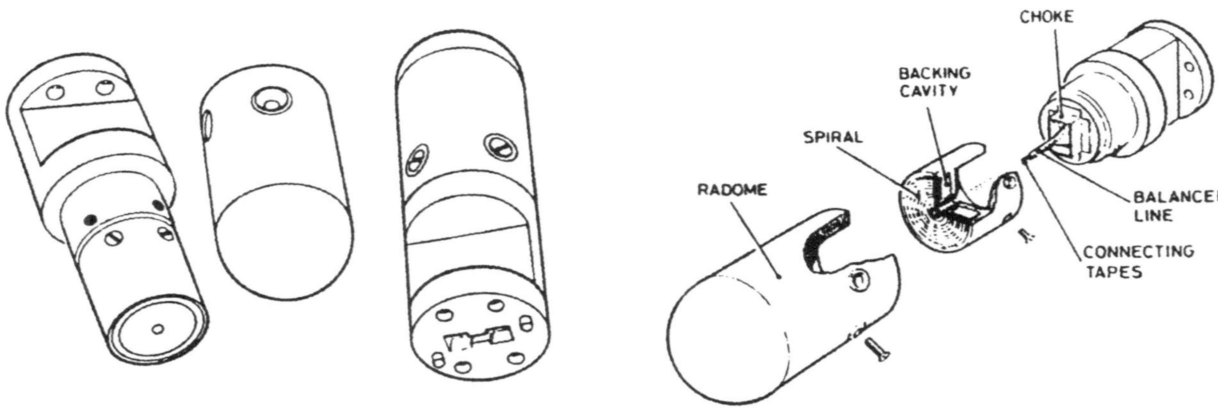

Figure 4.39 Spiral antenna developed in the frequency range of 18 to 40 GHz (from [45])

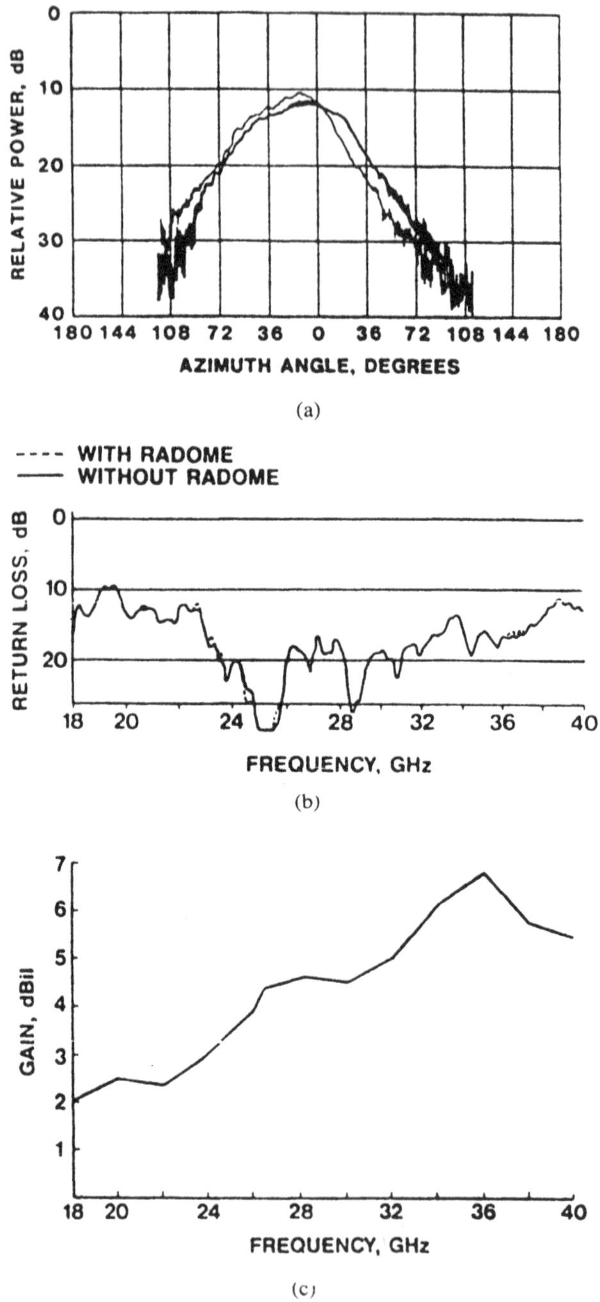

Figure 4.40(a) Radiation pattern of the antenna without choke (from [45])
(b) Return loss *versus* frequency of the antenna, with and without radome (from [45])
(c) Gain *versus* frequency of the antenna (from [45])

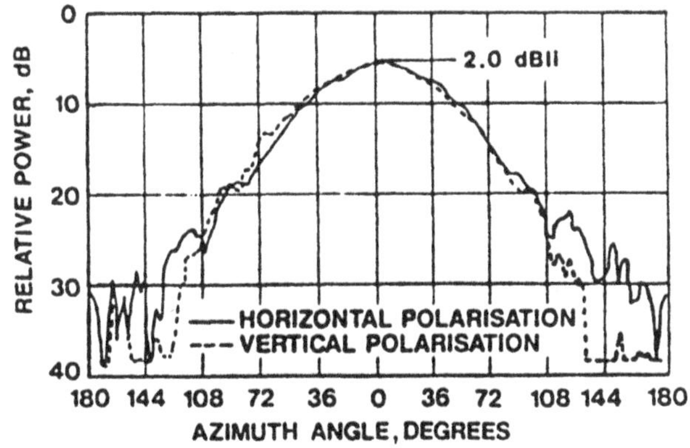

(d)

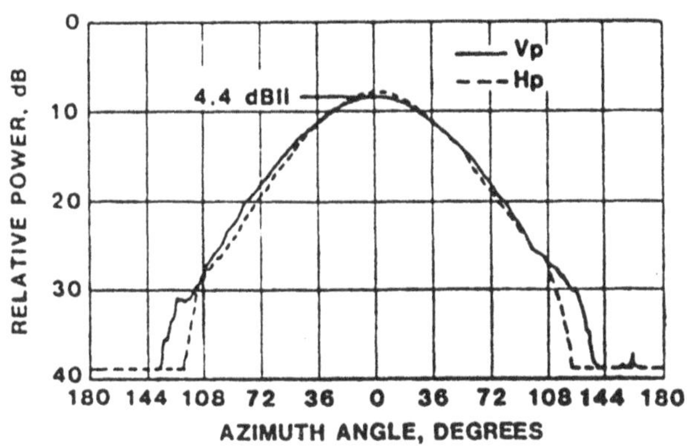

(e)

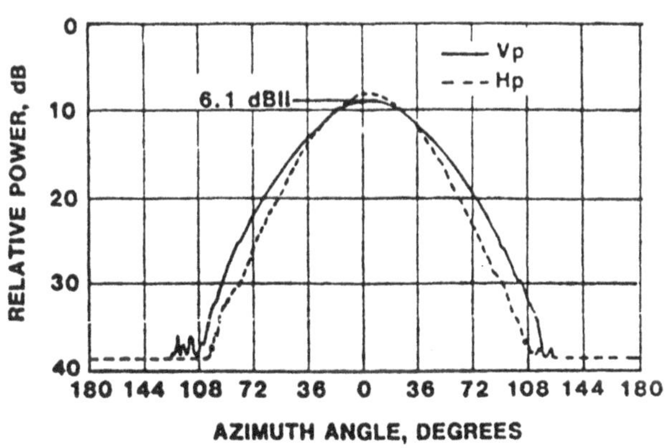

(f)

Figure 4.40(d) Radiation pattern at 18 GHz (from [45])
(e) Radiation pattern at 26.5 GHz (from [45])
(f) Radiation pattern at 34 GHz (from [45])

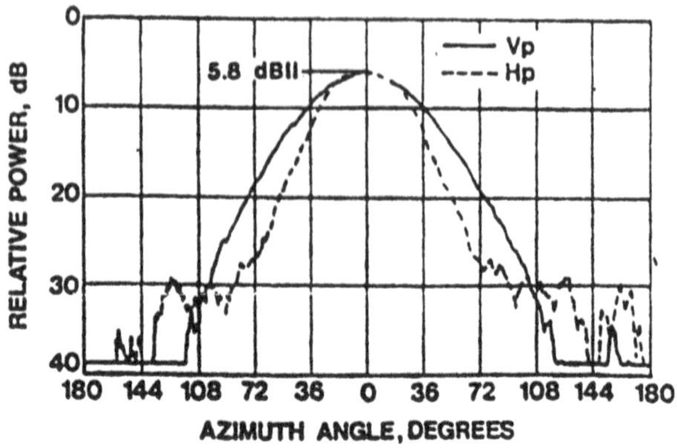

(g)

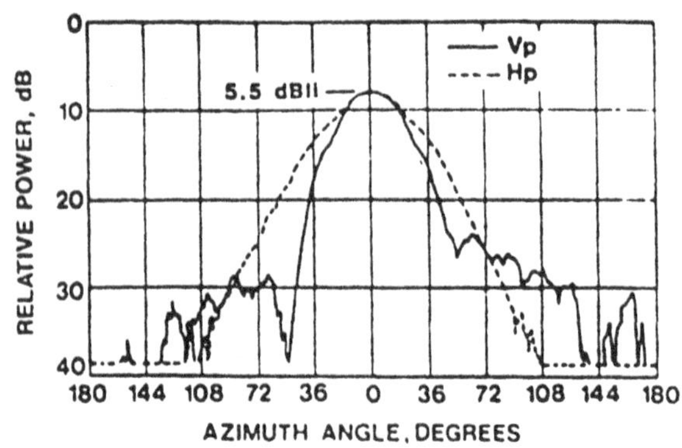

(h)

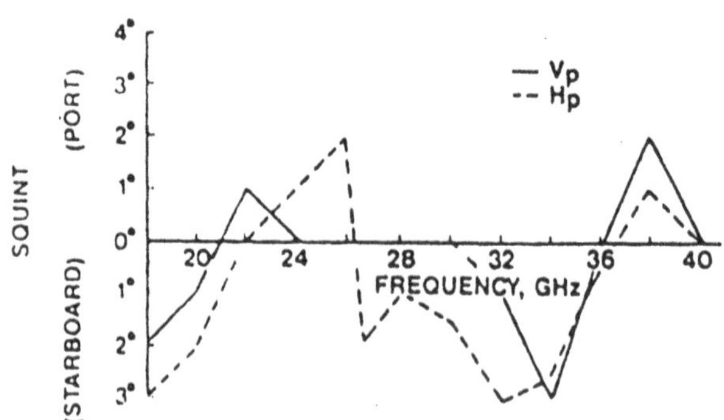

(i)

Figure 4.40(g) Radiation pattern at 38 GHz (from [45])
(h) Radiation pattern at 40 GHz (from [45])
(i) Squint in radiation *versus* frequency (from [45])

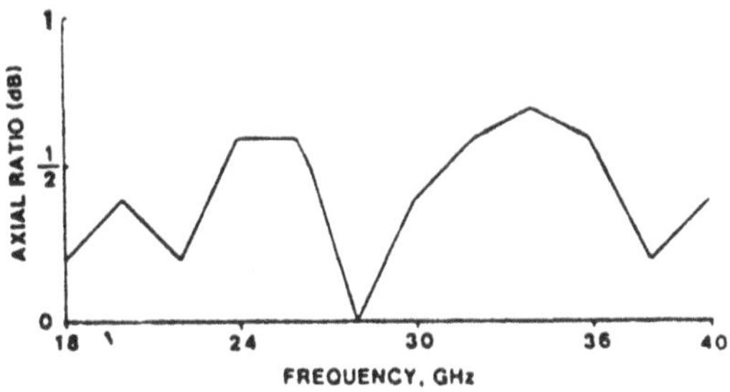

(j)

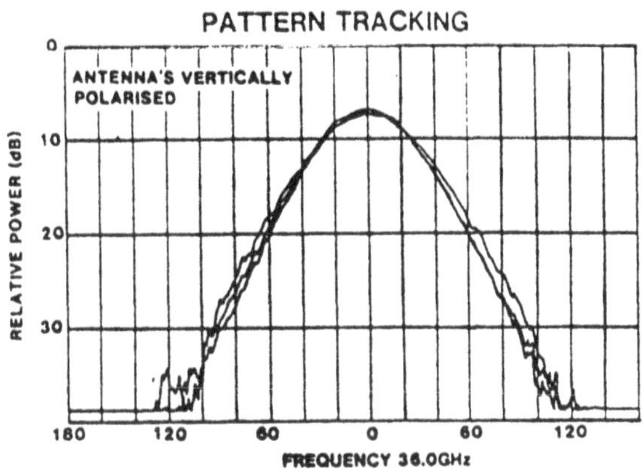

(k)

Figure 4.40(j) Axial ratio *versus* frequency (from [45])
(k) Radiation patterns of three antennas at 36 GHz (from [45])

and microwave equipment in space-borne payloads. In multipactor breakdown, the pressure is sufficiently low so that the mean free path is longer than the electrode separation distance. If sufficient energetic electrons collide with the electrode surfaces that have a secondary emission coefficient greater than one, then secondary electrons will be released. If this occurs as the electric field passes through zero, the reversed electric field will accelerate the electrons back across the gap. If the transit time of the electrons across the gap is one half-cycle of the RF field, the secondary electrons formed by the initial electrons become primary electrons in the next half-cycle, to form another group of secondary electrons. In this way, large electron densities rapidly build up in the gap and breakdown results. Multipactor also occurs if the transit time of the electrons is an odd integral multiple of a one-half cycle. For this discharge to occur, certain conditions must be fulfilled:

1. The multipactor gap must be in a vacuum where the electron mean free paths are much greater than the gap dimensions.

2. The secondary electron emission coefficient must be greater than unity for the gap surface. This depends on the type of surface and its cleanliness.
3. The RF voltage, gap spacing (d), and field frequency (f) must be such as to ensure resonance between the electron motion and the field.

4.4.2 Multipactor Regions in Cup Dipole Antennas

The regions most critical for multipactor breakdown in a crossed-dipole fed cup or horn antenna are shown in Figure 4.41. These places are: (1) The 50 ohm coaxial line and in the matching step transformer; (2) the slot gap; and (3) the gap between the tip of the longest dipole and the cup.

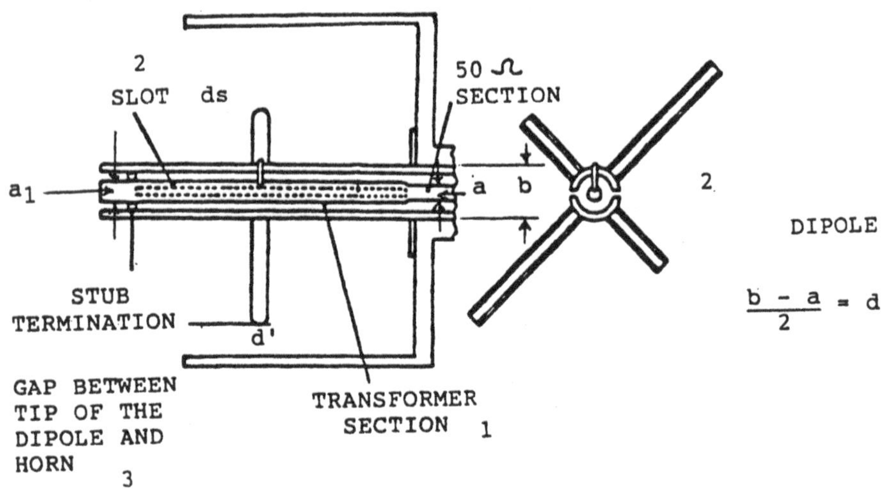

Figure 4.41 Cross-dipole fed cup or horn antenna with possible multipactor regions

(1) Multipacting in the 50 ohm coaxial line and the step matching transformer

Woo [46,47] has shown experimentally that under certain conditions, gaps in coaxial lines with similar geometries exhibit identical breakdown characteristics. Francis and Von Engel [48] have described the equations of motion in the electron gap. If $V/(fd)^2$, B/f, and V/fd are invariant (where V = RF voltage, B = magnetic flux density, f = frequency, d = gap in the antenna components, and V_f = electron velocity on impact) and if multipactor occurs for such a configuration, it will occur for all other configurations in that family. This principle is most useful when applied to geometries in which the electric field is nonuniform. Experimental data obtained for a 50 ohm coaxial transmission line [46,47] is summarized in Fig. 4.42. The multipactor calculations for several possible designs of the coaxial structure in the feed are shown in Table 4.4. In the calculations, the following assumptions have been made:

1. Multipactor breakdown power in coaxial cables occurs at pressure (P) of 10^{-1} torr or less, corresponding to p/f of 10^{-5} torr/MHz or less (f = frequency in MHz).
2. In Fig. 4.42, the average breakdown power given by Woo [46,47] has been converted to peak power.
3. In the case of Z = 50 ohm, the values are corrected by 10 log (Z/50).
4. The breakdown for f_d higher than 450 MHz-cm is given by $P = (f_1 d_1/f_d)^2 P_1$, where P = multipactor breakdown power at $f_1 d_1$, f_d = 450 MHz-cm, and P_1 = peak multipactor breakdown power at f_d.
5. Peak power level in the feed is taken to be 600 W (27.8 dBW). The margin for multipactor breakdown is: margin = breakdown peak power (dBW) − 27.8 (dBW).
6. In the present case, 50 ohm impedance is not maintained by the gap (near transformer), and therefore the breakdown power must be derated by a correction factor 10 log 10 (Z/50).

The results in Table 4.4 [49,50] indicate that it will be necessary to use design option 5 to obtain greater than a 6 dB margin for multipaction.

(2) Multipactor in the Slot Gap

Although the field in the slot gap will not be across the slot in normal operation and thus will not cause any multipacting problems, our analysis assumes a worst case, where the field is orthogonal to the slot. Multipactor breakdown power in the slotgap is given by

$$P = V_{max}^2 G$$

where V_{max} is the multipactor breakdown voltage and G is the admittance of the slot [51]. A typical value of the admittance of the slot is [22]: G = 0.00205 mho or Z = 487.8 ohm.

The formula used to calculate the multipactor for the present slot is not very accurate; therefore, it may be necessary to establish it by test. A dielectric plug can be used in the slot to stop multipactor breakdown.

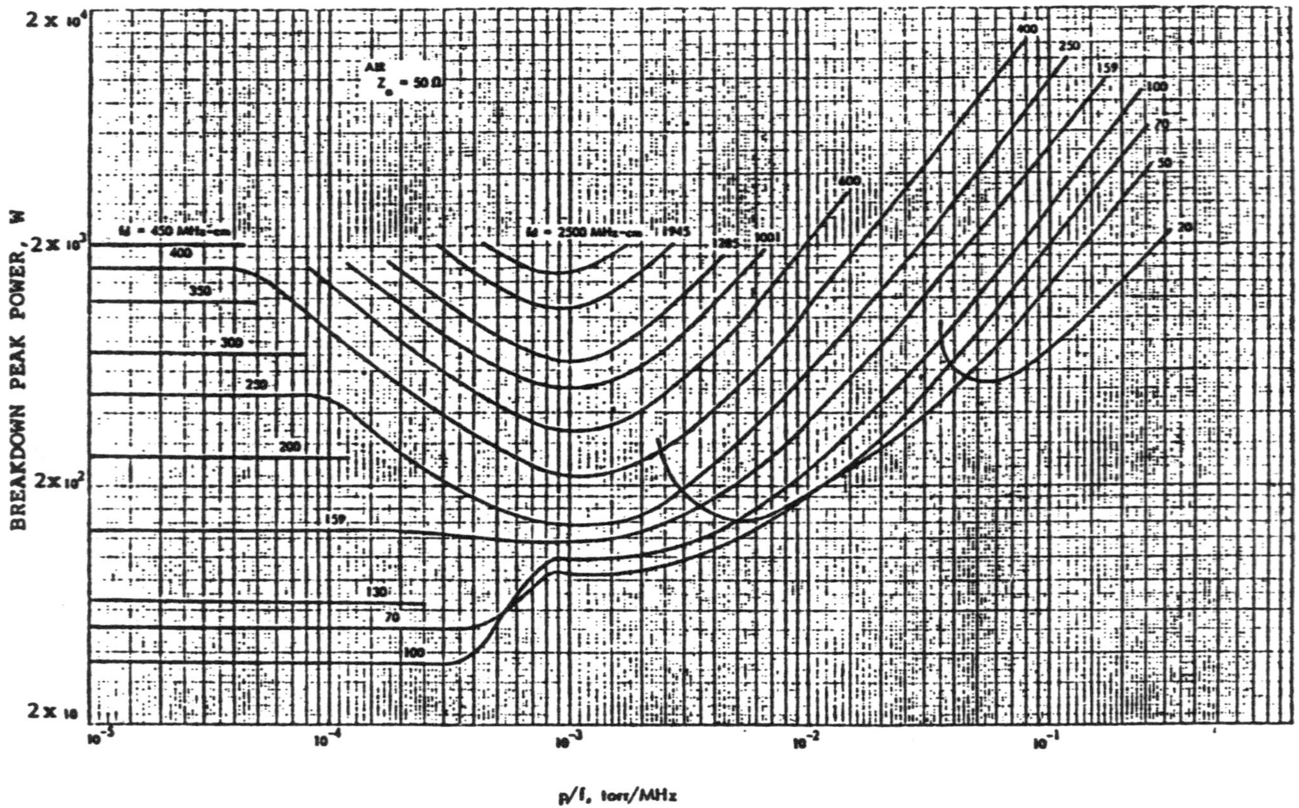

Figure 4.42 *RF* voltage breakdown in coaxial transmission lines (P = pressure in Torr, f = frequency in MHz, d = distance between inner to outer conductor in cm) (from [47])

Table 4.4 Multipactor Breakdown in 50 ohm Coax and in Transformer
f = 1550 MHz, Peak Power = 600 Watts (27.8 dBW)

Design Option	*a or a_l in cm*	*b in cm*	*Imp. in ohm*	*f × d MHz. cm*	*Breakdown Peak Power (From Figure 2 in Watts and dBW)*	*Margin dB*
1. Coax	0.5	1.15	50	1550 × 0.325 = 503.75	2506W (33.99)	6.2
1. Trans	0.64	1.15	35	1550 × 0.255 = 395.25	1542.9W (31.88 − 1.5) 1096.5 = 30.4	2.6
2. Coax	0.565	1.3	50	1550 × 0.3675 = 570	3208.89 (35)	7.2
2. Trans	0.73	1.3	35	1550 × 0.285 = 441.75	1927.34 (32.9 − 1.5) = 31.4 = 1380.4W	3.6
3. Coax	0.65	1.5	50	1550 × 0.425 = 658.75	4285.9W (36.3 dBW)	8.5
3. Trans	0.84	1.5	35	1550 × 0.33 =511.5	2584.02W (34.1 − 1.5) (1819.7W) = 32.6	4.8
4. Coax	0.695	1.6	50	1550 × 0.4525 =701.4	4858.9W (36.8 dBW)	9
4. Trans	0.89	1.6	35	1550 × 0.355 = 550.3	2990.9W (34.8 − 1.5) 2138 = 33.3	5.5
5. Coax	0.78	1.8	50	1550 × 0.51 =790.5	6171.8W (37.9)	10.1
5. Trans	1.0	1.8	35	1550 × 0.4 = 620.0	3796.5W (35.8 − 1.5) 2691.5 = 34.3	6.5
6. Coax	0.87	2.0	50	1550 × 0.565 = 875.8	7574.7 (38.8)	11.0
6. Trans	1.12	2.0	35	1550 × 0.44 = 682.00	4593.8 (36621) = 35.1 = 3235.9	7.3
7. Inmarsat Coax	0.198	0.64	70.4	1550 × 0.221 = 342	1155W (30.671.49) = 32.09	4.39
7. Trans	0.277	0.64	50	1550 × 0.1565 = 281	781W (28.9)	1.43

(3) Multipacting Between the Tip of the Longest Dipole and the Cup

Figure 4.41(a) shows the gap between the tip of the longest dipole and the cup. To calculate multipactor breakdown in a vacuum, we assume that the tip of the dipole and the cup wall act like parallel plates. We can use Clancy's equation [51] to calculate the breakdown power in the gap:

$$P(\text{dBW}) = 6.5 + 20 \log 10\,(f_d)$$

Table 4.4 shows a summary of the multipactor breakdown power for different gaps at 1.55 GHz.

4.4.3 Multipactor Regions in Cavity-Backed Helical and Spiral Antennas

There is a possibility of multipacting at the feeding point of the helix in the cavity-backed helical and spiral antennas. The calculation of the multipactor in an antenna can be made using the same formulas as described in Section 4.4.2.

4.5 APPLICATIONS OF CAVITY-BACKED ANTENNAS

4.5.1 Cup Dipole Cavity-Backed Antennas

The use of a crossed-dipole feed with cup shaped ground plane provides a simple and efficient method for achieving CP with a low axial ratio. This type of feed is very efficient for illuminating small front-fed parabolas for point-to-point communications. Silberberg [3] has described the application of this type of antenna as a feed. Figure 4.2 shows the split-coaxial balun with circularly-polarized dipoles. This feed is optimized for a reflector antenna operating in the frequency range 2200 to 2300 MHz. It is also optimized to produce 1 dB axial ratio over 5% of the frequency bandwidth.

Good matching is obtained by the use of step transformers located on the inner conductor of the coaxial line. The impedance matching is very sensitive to the length of the crossed dipoles. The optimized axial ratio of the cup dipole feed is better than 0.9 dB, as shown in Figure 4.43(a). The axial ratio measurement is taken at 2250 MHz.

Figure 4.43(b) shows a high-gain antenna (feed and reflector) assembly. The reflector diameter is 61 cm, with an f/D (focal/diameter) ratio of 0.4. The diameter of the cup is 12.7 cm. In the figure, the cup is located a quarter-wavelength behind the crossed dipoles, which reflect the electromagnetic wave incident upon it back into the parabolic reflector in phase with the dipole radiation in that direction. The shape and size of the cup control the primary illumination taper across the surface of the reflector to obtain nearly equal E-plane and H-plane beamwidths.

In the presence of a reflector, the crossed dipoles are returned to produce a good axial ratio. A pair of cone-shaped disks are placed on the feed support to achieve an antenna system with a low axial ratio. The larger disk (closest to the paraboloid) is used to “tune” the longer dipole; the smaller disk affects the shorter dipole. The two disks, 0.1 cm thick with diameters 11.43 cm and 6.99 cm, are separated by a distance of 0.5 λ_0. The present technique broadens the axial ratio by reducing the mismatch introduced in the cup dipole feed by the reflector. A typical radiation pattern for a parabolic reflector antenna of 61 cm diameter is shown in Fig. 4.43(c). Figure 4.43(d) shows the measured axial ratio pattern at 2250 MHz.

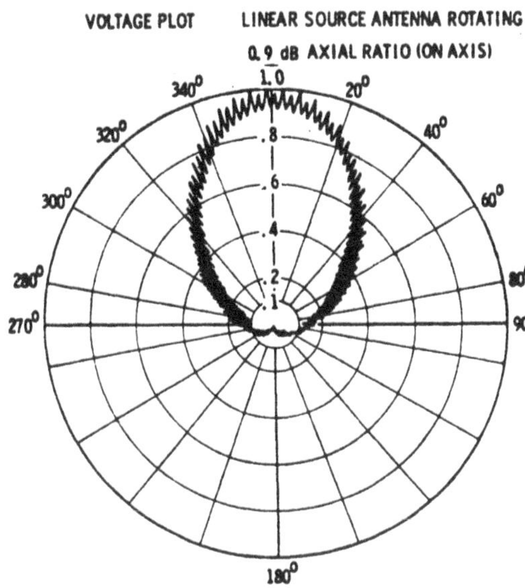

Figure 4.43(a) Axial ratio azimuthal pattern of the cup dipole antenna at 2250 MHz (from [3])

Figure 4.43(e) shows a typical radiation pattern of the antenna without disks on the support rod. A good axial ratio is achieved, but at the expense of an increase in sidelobe level and decrease in gain. The gain of the antenna is 20.3 dB at 2200 MHz, measured by comparison to a standard gain horn antenna. This corresponds to 55% efficiency. The input VSWR of the antenna is less than 1.5 over the frequency band, as shown in Figure 4.43(f). The bandwidth of the antenna can be increased by Kumar's technique [7]. An axial ratio of less than 1 dB over a 5% frequency band can be obtained by experimental adjustment of the diameters and positions of the disks. Figure 4.44(a) shows a typical cup dipole feed used in present microwave point-to-point communication antennas. A complete antenna is shown in Figure 4.44(b), where the reflector is covered by a protective dielectric layer. Table 4.5 shows various antenna sizes and frequencies, along with their electrical characteristics.

Fisk and Donovan [12] described a cavity-backed CP radiator which meets the requirements for TV broadcast at VHF. The design and requirements for this type

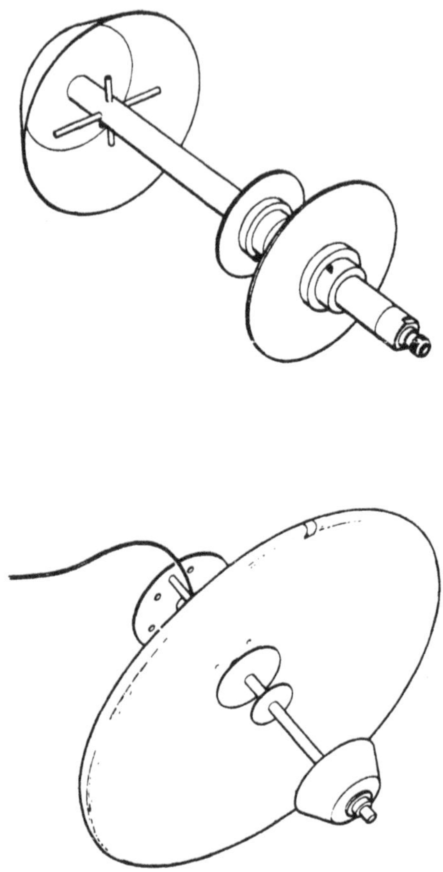

Figure 4.43(b) Antenna assembly, including cup dipole feed and support tube (from [3])

of antenna are discussed in the previous section. The radiation pattern characteristics of a typical 120° beamwidth unit, designed for use in a three-around configuration, is shown in Fig. 4.45(a), with the calculated omnidirectional pattern in Fig. 4.45(b). The radiation pattern bandwidth is uniform over several channels. Figure 4.45(c–d) shows similar radiation characteristics for the 90° beamwidth unit. Figure 4.45(e) shows the measured axial ratio *versus* off-axis angle of the antenna. Figure 4.45(f) shows the typical axial ratio pattern of the antenna. A single-element cavity-backed antenna mounted on a tower is shown in Figure 4.45(g). Figure 4.45(h and i) shows

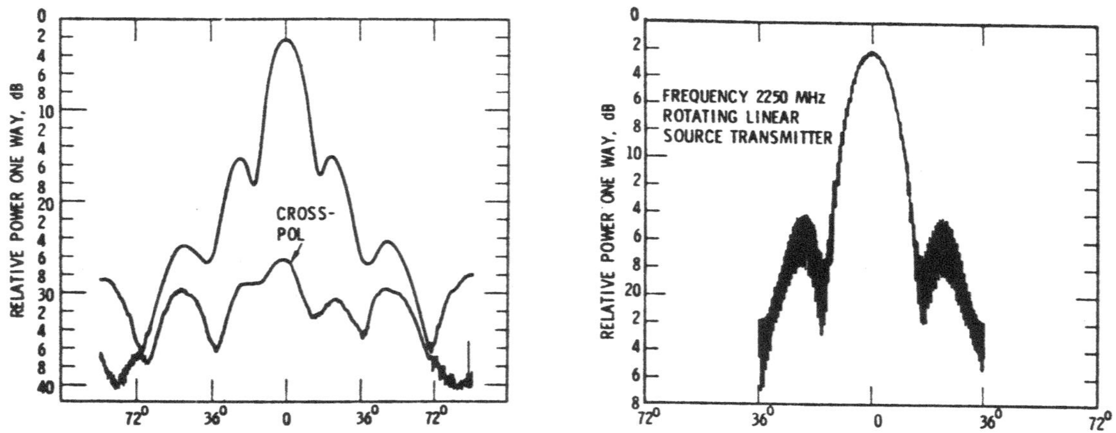

(c) (d)

Figure 4.43(c) Typical radiation pattern of the antenna (from [3])

(d) Axial ratio pattern of the antenna at 2250 MHz (from [3])

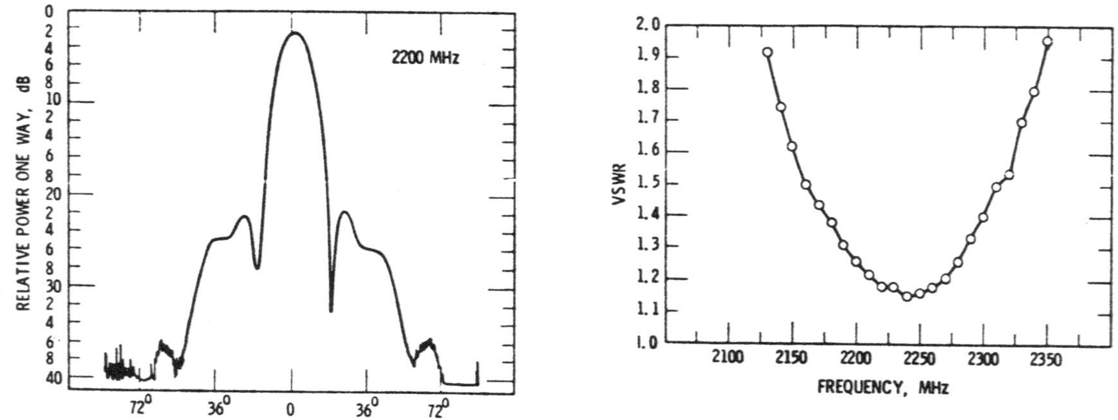

(e) (f)

Figure 4.43(e) Radiation pattern of the antenna without disks at the support rod (from [3])

(f) VSWR *versus* frequency of the antenna with disks on the support rod (from [3])

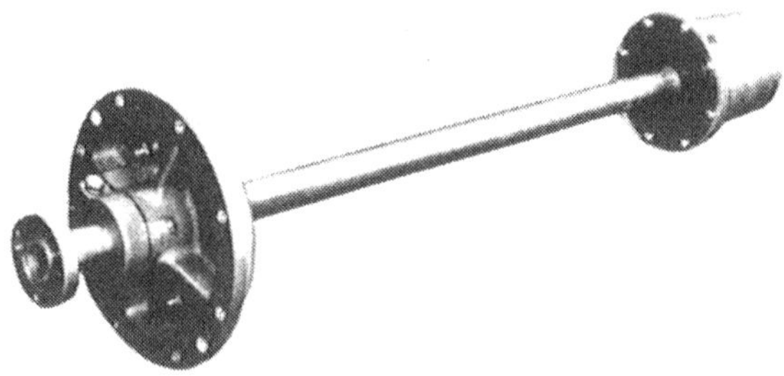

(a)

(b)

Figure 4.44(a) Photograph of a cup dipole feed used in communication (from [52])
(b) Photograph of a reflector using the feed (inside the cover) (from [52])

the expected circularity as a function of the tower face size in feet. Figure 4.45(j) shows the relative gain per bay *versus* the amount of null fill for a typical array of cavity-backed radiators. Null fill and beam tilt are controlled by the vertical phase distribution over the aperture, so that all radiators are fed equal power levels. In this case, the antenna is used with a number of vertically-stacked bays to provide desired power gain. Table 4.6 shows typical characteristics for high-channel arrays.

Table 4.5 Reflectors and Cup Dipole Antennas Used in Microwave Communications

Frequency GHz	*Model Number*	*Reflector Diameter Ft. (Meter)*	*U.S. F.C.C. Category*	*Low*	*Gain at Midband dBi*	*High*	*Beam-width*	*F/B Ratio*	*Maximum VSWR*
PLANE POLARIZED—LOW VSWR									
1.700–1.850	RF4P-2J17	4 (1.2)	n/a	24.2	24.5	24.7	9.4°	30 dB	1.15
	RF6P-2J17	6 (1.8)	n/a	27.6	27.9	28.2	6.3°	35 dB	1.10
	RF8P-2J17	8 (2.4)	n/a	30.1	30.4	30.7	4.7°	39 dB	1.06
	RF10P-2J17	10 (3.0)	n/a	31.9	32.3	32.5	3.7°	42 dB	1.06
	RF12P-2J17	12 (3.7)	n/a	33.5	33.9	34.2	3.1°	45 dB	1.06
	RF15P-2J17	15 (4.6)	n/a	35.3	35.7	36.0	2.6°	48 dB	1.06
1.850–1.990	RF4P-2J18	4 (1.2)		24.7	25.0	25.3	8.9°	30 dB	1.15
	RF6P-2J18	6 (1.8)	B	28.7	29.0	29.3	5.9°	35 dB	1.10
	RF8P-2J18	8 (2.4)	A	30.7	31.0	31.3	4.4°	39 dB	1.06
	RF10P-2J18	10 (3.0)	A	33.2	33.5	33.8	3.6°	42 dB	1.06
	RF12P-2J18	12 (3.7)	A	34.7	35.0	35.3	3.0°	45 dB	1.06
	RF15P-2J18	15 (4.6)	A	36.2	36.5	36.8	2.4°	48 dB	1.06
1.990–2.200†	RF4P-2J19	4 (1.2)		25.3	25.7	26.1	8.5°	34 dB	1.15
	RF6P-2J19	6 (1.8)	B	28.9	29.3	29.7	5.2°	37 dB	1.10
	RF8P-2J19	8 (2.4)	A	31.3	31.7	32.1	4.1°	40 dB	1.06
	RF10P-2J19	10 (3.0)	A	33.3	33.7	34.1	3.3°	44 dB	1.06
	RF12P-2J19	12 (3.7)	A	35.0	35.4	35.8	2.8°	46 dB	1.06
	RF15P-2J19	15 (4.6)	A	37.0	37.4	37.8	2.2°	50 dB	1.06
2.100–2.200	RF6P-2J21	6 (1.8)	A	29.8	30.0	30.2	5.0°	38 dB	1.10

For the Landsat-D Earth Resources Satellite, TRW designed an *X*-band shaped beam data link antenna. A cupped crossed-dipole feed was designed as a feed for the shaped reflector [53]. The dipoles were of unequal length and attached to a coaxial split balun (Fig. 4.2). Figure 4.46(a) is the assembled prototype of the shaped-beam antenna. The measured radiation patterns of the antenna are shown in Fig. 4.46(b). The peak gain of the antenna is 8.2 dBi at ±63 degrees; beyond this angle, the energy falls rapidly. Figure 4.46(c) shows the Landsat-D Delta interface [54], which includes a highly-shaped beam telemetry antenna fed by a cup dipole antenna.

The feasibility of wideband low-profile aircraft antenna design has been established for aircraft-to-satellite communication systems. A shallow, 10.16 cm deep, ridged-cavity, crossed slot antenna and a 15.24 cm deep, cavity-backed sleeve-turnstile dipole antenna were developed for operation in the 240 to 400 MHz band. The turnstile dipole antenna radiation characteristics were evaluated in a satellite-to-KC135 aircraft communication link [55,56]. Figure 4.47 shows a photograph of the cavity-dipole antenna on the aircraft.

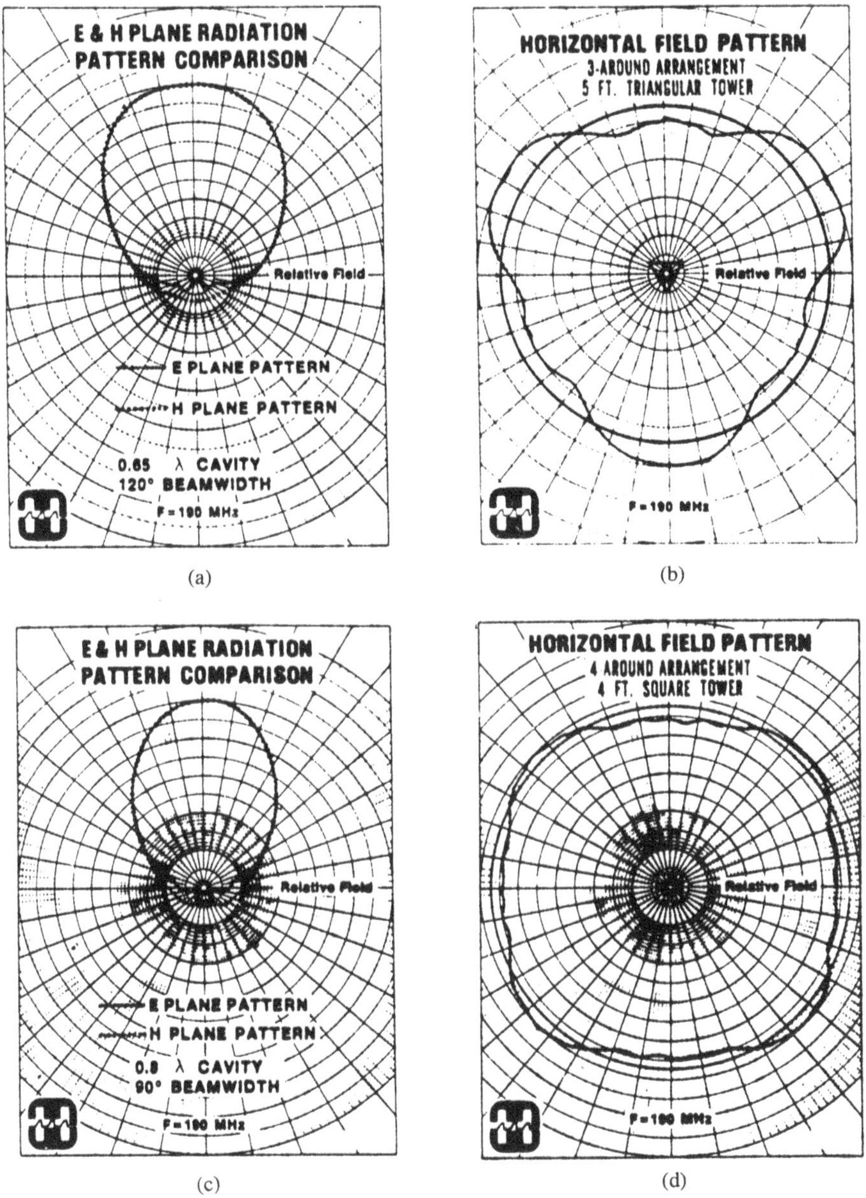

Figure 4.45(a) *E*-plane and *H*-plane pattern comparison for 0.65 λ_0 cavity at 190 MHz and 120° beamwidth (from [12])

(b) Horizontal field (omnidirectional) pattern on a 1.524 m triangular tower (from [12])

(c) *E*-plane and *H*-plane patterns for 0.8 λ_0 cavity at 190 MHz (from [12])

(d) Horizontal field (omnidirectional) pattern on a 1.22 m square tower (from [12])

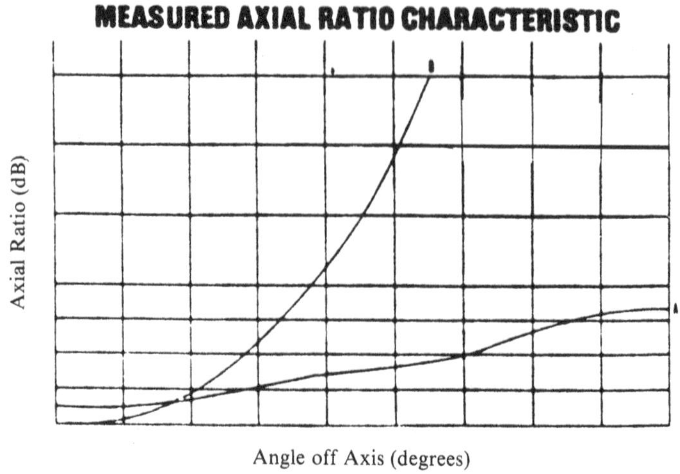

(e)

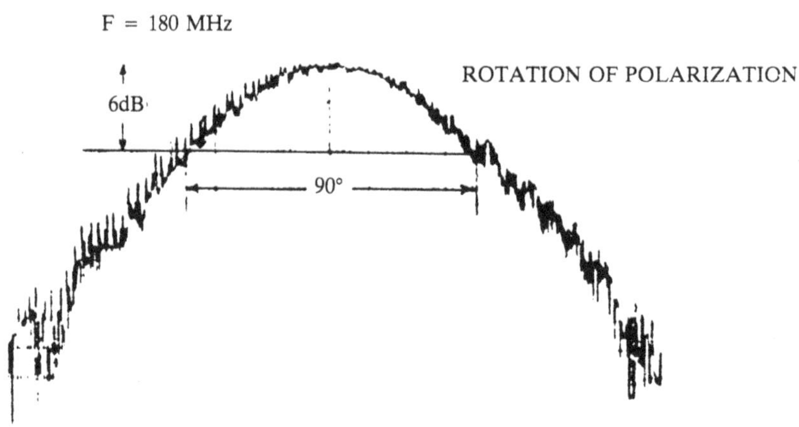

(f)

Figure 4.45(e) Axial ratio *versus* off-axis angle (from [12])
(f) Typical axial ratio pattern of the antenna (from [12])

A cavity-backed antenna, which was described in Section 4.2, has been used for low-altitude orbiting satellites operating in the 225 to 400 MHz band. The antenna consists of an open-sleeve dipole in a 1.4224 m diameter cavity, and is circularly polarized. It provides a nearly constant gain of 10.5 ± 1 dB over the entire operating frequency range. A photograph of the antenna is shown in Figure 4.15.

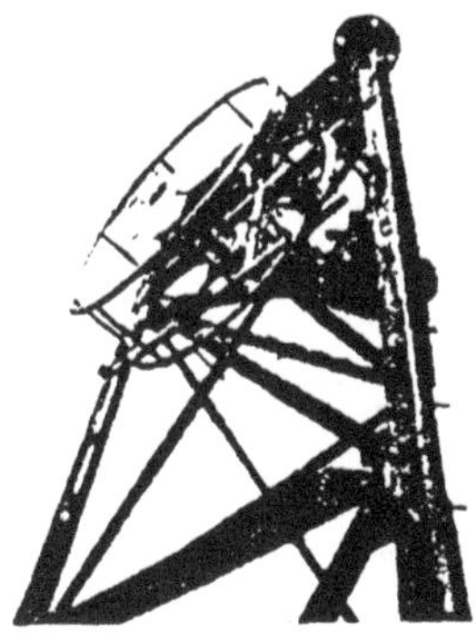

(g)

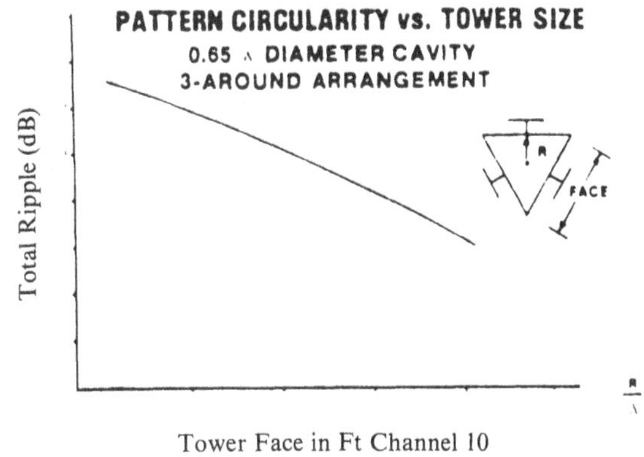

(h)

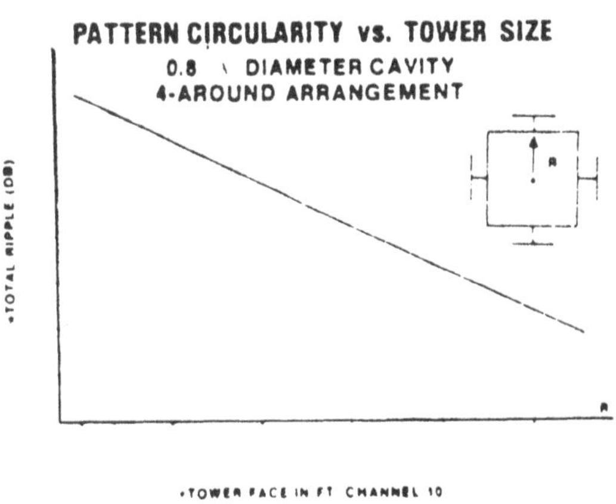

(i)

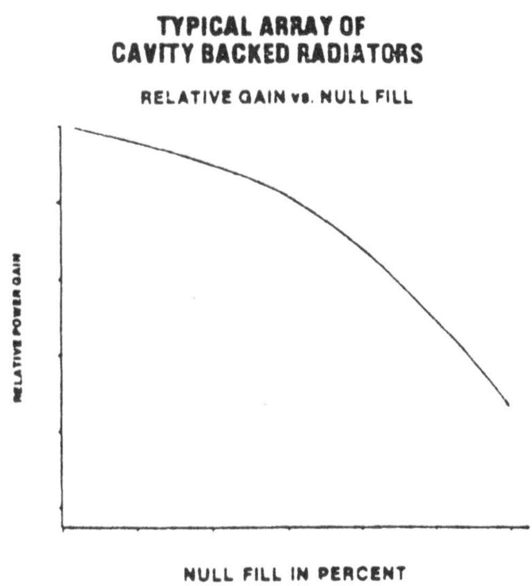

(j)

Figure 4.45(g) Single element cavity-backed antenna on a tower (from [12])

(h) Pattern circularity *versus* tower size for 0.65 λ_0 cavity (from [12])

(i) Pattern circularity *versus* tower size for 0.8 λ_0 cavity (from [12])

(j) Relative gain *versus* percentage null fill for a typical array of cavity-backed radiators (from [12])

Table 4.6 Tentative Performance Specifications

CP Cavity-Backed Radiator Antenna
Channels 7–13

	8 BAY	12 BAY	16 BAY
VSWR	≤11	≤11	≤11
GAIN TOTAL	7.92	11.88	15.84
HORIZ	3.96	5.90	7.92
VERT	3.96	5.90	7.92
CIRCULARITY	±2.0db	±2.0db	±2.0db
AXIAL RATIO	≤2.0dB	≤2.0dB	≤2.0dB
SHEAR AT 50 PSF	5000 lbs	7500 lbs	10,000 lbs
APERTURE	41′	60.6′	80°

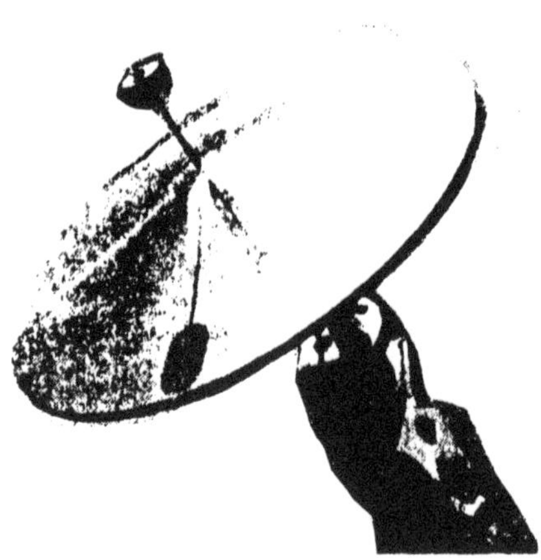

Figure 4.46(a) Antenna assembly using crossed-dipole feed (from [53])

The telemetry antenna developed for the SPOT satellite payload, functioning in the frequency band 8.20 to 8.31 GHz, is a highly-shaped beam antenna, consisting of a symmetric reflector of shaped profile, and illuminated by a crossed-dipole-fed horn. Figure 4.48(a) shows the cross section of the crossed-dipole feed used in this telemetry antenna. Figure 4.48(b) shows the radiation pattern of the feed in the presence of axial monopod and horn. The telemetry antenna assembly is shown in Fig.

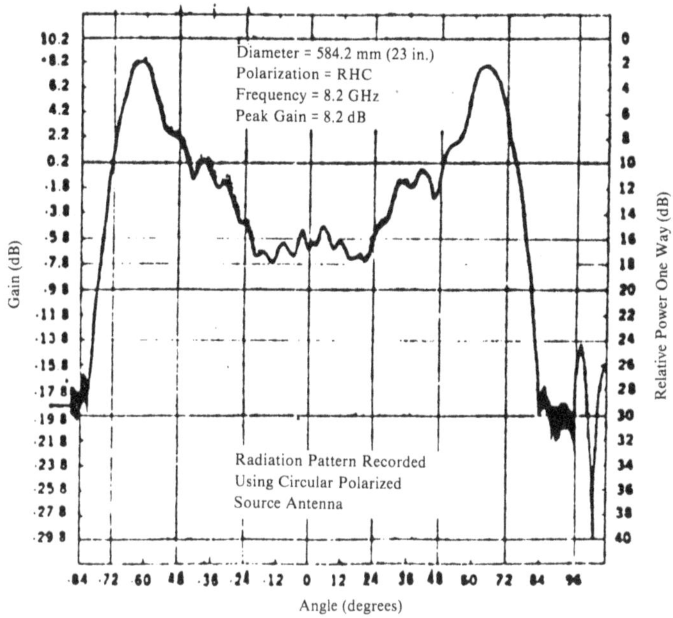

Figure 4.46(b) Radiation pattern of the antenna (from [53])

4.48(c). The diameter of the reflector is 0.8 m. The radiation pattern of the antenna is shown in Fig. 4.48(d).

Recently, a telemetry antenna was developed by Kumar [58–61] for the *European Space Agency* (ESA) ERS-1 satellite. The ESA ERS-1 satellite will map a portion of the earth's surface by synthetic aperture radar techniques. The radar beam itself is generated by a large planar array antenna. At *X*-band (8.03 to 8.17 GHz), the right-hand circularly-polarized shaped-beam telemetry antenna is suitable for retransmitting radar data back to an earth terminal. The antenna is comprised of a center-fed shaped-beam reflector and a cupped horn fed by a crossed dipole. Figure 4.49(a) shows the reflector, feed, and support rod assembly. The feed design is based on the approach developed by Silberberg [3] and Kumar [58]. It comprises a crossed dipole mounted in a dual-mode horn. The dipole lengths and cup geometry are optimized to obtain the best feed for the highly-shaped beam reflector. The shaped reflector is 0.60 m in diameter; the cup has a maximum diameter of 5.067 cm. The horn and the dipoles are supported by a coaxial rod, as shown in Fig. 4.49(b). Figure 4.49(c) shows how superposition of the dominant TE_{11} and TM_{11} modes leads to a field configuration at the input of the horn that impresses a voltage across the slot, and subsequently excites the dipoles. The step of the cup generates $TE_{11} + TM_{11}$ modes, which propagate through the aperture of the stepped horn. The shape and

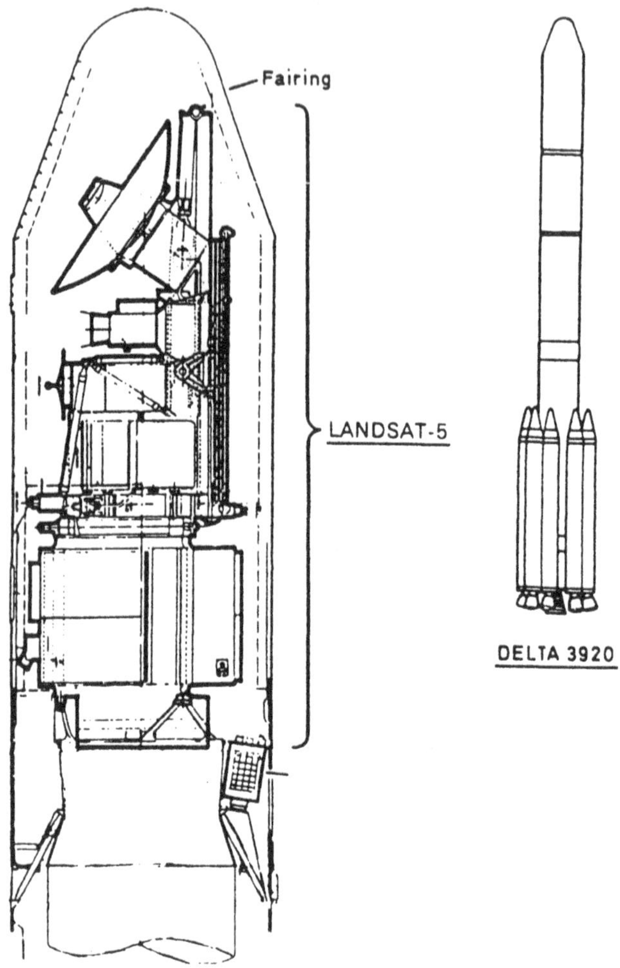

Figure 4.46(c) Landsat-D Delta interface, including telemetry antenna (from [54])

size of the cup control the primary illumination taper across the surface of the reflector. The coaxial line is fed from the back of the horn, and a short circuit is placed after the slot. A matching transformer is placed at the input coaxial line of the feed.

The test feed is fixed at the receiving end, and a rotating, linearly-polarized rectangular horn is used as a transmitter. The feed dipoles are adjusted to produce the best axial ratio. The measured feed pattern at 8.1 GHz is shown in Fig. 4.49(d). The axial ratio is better than 0.4 dB on the axis in the frequency range 8.0 to 8.2 GHz, and the return loss is better than -19 dB. Figure 4.41(e) shows the measured

Figure 4.47 Cavity dipole antenna on aircraft (from [55])

phase center of the feed at 8.1 GHz. The phase center is 0.1 cm behind the aperture of the horn. The axial ratio deteriorates with an increase in the thickness of the support rod [58,59]. A photograph of the simulated panel, sensors, sensor brackets, other hardware, and the antenna is shown in Fig. 4.49(f). For the co-polar measurement, the polar axis of the test antenna is driven by a motor at a rapid speed, while the azimuth position is turned simultaneously at a much slower rate. The co-polar pattern of the antenna is shown in Fig. 4.49(g).

The ERS-1 satellite payload has been integrated successfully at Fokker's Schiphol Integration Facility, which is shown in Figure 4.49(h). The satellite was vibrated using low-level excitations, to identify the eigen-modes of the satellite and to successfully qualify for launch on an Ariane-4 [62]. The Inmarsat antennas are designed to provide global coverage from a geostationary satellite. Each antenna is an array of cup dipole radiating elements, which produce a circularly-polarized earth coverage pattern. The radiating element is a crossed-dipole installed in a shallow circular cup. The dipole and cup dimensions are designed to maximize the antenna element radiating efficiency with good axial ratio. The antenna details are summarized in Table 4.7.

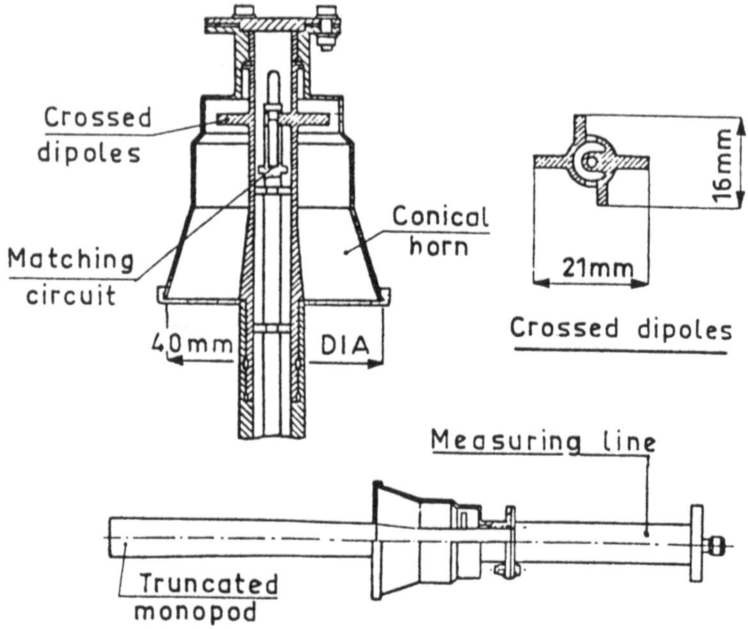

(a)

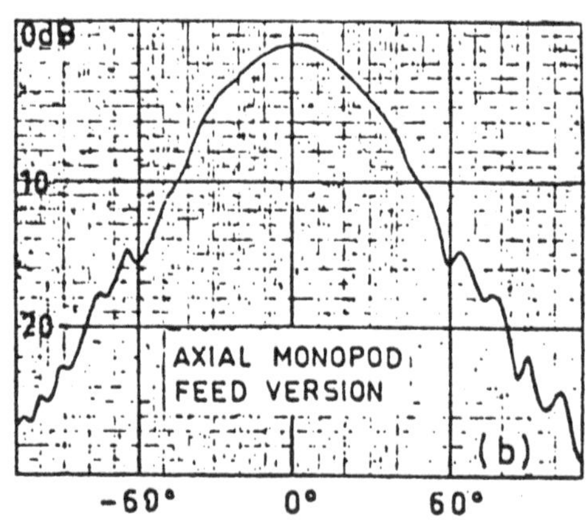

(b)

Figure 4.48(a) Crossed-dipole-fed horn with truncated monopod (from [57])
(b) Radiation pattern of the crossed-dipole-fed horn (from [57])

Figure 4.48(c) Telemetry antenna for the SPOT satellite payload (from [57a])

The *L*-band transmitting antenna is a 61-element array of cup dipoles each of 0.95 wavelength diameter, as shown in Fig. 4.50(a). Figure 4.50(b) shows a photograph of the cup dipole used in the array. The antenna gain *versus* cup depth for a single element is shown in Fig. 4.50(c). The axial ratio pattern of the cup dipole antenna is shown in Fig. 4.50(d).

For the *L*-band receiving antenna, the cup diameter for the seven-element array is 1.07 λ_0. A cup of this size has good axial ratio and an efficiency of about 92%. The efficiency can be improved by elongating the cup length, but elongation of the cup length is not acceptable for mechanical reasons. A nine-element array becomes attractive in this case, because the element size for the outside ring is 0.88 λ_0, and is 1.42 λ_0 for the center. Cups of these sizes are found to have nearly 100% efficiency. Figure 4.50(e) shows the seven-element *C*-band receiving and transmitting

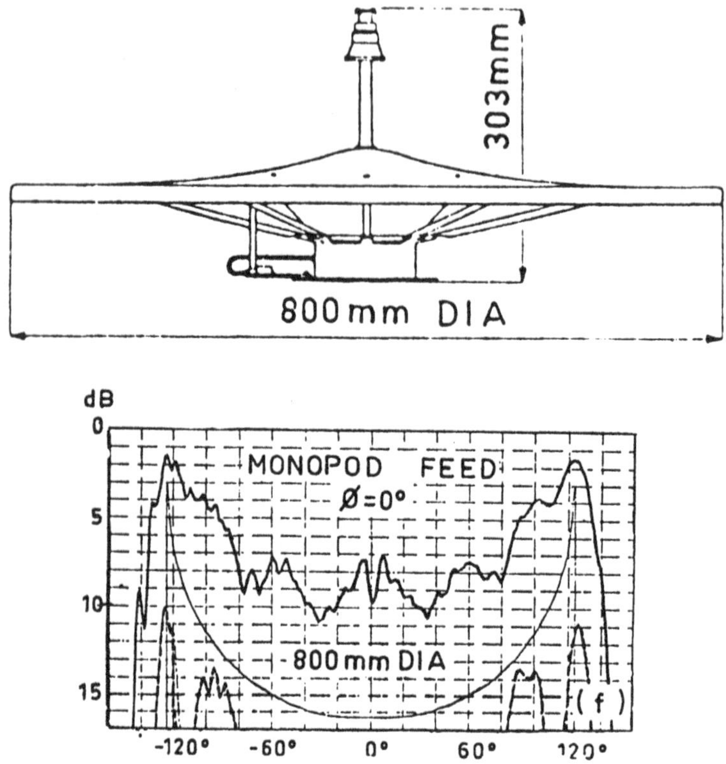

Figure 4.48(d) Radiation pattern of the antenna in the presence of the satellite body (from [57])

array configurations, and the nine-element L-band receiving array geometry [63]. Table 4.8 summarizes the edge gain coverage of these Inmarsat antennas. The cup dipole antenna is a major candidate for future satellite programs such as M-Sat, Intelsat, Inmarsat, and military applications [64].

4.5.2 Helical Antennas

Helical antennas are popular for the reception of spacecraft telemetry signals. They can be made cheaply and simply, and can be easily protected from shipboard environment by enclosing the helix in a fiberglass cylinder. Helices are also commonly used in three- or four-element planar arrays for spacecraft tracking.

The design of the quasitapered helical cavity-backed antenna used in the FLTSATCOM satellite was developed when the intermodulation problem on that satellite was identified. A quasitapered helical antenna, consisting of two or more

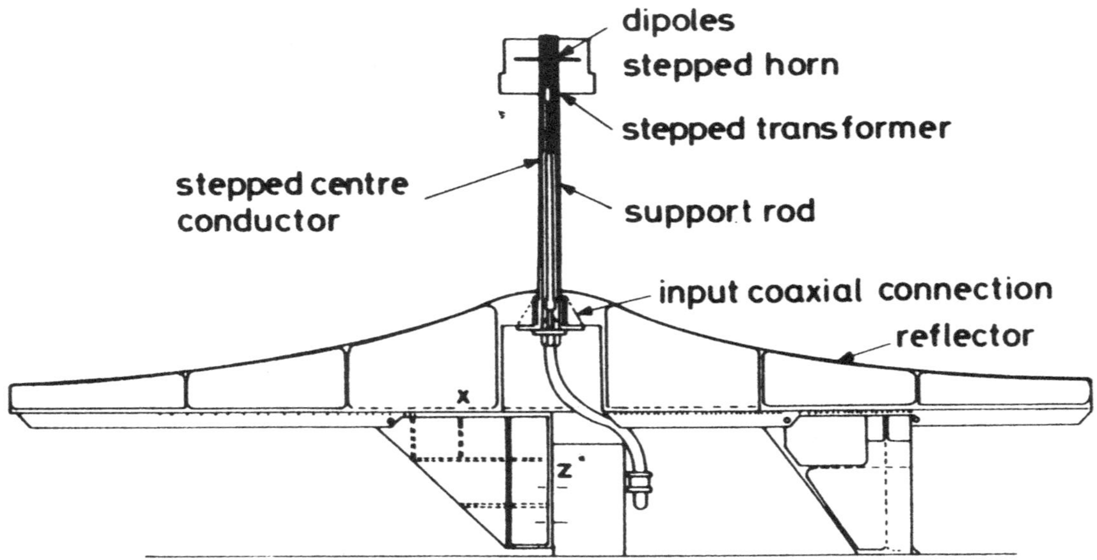

Figure 4.49(a) Sketch of an ERS-1 telemetry antenna (from [58–61])

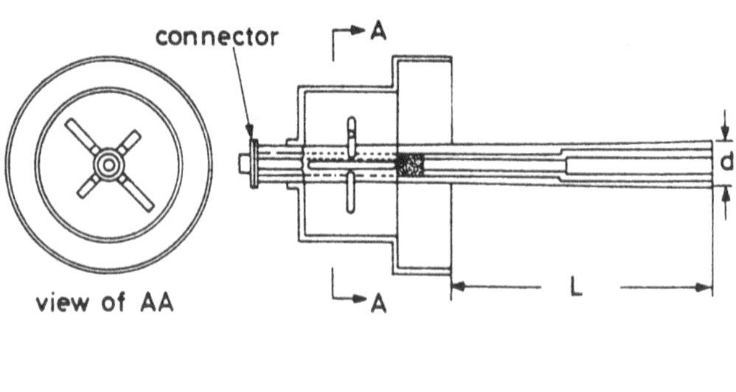

$L = 0.0953$ m; $d_1 = 0.0114$ m

Figure 4.49(b) Feed assembly (from [61a])

(c)

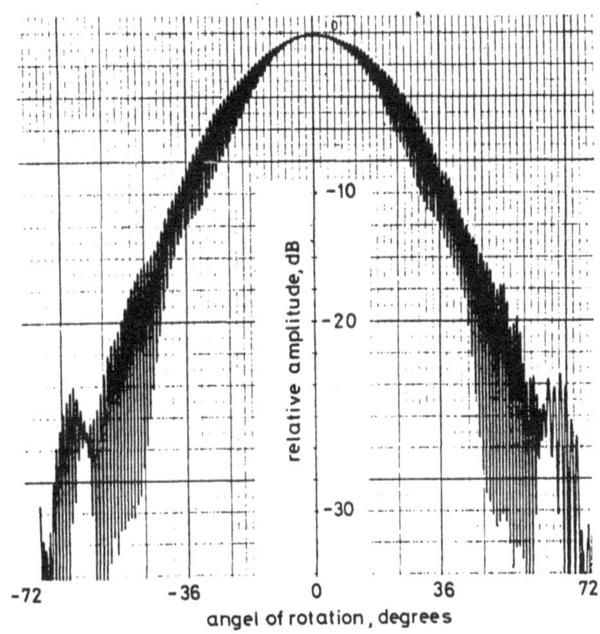

(d)

Figure 4.49(c) Superposition of TE_{11} and TM_{11} modes (from [61])
(d) Axial ratio pattern of the feed at 8.10 GHz (from [61])

uniform helical sections of different diameters, was joined (using short tapered transitions) to form a single radiating structure. This nonuniform configuration is a unique approach for widening the bandwidth of a cavity-backed helical antenna and improving gain, pattern, and axial ratio characteristics. The bandwidth is twice that of

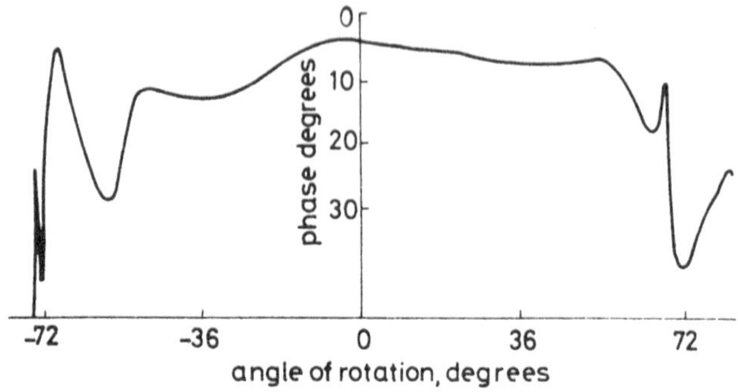

(e)

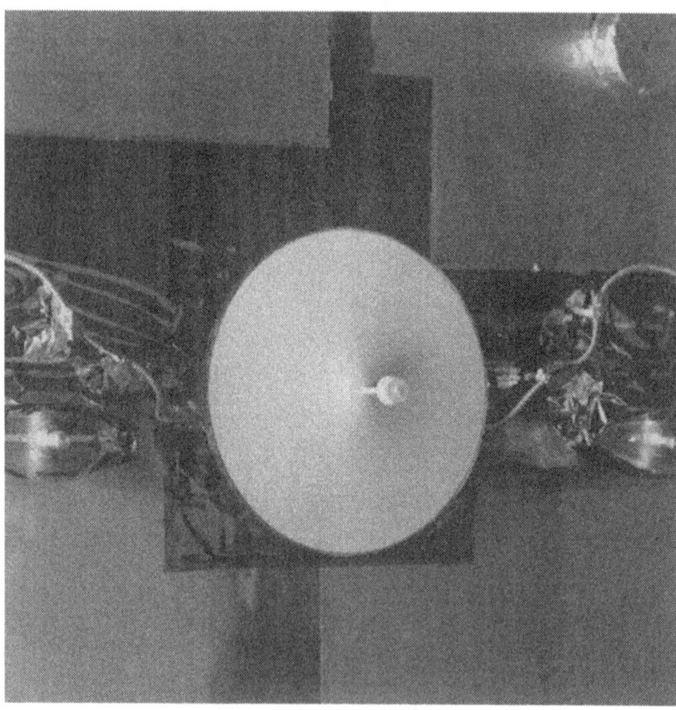

(f)

Figure 4.49(e) Phase against angle of rotation (from [61])
(f) Photograph of the antenna, including a simulated satellite body (from [61])

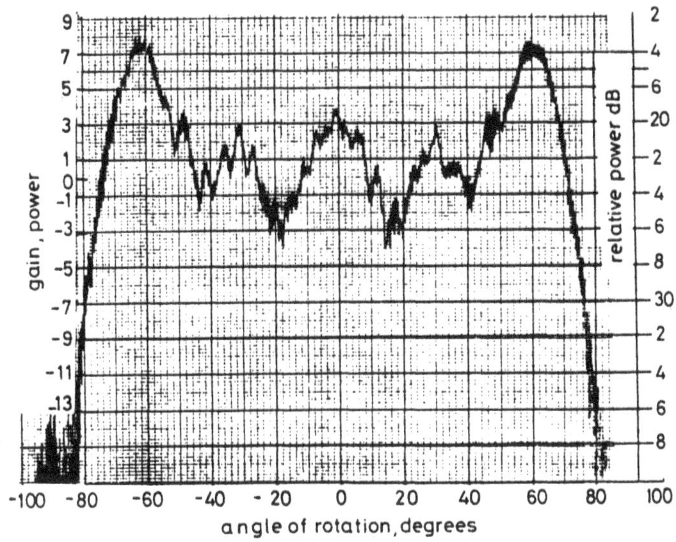

(g)

Figure 4.49(g) Co-polar pattern of the antenna, in the presence of a simulated satellite body, at 8.1 GHz (from [61])

a conventional uniform diameter helix. In addition, it provides the capability of shaping the gain-frequency response to the desired characteristics, as described in Section 4.4.1. The helix is used as a receiving antenna, operating in the 290 to 400 MHz range, with 12.8 to 14 dB edge-of-earth coverage gain. Figure 4.51(a) shows a 3/8 scale model FLTSATCOM test antenna mounted adjacent to the transmitting antenna [65]. A detailed description of the transmitting antenna feed will be given in the next section. Isolation and interaction between the transmitting antenna and the receiving antenna have already been discussed.

Isolation and interaction effects between the transmitting antenna (reflector fed by backfire helix) and the receiving (nonuniform cavity-backed helix) antenna were evaluated by swept frequency measurements. The measurements were made on the 3/8 scale antennas with a center-to-center spacing of 1.3716 m (corresponding to 3.6576 m full scale). Figure 4.51(b) shows the measured isolation (45 + 2 dB) between transmitting and receiving antennas. The relative gain and axial ratio of these antennas are shown in Figure 4.51(c and d). The axial ratio envelopes are relatively unchanged. The edge-of-earth coverage gain variation of the transmitting antenna at 9 degrees is within ±0.1 dB, while the gain change of the receiving helix

Figure 4.49(h) Photograph of the highly-shaped beam telemetry antenna assembled on the ERS-1 satellite payload (from [62])

Table 4.7 Cup Dipole Array Antenna Details

Antenna	*Number of radiators*	*Aperture shape*
L-band transmitting	61	hexagonal
L-band receiving	9	octagonal
C-band transmitting	7	hexagonal
C-band receiving	7	hexagonal

The radiators are crossed-dipoles mounted in circular cups.

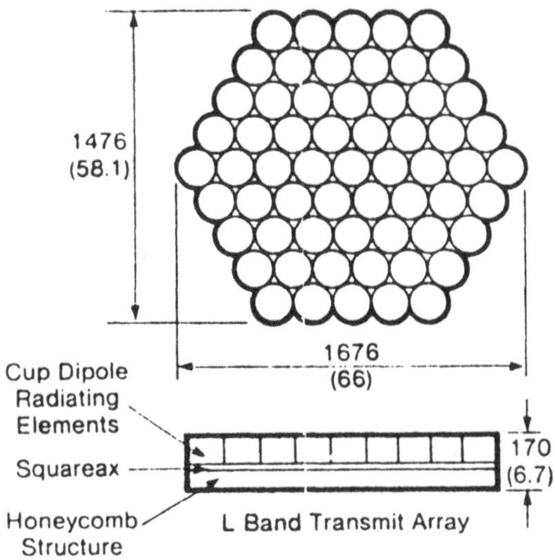

(a)

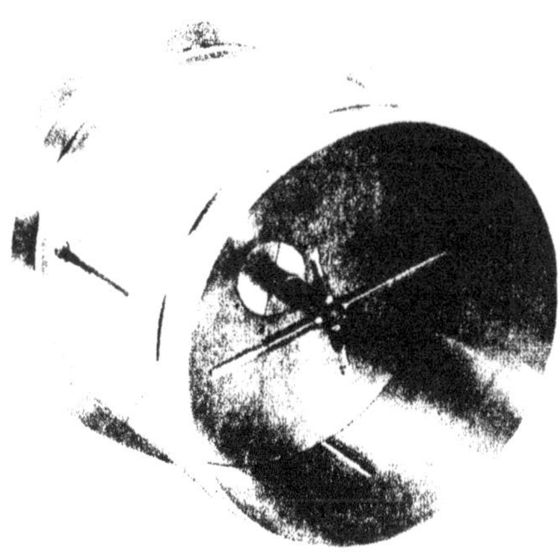

(b)

Figure 4.50(a) *L*-band transmitting array configuration (from [63])
(b) Photograph of a cup dipole antenna (from [63])

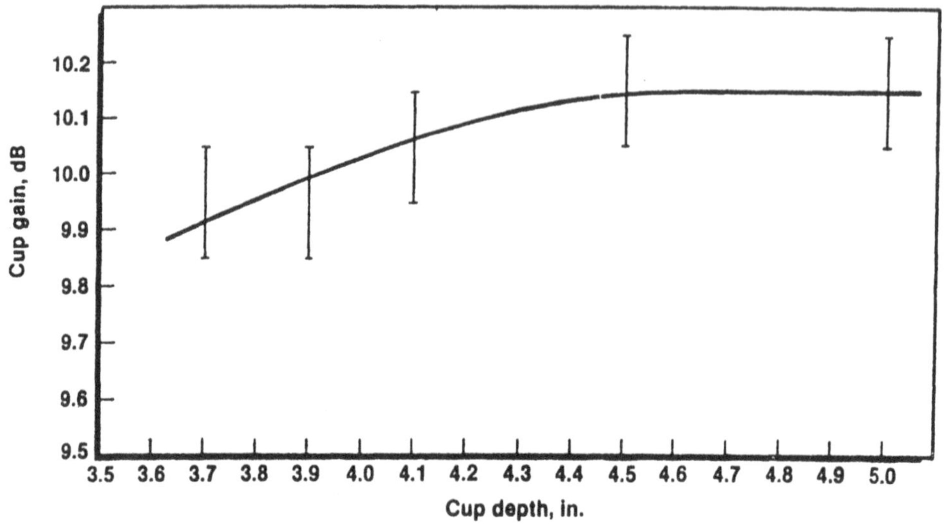

(c)

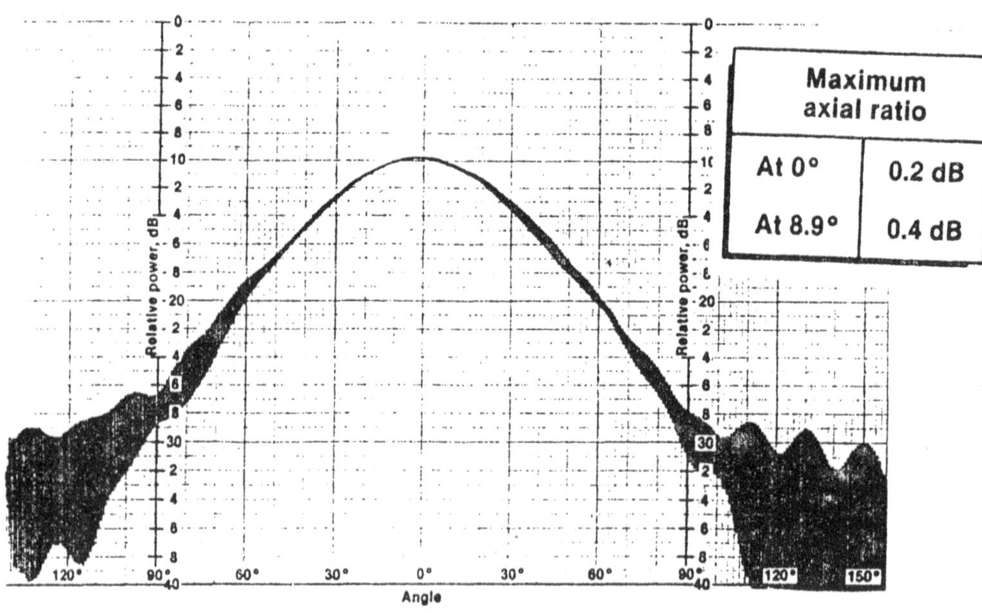

(d)

Figure 4.50(c) Cup dipole antenna gain *versus* cup depth (0.95 λ_0 cup) (from [63])
(d) Axial ratio pattern of 0.95 λ_0 cup dipole antenna (from [63])

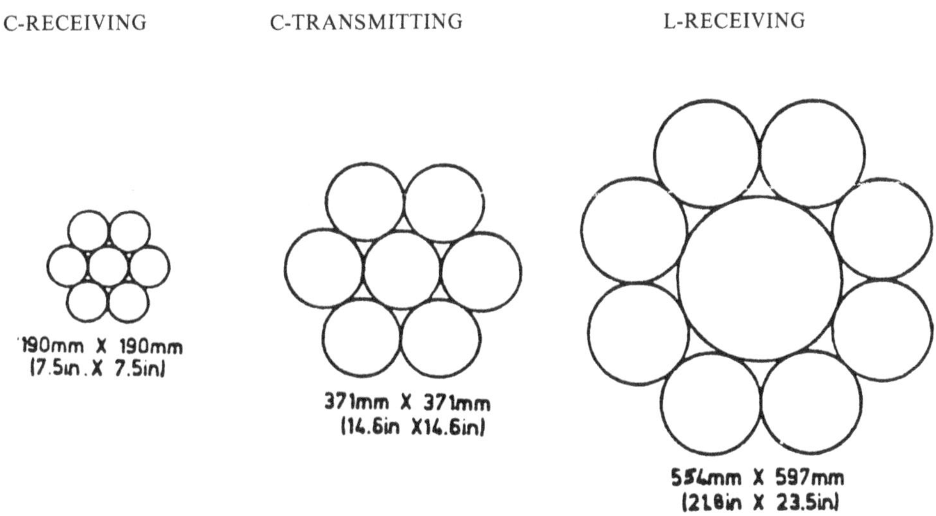

Figure 4.50(e) *C*-band receiving and transmitting, and *L*-band receiving array geometries (from [63,64])

Table 4.8 Edge Gain Coverage of the Inmarsat Cup Dipole Array Antennas

Antenna	*Edge gain coverage, dB* (*±9 degrees*)
L-band transmitting	18.2
L-band receiving	16.0
C-band transmitting	17.0
C-band receiving	16.7

at 10 degrees is within +0.2 dB. We conclude, therefore, that the interaction effects are relatively small.

Figure 4.52 shows the *Tracking and Data Relay Satellite System* (TDRSS) spacecraft developed by TRW. Cavity-backed helical phased array antennas have been used for TDRSS multiple-access services at *S*-band. A detailed description of the antenna was given in Section 4.4.1. Figure 4.53(a) shows the Skynet 4 satellite configuration. The layout of the antennas is shown in Fig. 4.53(b). A cavity-backed helix antenna is used for earth coverage in the UHF band. In the UHF antenna, the objective has been to achieve the maximum gain for the single transmit-receive assembly from the volume available: the antenna must provide Earth cover at relatively widely-separated frequency bands with RHCP. The solution is a single helix of about 2.5 m in length, which is stowed during the launch phase.

(a)

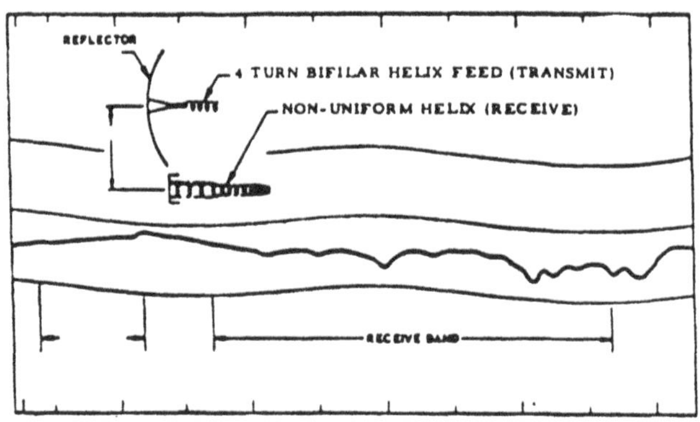

Isolation, dB

Scale Frequency, MHz

(b)

Figure 4.51(a) A 3/8 scale antenna (from [65])

(b) Isolation between transmitting reflector and receiving nonuniform helical antennas (from [65])

The structural details of the cavity-backed quad-helical array antenna are shown in Fig. 4.54. In the figure, the reflector diameter, antenna height, element spacing, and cavity rim length are about 1.7 λ_0, 0.425 λ_0, 0.7 λ_0, and 0.25 λ_0, respectively. A power gain of 13 dB, an axial ratio of better than 1 dB, a beamwidth of about

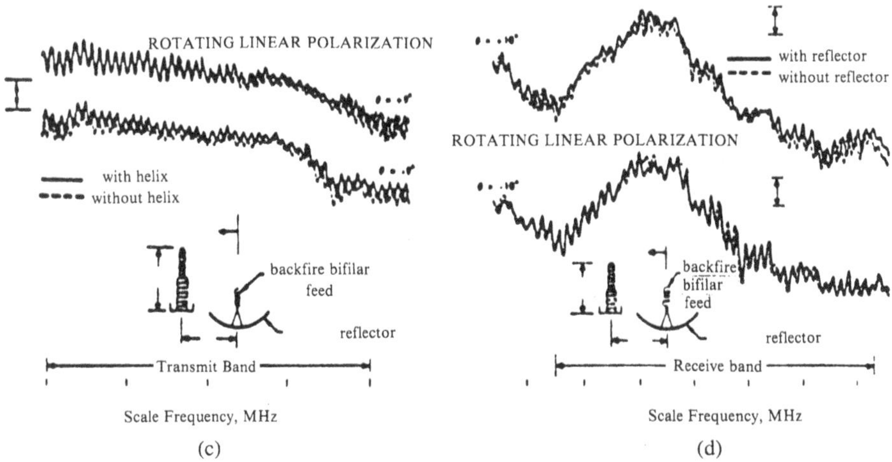

Figure 4.51(c) Relative gain and axial ratio of a transmitting reflector antenna with and without side-mounted helix (from [65])

(d) Relative gain of a receiving helical antenna with and without transmitting reflector antenna (from [65])

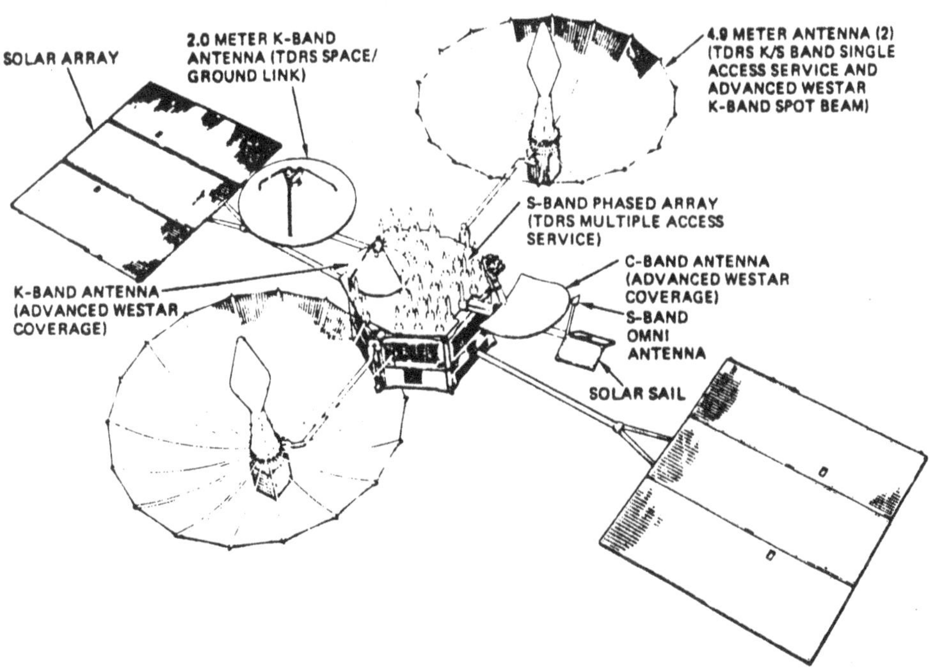

Figure 4.52 TDRSS antennas, including solar panel (from [66])

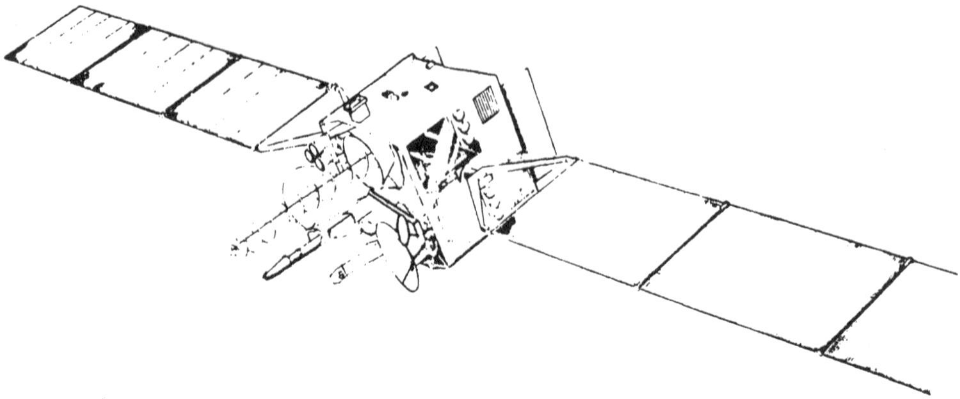

Figure 4.53(a) Skynet 4 United Kingdom defence communication satellite (from [67])

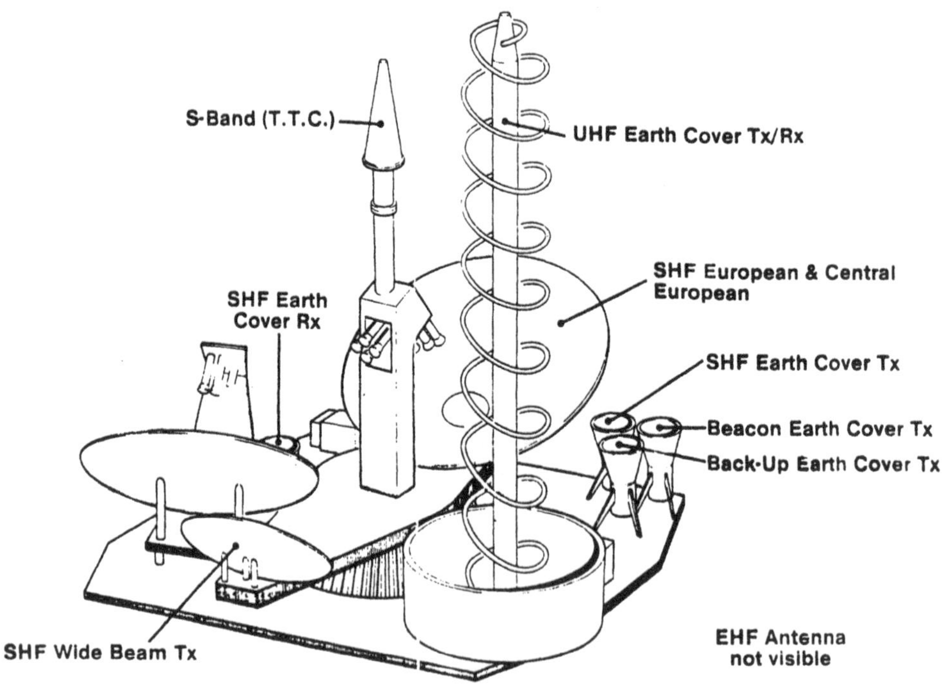

Figure 4.53(b) Antenna layout on the Skynet 4 [67]

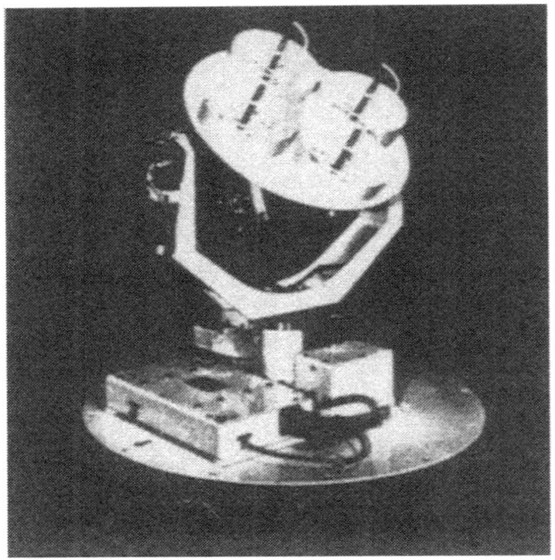

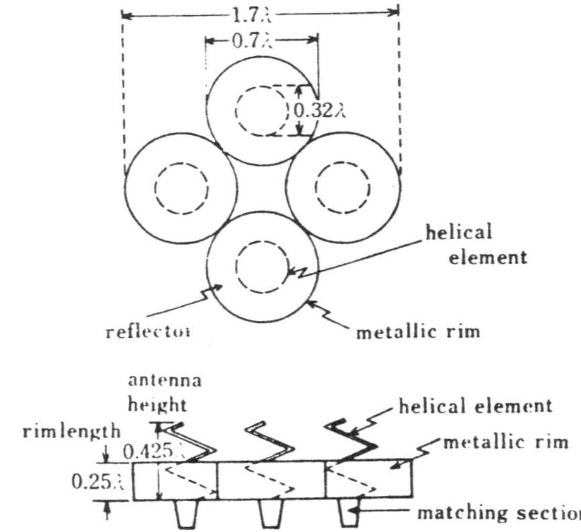

Figure 4.54 Structural details of shipborne helical array antenna (from [68])

40°, and an aperture efficiency of approximately 100% are obtained. The performance of this antenna satisfies the standard-B ship station ($G/T \geq -12$ dB/K) for the Inmarsat system.

REFERENCES

1. Bolster, M.F., "A New Type of Circular Polarizer Using Crossed Dipoles," *IRE Trans. Microwave Theory and Techniques,* Vol. MTT-9, No. 5, September 1961, pp. 385–388.
2. King, R.W.P., *The Theory of Linear Antennas,* Cambridge, MA: Harvard University Press, 1956, p. 172.
3. Silberberg, R.W., "The Paradisc Antenna—A Novel Technique to Improve the Axial Ratio of A Circularly Polarized High Gain Antenna System," *IEEE Trans. Antennas and Propagation,* Vol. AP-21, No. 1, January 1973, pp. 108–110.
4. Katulski, R.J., "The Optimization of the Crossed Dipoles as a Radiating Source in the Short Backfire Antenna," *Proc. Third International Conference on Antennas and Propagation,* ICAP 83, IEE Publication No. 219, University of East Anglia, Norwich, U.K., April 1983, pp. 412–416.
5. McLaughlin, J.W., D.A. Dunn, and R.W. Grow, "A Wide-band balun," *IRE Trans. Microwave Theory and Techniques,* Vol. MTT-6, No. 3, July 1958, pp. 314–316.
6. Oltman, H.G. Jr., "The Compensated Balun," *IRE Trans. Microwave Theory and Techniques,* Vol. MTT-14, No. 3, March 1960, pp. 112–119.
7. Kumar, A., "Cavity-backed Cross Dipole Antenna," *Research Report No. 2-77,* E.E.T.C. Coventry, U.K., 1977.
8. Hong, M.H., D.P. Nyquist, and K.M. Chen, "Radiation Fields of Open-Cavity Radiators and a Backfire Antenna," *IEEE Trans. Antennas and Propagation,* Vol. AP-18, No. 6, November 1970, pp. 813–815.
9. King, H.E., and J.L. Wong, "An Experimental Study of a Balun-fed Open-sleeve Dipole in Front of a Metallic Reflector," *IEEE Trans. Antennas and Propagation,* Vol. AP-20, No. 2, March 1972, pp. 201–204.
10. Wong, J.L., and H.E. King, "A Cavity-Backed Dipole Antenna with Wide-Beamwidth Characteristics," *IEEE Trans. Antennas and Propagation,* Vol. AP-21, No. 5, September 1973, pp. 725–727.
11. Ehrenspeck, H.W., "A New Class of Medium-Size High-Efficiency Reflector Antennas," *IEEE Trans. Antennas and Propagation,* Vol. AP-22, No. 2, March 1974, pp. 329–332.
12. Fisk, R.E., and J.A. Donovan, "A New CP Antenna for Television Broadcast Service," *IEEE Trans. Broadcasting,* Vol. BC-22, No. 3, September, 1976, pp. 91–96.

12a. Duncan, J.W., S.J. Hamada, and P.G. Ingerson, "Dual Polarization Multiple Beam Antenna for Frequency Reuse Satellites," *AIAA/CAST 6th Communications Satellite Systems Conference,* AIAA Paper No. 76-248, Montreal, Quebec, Canada, April 5–8, 1976, pp. 1–8.

13. Kumar, A., "The Short Backfire Antenna: A Theoeretical Approach," *Proc. IEEE Electrocom 85,* Toronto, Canada, September 1985.
14. Sinha, N., "Conical Cup Dipole Antennas," *Research Report No. 5-77,* E.E.T.C. Coventry, U.K., 1977.
15. Kumar, A., "Short Backfire Antenna Arrays," *IEEE APS/URSI Symposium,* US, 1984, p. 20.
16. Love, A.W., A.W. Rudge, and A.D. Olver, "Primary Feed Antennas," (in *Handbook of Antenna Design,* Rudge, A.W., K. Milne, A.D. Olver, and P. Knight, Eds., Vol. 1) London, Peter Pereriuns Press, 1986, pp. 386–397.
17. Cohn, S.B., "Flare-angle Changes in Horn as Means of Pattern Control," *Microwave Journal,* Vol. 13, No. 10, October 1970, pp. 41–46.

18. Wong, G., "A Compact Corrugated Horn Antenna," *IEEE Montech '86 Conference on Antennas and Communications,* Montreal, Quebec, Canada, IEEE Cat No. TH0156-0, September–October, 1986, pp. 33–35.
19. Kumar, A., "Circularly Polarized Square Horn Feed," *IEEE APS Symposium,* Blacksburg, Virginia, IEEE Cat. No. 87 CH 2435–6, Vol. 2, June 1987, pp. 668–671.
20. Kumar, A., "*L*-Band Radiator Design/Test Cup Dipole," *Report No. RML-009-87-109,* Spar Aerospace Ltd., Ste-Anne-de-Bellevue, Quebec, Canada, August 1987.
20a. Kumar, A., "*L*-Band Radiator Design," *Research Report,* EETC Coventry, UK, 1982.
21. Kraus, J.D., "Helical Beam Antennas for Wide Band Applications," *Proc. IRE,* Vol. 36, October 1948, pp. 1236–1242.
22. Silver, S., *Microwave Antenna Theory and Design,* MIT Rad. Lab. Series, Vol. 12, New York: McGraw-Hill, 1949.
23. Kraus, J.D., *Antennas,* New York: McGraw-Hill, 1950, Chapter 7.
24. Henery, J., *Antenna Engineering Hand Book,* New York: McGraw-Hill, 1984.
25. Andersen, J.B., "Low- and Medium-Gain Microwave Antennas," (*The Handbook of Antenna Design,* Rudge, A.W., K. Milne, A.D. Olver, and P. Knight, Eds.), Chapter 7, U.K., Peter Peregrinus Ltd., 1986.
26. Bystrom, J., and D.G. Berntsen, "An Experimental Investigation of Cavity Mounted Helical Antennas," *IRE Trans. Antennas and Propagation,* Vol. AP-4, No. 1, January 1966, pp. 53–58.
26a. Heckert, G.P., "Investigation of *L*-Band Shipboard Antenna for Maritime Satellite Applications," *Report No. N72-24161,* Automated Marine International, Newport Beach, California, US, February 1972, pp. 225–227.
27. Wong, J.L., and H.E. King., "Broadband Quasi-Taper Helical Antennas," *IEEE Trans. Antennas and Propagation,* Vol. AP-27, No. 1, January 1979, pp. 72–78.
28. King, H.E., and J.L. Wong, "Characteristics of 1 to 8 Wavelength Uniform Helical Antennas," *IEEE Trans. Antennas and Propagation,* Vol. AP-28, No. 2, March 1980, pp. 291–296.
29. Wong, J.L., and H.E. King, "Empirical Helix Antenna Design," *IEEE APS Symposium,* Albuquerque, New Mexico, 1982, pp. 366–369.
30. Kraus, J.D., "A 50 Ohm Input Impedance for Helical Beam Antennas," *IEEE Trans. Antennas and Propagation,* Vol. AP-25, No. 6, November 1977, p. 913.
31. Kumar, A., "*C*-Band Rigid Helical Antenna," *1986 National Radio Science Meeting* (URSI), Philadelphia, June 1986, p. 30.
32. Agrawal, V.D., and G.G. Wong, "A High Performance Helical Element for Multiple Access Array on TDRSS Spacecraft," *IEEE APS Symposium,* 1979, pp. 481–484.
33. Marsh, J.A., "Current Distribution on Helical Antennas," *Proc. IRE,* Vol. 39, June 1951, pp. 668–675.
34. Stratoti, A.R., and E.J. Wilkinson, "An Investigation of the Complex Mutual Impedance between Short Helical Array Elements," *IRE Trans. Antennas and Propagation,* Vol. AP-7, July 1959, pp. 279–280.
35. Yen, J.L., "Coupled Surface Waves and Broadside Arrays and Endfire Antennas," *IRE Trans. Antennas and Propagation,* Vol. AP-9, May 1961, pp. 296–304.
36. Shiokawa, T., Y. Karasawa, and H. Yokoi, "A Ship-born Helical Array Antenna for Maritime Satellite Communication," *IEE ICAP 81,* IEE Conference Publication No. 195, London, U.K., 1981, pp. 303–307.
37. Bawer, J., and J. Wolfe, "The Spiral Antenna," *IRE National Convention Record,* Part I, 1960, pp. 84–95.
38. Nakano, H., K. Nogami, S. Arais, H. Mimaki, and J. Yamauchi, "A Spiral Antenna Backed by a Conducting Plane Reflector," *IEEE Trans. Antennas and Propagation,* Vol. AP-34, No. 6, June 1986, pp. 791–796.

39. Dyson, J., "The Equiangular Spiral," *IRE Trans. Antennas and Propagation,* Vol. AP-7, No. 4, October 1959, pp. 329–334.
40. US Patent No. 4,260,964.
41. Marchant, N. "Transmission Line Conversions," *Electronics,* December 1944, pp. 142–145.
42. Duncan, J.W., and V.P. Minerva, "100:1 Bandwidth Balun Transformer," *Proc. IRE,* Vol. 48, February 1960, pp. 156–164.
43. Morgan, T.E., "Spiral Antennas for ESM," *Proc. IEE,* Pt. F, Vol. 132, No. 4, July 1985.
44. Saul, D.L., "Cavity-Backed Spiral Has Microstrip Feed," *Microwaves,* Vol. 20, No. 10, October 1981, pp. 88–89.
45. Morgan, T.E., "18-40 GHz Spiral Antenna," *17th European Microwave Conference,* Italy, September 1987, pp. 195–200.
46. Woo, R., "Multipactor Breakdown in Coaxial Transmission Lines," *Proc. IEEE,* Vol. 56, No. 4, April 1968, pp. 776–777.
47. Woo, R., "Final Report on RF Voltage Breakdown in Coaxial Lines," *Tech. Report 32-1500,* JPL, California, US, October, 1970.
48. Francis, G., and A. Von Egel, "The Growth of High Frequency Electrodeless Discharge," *Trans. Royal Society of London,* Vol. A-246, 1953, p. 19.
49. Kumar, A., "Multipactor Analysis for Cross-Dipole Feed Horn," *IEEE APS/URSI Symposium,* Syracuse, N.Y., June 1988.
50. Kumar, A., "Cross-dipole fed square Horn," *Int. Jour. Electronics,* Vol. 63, 1988.
51. Clancy, P.F., "Multipactor Control in Microwave Space Systems," *Microwave Journal,* Vol. 21, No. 3, March 1978, pp. 77–83.
52. Antenna and Transmission Line Systems, *Catalog 871,* Gabriel Electronics Incorporated, Scarborough, Maine, US, 1987, pp. 16–17.
53. Moskowitz, S., "Shaped-Beam Antenna Spreads Uniform Signal, Overearth," *Microwaves,* Vol. 18, No. 3, March 1979, pp. 10–11.
54. Sweeting, M.N., "The University of Surrey UOSAT-2 Space-Craft Mission," *J. IERE Supplement to Vol. 57,* No. 5, September–October 1987, pp. S99–S115.
55. Nawman, L.R., and H.E. King, "Flight Test Evaluation of a 240 to 400 MHz Cavity-Backed Cross Open-Sleeve Dipole Antenna," *The Aerospace Corporation Technical Report No. TR-0076(6401)-1 (AD A017298),* October 10, 1975.
56. Wong, J.L., and H.E. King, "A Shallow Ridge-Cavity Crossed-Slot Antenna for the 240 to 400 MHz Frequency Range," *IEEE Trans. Antennas and Propagation,* Vol. AP-21, September 1973, pp. 725–727.
57. Do Boi, H., "The Highly Shaped Beam Telemetry Antenna for the Spot Satellite," *14th European Microwave Conference,* Belgium, September 1984, pp. 234–240.
57a. Matra Espace, France, Private Communication.
58. Kumar, A., "IDHT Antenna for ERS-1 Satellite," *IEEE IGARSS'85 Conference,* University of Massachusetts, Amherst, MA, IEEE Cat. No. 85CH2162-6, October 1985.
59. Kumar, A., "ERS-1 X-Band IDTS Antenna," *IEEE APS Symposium,* Philadelphia, IEEE Cat. No. 86CH2325-9, Vol. 2, June 1986, pp. 739–742.
60. Kumar, A., "Highly-Shaped Beam Telemetry Antenna for the ERS-1 Satellite," *IEEE Montech'86 Conference on Antennas and Communications,* Montreal, Quebec, Canada, IEEE Cat. No. THO156-0, September–October 1986, pp. 46–49.
61. Kumar, A., "Highly Shaped Beam Telemetry Antenna for the ERS-1 Satellite," *Proc. IEE,* Pt. H, Vol. 134, No. 1, February 1987, pp. 106–108.
61a. Kumar, A., "Shape Beam Antenna for RADARSAT," Seventh Annual Benjamin Franklin Symposium, Cherry Hill, New Jersey, March 1989.
62. "Last News: ERS-1 on Tests," *Earth Observation Quarterly* No. 17, March 1987.
63. Hughes Aircraft Company, LA, US, Capability Voucher on Inmarsat Satellite Antennas, 1985.

64. Kumar, A., *Mutual Coupling of a Closely Packed Array,* National Radio Science Meeting (URSI), Colorado, January 1989.
65. King, H.E., and J.L. Wong, "240–400 MHz Antenna System for the FLEETSATCOM Satellites," *IEEE APS Symposium Digest,* June 20–22, 1977, pp. 349–352.
66. Smith, B.A., "Data Relay System Made Ready for January Launch," *Aviation Week & Space Technology,* Vol. 117, No. 2, 12th July 1982, pp. 67–68.
67. Dawson, T.W.G., D.E. McLaurin, and M.J. Painter, "Skynet 4 Defence Communication Satellite Project," *Conference Proc. Military Microwaves,* October 20–22, 1982, London, pp. 94–103.
68. Shiokawa, T., Y. Karasawa, and M. Yamada, "Compact Antenna Systems for Inmarsat Standard-B Ship Earth Station," *IEE 3rd Int. Conf. Satellite Syst. Mobile Comm. and Nav.,* IEE Conference Publication No. 222, London, 1983, pp. 95–99.

Chapter 5
Short Backfire Antennas

The SBFA is a highly-efficient radiator of simple and compact construction. It contains two parallel reflectors of unequal apertures, spaced approximately a half-wavelength apart and fed by either a dipole placed normal to the axis of the reflectors, or an open-ended waveguide coaxial with the reflector axis. The dipole is placed between the reflectors. Such a system is characterized by multiple reflections of electromagnetic waves, and maximum radiation occurs along the outward edge from the large reflector.

Multiple reflections of electromagnetic waves between two planes have been studied. The increase in directivity that results from placing a partially reflecting sheet in front of an antenna with a reflecting screen has been investigated at a wavelength of 3.2 cm by Trentini [1].

SBFAs were developed experimentally by Ehrenspeck in the 1960s. A backfire antenna is an ordinary endfire slow-wave antenna (e.g., a Yagi, helical, or dielectric rod antenna), terminated by a large reflecting plane, with the slow-wave reflector normal to its axis. The reflector backfires the electromagnetic waves onto the slow-wave structure, thus causing the radiation to travel the physical length of the structure twice. The measured directivity of such antennas is approximately 6 dB higher than that of corresponding endfire antennas. SBFAs are used as replacements for conventional dipoles, slot radiators, and Yagis, because they offer several advantages: a high-gain element; a lightweight antenna element; reduction in the number of feed elements in an array environment to between one-fourth and one-sixth that of slots and dipoles, without any loss in directivity or pattern quality; compactness and ease of manufacture, including low construction cost; reduced insertion losses because of fewer elements, and therefore increased overall gain; good axial ratio for CP applications; easy to adopt dual-polarization characteristics; ease of matching; low sidelobes and backlobes; ease of design for high-power applications; easy adoption of environmental protection requirements; and ease of mechanical design.

Because of favorable pattern characteristics, SBFAs lend themselves to the array techniques of high-gain antennas used for satellite communication, tracking, and telemetry.

5.1 EXPERIMENTAL CONSIDERATIONS OF A SBFA FED BY A CROSSED DIPOLE

The SBFA consists of a leaky cavity resonator formed from two plane reflectors of different diameters, spaced a half-wavelength apart, with a source (feed) placed between them, as shown in Figure 5.1. The SBFA was first developed by Ehrenspeck in 1965 [2]. A photograph of Ehrenspeck's S-band rim-edged SBFA model for 3 GHz is shown in Figure 5.2. Table 5.1 summarizes the dimensions of Ehrenspeck's SBFA.

Figure 5.1 Geometry of Ehrenspeck's SBF antenna

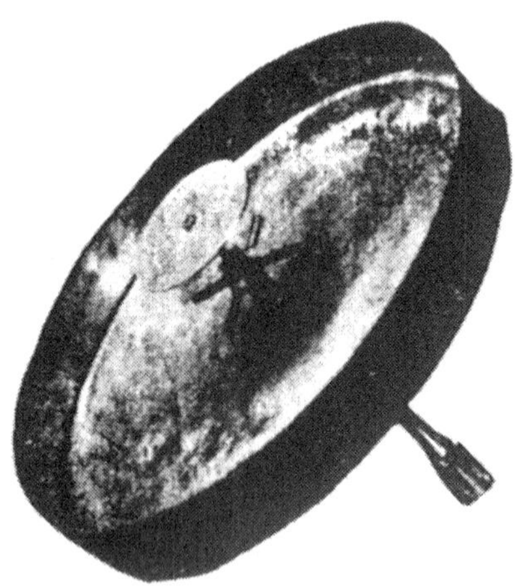

Figure 5.2 Photograph of Ehrenspeck's SBF antenna (from [2])

Table 5.1 Dimensions of Ehrenspeck's SBFA (λ_0 = 10 cm)

SBF antenna components	*Dimensions*
Large reflector	2.00 λ_0 in diameter
Small reflector	0.40 λ_0 in diameter
Spacing between the large and small reflector	0.50 λ_0
Width of the rim surrounding the large reflector	0.25 λ_0
Dipole or source feed point from the large and small reflector	0.25 λ_0

The measured *E*-plane and *H*-plane patterns of the *S*-band SBF antenna are shown in Figure 5.3(a) and (b). The main characteristics of Ehrenspeck's antenna are summarized in Table 5.2.

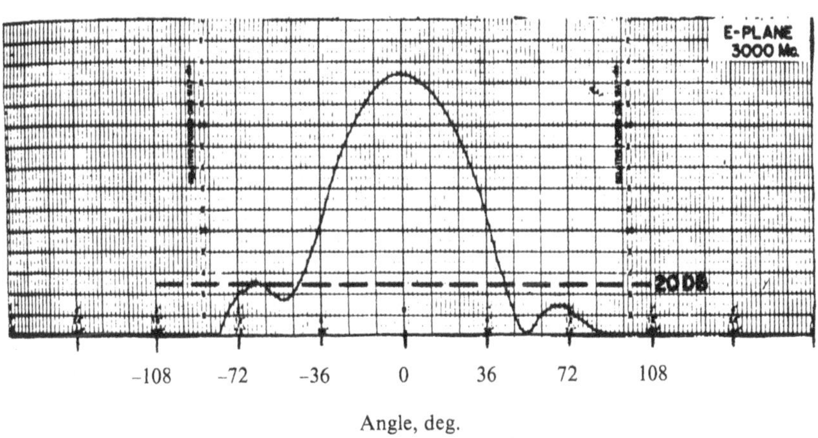

Figure 5.3(a) Measured *E*-plane radiation pattern of Ehrenspeck's SBFA (from [2])

Ehrenspeck compared his antenna with the Yagi endfire antenna. A conventional Yagi antenna with a gain of 15 dBi has to be about 4 λ_0 long, and needs 15 to 25 dipole elements. The sidelobe level of a Yagi antenna is higher than the SBFA. To obtain radiation patterns comparable to the low sidelobe and backlobe patterns

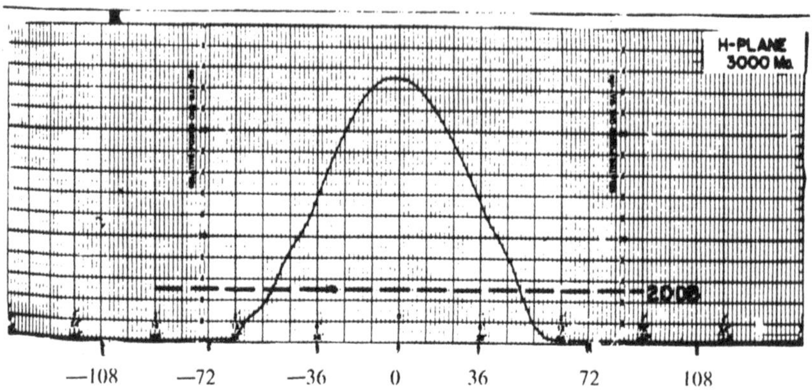

Figure 5.3(b) Measured *H*-plane radiation pattern of Ehrenspeck's SBFA (from [2])

Table 5.2 Characteristics of Ehrenspeck's SBFA

1. Frequency of operation:	3 GHz
2. Measured gain above a dipole feed:	13.1 dB
3. Measured gain above isotropic:	15.2 dBi
4. *E*-plane 1st sidelobe level:	20 dB down from peak
5. *H*-plane 1st sidelobe level:	25 dB down from peak
6. Backlobe at ±180 degrees:	More than 25 dB down from peak, latter confirmed 30 dB down.

of the SBFA, the Yagi needs more than 27 elements and an axial length of the more than 5.5 λ_0.

In 1966, Dod [3] experimentally investigated the SBFA's properties as functions of the antenna dimensions. Figure 5.4 (a) and (b) shows a sketch and the physical dimensions of the antenna in terms of wavelength. The antenna consists of two planar reflectors of different diameters, separated by a half-wavelength, forming a leaky cavity resonator with radiation normal to the smaller reflector. The antenna is fed by orthogonal sleeve dipoles at the midpoint between the two reflectors, with these dipoles fed by a dual-line balun. The mechanical tolerance in construction of the antenna is better than ±0.010 cm.

The effect of changes in the size of the smaller reflector on half-power beamwidth and sidelobe levels is shown in Figure 5.5. This figure suggests that the half-

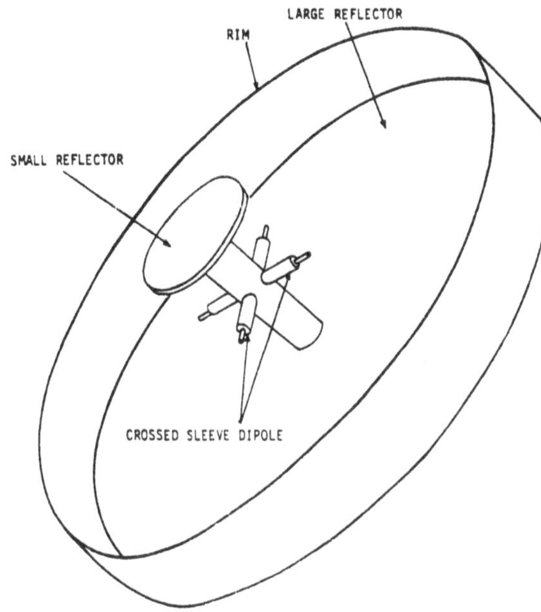

Figure 5.4(a) Sketch of an SBFA (from [3])

power beamwidth is reduced as the diameter of the small reflector is increased. Sidelobe level increases as the diameter of the small reflector increases, but remains at reasonable levels for the range of diameters (0.3 to 0.65 λ_0).

The radiation pattern measurements in the E-plane and H-plane were repeated for the large reflector diameter reduced to 1.5 λ_0, and the small reflector diameters, which varied from 0.30 to 0.65 λ_0 in steps of 0.05 λ_0. The height of the rim on the large reflector was fixed at 0.25 λ_0. The effect that varying the small reflector diameter has on half-power beamwidth and sidelobe level is shown in Figure 5.6. Reducing the diameter of the large reflector from 2 λ_0 to 1.5 λ_0 causes higher sidelobe levels in the E-plane and broader half-power beamwidths.

The E-plane and H-plane patterns of the antenna without a rim on the large reflector show that the removal of the rim results in higher backlobes and sidelobes, because the resonator shielding is reduced.

The maximum measured gain of the antenna is 14.9 dB, with 2.0 λ_0 large reflector, with and without rim, and a 0.65 λ_0 diameter small reflector, as shown in Figure 5.7. The addition of the rim increases the gain by 1 dB. The width of the main lobe increases as the operating frequency is reduced below the design frequency (1.5 GHz).

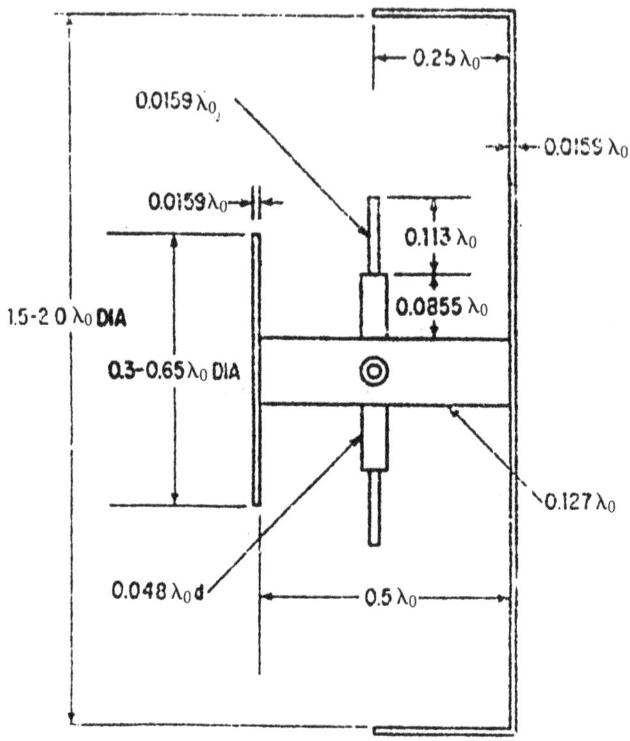

Figure 5.4(b) Dimensional sketch of an SBFA (from [3])

Figure 5.8 shows the plot of the input VSWR of the antenna *versus* the frequency of a 2.0 λ_0 diameter large reflector with 0.25 λ_0 rim and 0.50 λ_0 diameter small reflector. The antenna has an impedance matching bandwidth of 6.7% for a VSWR of less than 3 to 1. The leaky cavity is a high-Q device that requires broadband impedance matching techniques. It is possible to extend the impedance bandwidth by the proper choice of compensated balun (discussed in Chapter 4) and sleeve dipole parameters. In the present configuration, the antenna is matched to less than 1.1 to 1 VSWR at 1.5 GHz by adjusting the lengths of the dipole sleeves and the dipole elements.

In 1970, Nielsen [4] made numerical and experimental investigations on the size of the large and small reflector plate, the length of the feed dipole, the position of the dipole between the two reflector plates, the distance between those plates, and the influence of a frequency variation.

In 1974, Ehrenspeck [5] presented an empirical theory for optimizing the diameter and rim width of cup dipole and SBFAs. The antenna was optimized by

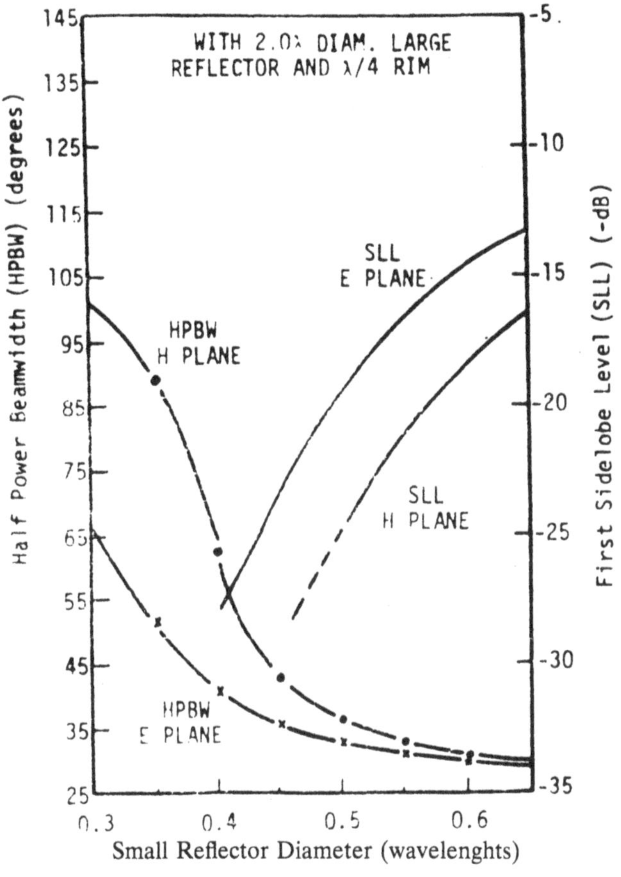

Figure 5.5 Half-power beamwidth and sidelobe level *versus* small reflector diameters, with the diameter of the large reflector $2.0\lambda_0$ and rim on the large reflector $0.25\lambda_0$ deep (from [3])

varying the frequency and rim width to find its highest directive gain. The directive gain of the antenna was 18.1 dB at 3.4 GHz, with the large reflector diameter 2.35 λ_0 and the rim width adjusted to 0.57 λ_0. Figure 5.9 shows the resulting directive gain as a function of frequency for the antenna. The directive gain is given by:

$$G = \eta[(4\pi A/\lambda_0^2)] \tag{5.1}$$

where A is the area of the large reflector and η is the aperture illumination efficiency.

The influence of the rim width adjustment on the directive gain of a cup dipole antenna or an SBFA can be most clearly demonstrated with those antennas in which

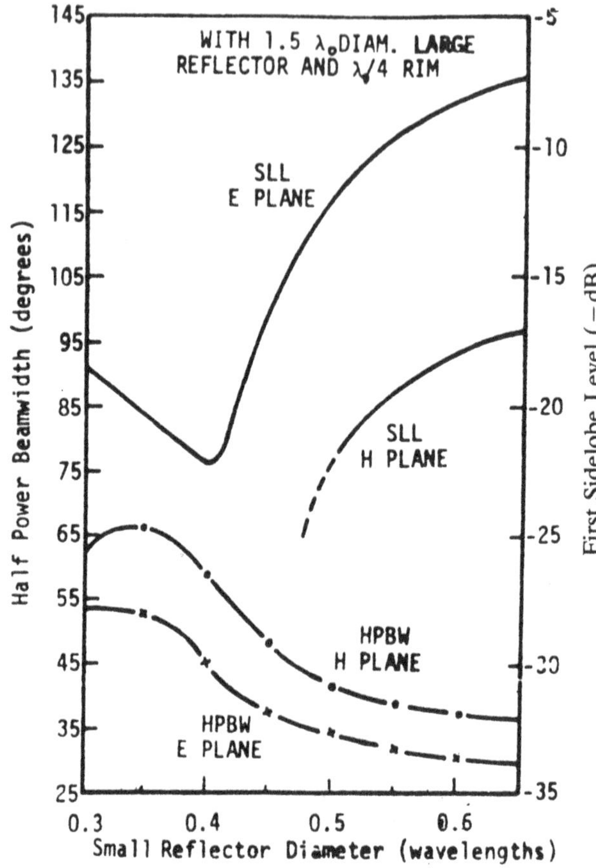

Figure 5.6 Half-power beamwidth and sidelobe level of the antenna *versus* diameters of the small reflector, with the large reflector diameter 1.5 λ_0 (from [3])

dimensions yield the smallest directive gain because of their lowest aperture efficiency. Figure 5.10 shows the effect of rim width on the directive gain of cup dipole and SBF antennas. This effect is caused mainly by phase changes between the far-field contributions of the edge and feed radiators. We can assume that the low directive gain of the rimless antenna indicates a phase difference of approximately 180 degrees between the two radiators.

Experimental measurement is performed with an SBFA, which has a large reflector diameter of 2.35 λ_0 and aperture area of 4.32 λ_0^2. The measured gain of the antenna is 18.1 dB, which indicates that the aperture efficiency of the antenna is higher than 100%. The half-power beamwidths are 24.5° and 23.5° in the *E*-plane

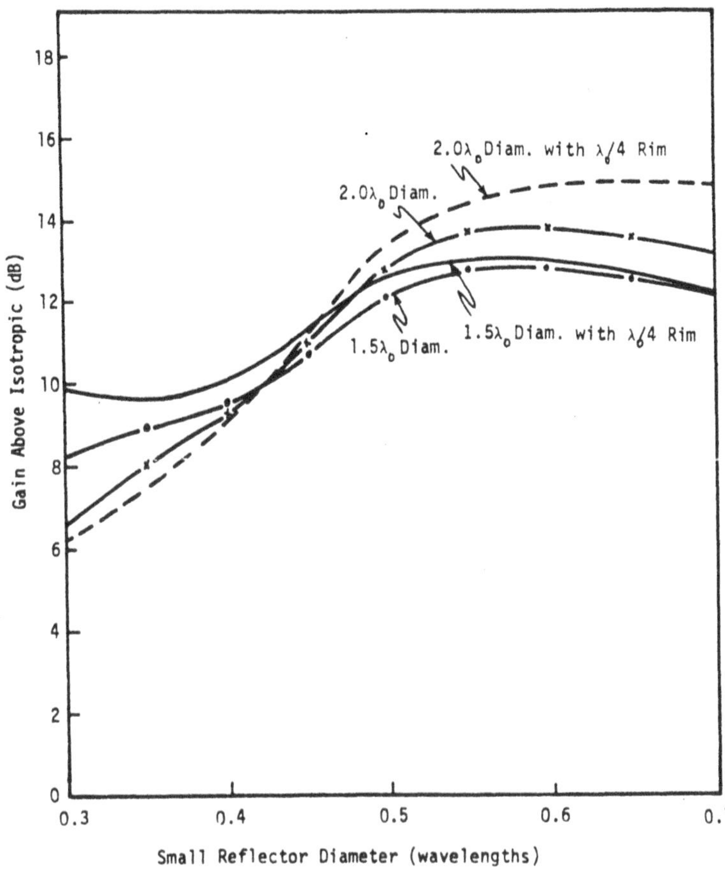

Figure 5.7 Measured gain of the antennas *versus* diameter of the small reflector for two sizes of the large reflector with and without rim (from [3])

and H-plane patterns, respectively, and the directive gain is 2.9 dB above that, calculated for $W_B = 0.65\ \lambda_0$ [5]. Ehrenspeck and Strom [6,7] described an optimized SBFA on the basis of Ehrenspeck's previous work [2]. Figure 5.11 shows an SBFA which has a small reflector diameter of $0.5\ \lambda_0$. Table 5.3 describes the dimensions and characteristics of the antenna at 3 GHz. The E- and H-plane patterns of the antenna are shown in Figure 5.12.

A simplified SBFA described by Woodward [8] provides both sum and error radiation patterns in the H-plane. The antenna is comprised of a pan-type large reflector, a small reflector disk and three parallel dipoles. The center dipole is excited to produce a sum pattern and the two end dipoles are excited out of phase to provide

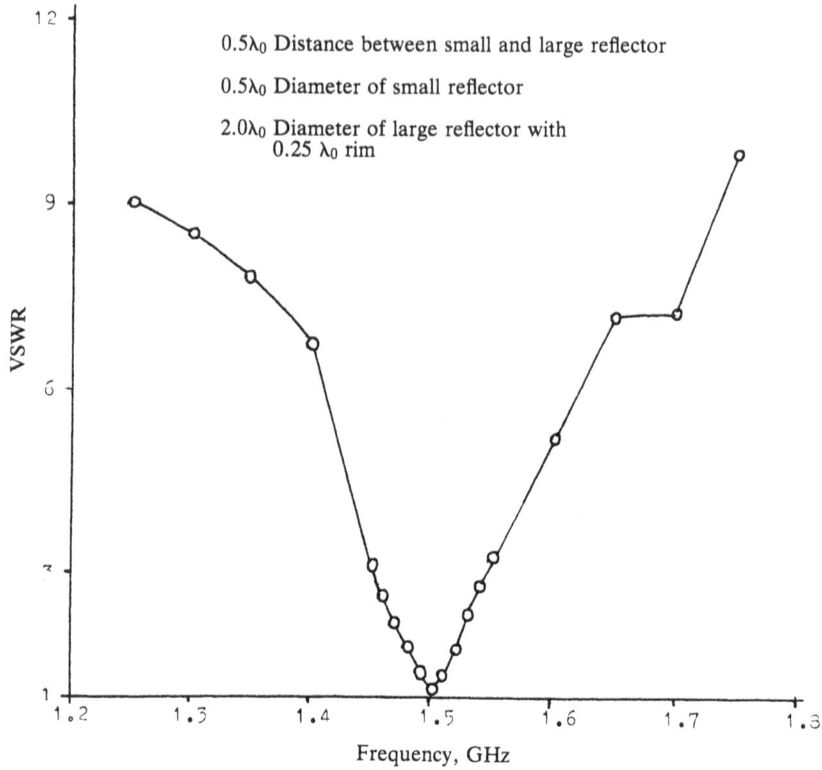

Figure 5.8 VSWR of the antenna *versus* frequency (from [3])

an error pattern, with a broadside null in the H-plane. Figure 5.13 shows a pan-type octagonal reflector P, with a width of approximately 2 λ_0, which includes a vertical rim of 0.25 λ_0 extending about the perimeter of the reflector. A small metal disk of approximately 0.4 λ_0 in width is positioned centrally and parallel to the larger pan reflector and spaced approximately 0.5 λ_0 above the pan reflector along a common vertical axis to form a resonant cavity in between.

Excitation of the central dipole produces like sum patterns in the E-plane and H-plane as shown in Figure 5.14 ((a) and (b)), with typical operating frequencies of 1.03 and 1.09 GHz. A conical beam having a gain of about 15 dBi is obtained, with very low sidelobes and backlobes. With the two outer dipoles fed 180° out of phase to each other using a tee junction with lines of equal length, as shown in Figure 5.15, an error pattern having a broadside null in the H-plane is obtained (Figure 5.14(b)). The lengths of outer dipole lines are selected to detune the outer dipoles to avoid pattern interaction in the sum mode of operation. The effect of radiation on

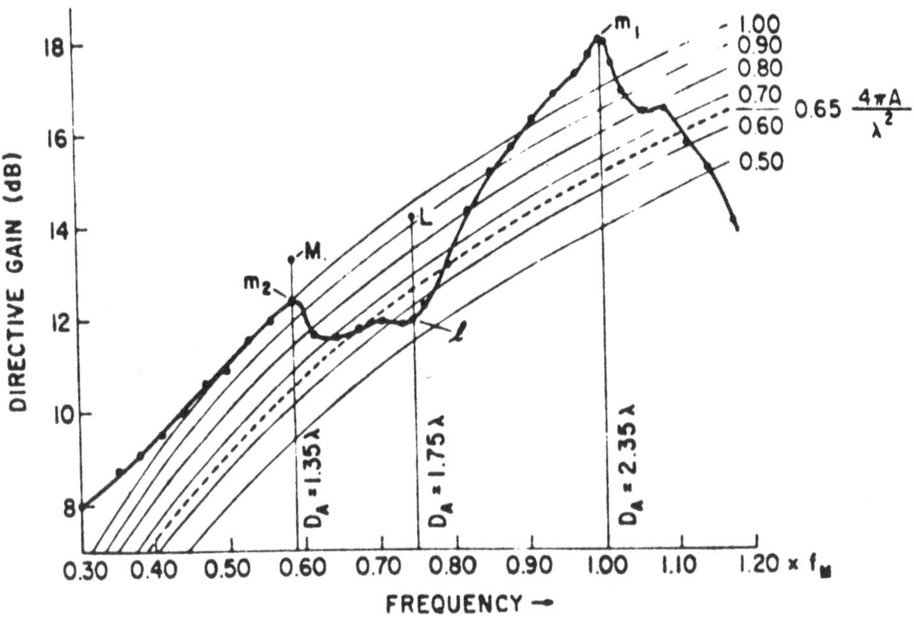

Figure 5.9 Directive gain *versus* frequency curve for cup dipole and SBF antennas (from [5])

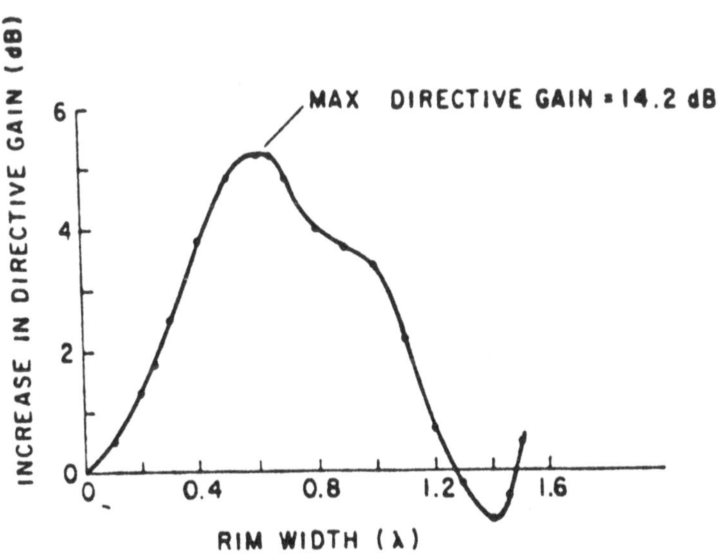

Figure 5.10 Effect of rim width on directive gain for cavity-backed antennas (from [5])

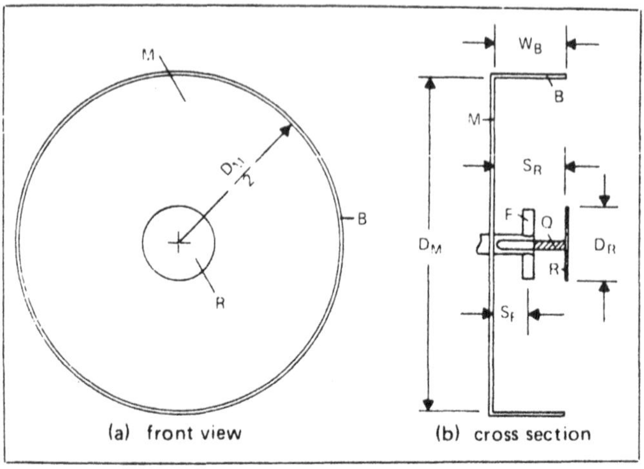

Figure 5.11 3 GHz SBFA model (from [6,7])

Table 5.3 Experimental Dimensions and Characteristics of the SBFA at 3 GHz

Diameter of the large reflector (D_M)	2.0 λ_0
Diameter of the small reflector (D_R)	0.5 λ_0
Distance between slotted dipole feed F and the large reflector $M(S_F)$	0.25 λ_0
Distance between the large and the small reflector (S_R)	0.5 λ_0
Depth of the rim (W_B)	0.5 λ_0
Aperture area of the antenna	3.14 λ_0^2
Aperture efficiency	83%
Directive gain at 3 GHz	15.1 dB
First sidelobe level	<-22 dB
Far sidelobe level	<-30 dB
Half-power beamwidth	34°

the center dipole is thus reduced. The sum pattern amplitude should exceed the error pattern only in the limited angular region of the error pattern null directed broadside to the large reflector, at least over the major portion of the forward hemisphere. This permits a relatively small aperture to have a sharp resolution equivalent to that of a larger antenna. The combination of sum and error patterns provides greater azimuthal information for improved target location. Optimum pattern shaping and gain can be achieved by choosing the optimum large reflector, the small reflector, and dipoles

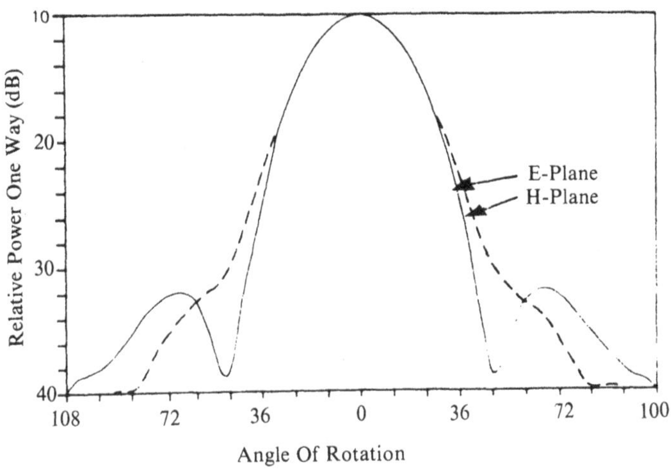

Figure 5.12 *E*-plane and *H*-plane patterns of the antenna (from [6,7])

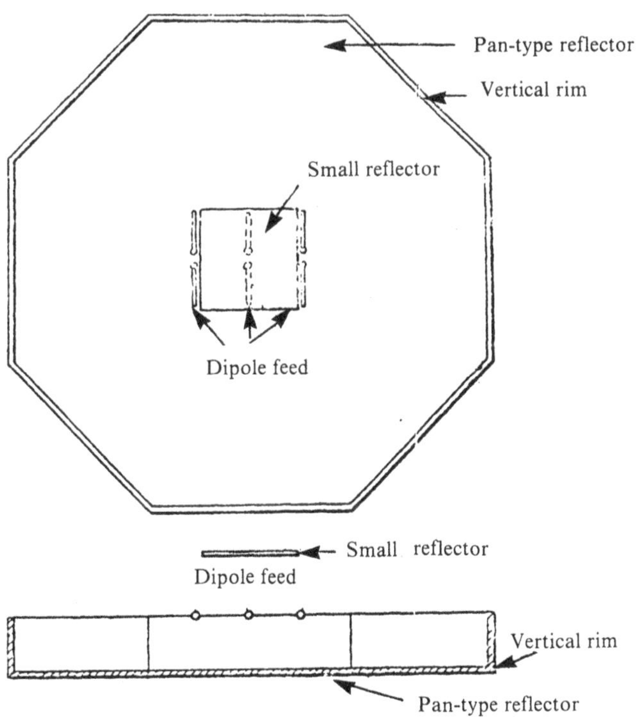

Figure 5.13 Sketch of an SBFA with octagonal-shaped large reflector, three dipoles, and a small reflector (from [8])

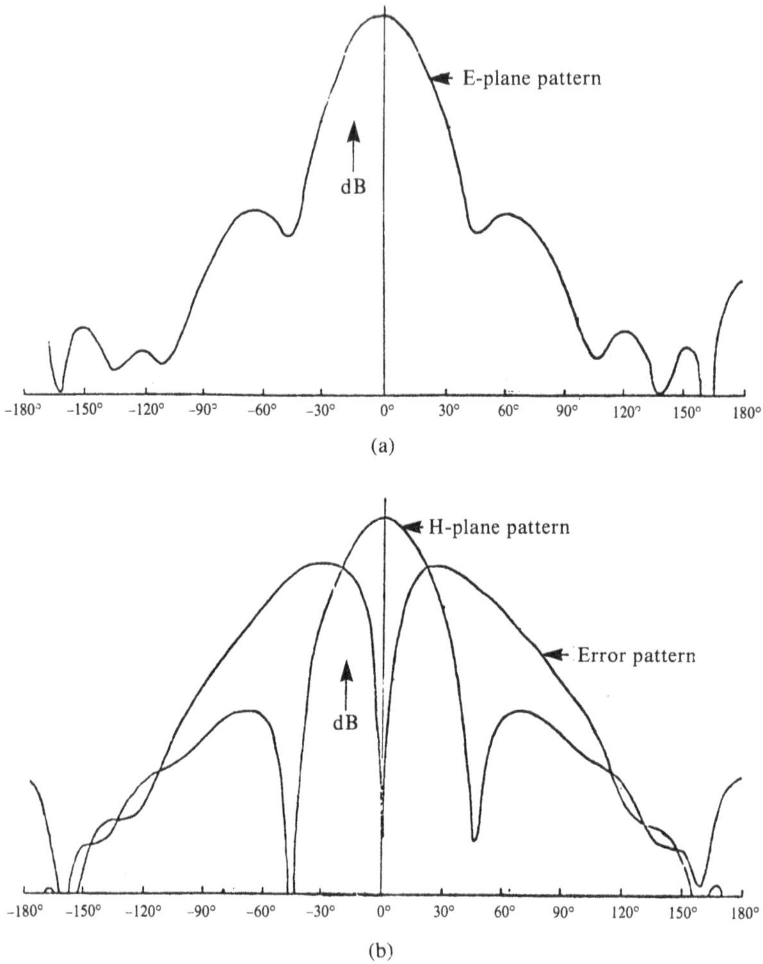

Figure 5.14 (a) *E*-plane and (b) *H*-plane patterns (from [8])

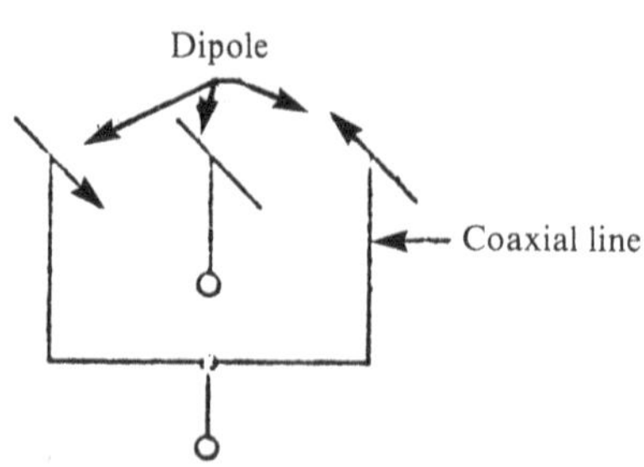

Figure 5.15 Three-dipole exciter (from [8])

(including spacings). A typical balun support structure for the center dipole and small reflector is shown in Figure 5.16. Figure 5.17 shows an alternate feed system, which employs two parallel in-phase dipoles and an intermediate hybrid ring rat-race with equal feed lines to generate both sum and error patterns in the H-plane.

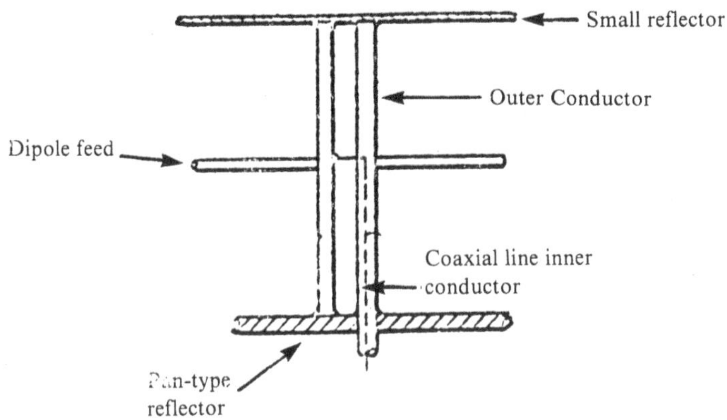

Figure 5.16 A typical balun support structure (from [8])

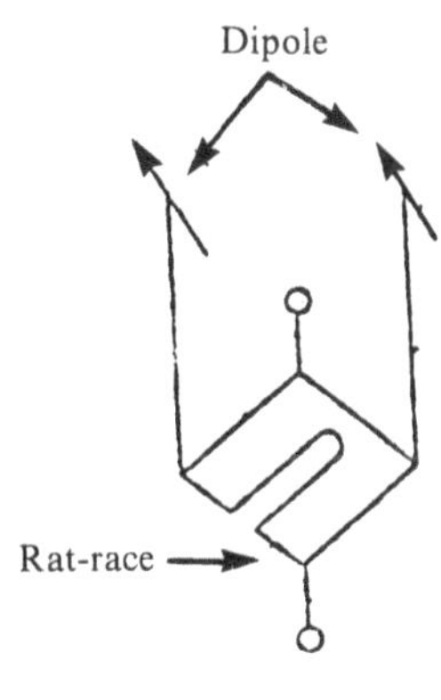

Figure 5.17 Alternate method to feed two-dipole exciter (from [8])

Kooi *et al.* [9] have described a dipole-excited SBFA with corrugated rim at X-band. Figure 5.18 shows the sketch of the dipole-fed SBFA. The measured E-plane and H-plane radiation patterns are shown in Figure 5.19 ((a) and (b)) in the frequency range of 9.5 to 10.5 GHz. Also shown in the figure (for comparison), are the corresponding patterns for the SBFA having no corrugations on the rim. Tables 5.4 and 5.5 show the measured 10 dB beamwidths and directive gain of the antenna with and without corrugations, respectively. The results obtained show that the presence of corrugations can lead to improvement in directive gain, sidelobe levels and

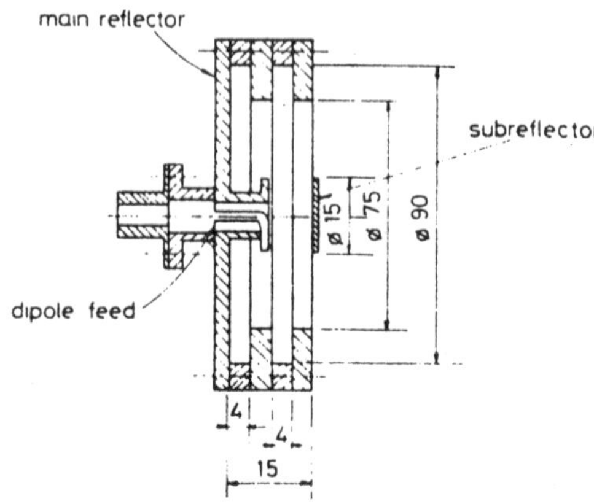

Figure 5.18 Crossed-dipole-fed SBFA with corrugated rim (from [9])

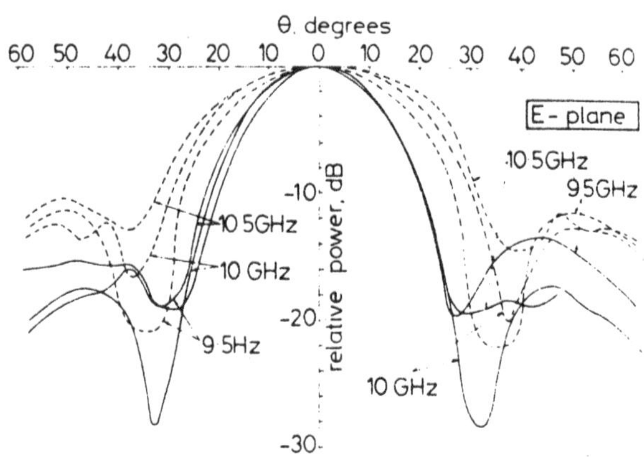

Figure 5.19(a) Measured *E*-plane patterns of crossed-dipole-fed SBFA with (———) and without (-----) corrugations at 9.5, 10.0, and 10.5 GHz (from [9])

impedance matching of the antenna. Ohmori *et al.* [10] described an improved version of the SBFA. Figure 5.20 ((a) and (b)) shows both a conventional and an improved version of the SBFA. Tables 5.6 and 5.7 show the physical dimensions and the electrical properties of both a conventional and an improved SBFA. Figure 5.21 (a) and (b) shows the radiation patterns of both a conventional and an improved SBFA at 1.54 GHz. Figure 5.22 shows the input VSWR of a conventional SBFA with a large flat reflector, an improved SBFA with a large conical reflector (α = 15°), and an improved SBFA with a large conical reflector and a second smaller reflector. Electrical characteristics of a conventional SBFA have been improved by

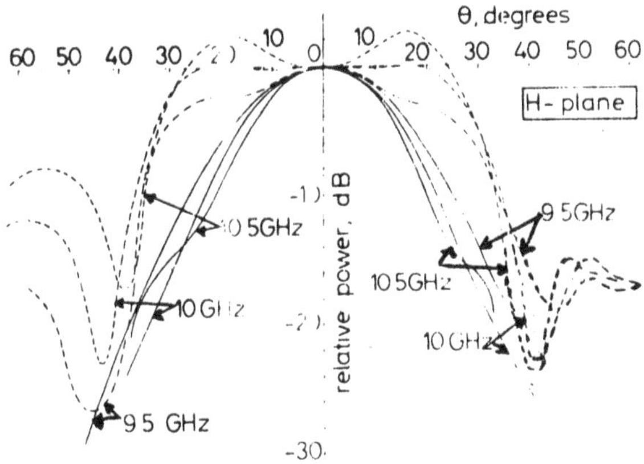

Figure 5.19(b) Measured *H*-plane patterns of crossed-dipole-fed SBFA with (———) and without (------) corrugations at 9.5, 10.0, and 10.5 GHz (from [9])

Table 5.4 Measured 10 dB Beamwidths of Dipole-Excited Antenna With and Without Corrugations

Frequency, GHz	*10 dB beamwidth, degrees*			
	Without corrugations		*With corrugations*	
	E-plane	*H-plane*	*E-plane*	*H-plane*
9.5	53.5	67	42.5	51.5
10.0	59.5	71.5	42.5	45.5
10.5	64	68	40	68

Table 5.5 Measured Directive Gain of Dipole-Excited Antenna With and Without Corrugations

Frequency, GHz	*Directive gain, dB*	
	Without corrugations	*With corrugations*
9.5	11.7	14.9
10.0	11.6	15.4
10.5	10.7	16.2

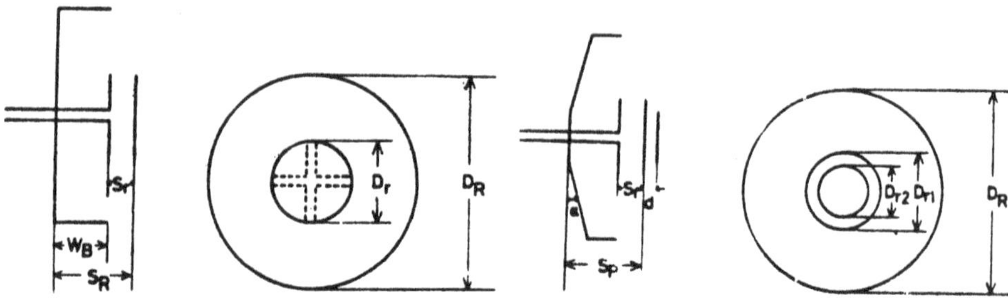

(a) (b)

Figure 5.20 Sketch of SBFAs: (a) conventional and (b) improved version (from [10])

Table 5.6 Physical Dimensions of a Conventional SBFA and an Improved SBFA

	Plane reflector SBF Antenna	*Conical reflector SBF Antenna*
Diameter of a large reflector; D_R (cm)	40 (2.05λ)	40 (2.05λ)
Diameter of a small reflector; D_{r1} (cm)	9 (0.46λ)	9 (0.46λ)
D_{r2} (cm)	—	8 (0.41λ)
Width of a rim; W_B (cm)	4.9 (0.25λ)	4.9 (0.25λ)
Distance between large and small reflectors; S_R (cm)	9.7 (0.5λ)	12.9 (0.66λ)
Distance between an exciter and a small reflector; S_r (cm)	4.9 (0.25λ)	5.7 (0.29λ)
Distance between small and second small reflectors; d (cm)	—	1.8 (0.09λ)
Slanting angle of a large reflector; α (degrees)	0°	15°

(λ; Wave length at 1540MHz)

Table 5.7 Electrical Characteristics of both a Conventional and an Improved SBFA

	Plane reflector SBF Antenna	*Conical reflector SBF Antenna*
Effective gain (dB)	14.3	15.0
Half power beam width (degrees)	34*	34*
Directive gain (dB)	14.8	15.5
1st sidelobe level (dB)	−21.0	−22.5

Table 5.7 (cont'd)

	Plane reflector SBF Antenna	*Conical reflector SBF Antenna*
Axial ratio (dB)	1.28	1.1
Aperture efficiency with respect to		
effective gain	65%	76%
directive gain	73%	85%
Frequency bandwidth for VSWR under 1.5	3%	9%
		20% (with a second small reflector)

adopting a large conical reflector and a second small reflector. Directive gain has also been increased (by 10 to 15%) from a conventional SBFA. There is an improvement in the VSWR of the SBFA with a conical reflector and second small reflector.

Katulski [11] reported on the optimization of crossed dipoles as a radiating source in the SBFA. Table 5.8 shows the lengths of the dipoles reported here. The polarization ratios of the antenna for all considered cases of the crossed dipoles are shown in Figure 5.23(a). The experimental results of the SBFA VSWR *versus* frequency as a function of dipole lengths are shown in Figure 5.23(b). The VSWR characteristic can be improved using a step transformer, as discussed in Chapter 4.

Keen [11a] has shown that if the quadrature hybrid in Figure 5.23(c) is perfect and the feeder cables are exactly the same length, the polarization response in the antenna boresight direction is perfectly circular. However, if the hybrid does not provide exactly equal amplitude with an exactly 90° phase difference at its output ports, and if there is an additional phase error in the coaxial feeder cables between the hybrid and the antenna, elliptical, rather than circular polarization results. Figure 5.23(d) shows a useful set of curves for axial ratio at the boresight of the antenna as a function of amplitude and phase errors in hybrid and feeder.

5.2 EXPERIMENTAL INVESTIGATIONS ON SBFAs FED BY A WAVEGUIDE

SBFAs excited by simple open-ended waveguides with rectangular or square cross sections were reported by Hristov *et al.* [12–14]. The size of the aperture is approximately 2.6 λ_0^2. It consists of two plane rectangular or square reflectors and a

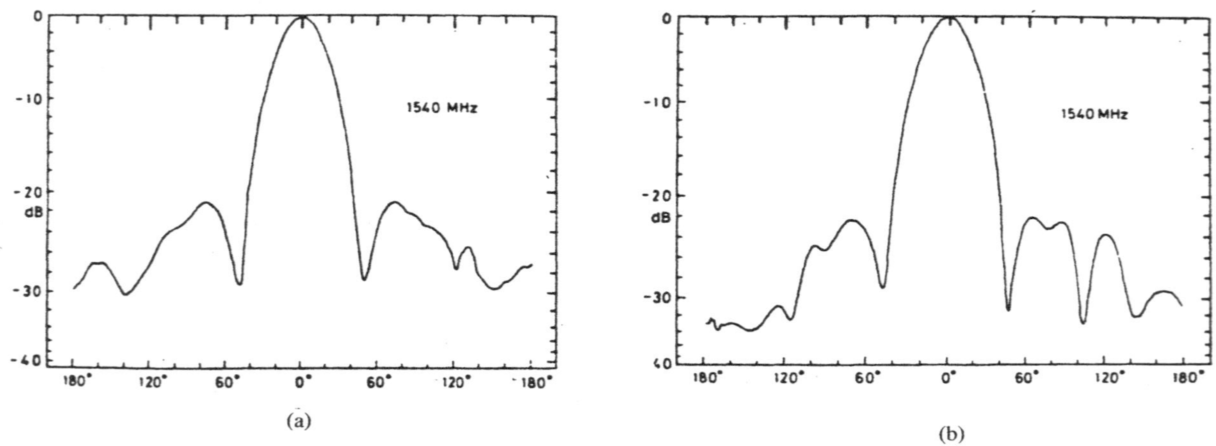

Figure 5.21 Radiation patterns of SBFAs at 1.54 GHz: (a) conventional and (b) improved (from [10])

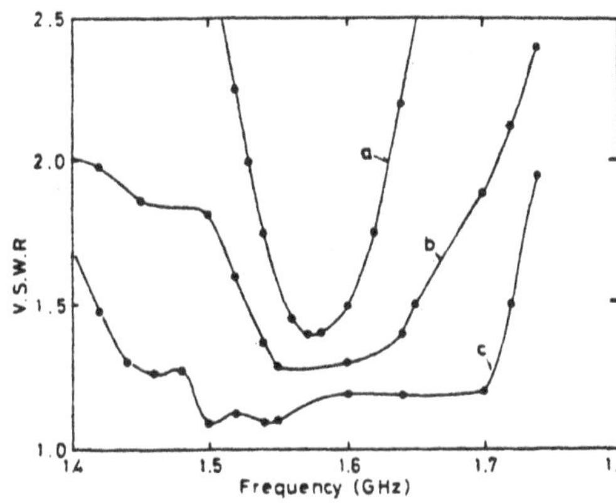

Figure 5.22 VSWR of three types of SBFAs: (a) Conventional with large flat reflector, (b) improved with large conical reflector and (c) improved with large conical reflector and second small reflector (from [10])

Table 5.8 Lengths of the Investigated Dipoles

i (Number of Dipoles, Pair)	l_{xi} *(mm)*	l_{yi} *(mm)*
1	82	46
2	84	48
3	86	49.5
5	88	51

dielectric structure as a rectangular waveguide prolongation. This antenna construction was compared with those of ordinary dielectric and horn antennas, which are more than twice the length of the BFA. Simple expressions for the far-field pattern were derived in terms of the near-field amplitude distribution. A detailed discussion of the antenna is presented in Chapter 6.

The use of a waveguide feed has several important advantages over a dipole feed design, particularly at *C*-band and *X*-band: high-power operation; simple construction; low feed system loss; improved impedance bandwidth over a dipole feed; ease of design at high frequency; elimination of the high mechanical tolerance of the dipole; ease of installing weather and environmental sealing devices into the

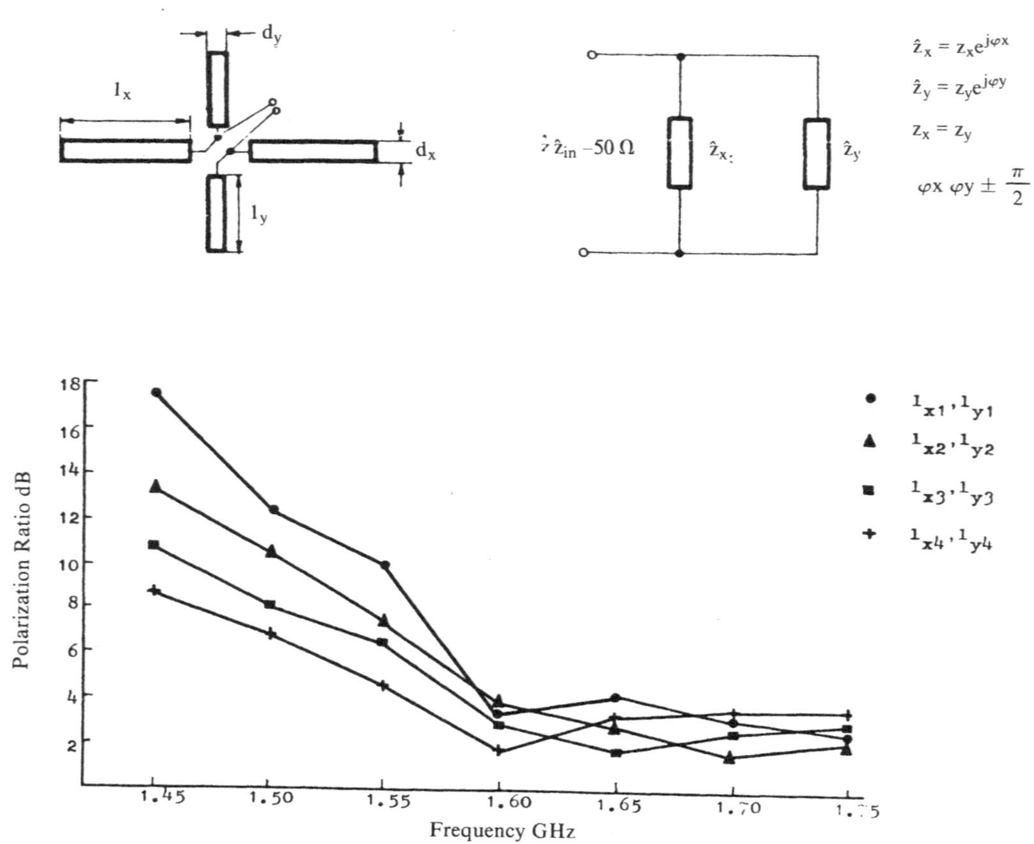

Figure 5.23(a) Polarization ratio *versus* frequency as a function of dipole lengths (from [11])

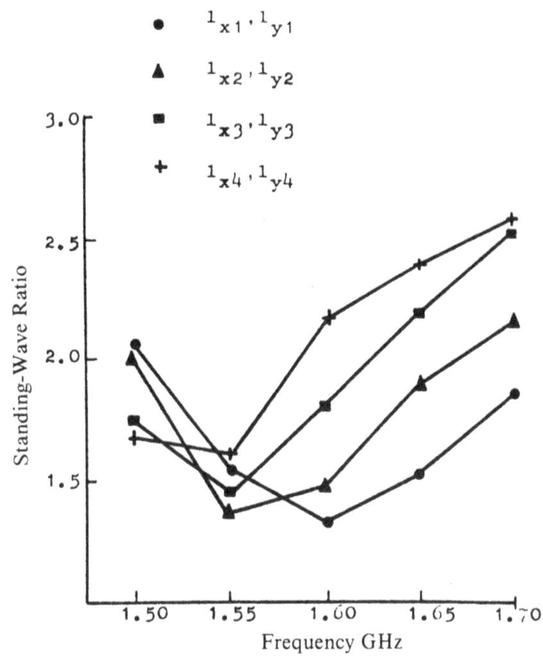

(b)

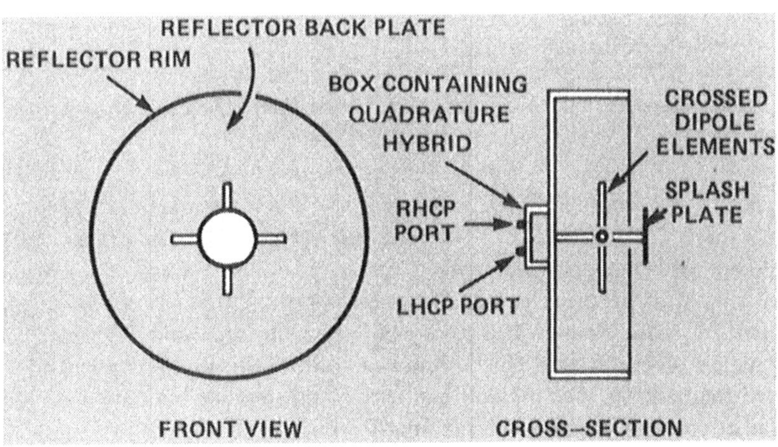

(c)

Figure 5.23(b) VSWR *versus* frequency of the SBFA as a function of dipole lengths (from [11])
(c) SBFA fed with a quadrature hybrid for circular polarization (from [11a])

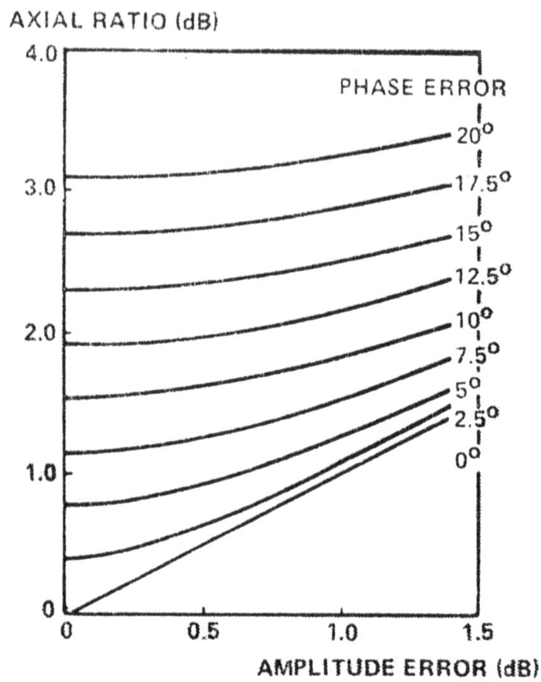

Figure 5.23(d) Antenna boresight axial ratio as a function of amplitude error and phase error of hybrid and feeder lines (from [11a])

antenna; better reliability than dipoles; and better than dipoles in thermal and structural properties.

Large [15–18] made experimental investigations of SBFAs with waveguide and linear fields. Two versions of the antenna using a rectangular waveguide feed are shown in Figure 5.24 (a) and (b). The simplest version of the antenna consists of a large reflector without a rim but with a centrally-mounted flush feed. A small reflector disk, spaced from the center of the large reflector by approximately 0.5 λ_0, is supported by foam plastic. The cavity between the large and the small plate can be filled with a block of rigid plastic foam or solid dielectric material of low loss at microwave frequencies. The dielectric block is held in place by a low-loss adhesive, and can also act as a support for the small reflector, as shown in Figure 5.25(a). Figure 5.25(b) shows a method of providing weather protection for the antenna. A thermoplastic or reinforced plastic molding is attached to the rim, providing a cover over the antenna. This type of weather protection radome can also act as an air pressure seal if the waveguide needs pressurization for use at high microwave powers.

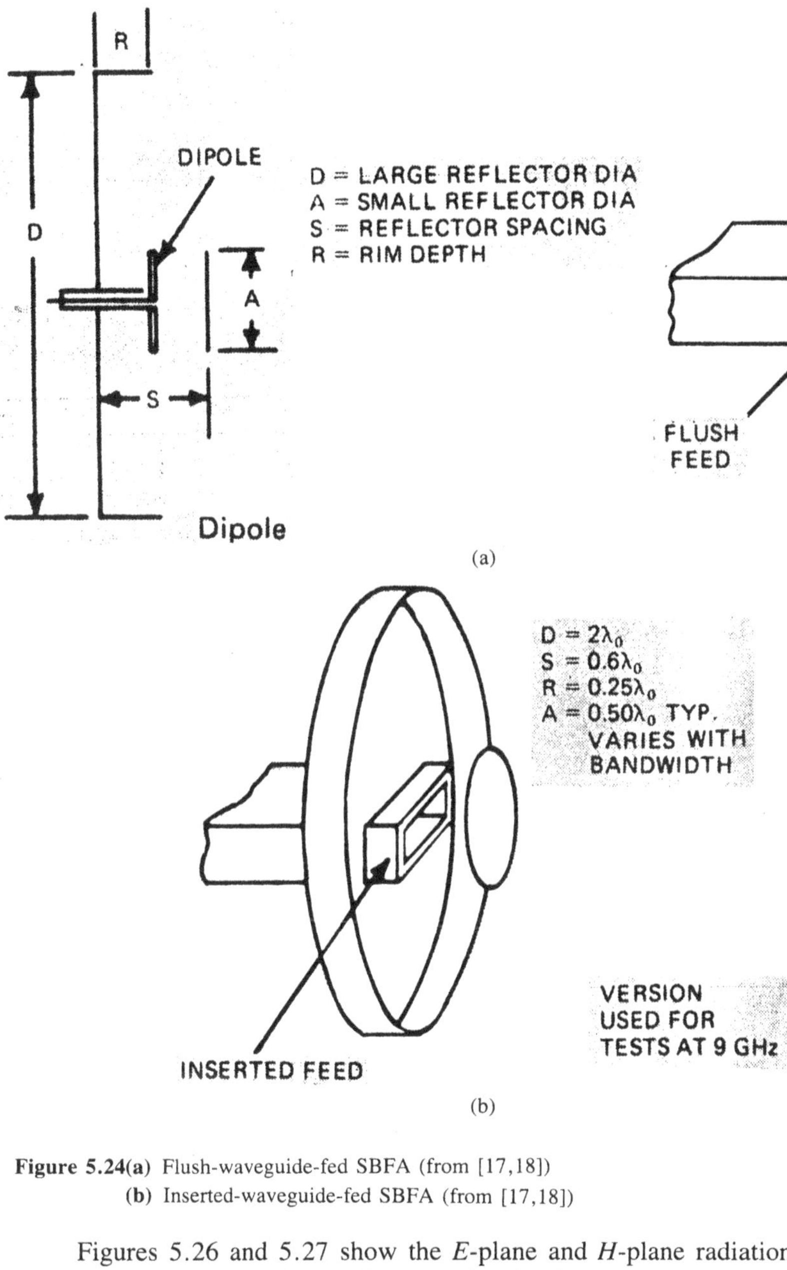

Figure 5.24(a) Flush-waveguide-fed SBFA (from [17,18])
(b) Inserted-waveguide-fed SBFA (from [17,18])

Figures 5.26 and 5.27 show the *E*-plane and *H*-plane radiation properties of flush-feed and inserted-feed SBFAs at *X*-band. These measurements were performed for circular small reflector plates of $0.5\lambda_0$ in diameter, because radiation patterns

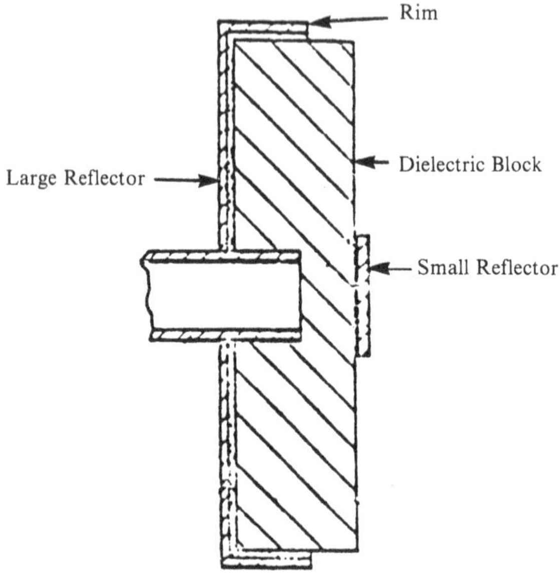

(a)

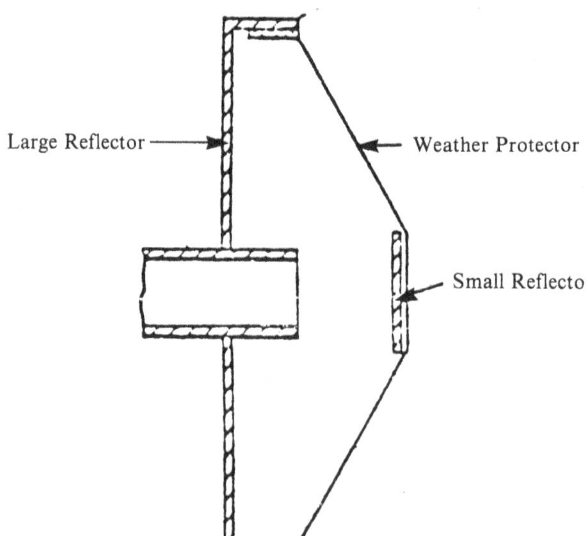

(b)

Figure 5.25(a) SBFA with low-loss dielectric material (from [17])
(b) SBFA with weather protection radome (from [17])

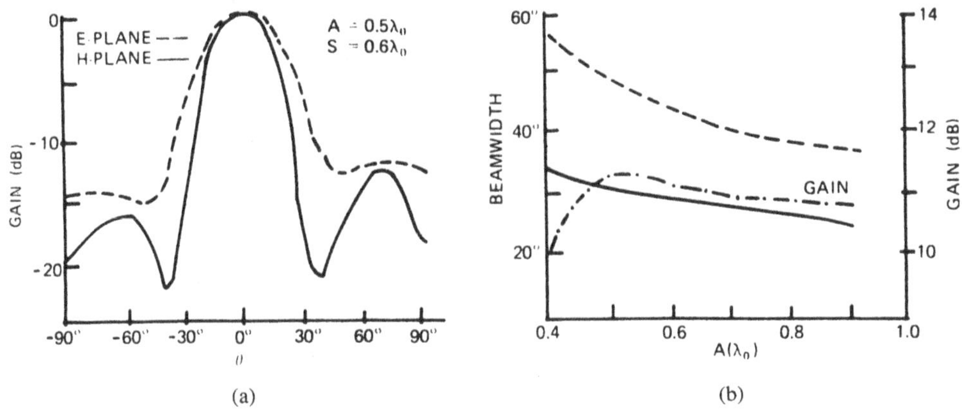

Figure 5.26 Radiation properties of flush-feed waveguide-fed SBFA (from [18]): (a) *E*-plane and *H*-plane radiation patterns; (b) Plots of 3 dB beamwidth and gain for various small reflectors (3 dB beamwidth *E*-plane ----, *H*-plane ———, Gain -·-·-·-·).

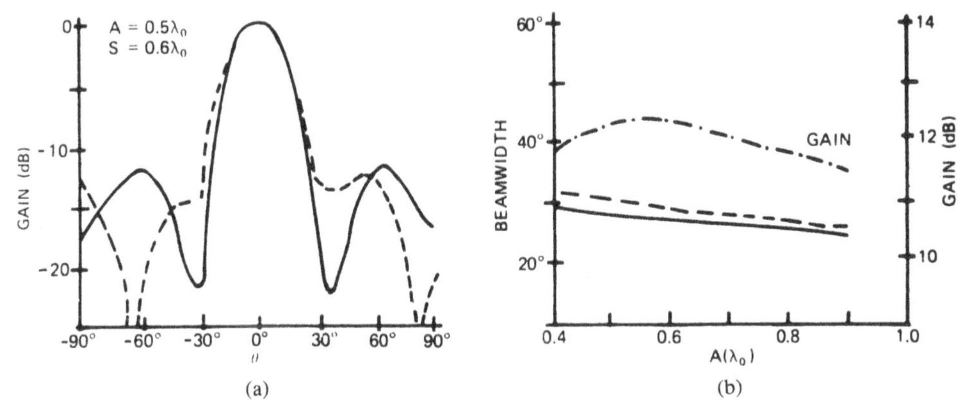

Figure 5.27 Radiation properties of inserted-feed waveguide-fed (from SBFA [18]): (a) *E*-plane and *H*-plane patterns; (b) Plot of 3 dB beamwidth and gain for various small reflectors (*E*-plane ----, *H*-plane ———, Gain -·-·-·-·-·).

and gain were measured for reflector plates of $0.4\lambda_0$ to $0.9\lambda_0$ in diameter, and it was found that a plate of $0.5\lambda_0$ produced an optimum radiation pattern. A summary of measured properties are shown in Table 5.9.

Figure 5.28 shows an SBFA with a small rectangular reflector. Measured radiation patterns of the antenna (for an optimized small rectangular reflector of size 0.4 by $0.9\lambda_0$) is shown in Figure 5.29. Figure 5.30 shows the plot of VSWR *versus*

Table 5.9 Summary of Measured Radiation Properties

Beam *Width* (3dB)	*Flush* *Feed*	*Inserted* *Feed*
E-plane	36°	29°
H-plane	32.5°	31°
Average side lobes (dB)		
E-plane	−15	−16
H-plane	>−30	>−30
Gain (dB)	12.5	13.5

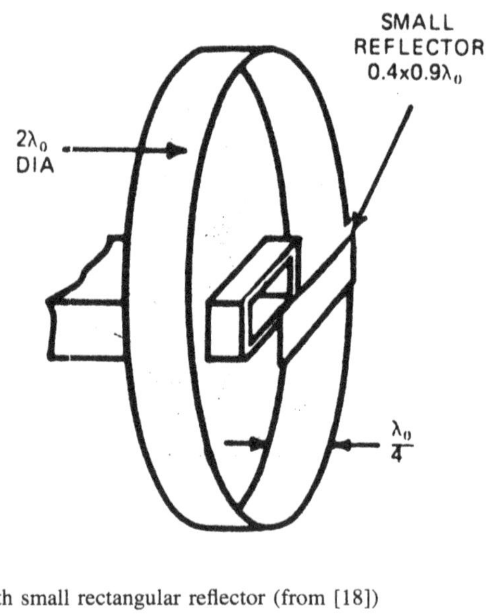

Figure 5.28 Antenna with small rectangular reflector (from [18])

frequency of flush-feed and inserted-feed waveguide-fed SBFAs with a small reflector of 0.4 by 0.9 λ_0. The VSWR for both feed positions is less than 2 over a 10% band, and less than 1.5 over an 8% band centered on 9 GHz, for flush feed. The matching can be improved by using irises with the waveguide aperture.

Leong and Kooi [19] have shown that the corrugated rim on the large reflector influences the directive gain, sidelobe level, and input impedance in the vicinity of the design frequency of an SBFA. Figure 5.31 shows a rectangular-waveguide-excited SBFA with a corrugated rim at 10 GHz. The measured *E*-plane and *H*-plane patterns (with and without corrugations on the rim) are shown in Figure 5.32. These

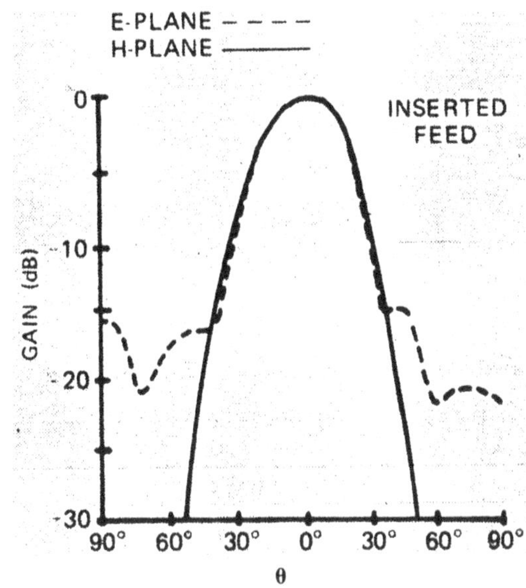

Figure 5.29 Radiation patterns of the SBFA with small rectangular reflector (from [17,18])

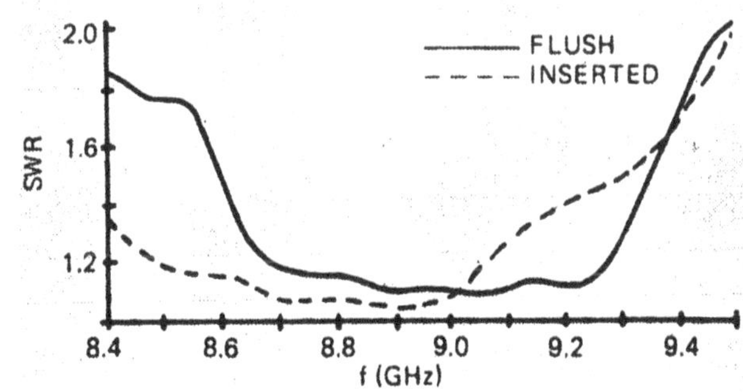

Figure 5.30 Plot of VSWR *versus* frequency (from [18])

measurements were taken at 10.0, 10.5 and 11.0 GHz. Table 5.10 summarizes the measured 10 dB beamwidths and sidelobe level of the antenna with and without corrugations. Table 5.11 shows the measured directive gain of a rectangular-waveguide-fed SBFA with and without corrugations.

The input impedance measured at the feeder waveguide aperture plane as a function of frequency is shown in Figure 5.33 for the 1.5-cm diameter small reflector. The presence of corrugations improves the VSWR from 1.6 to 1.1 at the

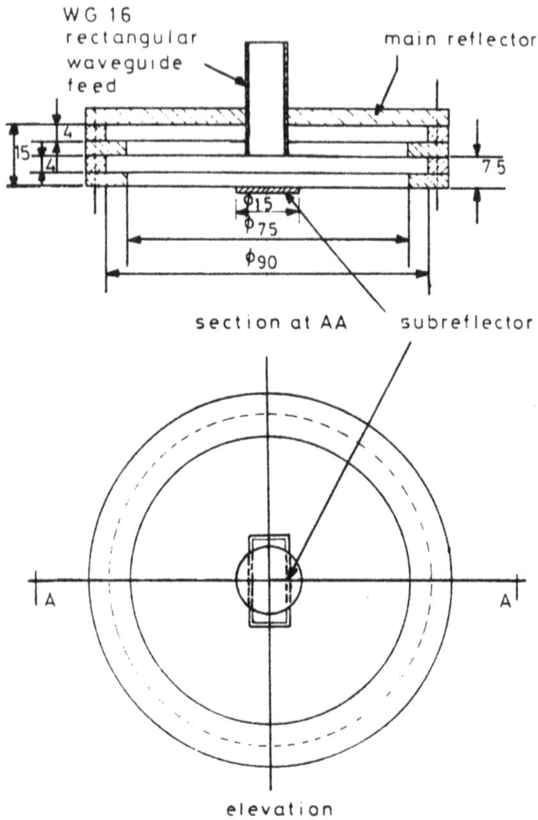

Figure 5.31 Rectangular-waveguide-excited SBFA with corrugated rim (from [19])

design frequency of 10 GHz. The rectangular-waveguide-fed SBFA with a corrugated rim produces a directive gain of 20 dB, a sidelobe level of −17 dB, and a VSWR in the range of 1.1 to 2 at *X*-band.

Kooi and Leong [20] and Leong and Kooi [62] described experimental works at *Q*-band for waveguide-fed SBFAs. Figures 5.34 and 5.35 show the dimensions of antennas with a corrugated rim and a dielectric-lined rim, respectively. In Figure 5.34, the large reflector has a diameter D of 4 λ_0, the small reflector has a diameter of 0.5 λ_0, and the rim has two corrugations per half-wavelength of 34 GHz. The aperture of the rectangular waveguide feed is at a distance of 0.25 λ_0 from the small reflector.

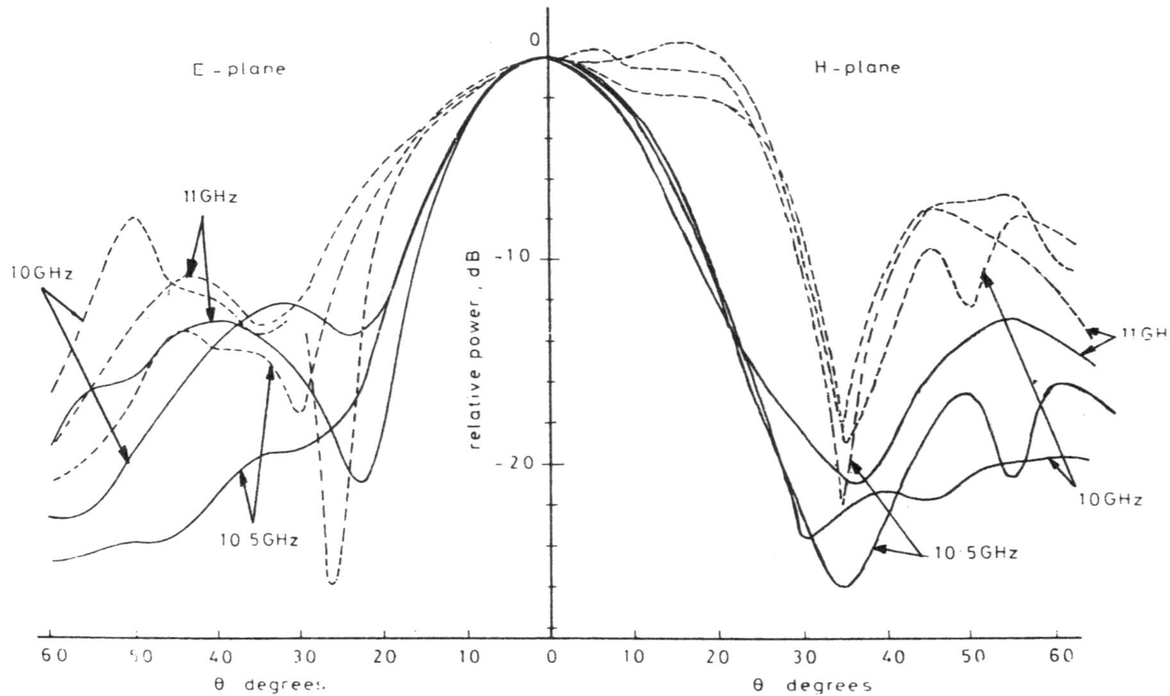

Figure 5.32 Measured E-plane and H-plane patterns with and without corrugations at 10.0, 10.5 and 11.0 GHz (from [19]): with corrugations ———, without corrugations ------

Table 5.10 Measured 10 dB Beamwidths and 1st Sidelobe Level of Rectangular-Waveguide-Excited SBFA With and Without Corrugations (from [19])

Frequency GHz	10 dB Beamwidth, Degrees			
	Without Corrugations		*With Corrugations*	
	E-plane	*H-plane*	*E-plane*	*H-plane*
10.0	42	62	35	39
10.5	50	59	35	38
11.0	55	60	31	36

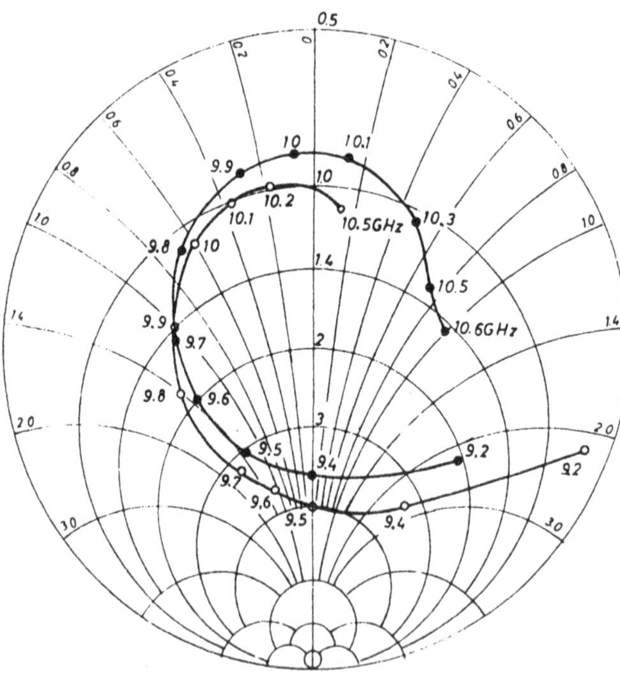

Figure 5.33 Measured input impedance of rectangular-waveguide-fed SBFA (from [19]): without corrugations ●-●-●, with corrugations ○-○-○

Figure 5.35 shows the large reflector rim with a diameter of 3 λ_0, and a small reflector diameter of 0.5 λ_0. The rim is lined with a 1 mm thick uniform layer of Emerson & Cuming Eccosil 4122. Eccosil 4122 [21] is a white, low-viscosity silicon

Table 5.11 Measured Directive Gain of Rectangular-Waveguide-Fed SBFA With and Without Corrugations (from [19])

Frequency GHz	*Directive Gain, dB*	
	Without Corrugations	*With Corrugations*
10.0	15.4	20.2
10.5	15.0	20.1
11.0	14.7	20.8

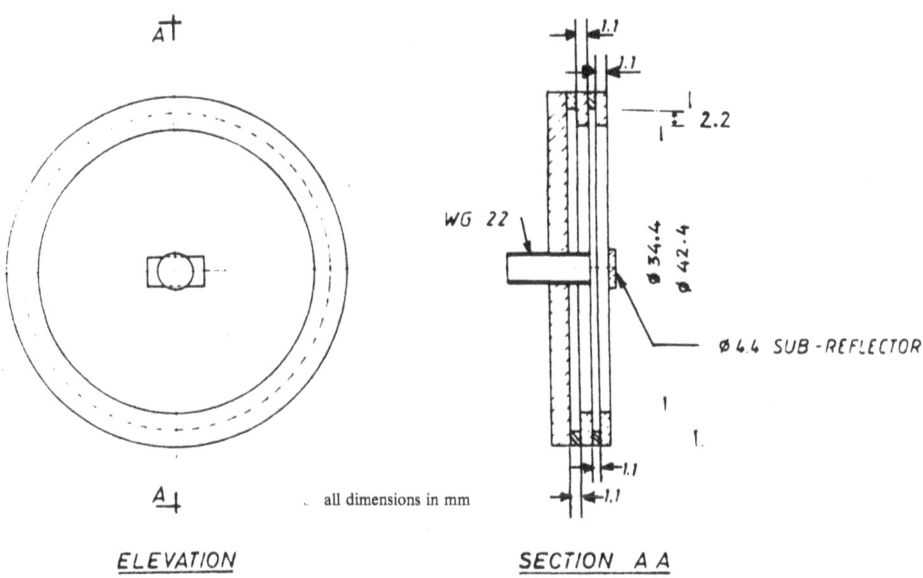

Figure 5.34 Sketch of an SBFA with a corrugated rim on the large reflector (from [20])

resin. Table 5.12 shows typical properties of the material. Eccosil 4122 cures at room temperature when a small amount of Emerson & Cuming Catalyst 50 is added to it. Complete cure at room temperature requires five to six days, but strength properties will reach about 80% of their ultimate value in 24 hours. To speed cure, temperatures up to about 66° C may be used.

In both cases, the small reflector is positioned by means of a low permittivity dielectric slab inserted into the feed waveguide as suggested by Hristov, *et al.* [13] and Large [15–18]. Recently, Kumar [22] has suggested use of rohacell foam, which

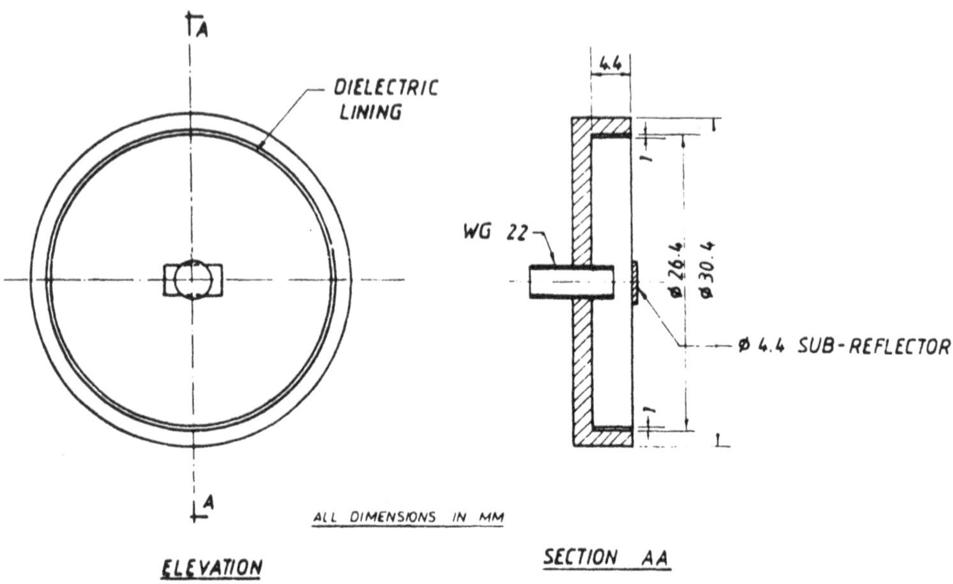

Figure 5.35 Sketch of an SBFA with dielectric-lined rim on the large reflector (from [62])

Table 5.12 Typical Properties of Emerson & Cuming Eccosil 4122 (from [21])

In the Uncured State		*In the Cured State*	
Color	White	Specific Gravity	1.15
Viscosity at 25° C	10,000 cps	Tensile Strength, psi (kg/cm^2)	400 (28)
Shelf Life	6 months	Tear Resistance, lbs/in (kg/cm)	18 (3.2)
		Hardness, Shore A	45
		Elongation, %	180
		Dielectric Constant, at 1 MHz	3.1
		Loss Tangent at 1 MHz	0.004
		Dielectric Strength, V/mil (kV/mm)	500 (19.7)

has good electrical and mechanical properties for space applications. Figure 5.36 shows the measured E-plane and H-plane patterns at 34 and 35 GHz for the SBFA, with and without corrugations on the rim of the large reflector. There is an improvement of approximately 5 dB in gain and 4 dB in sidelobe level due to corrugations on the rim.

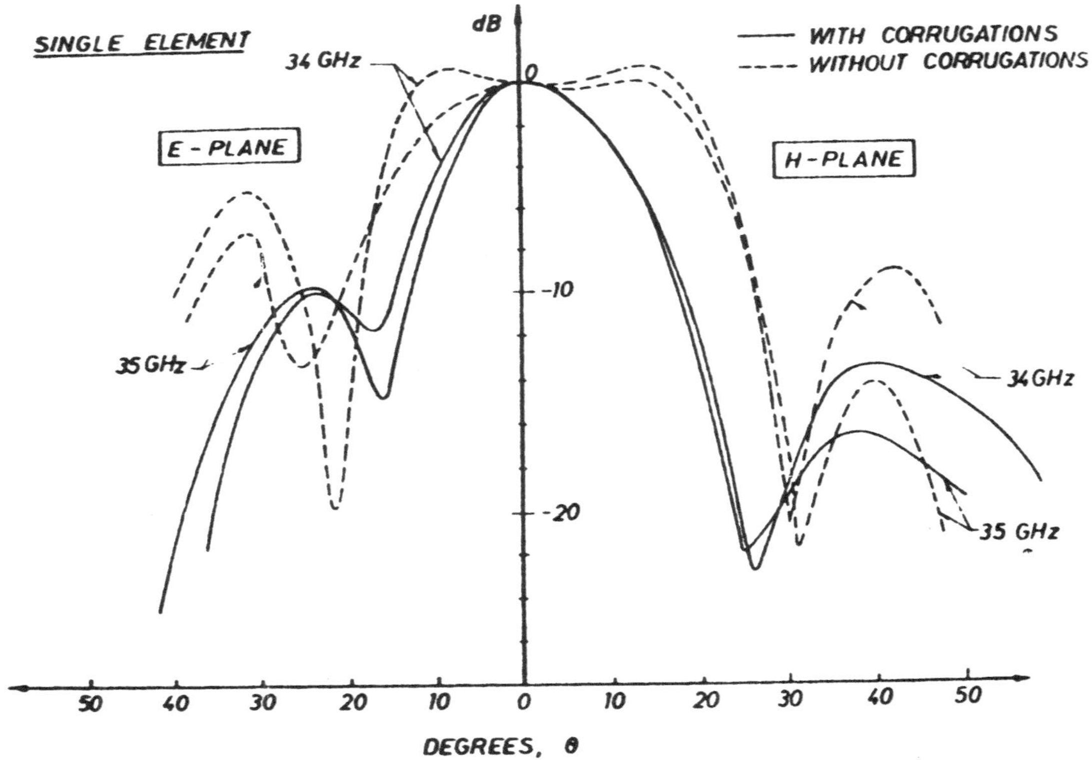

Figure 5.36 Measured E-plane and H-plane patterns of the SBFA, with and without corrugations on the rim of the large reflector at 34 and 35 GHz (from [20])

The measured E-plane and H-plane patterns at 33 and 34 GHz for the SBFA element, with and without a dielectric-lined rim on the large reflector, are shown in Figure 5.37. There is an improvement of approximately 3.5 dB in gain and a suppression of approximately 7 dB in the sidelobes at 33 GHz, due to the dielectric lining on the rim. The choice of the dielectric instead of quarter-wavelength corrugations can offer a less expensive alternative; however, corrugations on the rim produce a 1.5 dB higher gain than does the dielectric lining.

Leong and Kooi [23–24] used a conical large reflector excited by a rectangular waveguide. This type of BFA was discussed by Hristov and Kumar [25], and a similar idea was investigated by Ohmori, *et al.* [10], which was discussed in Section 5.1. Figure 5.38 shows an antenna fed by a rectangular waveguide. The conical section is constructed in the form of an attachment to the large reflector plate. Conical sections with slant angles θ = 5, 10, 15 and 20° have been constructed to determine the optimum radiation properties of the SBFA as a function of slant angle.

Figure 5.39 shows the measured E-plane and H-plane patterns at 10 GHz of the SBFA as a function of slant angle θ of the conical large reflector. The H-plane beamwidth is virtually unaffected by the slant angle of the conical large reflector of the antenna. However, there is an optimum angle θ = 10 to 15°, beyond which beamwidths become wider and the sidelobe levels rise, resulting in reduced gain. Figure 5.40 (a) and (b) shows the representative radiation patterns of SBFAs with a conical large reflector of 15 degrees slant angle, and with a large plane reflector in the frequency range 9 to 11 GHz. The measured 3 dB and 10 dB beamwidths, the first sidelobe level, and functions of frequency are summarized in Table 5.13.

Table 5.14 shows the calculated directive gain of the SBFA with conical (θ = 15°) and plane (θ = 0°) large reflector at 9, 10, and 11 GHz. A directive gain of 16.7 to 18.3 dB is shown for the antenna over a 20% bandwidth. The VSWR *versus* frequency curves are plotted (as a function of large reflector slant angle) in Figure 5.41. The VSWR is better than 2.5 for large reflector slant angle θ = 10–15° over the frequency range of 9 to 11 GHz. The use of a conical large reflector improves bandwidth and impedance matching, and narrows the beamwidth of the SBFA, resulting in a higher directivity.

5.3 THEORETICAL STUDIES OF SHORT BACKFIRE ANTENNAS

A theoretical analysis of the SBFA was developed by Chen, *et al.* [26] in 1968. They proposed an approximate method for calculating the radiation pattern of the antenna from an assumed cosinusoidal aperture field distribution, which is deduced from the results of the near-zone field measurements. The configuration of the SBFA is shown in Figure 5.42(a) and a detailed description of it is given in Section 5.1.

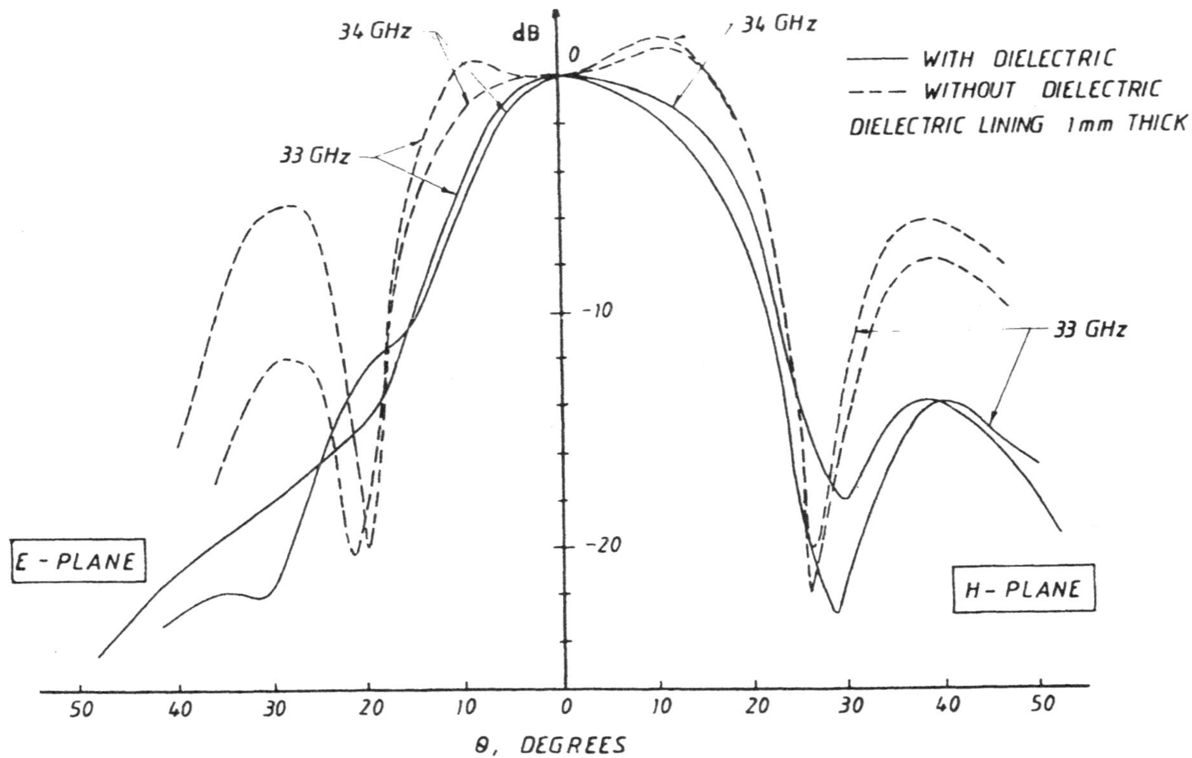

Figure 5.37 Measured E-plane and H-plane patterns of the SBFA, with and without dielectric lining on the rim of the large reflector (from [62])

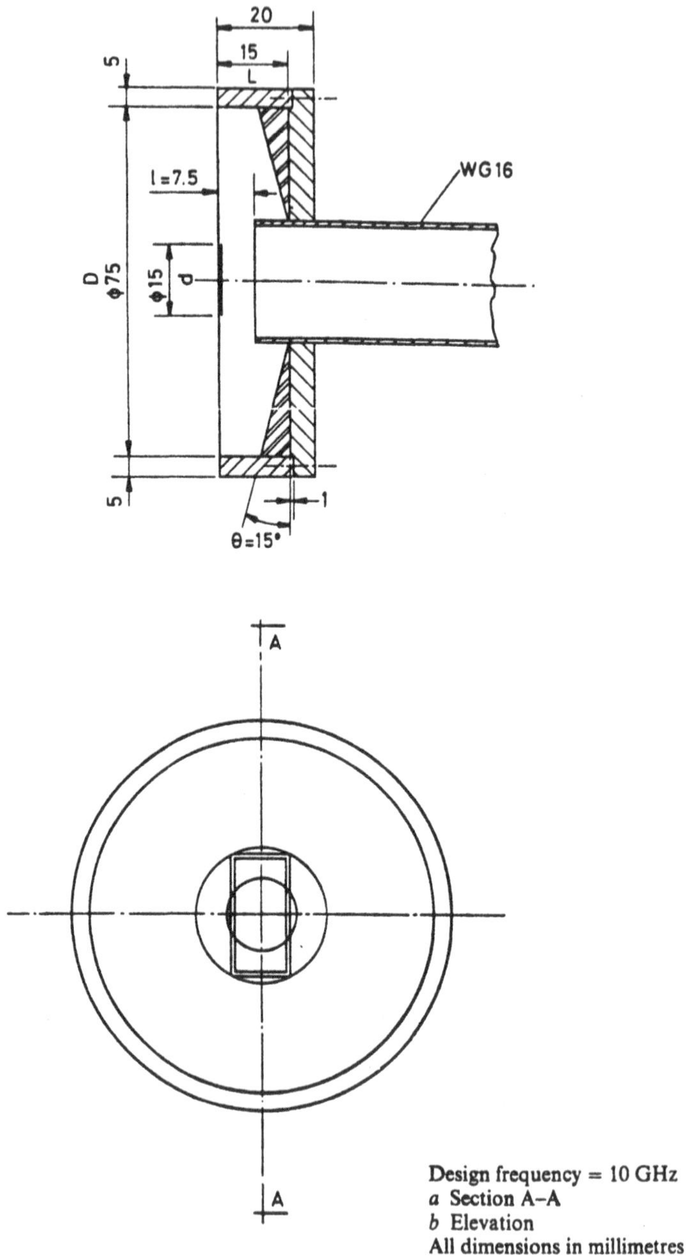

Figure 5.38 SBFA fed by a rectangular waveguide, with conical large reflector (from [23,24])

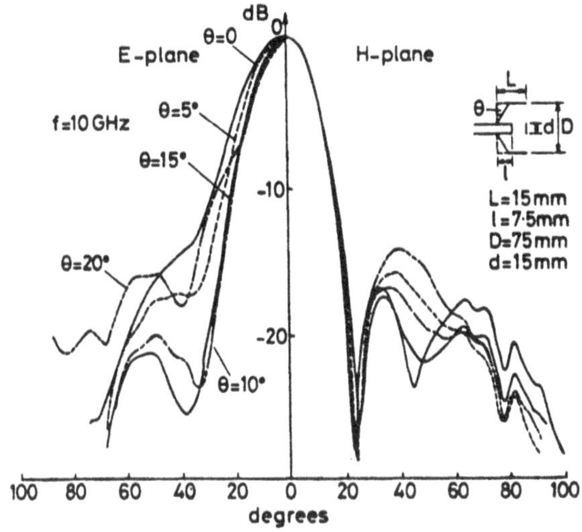

Figure 5.39 Measured *E*-plane and *H*-plane patterns of rectangular waveguide-fed SBFA at 10 GHz, with conical large reflector of slant angles θ = 0, 5, 10, 15 and 20° (from [23,24])

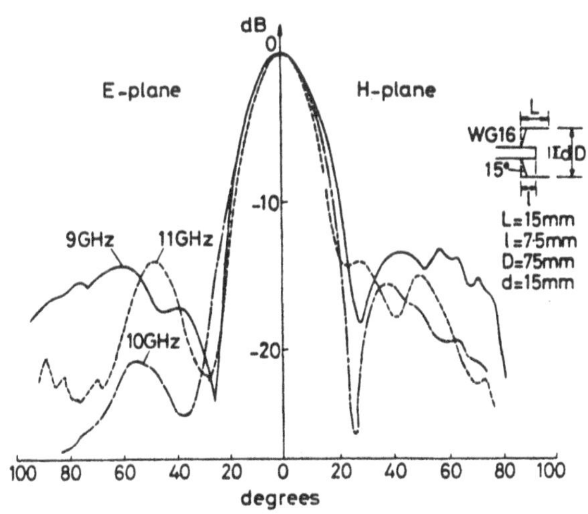

Figure 5.40(a) Measured *E*-plane and *H*-plane patterns of rectangular-waveguide-fed SBFA, with conical large reflector (θ = 15°) (from [23])

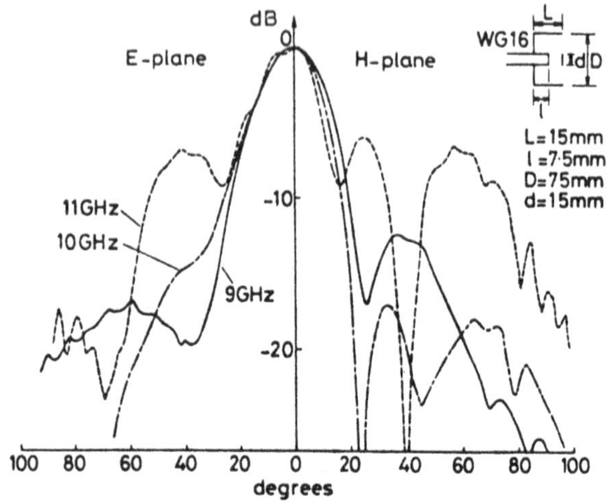

Figure 5.40(b) Measured E-plane and H-plane patterns of rectangular-waveguide-fed SBFA with large plane reflector ($\theta = 0°$) (from [23])

Table 5.13 Measured 3 and 10 dB Beamwidths of Rectangular Waveguide-Excited Short Backfire Antenna with Plane ($\theta = 0°$) and Conical ($\theta = 15°$) Main Reflectors (from [23])

	Plane Main Reflector (θ = 0°)						*Conical Main Reflector (θ = 15°)*					
	E-Plane			*H-Plane*			*E-Plane*			*H-Plane*		
	Beamwidth (Deg.)		*1st Sidelobe, dB*	*Beamwidth (Deg.)*		*1st Sidelobe, dB*	*Beamwidth (Deg.)*		*1st Sidelobe, dB*	*Beamwidth (Deg.)*		*1st Sidelobe, dB*
Frequency, GHz	*3 dB*	*10 dB*		*3 dB*	*10 dB*		*3 dB*	*10 dB*		*3 dB*	*10 dB*	
9	25.8	47.1	−16.2	22.2	40	−11.6	22.2	37.3	−13.8	24	40.9	−13.5
10	25.8	52.4	—	20.4	33.8	−16.5	23.1	40.9	−20	21.3	36.4	−15.4
11	24	—	−6.7	16.9	—	−5.7	19.6	35.6	−13.8	18.7	34.7	−13.8

D = 75 mm; d = 15 mm; L = 15 mm; l = 7.5 mm

The measured and theoretical (consinusoidal) field distributions in $Y = 0$ (the plane of aperture) are shown in Figure 5.42(b). The radiation fields in the E-plane ($\theta = \pi/2$) are calculated by the expression:

$$E_{\phi}^{r} = \frac{j4E_0\lambda_0}{\pi}\frac{e^{-jk_0 r}}{r} I_E(A,\phi) \tag{5.2}$$

Table 5.14 Directive Gain as a Function of Frequency, for SBFA with Plane ($\theta = 0°$) and Conical ($\theta = 15°$) Main Reflectors (from [23])

	Directive Gain, dB	
Frequency, GHz	*Plane Main Reflector ($\theta = 0°$)*	*Conical Main Reflector ($\theta = 15°$)*
9.0	16.4	16.7
10.0	16.8	17.1
11.0	17.9	18.3

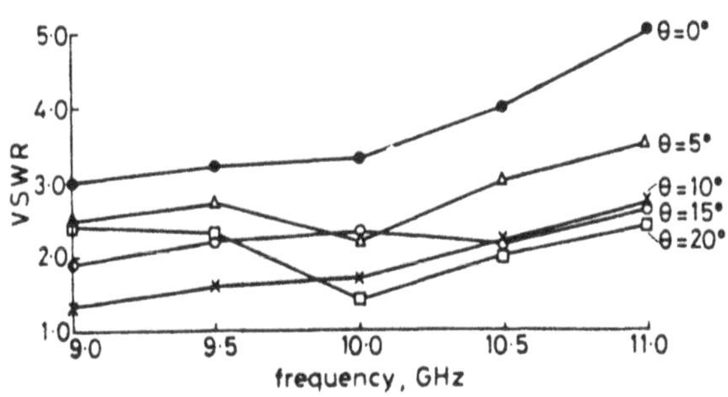

Figure 5.41 Measured VSWR *versus* frequency of rectangular-waveguide-fed SBFA with conical large reflector of slant angle θ = 0, 5, 10, 15, and 20° (from [23]): $\theta = 0°$ ●—●—●—●—●; $\theta = 5°$ △—△—△—△—△; $\theta = 10°$ ○—○—○—○—○; $\theta = 15°$ ★—★—★—★—★; $\theta = 20°$ □—□—□—□

The radiation fields in the H-plane ($\phi = 0$) are given by:

$$E_\phi^r = \frac{j4E_0\lambda_0}{\pi}\frac{e^{-jk_0r}}{r}I_H(A,\theta) \tag{5.3}$$

where

λ_0 = free-space wavelength,
k_0 = the wave number,
r = the observation point in far-field,
a = the aperture radius of the SBFA, and
A = a/λ_0, the aperture radius in wavelength.

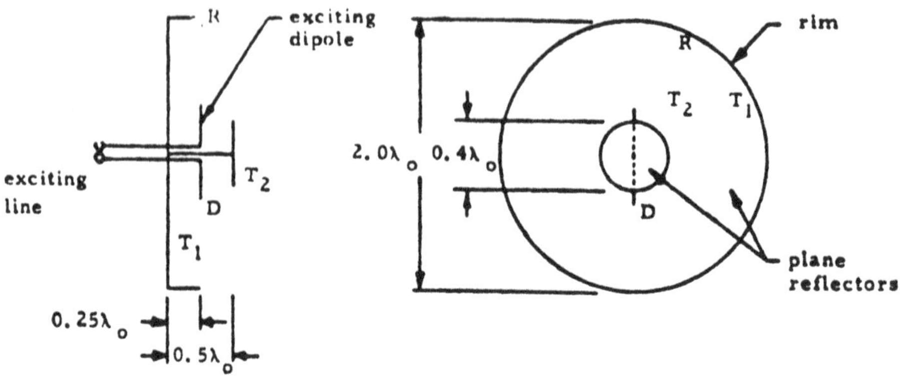

Figure 5.42(a) Configuration of the SBFA (from [26])

$E(Z)$, E_0 and Z are shown in Figure 5.42 (b), functions $I_E(A,\phi)$ and $I_H(A,\theta)$ are given by:

$$I_E(A,\phi) = (1 + \cos\phi)\int_0^A \frac{\sqrt{A^2 - Z^2}}{1 - (4\sin\phi\sqrt{A^2 - Z^2})^2} \cdot \cos\frac{\pi Z}{2A}\cos(2\pi\sin\phi\sqrt{A^2 - Z^2})\,dZ \tag{5.4}$$

and

$$I_H(A,\theta) = (1 + \sin\theta)\int_0^A \sqrt{A^2 - Z^2} \cdot \cos\frac{\pi Z}{2A}\cos(2\pi Z\cos\theta)\,dZ \tag{5.5}$$

Chen, Nyquist, and Lin [26] plotted two sets of patterns for $a/\lambda_0 = 1.0$ and $a/\lambda_0 = 1.1$; these patterns are shown in Figures 5.43(a) (E-plane) and 5.43(b) (H-plane). The case of $a/\lambda_0 = 1.1$ shows good agreement between theoretical and experimental E-plane and H-plane patterns for major lobe shapes, which are nearly identical up to 3 dB beamwidth.

In 1970, Loh and Leung [27] derived a mathematical expression for the radiated field of a rimless SBFA by assuming that images produced by two plane reflectors are infinite in extent. Their model of a rimless SBFA, shown in Figure 5.44, essentially consists of two plane reflectors (L and S), of unequal apertures,

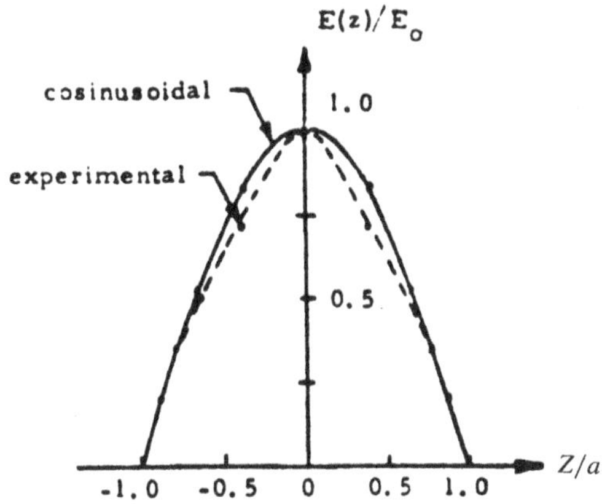

Figure 5.42(b) Theoretical (cosinusoidal) and experimental field at the aperture of an SBFA (from [26])

arranged parallel to each other. A dipole feed A_0 is placed midway between the reflectors, and transverse to the axis of the reflectors. Such a system is characterized by multiple images produced by reflections between the two conducting planes. Figure 5.45 shows images $A_1, A_2, A_3, \ldots A_n$ and $A_1', A_2', A_3', \ldots, A_n'$ of dipole feed A_0. The presence of the large reflector prevents radiation behind it, but the small reflector allows radiation beyond itself due to its comparatively small size; therefore, only images $A_1, A_2, A_3, \ldots A_n$ are considered in this analysis.

Loh and Leung [27] obtained the approximate expression for the electric field intensity in the far-field of the SBFA as follows:

$$E_{N\theta} = \frac{jZ_0 I_m \exp(-jkr)}{r} \left[\frac{\cos\left(\frac{1}{2}\pi\cos\theta\right)}{\sin\theta} \right] \times \left\{ \frac{1-(a)^{N+1}\exp[-j(N+1)(\pi\sin\theta\cos\phi+\alpha)]}{1-a\exp[-j(\pi\sin\theta\cos\phi+\alpha)]} \right\} \tag{5.6}$$

where

Z_0 = intrinsic impedance of free space,
I_m = maximum current amplitude at the dipole,
k = wave number,
A_n = $a_n \exp(jn\alpha)$, the reflection coefficient, and
r = observation point in the far field.

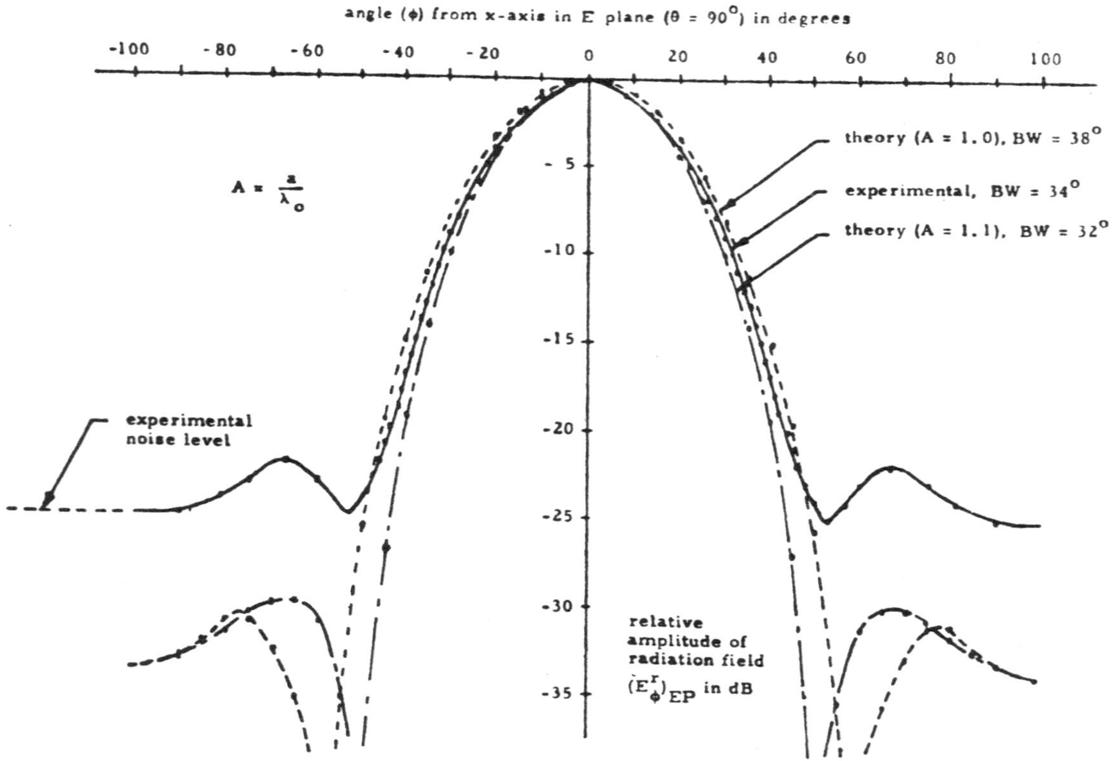

Figure 5.43(a) Radiation patterns in *E*-plane of an SBFA (from [26])

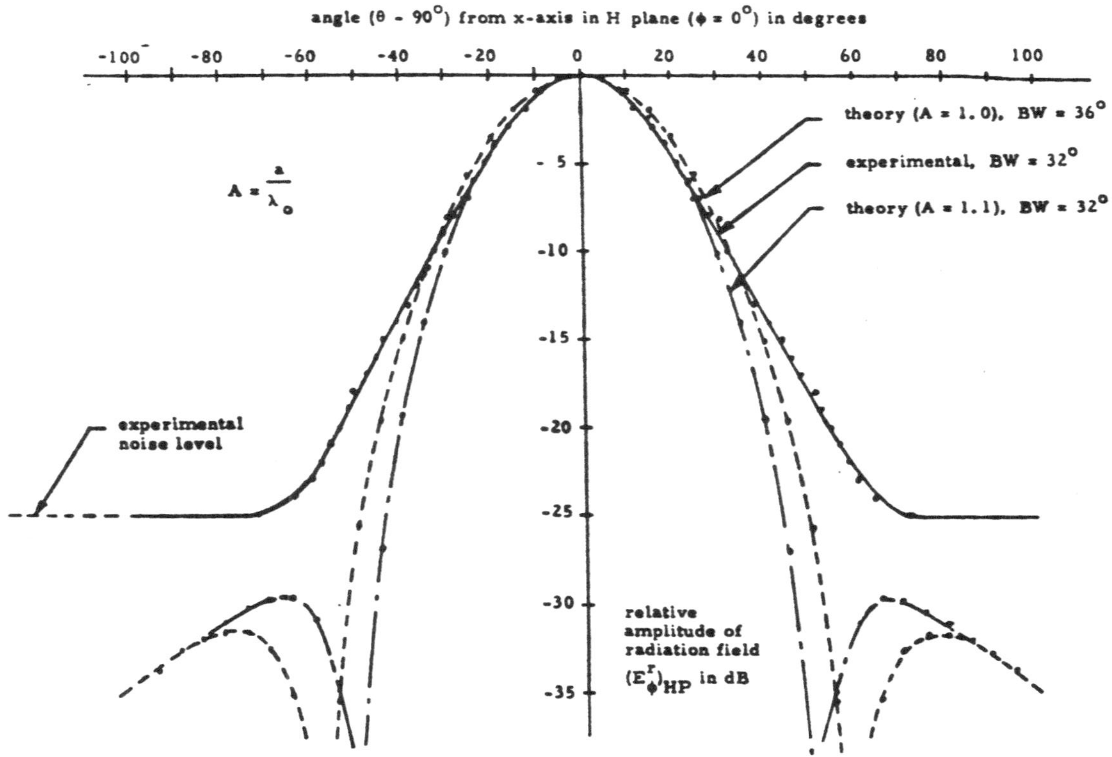

Figure 5.43(b) Radiation patterns in H-plane of an SBFA (from [26])

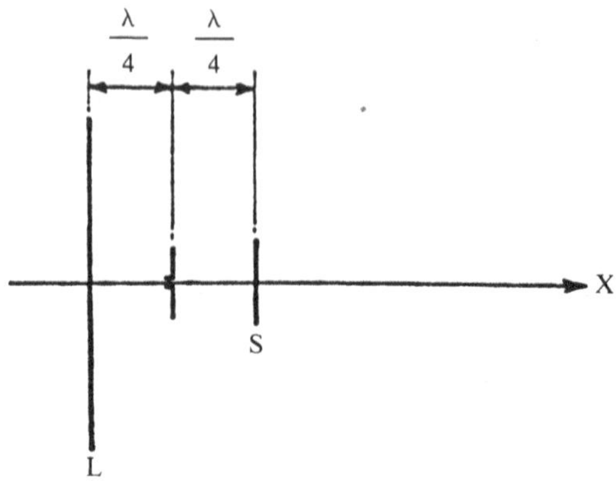

Figure 5.44 SBFA structure (from [27])

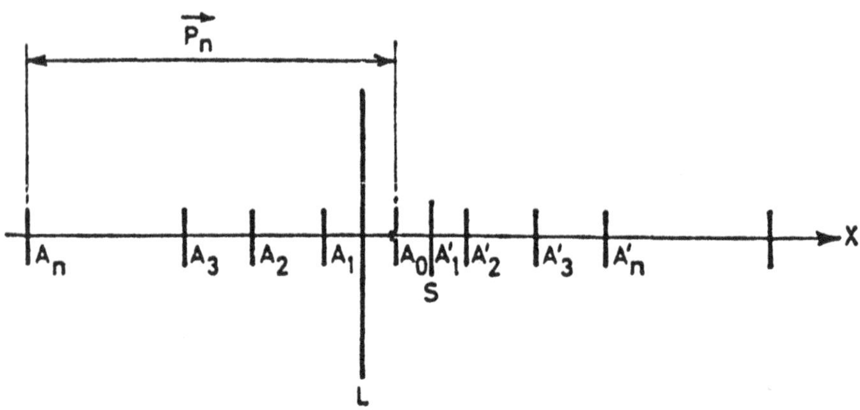

Figure 5.45 Images of a dipole feed placed between two infinite conducting planes [27]

Figure 5.46 (a) and (b) shows the *E*-plane and *H*-plane patterns of the SBFA, respectively. The measured patterns are taken from Ehrenspeck's work and the theoretical patterns are computed from his equation for $N = 50$, $A = 0.85$ and $N = 14$, $A = 0.98$. Computed and measured patterns agree reasonably well (up to 5 dB below the maximum); however, the theoretical expression gives a large number of sidelobes, due to the finite dimensions of the reflectors.

Nielsen and Pontoppidan [28] published theoretical and experimental work for SBFA design in 1970. The radiated pattern is computed by superposing the field due to the dipoles and the field due to the surface current distribution on the large reflector. The induced currents on the dipoles are determined through the use of image theory and the assumption that the reflectors are infinite in extent. A detailed description of their theory is discussed in Chapter 6.

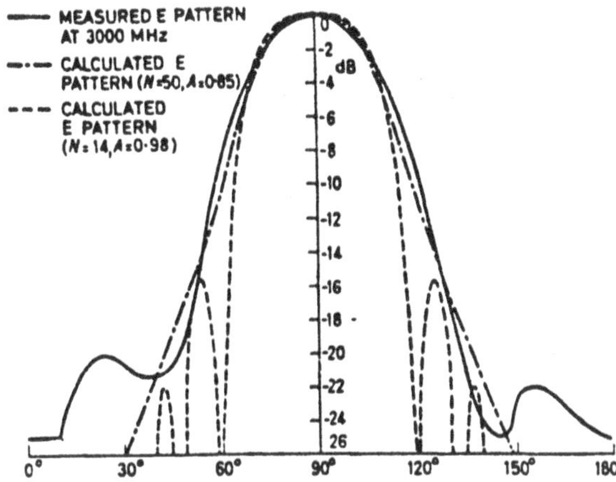

(a)

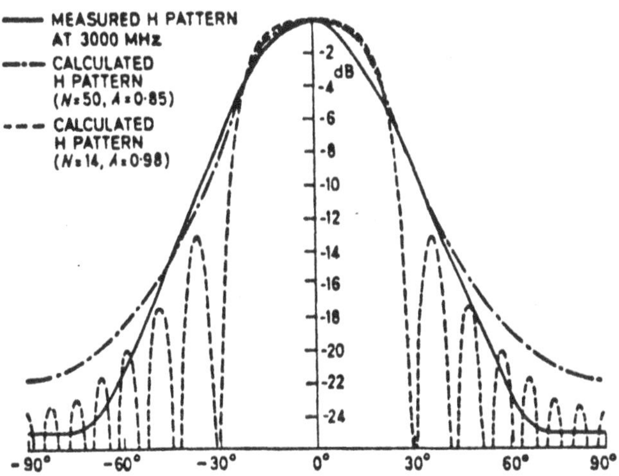

(b)

Figure 5.46(a) Computed and measured E-plane patterns of the SBFA (from [27])
(b) Computed and measured H-plane patterns of the SBFA (from [27])

The 1.5 GHz rimless SBFA shown in Figure 5.47 was investigated numerically using this theory. The feeder of this antenna is a slot-fed dipole of length of $0.47\lambda_0$ and radius $0.0085\lambda_0$; the small reflector has a diameter of $0.4\lambda_0$ and the large reflector a diameter of $2.0\lambda_0$. The large reflector is rimless and the dipole is centered between the two plates, the distance between which is $0.5\lambda_0$.

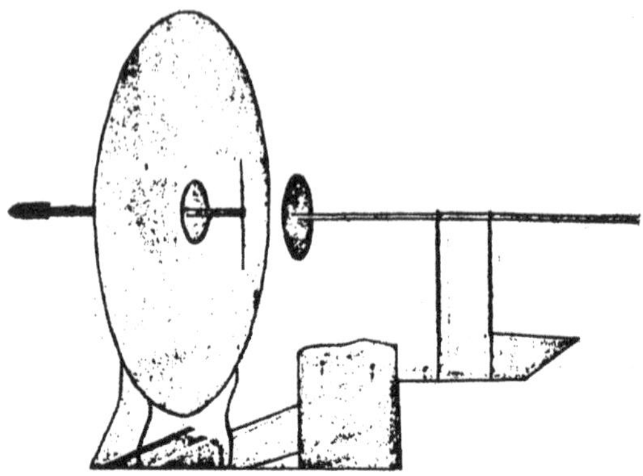

Figure 5.47 SBFA studied by Nielsen and Pontoppidan (from [28])

Two patterns of the SBFA are shown in Figure 5.48. Figure 5.48(a) shows the E-plane pattern, and 5.48(b) shows the H-plane pattern. These figures show good agreement between measured and computed results in both planes.

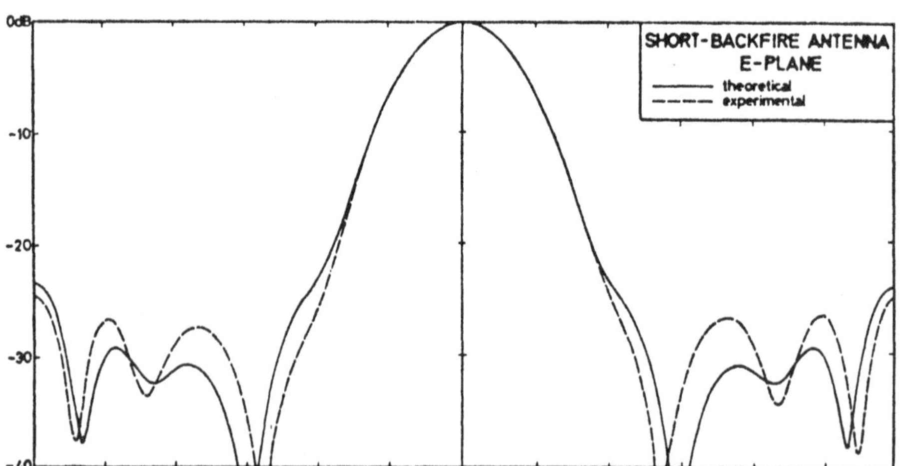

Figure 5.48(a) Comparison between the computed and measured E-plane patterns of an SBFA (from [28])

Nielsen [29] made theoretical and experimental investigations based on his previous works [30,31] on the rimless SBFA shown in Figure 5.49. Table 5.15 gives the parameters involved in the design of the antenna.

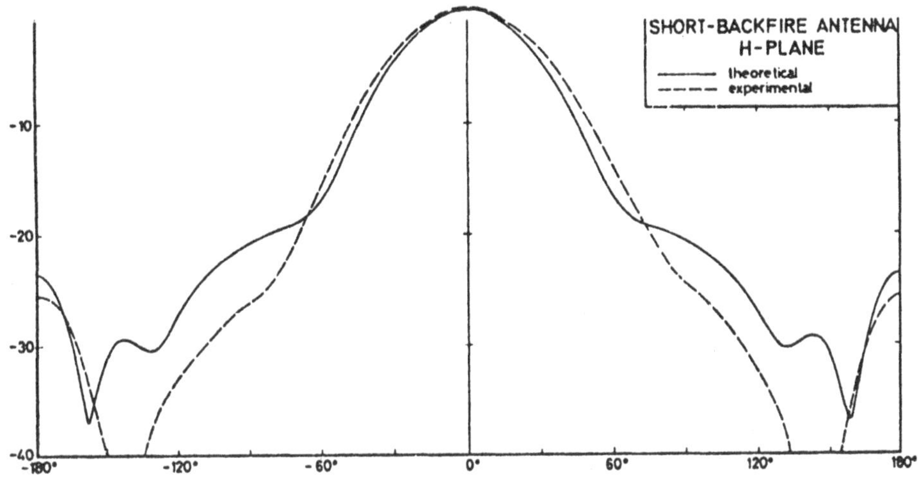

Figure 5.48(b) Comparison between the computed and measured H-plane patterns of an SBFA (from [28])

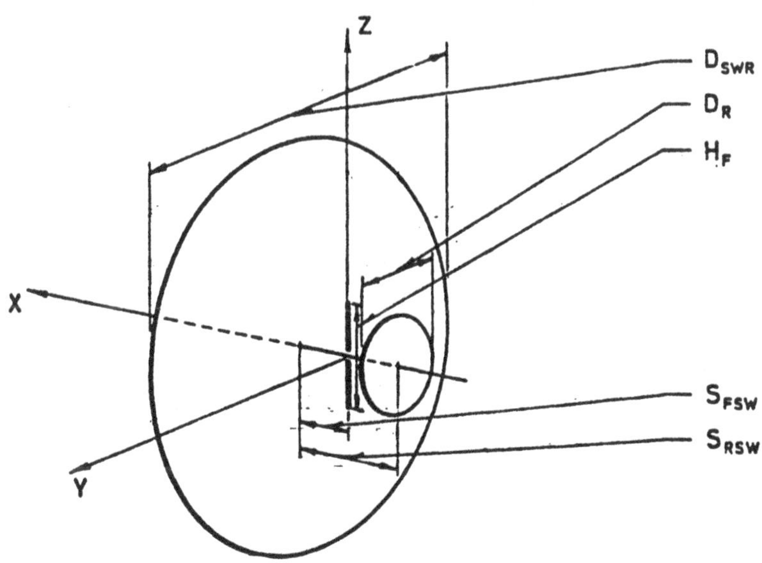

Figure 5.49 A rimless SBFA (from [29])

Figure 5.50 shows the effect that variation of the feed dipole length H_F has on the radiation properties of the antenna. The computed values in the figure show that the directivity and front-to-back ratio are constant with a variation of dipole length.

Table 5.15 Design Parameters of an SBFA

Parameters	*Notations and values*
Length of feed dipoles	$H_F = 0.47\lambda_0$
Distance from feed dipole to surface wave (large or main) reflector	$S_{FSW} = 0.25\lambda_0$
Distance between reflector plates	$S_{RSW} = 0.50\lambda_0$
Diameter of small reflector	$D_R = 0.40\lambda_0$
Diameter of large reflector	$D_{SWR} = 2.00\lambda_0$
Frequency ratio ($F = f/f_0$, $f_0 = c/\lambda_0$)	$F = 1.00$
Radius of the dipole elements	$d = 0.00634\lambda_0$

The input resistance of the dipole varies with the length, so that, for good matching, the length of the dipole is tuned. The E-plane and H-plane half-power beamwidths vary with the length of the dipole. Figure 5.51 shows the influence on the radiation properties of the antenna with the feed dipole positioned between the two reflector plates. This figure shows that a slight increase in directivity is obtained by moving the dipole from its centered position between the plates to a position nearer to the large reflector plate, and the properties (except front-to-back ratio) deteriorate significantly when the dipole is moved toward the small reflector.

The curves in Figure 5.52 show the influence that variation of the distance between the two reflectors S_{RSW} has on directivity, half-power beamwidth, front-to-back ratio, and input resistance when the dipole is kept in a center position between the reflectors. The radiation is optimal when the distance is about 0.5 λ_0. The effect of the small reflector diameter variation on directivity, half-power beamwidth, front-to-back ratio, and input resistance curves is shown in Figure 5.53. The directivity and half-power beamwidths are improved for $D_{SWR} = 0.44\ \lambda_0$; however, there is an increase in backlobe level. These curves show large variations in the radiation properties for $D_{SWR} = 0.46\ \lambda_0$, which is equal to the resonant length of the dipole element.

Figure 5.54 shows that the optimum diameter of an SBFA is 2.0 [1–3]. Figure 5.55 shows the computed effect of frequency variation on directivity, half-power beamwidth, backlobe, and input resistance of an SBFA. Optimum values for directivity and half-power beamwidths are obtained for $F = 1.07$, while a very low backlobe may be obtained for $F = 0.93$. The input resistance is approximately constant up to $F = 1.0$, but increases rapidly for frequencies f greater than f_0.

In 1971, Patel [32] gave an analytical expression of the far-field pattern of a rimless SBFA in terms of aperture illumination, assuming multiple reflection of electromagnetic waves between two conducting planes. Because the dipole feed is located midway between the reflectors, the multiple reflections can be approximated

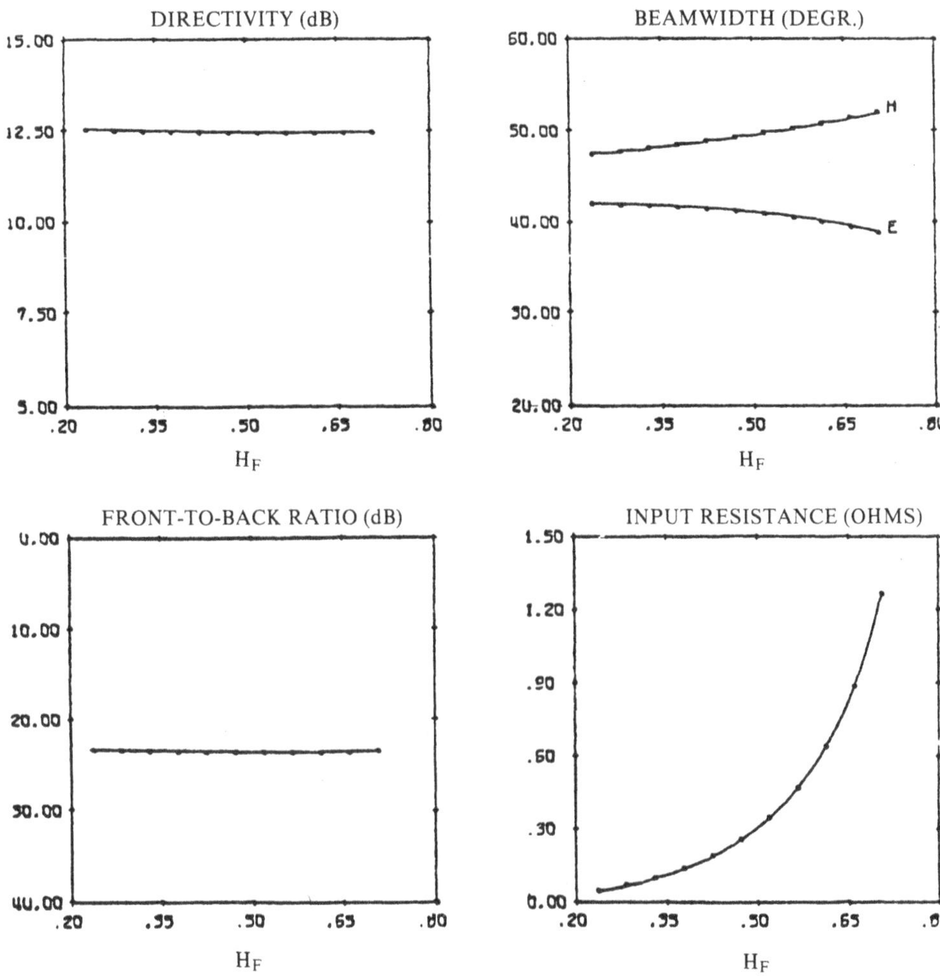

Figure 5.50 Effect due to variation of dipole lengths on directivity, half-power beamwidth, front-to-back ratio, and input resistance of an SBFA (from [29])

as an electromagnetic propagation $\cos(\omega x)$ originating from the feed, with frequency $\omega = \pi d$, x being the normalized distance along the abscissa with the origin at the feed, and d the separation between the reflectors. Such a wave generates an appropriate direction of current flow at each of the images, and the wave decays in geometrical progression. This may be approximated to a Gaussian function, $\exp(-bx^2)$, which superimposed on the wave $\cos(\omega x)$, corresponds to a continuous illumination.

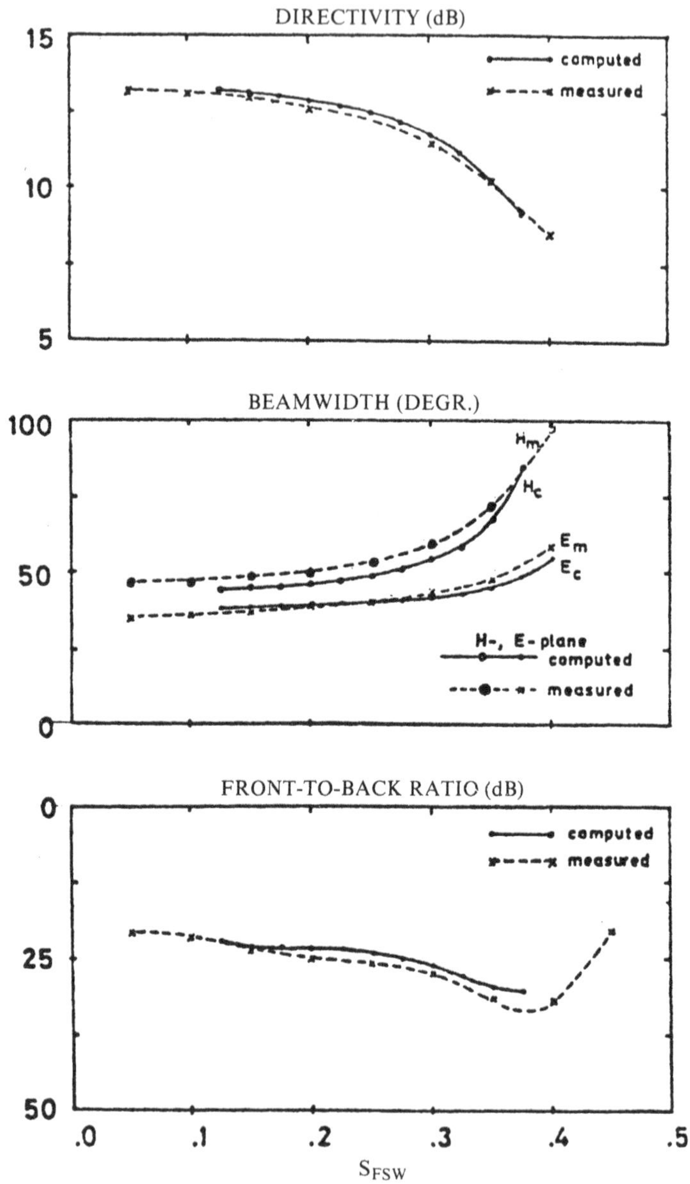

Figure 5.51 Influence on directivity, half-power beamwidth, front-to-back ratio, and input resistance due to variation of the distance S_{FSW} from the feed dipole to the large reflector (from [29])

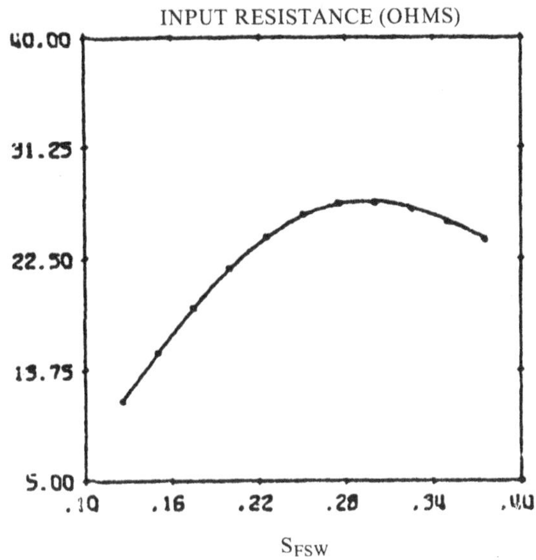

Figure 5.51 (cont'd)

The far-field pattern $F(u)$ is obtained by the Fourier transform of the field distribution across the aperture, with the final expression given by:

$$F(u) = 1/2\sqrt{\frac{\pi}{b}}\left[\exp\left(-\frac{q^2}{4b}\right)Re\,erf(\beta) + \exp\left(-\frac{r^2}{4b}\right)Re\,erf(\delta)\right] \quad (5.7)$$

where

$$q = u + \omega, \quad \beta = \sqrt{ba} + j\frac{q}{2\sqrt{b}}$$

$$r = u - \omega, \quad \delta = \sqrt{ba} + j\frac{r}{2\sqrt{b}}$$

and *Re erf* = real part of error function.

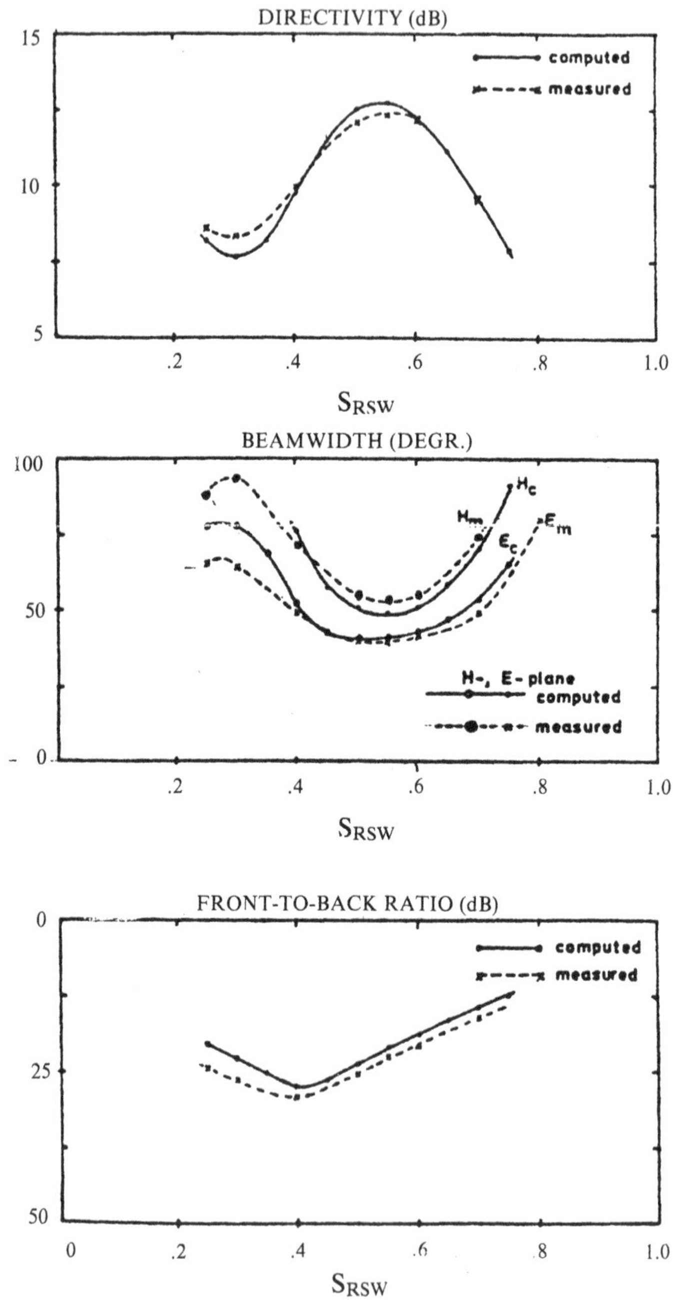

Figure 5.52 Influence of variation in the distance S_{RSW} between the two reflectors on directivity, half-power beamwidth, front-to-back ratio, and input resistance of an SBFA (from [29])

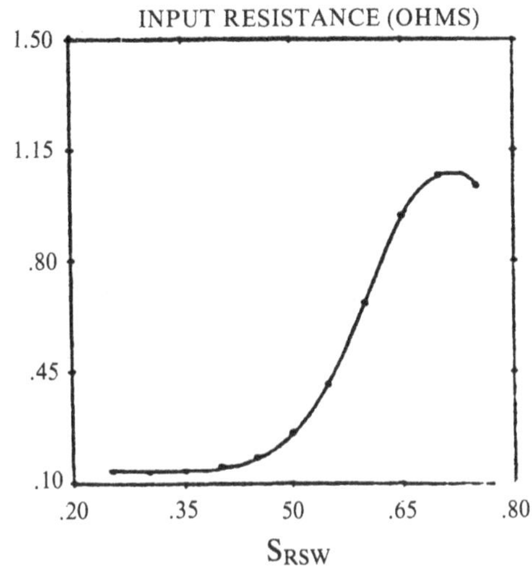

Figure 5.52 (cont'd)

Figure 5.56 shows the computed H-plane pattern from equation (5.7) and experimental data from Nielsen and Pontoppidan [29]. The results agree well with suitable choices of a and b.

Hsieh, Nyquist, and Chen [33], and Kumar [34] have developed a numerical-physical optics method to predict the characteristics of an SBFA. In this approach, the electromagnetic boundary value problem is formulated theoretically, and solved numerically to determine the induced current excited on the conducting surfaces of various components of the SBFA. The SBFA models studied in detail were a rimless large reflector SBFA, and an SBFA with a rim on the large reflector.

Figure 5.57 shows a sketch of a rimless SBFA model with typical dimensions. The radiator consists of a circular large reflector S of radius w and a small circular reflector R of radius b, arranged parallel to one another, with separation c. The planes of both reflectors are transverse to the longitudinal antenna axis. The dipole radiator D is centered along the antenna axis between the large and small reflectors at a distance d from the large reflector.

The center of the large circular reflector is placed at the origin of a Cartesian coordinate system that has its y axis lying along the SBFA axis. The dipole primary radiator is parallel to the z-axis, with half-length h and radius a, and its center is located at $y = d$ along the positive y-axis. The small reflector is parallel to the x-z plane, with its center along the positive y-axis at $y = c$.

A set of integral equations for the induced currents excited in the dipole and on the surface of the small reflector (as well as their images) is formulated, based on the boundary condition that the tangential electric field vanishes at all conducting

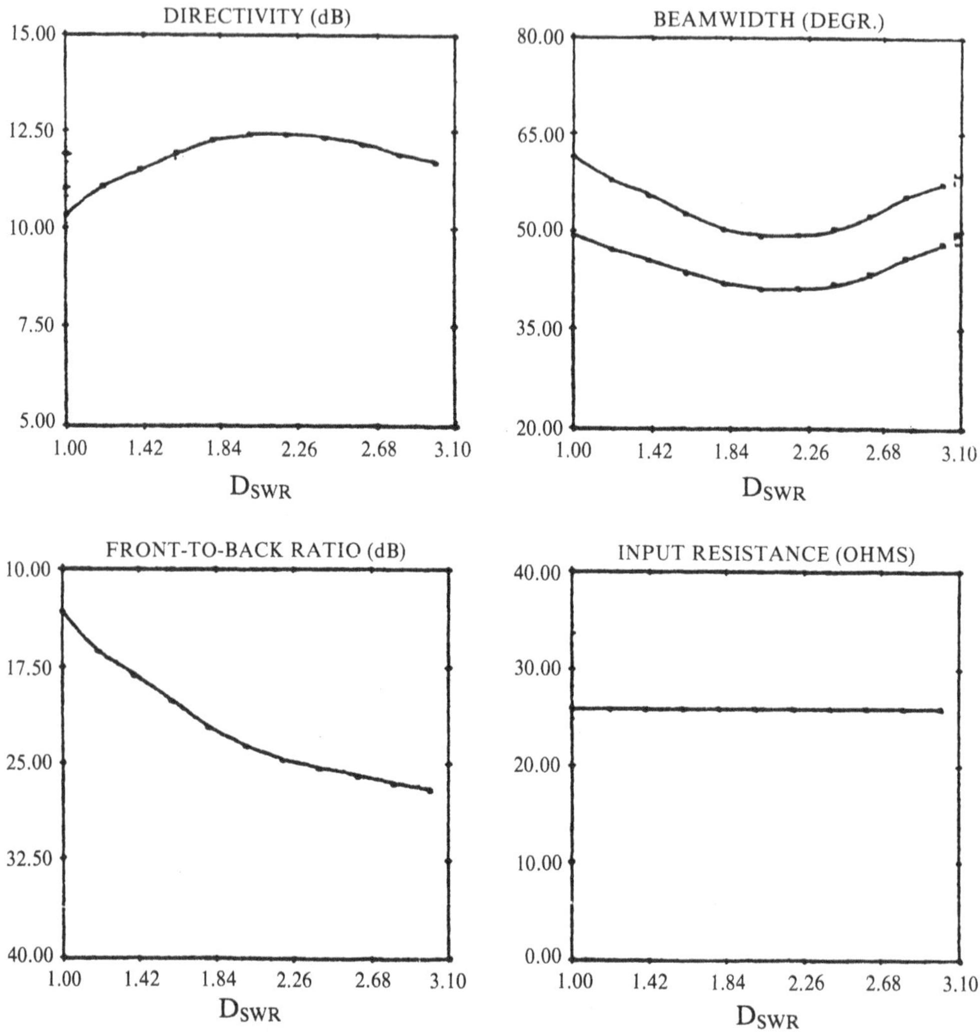

Figure 5.53 The effect of small reflector diameter variation (D_{SWR}) on directivity, 3 dB beamwidth, backlobe, and input resistance (from [29])

surfaces. In calculating the radiation field of the antenna, the large reflector currents are truncated by using the actual radius of the reflector. Image techniques determine the currents induced in the primary dipole radiator and on the surface of the small reflector. The large reflector is assumed to be infinite in extent. The induced field, $E(r)$ at any point r in space, maintained by all the currents in the antenna system, is given by:

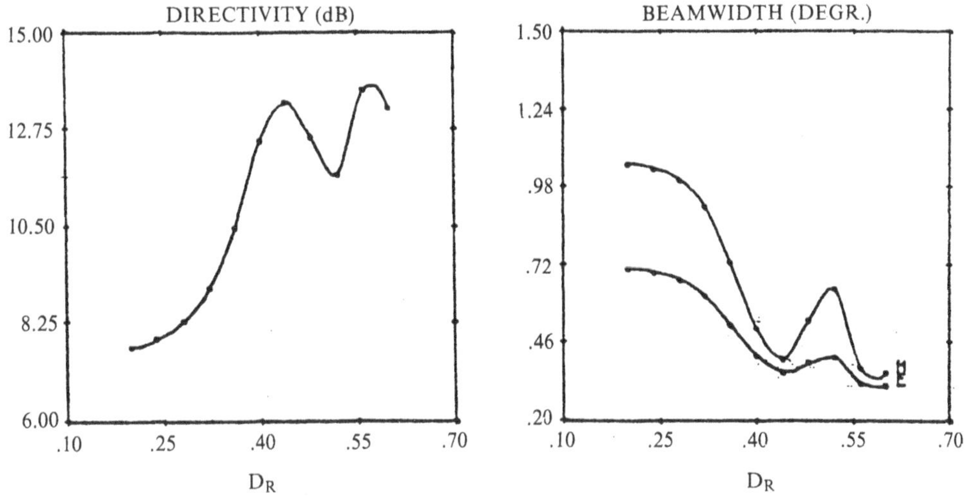

Figure 5.53 (cont'd)

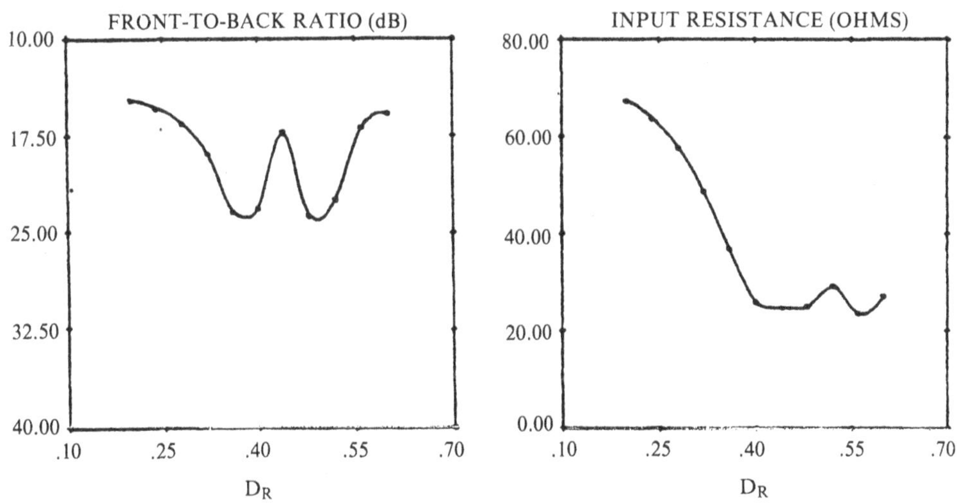

Figure 5.54 The effect of variation in the diameter of the large reflector (D_R), on directivity, 3 dB beamwidth, backlobe and input resistance (from [29])

$$\vec{E}(\vec{r}) = \frac{-j\zeta_0}{4\pi\beta_0}\left[\nabla \int_v \nabla' \cdot \vec{J}(\vec{r}')\, \frac{e_0^{-j\beta}R(\vec{r},\vec{r}')}{R(\vec{r},\vec{r}')}\, \mathrm{d}V' + \beta_0^2 \int_v \vec{J}(\vec{r}')\, \frac{e_0^{-j\beta}R(\vec{r},\vec{r}')}{R(\vec{r},\vec{r}')}\right] \mathrm{d}V' \tag{5.8}$$

where

$\beta_0 \quad = 2\pi/\lambda_0$ is the free-space wave number,

$\zeta_0 \quad = \sqrt{\mu_0/\varepsilon_0}$ is the free-space wave impedance,

$R(\vec{r},\vec{r}') = |\vec{r} - \vec{r}'|$ is the distance between the field point at r and a source point at r, and v is the volume, including all the currents in the antenna system.

The term $J(r)dV'$, in this case, consists of the current in the dipole, the surface current on the small reflector, and the effects of their images. The surface current excited on the infinite conducting plane is determined from the electromagnetic field maintained by the currents in the dipole and on the small reflector, and their images (Figure 5.58).

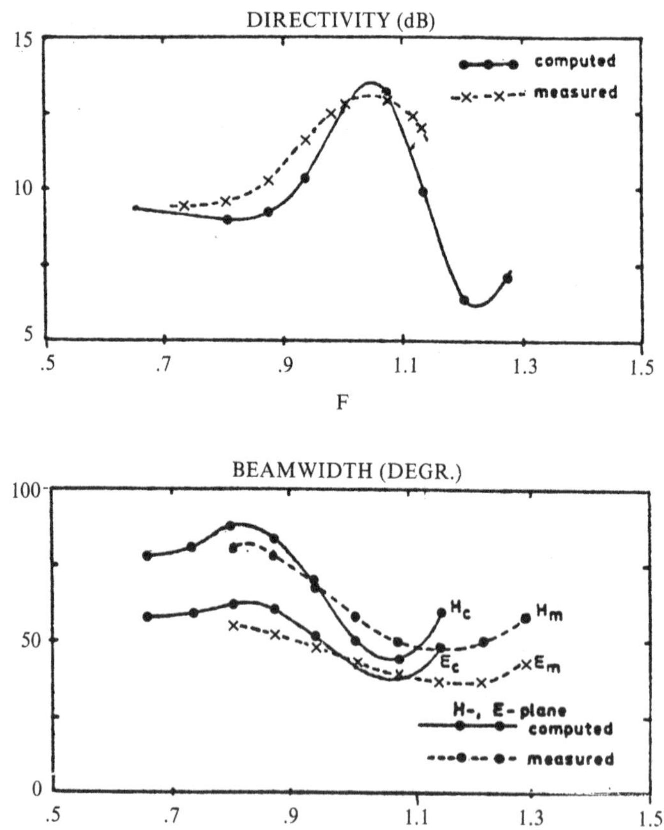

Figure 5.55 The effect of variation in frequency on directivity, half-power beamwidth, backlobe and input resistance (from [29])

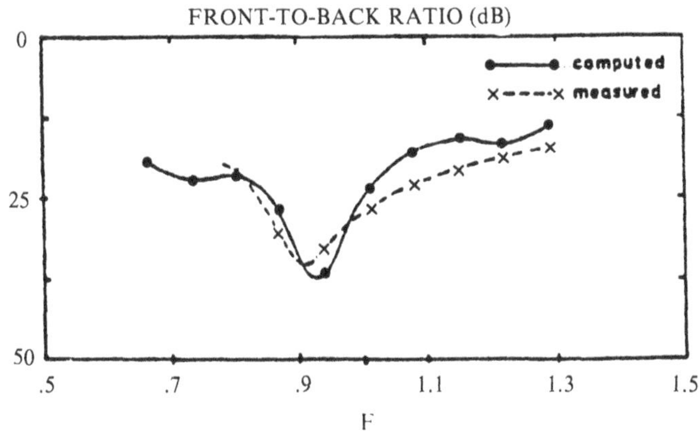

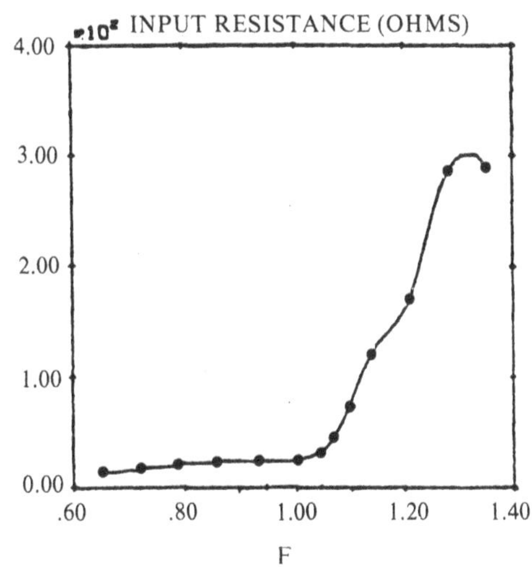

Figure 5.55 (cont'd)

In order to determine the induced current numerically, the dipole is partitioned into a number of subsections, and the small reflector is partitioned into a number of rectangular subareas. Within the subsections or subareas, the induced current is assumed to be approximately constant, and equal to its value at the center of the partition. In order to calculate the radiation field in terms of the various current distributions, the dipole exciter is again partitioned into a number of subsections, and the two reflectors are partitioned into a number of rectangular subareas. The induced current is assumed approximately constant over each partition, and equal to its value at the center partition. Figure 5.59 shows the geometry for the radiation field calculation.

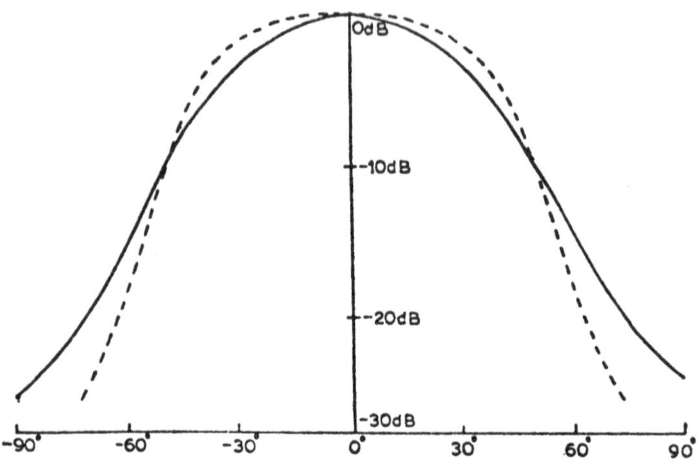

Figure 5.56 H-plane pattern of an SBFA: theoretical pattern for $a = 20$ and $b = 1.5$ ----------; experimental pattern of Nielsen and Pontoppidan (from [29]) ———

The radiation field maintained by the large reflector is given by:

$$\vec{E}^r(r,\theta,\phi) = \hat{\phi}\frac{j\zeta_0\beta_0}{4\pi}\frac{e^{-j\beta_0 r}}{r}\sin\theta\int_S K_{sz}(x',z')\cdot e^{j\beta_0(x'\sin\theta\cos\phi+z'\cos\theta)}\,dx'dz' \tag{5.9}$$

The radiation field maintained by the small reflector is given by [33]:

$$\vec{E}^r(r,\theta,\phi) = \hat{\phi}\frac{j\zeta_0\beta_0}{4\pi}\frac{e^{-j\beta_0 r}}{r}\sin\theta\int_R K_{sz}(x',z') \cdot e^{j\beta_0(x'\sin\theta\cos\phi+c\sin\theta\sin\phi+z'\cos\theta)}\,dx'dz' \tag{5.10}$$

The radiation field maintained by the dipole exciter is calculated by [33]:

$$\vec{E}^r(r,\theta,\phi) = \hat{\phi}\frac{j\zeta_0\beta_0}{4\pi}\frac{e^{-j\beta_0 r}}{r}\sin\theta\int_{-h}^{h} I_z(z')\,e^{j\beta_0(z'\cos\theta+d\sin\theta\sin\phi)}\,dz' \tag{5.11}$$

The total radiated field maintained by a rimless SBFA is the superposition of the fields described by (5.9), (5.10), and (5.11), and is expressed as [33]:

$$\vec{E}^r(r,\theta,\phi) = \hat{\theta}\frac{j\zeta_0\beta_0}{4\pi}\frac{e^{-j\beta_0 r}}{r}\sin\theta\left\{\int_S K_{sz}(x',z')\,e^{j\beta_0(x'\sin\theta\cos\phi z'\cos\theta)}\,dx'dz'\right.$$

$$+ \int_R K_{rz}(x',z')\, e^{j\beta_0(x'\sin\theta\cos\phi + c\sin\theta\sin\phi z'\cos\theta)}\, dx'dz'$$

$$+ \int_{-h}^{h} I_z(z')\, e^{j\beta_0(z'\cos\theta + d\sin\theta\sin\phi)}\, dz' \Bigg\} \tag{5.12}$$

The far-field radiation pattern is calculated from the above expression and given by [33]:

$$F(\theta,\phi) = \sin\theta \Bigg\{ \int_S K_{sz}(x',z') e^{j\beta_0(x'\sin\theta\cos\phi + z'\cos\theta)}\, dx'dz'$$

$$+ \int_R K_{rz}(x',z')\, e^{j\beta_0(x'\sin\theta\cos\phi + c\sin\theta\sin\phi + z'\cos\theta)}\, dx'dz'$$

$$+ \int_{-h}^{h} I_z(z')\, e^{j\beta_0(z'\cos\theta + d\sin\theta\sin\phi)}\, dz' \Bigg\} \tag{5.13}$$

From (5.12), the E-plane radiation is found by assuming $\phi = \pi/2$ in the y-z plane, and the H-plane is found by assuming $\theta = \pi/2$ in the x-y plane.

Hsieh, *et al.* [33] wrote software using these expressions and computed the E-plane and H-plane patterns of an SBFA. They used a typical SBFA developed by Ehrenspeck [5] to compare their theoretical results. The typical dimensions of Ehrenspeck's experimentally-optimized SBFA model at 3 GHz are:

length of dipole primary radiator ($2h$)	$= 0.5\,\lambda_0$
distance from dipole to large reflector (d)	$= 0.25\,\lambda_0$
distance between large and small reflectors (c)	$= 0.5\,\lambda_0$
diameter of rimless large reflector (w)	$= 1.0\,\lambda_0$
diameter of small reflector (b)	$= 0.2\,\lambda_0$
dipole (radius a) thickness parameter ($\Omega = 2\ln(2h/a)$)	$= 10.0$

Figure 5.60(a) and (b) shows the E-plane and H-plane radiation patterns of the rimless circular SBFA model. It shows good agreement between Ehrenspeck's experimental and theoretical (numerical-physical optics approach) patterns. In Figure 5.61(a) and (b) the E-plane and H-plane radiation patterns obtained by the numerical-physical optics technique for various small reflector radii b are presented. This figure shows that the optimal radius for the small reflector is $b = 0.2\lambda_0$, for which theoretical conclusion is in agreement with the optimum reflector radius determined by Ehrenspeck [2].

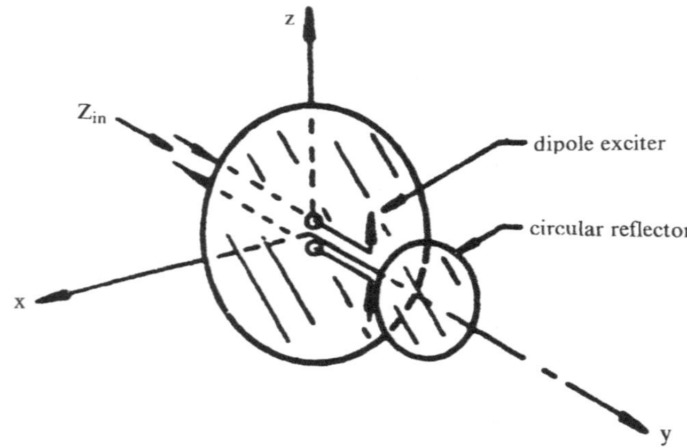

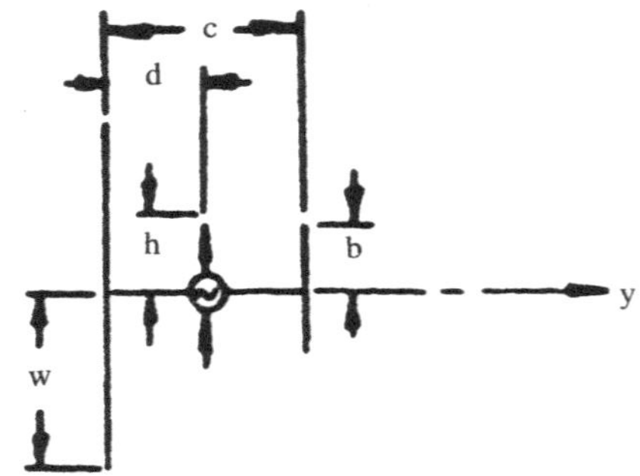

$w = 1.0\ \lambda_0$
$h = d = 0.25\ \lambda_0$
$b = 0.2\ \lambda_0$ typical
$c = 0.5\ \lambda_0$ dimensions
$\Omega = 2 \ln (2h/a) = 10.0$

Figure 5.57 Geometry of the rimless SBFA (from [33])

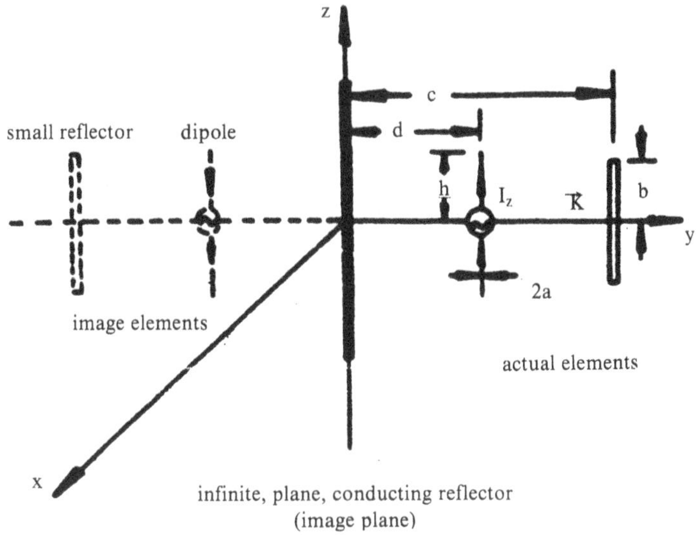

Figure 5.58 Infinite plane conducting reflector in the image plane (from [33–34])

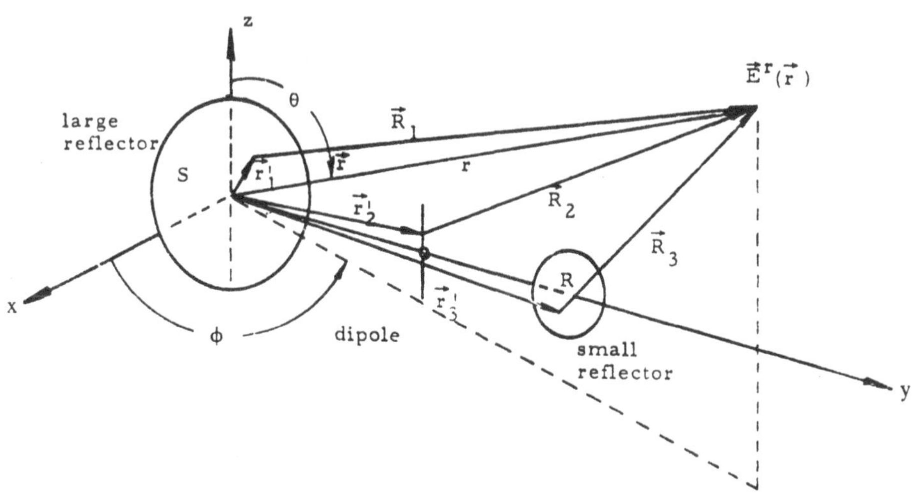

Figure 5.59 Geometry for radiation field calculation (from [33])

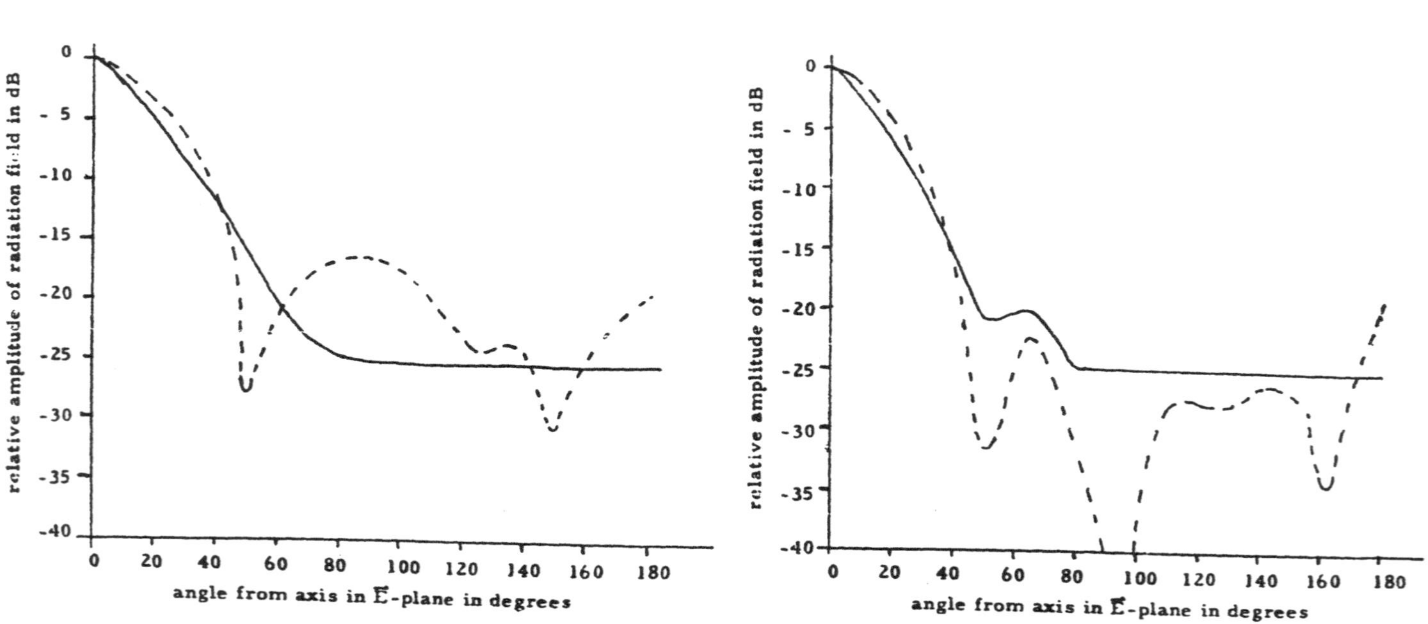

(a)

(b)

Figure 5.60(a) Theoretical and experimental E-plane patterns at 3 GHz (from [33]): ——— AFCRL SBFA; ---------- Numerical solution with 20-term current series for circular reflector (Z_{in} = 42.01 + j 144.0 ohms)

(b) Theoretical and experimental H-plane patterns at 3 GHz (from [33]): ——— AFCRL SBFA; ---------- Numerical solution with 20 term current series for circular reflector (Z_{in} = 42.01 + j 144.0 ohms)

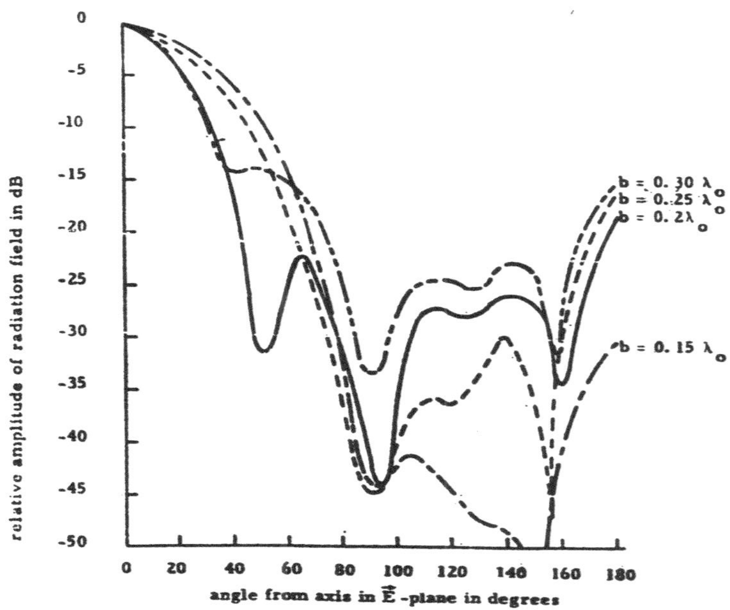

(a)

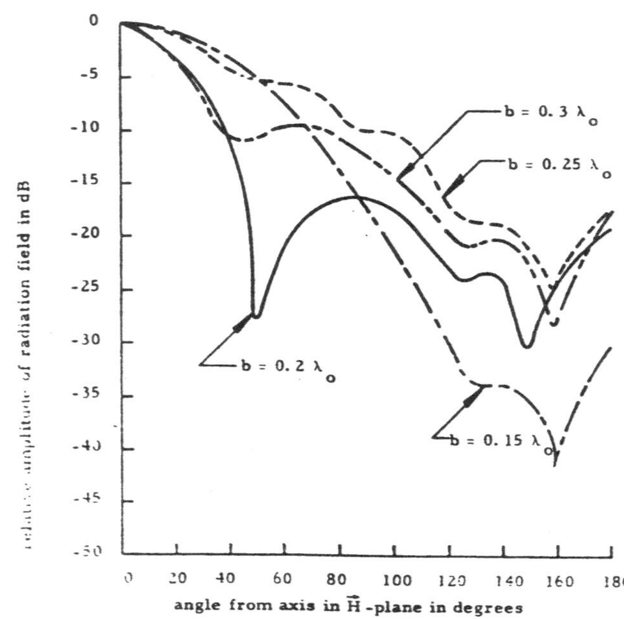

(b)

Figure 5.61(a) E-plane pattern of an SBFA obtained by the numerical-physical optics approach for various small reflector radii b (from [33])

(b) H-plane pattern of an SBFA obtained by the numerical-physical optics approach for various small reflector radii b (from [33])

Figure 5.62(a) and (b) shows the frequency dependence of the E-plane and H-plane radiation patterns. We find that the most desirable field patterns, with minimum beamwidth and sidelobe levels, are obtained at 3, the GHz frequency used by Ehrenspeck [2].

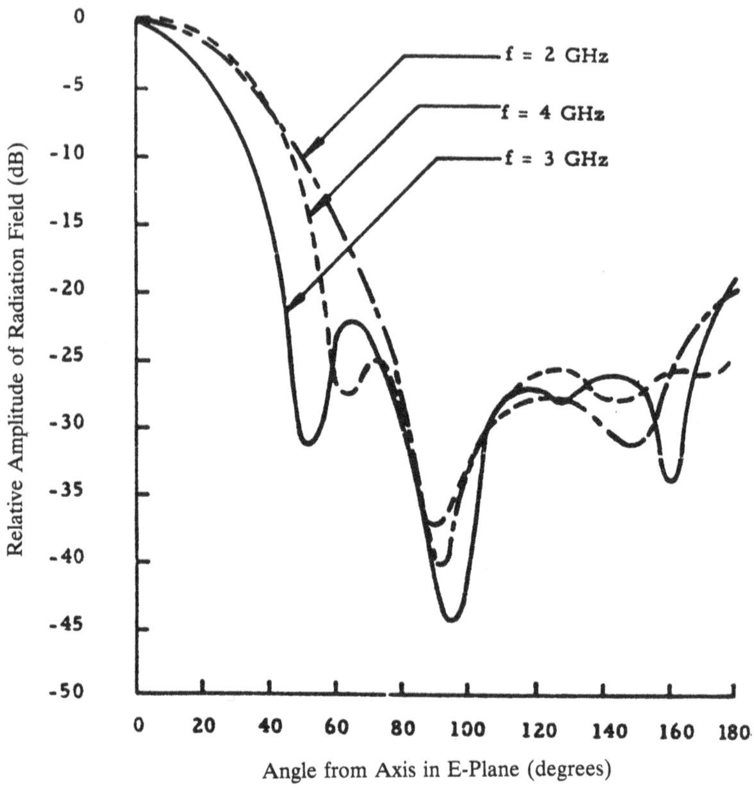

Figure 5.62(a) E-plane radiation pattern of an SBFA, obtained by numerical-physical optics approach for various frequencies (from [33])

Hsieh, *et al.* [33] and Kumar [34–35] have calculated the diffracted field from the rim of the antenna. The field maintained by the currents on the dipole, and the large and small reflectors is assumed to behave as a plane wave when it is incident at the rim edge, and to emerge from the center of the large reflector. This assumption is based on the experimental work by Ehrenspeck [3–5] at AFCRL. He showed that the phase center of the complicated near field of an SBFA is near the center of the large reflector. To calculate the field diffracted by the rim edge, a geometrical theory or diffraction [36] is used. Figure 5.63(a) and (b) shows the antenna geometry and the rays diffracted by the rim edge in the E-plane and the H-plane, respectively.

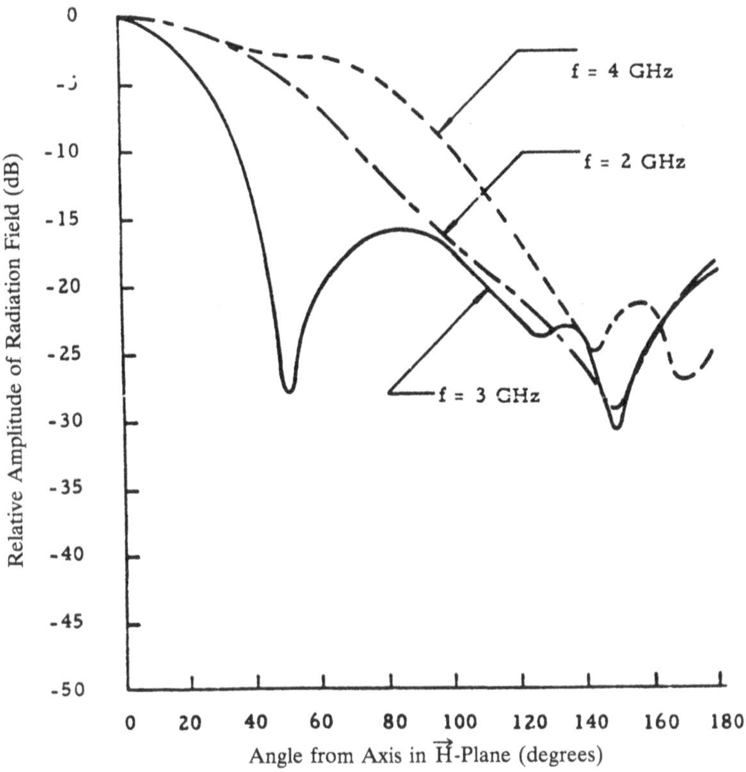

Figure 5.62(b) *H*-plane pattern of an SBFA obtained by the numerical-physical optics approach for various frequencies (from [33])

When the field is incident at Q_1, the diffracted ray propagates along the path r_1 to reach the observation point Q. Similarly, the diffracted ray from Q_2 propagates along the path r_2 to reach Q. The sum of these two diffracted rays gives the total diffracted field $\vec{E}^d(Q)$ by rim-edge at the observation point Q. The total diffracted field at point Q in the E-plane for $0° < \theta \leq 90°$ is given by:

$$\vec{E}^d(Q) = \hat{\theta} E_e^{inc} \left(\frac{-je^{-j\pi/4}}{\sqrt{\pi}} \right) \sqrt{ww'} \, \frac{e^{-j\beta_0 r}}{r} \cdot \left[\frac{\exp[j\beta_0(d\cos\theta' + w\sin\theta')]}{\sqrt{\sin\theta'}} D_h'(Q_1) + \frac{j\exp[j\beta_0(d\cos\theta' - w\sin\theta')]}{\sqrt{\sin\theta'}} D_h'(Q_2) \right] \tag{5.14a}$$

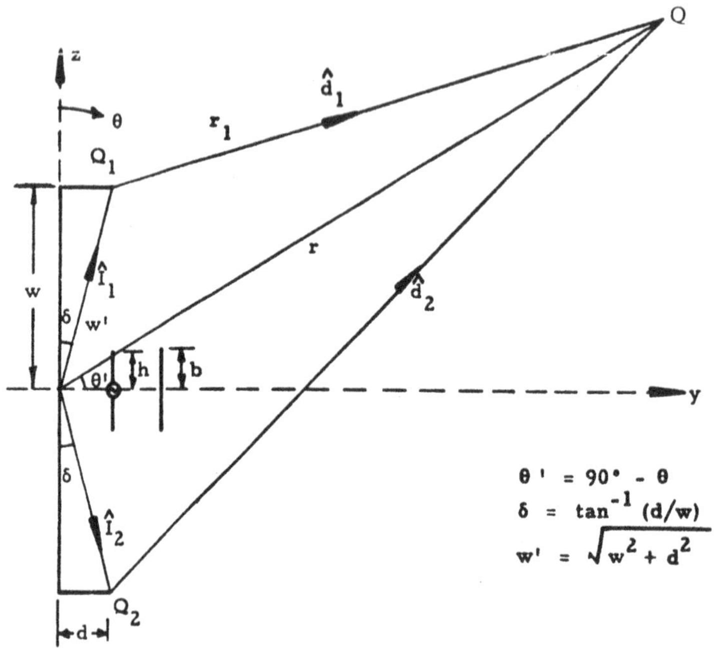

Figure 5.63(a) An SBFA geometry to calculate the diffracted field in E-plane (from [33–35])

For $90° \leq \theta < 180°$, the diffracted field in the E-plane is

$$\vec{E}^d(Q) = \hat{\theta} E_e^{inc} \frac{-je^{-j\pi/4}}{\sqrt{\pi}} \sqrt{ww'} \frac{e^{-j\beta_0 r}}{r} \cdot \left[\frac{\exp[j\beta_0(d \cos\theta' + w \sin\theta')]}{\sqrt{\sin\theta'}} D_h'(Q_1) \right] \tag{5.14b}$$

The total diffracted field at point Q in the H-plane for $0 < \phi' \leq 90°$ is given by

$$\vec{E}^d(Q) = \hat{z} E_h^{inc} \left(\frac{-je^{-j\pi/4}}{\sqrt{\pi}} \right) \sqrt{ww'} \frac{e^{-j\beta_0 r}}{r} \cdot \left[\frac{\exp[j\beta_0(d \cos\phi' + w \sin\phi')]}{\sqrt{\sin\phi'}} D_s'(Q_1) + \frac{j \exp[j\beta_0(d \cos\phi' - w \sin\phi')]}{\sqrt{\sin\phi'}} D_s'(Q_2) \right] \tag{5.15}$$

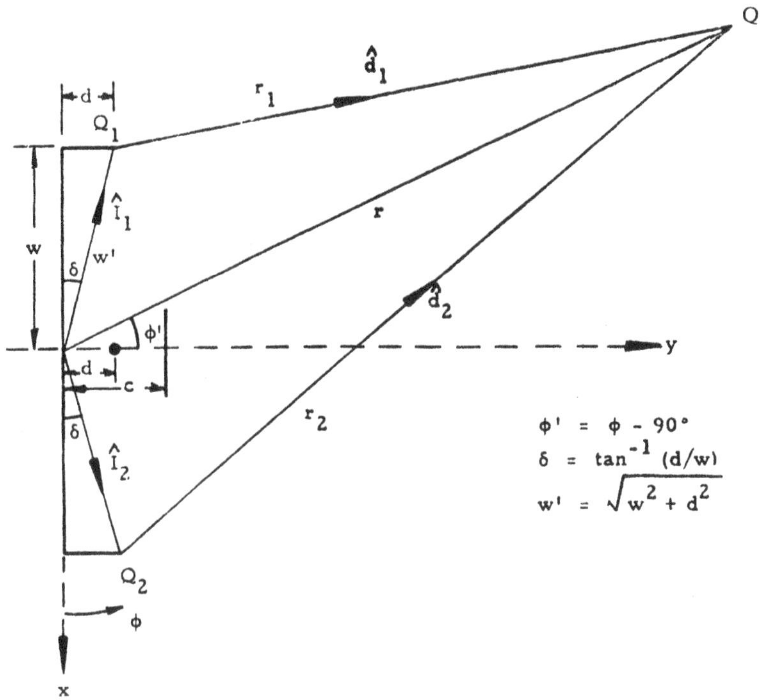

Figure 5.63(b) An SBFA geometry to calculate the diffracted field in H-plane (from [33–35])

For $90° \leq \phi' < 180°$ the field in the H-plane is given by:

$$\vec{E}^d(Q) = \hat{z}E_h^{inc}\left(\frac{-je^{-j\pi/4}}{\sqrt{\pi}}\right)\sqrt{ww'}\,\frac{e^{-j\beta_0 r}}{r} \cdot \left[\frac{\exp[j\beta_0(d\cos\phi' + w\sin\phi')]}{\sqrt{\sin\phi'}}D_s'(Q_1)\right] \tag{5.16}$$

where $D_s'(Q_1)$, $D_s'(Q_2)$, $D_h'(Q_1)$ and $D_h'(Q_2)$ are the normalized diffraction coefficients, E_e^{inc} is the electric field at Q_1 or Q_2 of the rim-edge of the E-plane, E_h^{inc} is the electric field at Q_1 or Q_2 of the rim-edge in the H-plane, and $\hat{Z}$ and $\hat{\theta}$ are the unit vectors in Z and θ directions. The E-plane and H-plane radiation patterns of a SBFA obtained at 3 GHz by the numerical physical optics method (including field diffracted by rim) for varions small reflector radii, b, are shown in Figures 5.64 and 5.65, respectively.

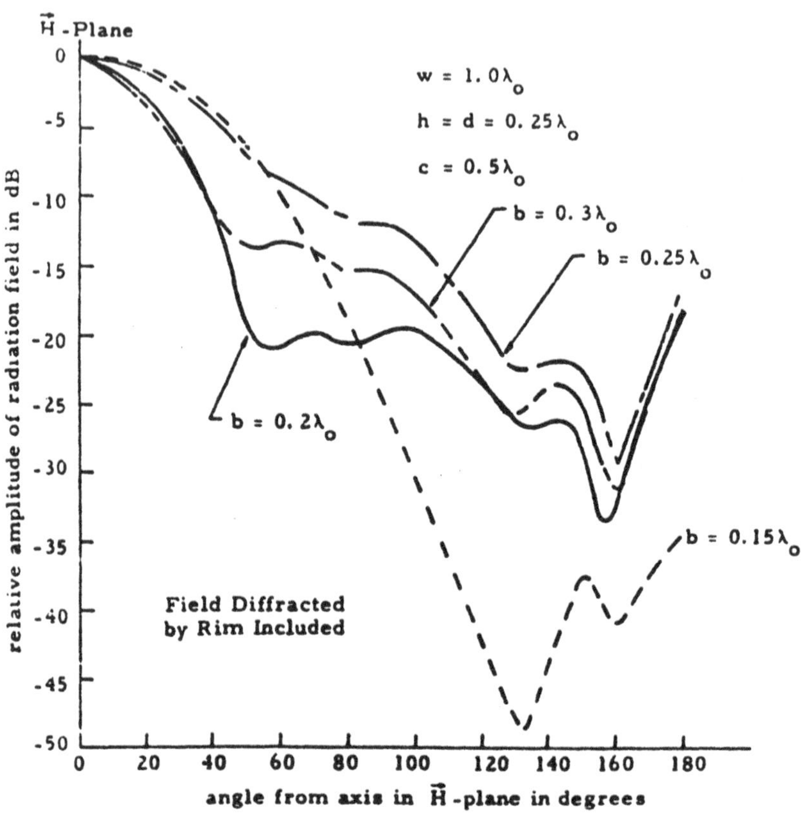

Figure 5.64 $\vec{H}$-plane (x-y plane) radiation patterns of an SBFA, obtained by the numerical-physical optics approach for various small reflector radii b (from [33])

The frequency dependence of the E-plane and H-plane radiation patterns is determined by the numerical physical optics method (PO), including correction due to rim-edge diffraction, as shown in Figure 5.66(a) and (b). The maximum directivity with minimum sidelobe level is achieved for the optimum size of small reflector ($b = 0.2\lambda_0$) and for the optimum frequency (3 GHz).

In 1976, Gibson, Govind, and Tsai [36] developed a simple analysis of a two-dimensional model of the SBFA by using the moment method [37]. The pattern sensitivity is considered a function of the subreflector size and location, lip protrusion of the large reflector, and location of the excitation of the line source. For simplicity of analysis, a TM problem (i.e., one with an electric line source excitation) is considered. The integral equation for the unknown currents $J_z(\bar{\rho})$ on the large reflector

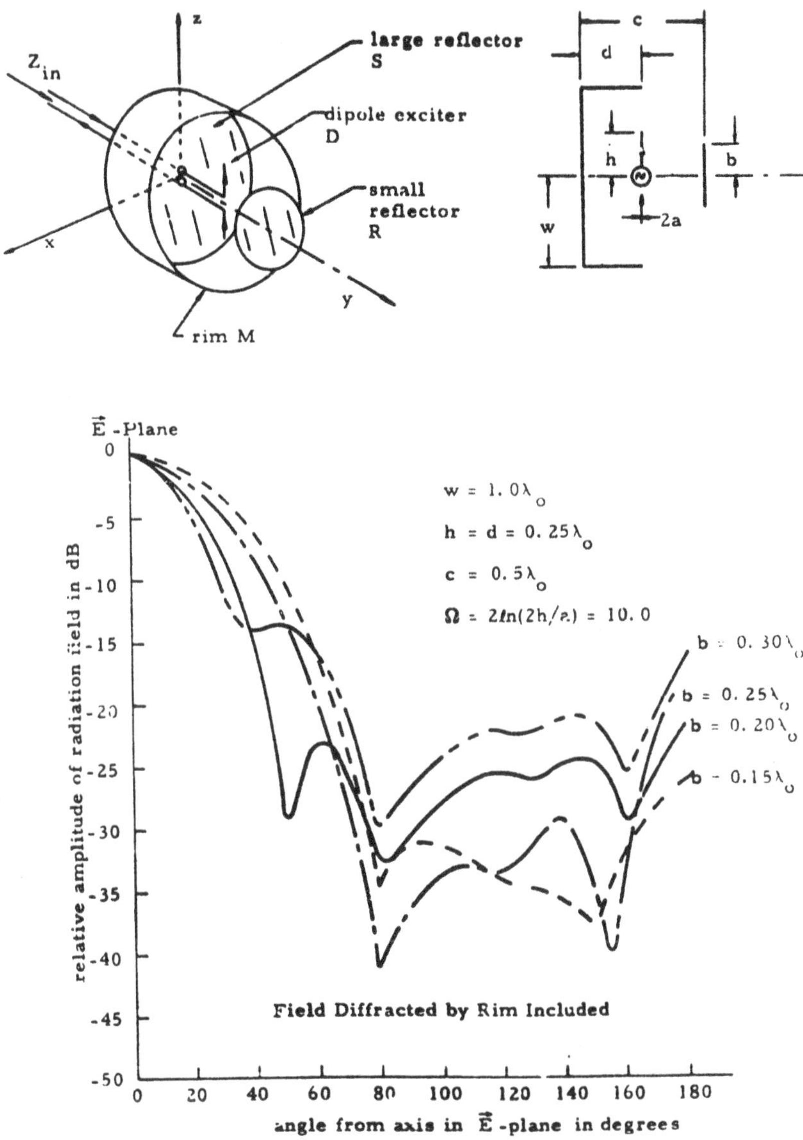

Figure 5.65 $\vec{E}$-plane (y-z plane) radiation patterns of an SBFA, obtained by the numerical-physical optics approach for various small reflector radii b (from [33])

is calculated by the moment method and given by:

$$\frac{k\eta}{4}\int_C J_z(\bar{\rho}')H_0^{(2)}(k|\bar{\rho}-\bar{\rho}'|)\,\mathrm{d}c' = E_z^i(\bar{\rho}) \tag{5.17}$$

where

k = $2\pi/\lambda_0$,
λ_0 = the free-space wavelength,
η = $120\ \pi$,
$H_0^{(2)}(k|x|)$ = the Hankel function of zero order and second kind, and
$E_z^i(\bar{\rho})$ = the line source excitation field.

The total far field is given by

$$E^t = E^i + E^s \tag{5.18}$$

where

E^i = the incident field and
E^s = the scattered field.

The far-field expression is derived by using the conventional large argument approximation for the Hankel function:

$$E_z^t = \sqrt{\frac{2j}{\pi k}}\frac{e^{-jk\rho}}{\sqrt{\rho}}\ 1 - \frac{k\eta}{4}\sum_{n=1}^{N} J_{z_n}e^{jk\rho_n\cos(\phi-\phi_n)}\Delta C_n \tag{5.19}$$

where ρ_n, ϕ_n are the coordinates of the currents J_{z_n}, and ΔC_n is the nth subdomain width.

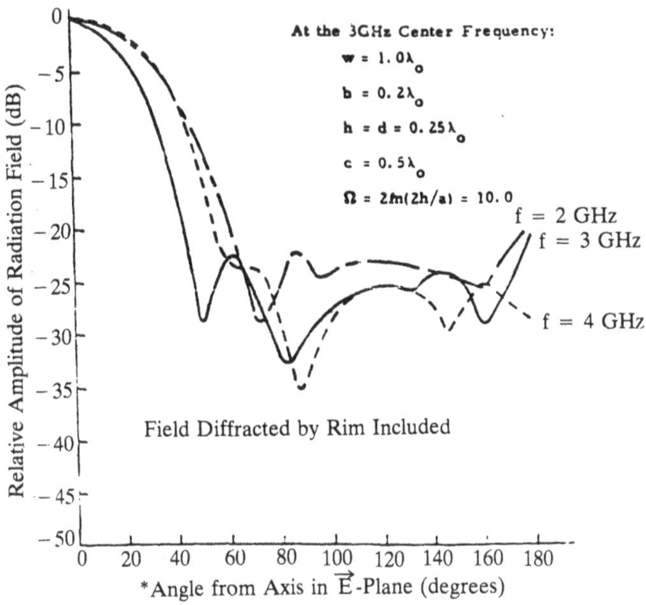

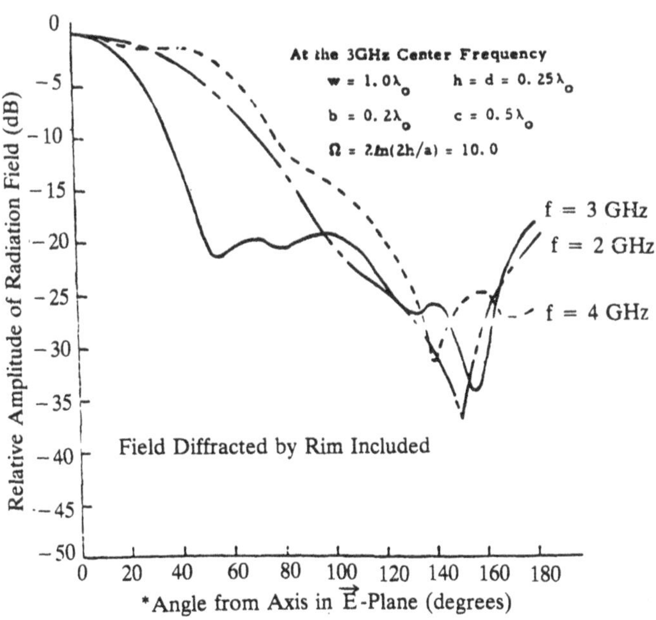

Figure 5.66(a) $\vec{E}$-plane (y-z plane) radiation patterns of an SBFA, obtained by the numerical-physical optics approach for various excitation frequencies (from [33])

(b) $\vec{H}$-plane (x-y plane) radiation patterns of an SBFA, obtained by the numerical-physical optics approach for various excitation frequencies (from [33])

Figure 5.67 shows the geometry of the SBFA model with a line source feed. The radiation pattern of the endfire antenna (i.e., without a small reflector) is shown in Figure 5.68. Figure 5.69(a) and (b) shows the dependence of the radiation pattern on the small reflector size d and lip protrusion p, respectively. The splitting off of the sidelobes and the increase in gain can be compared to Figure 5.68.

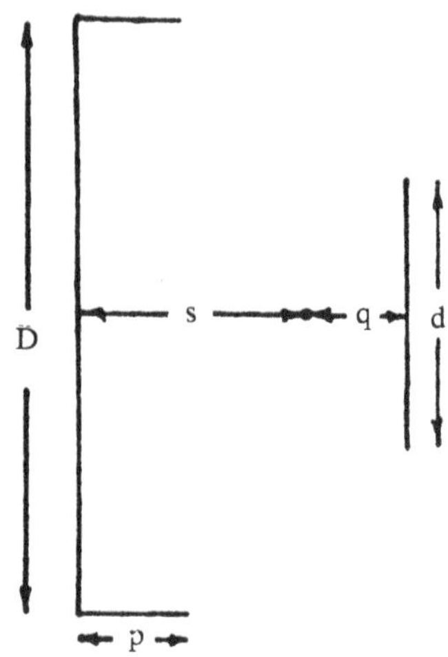

Figure 5.67 Geometry of the SBFA (from [36])

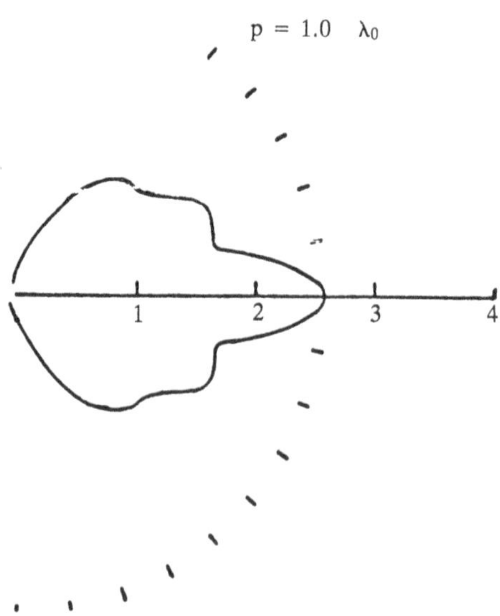

Figure 5.68 Far-field pattern of the SBFA, in the absence of a small reflector (from [36])

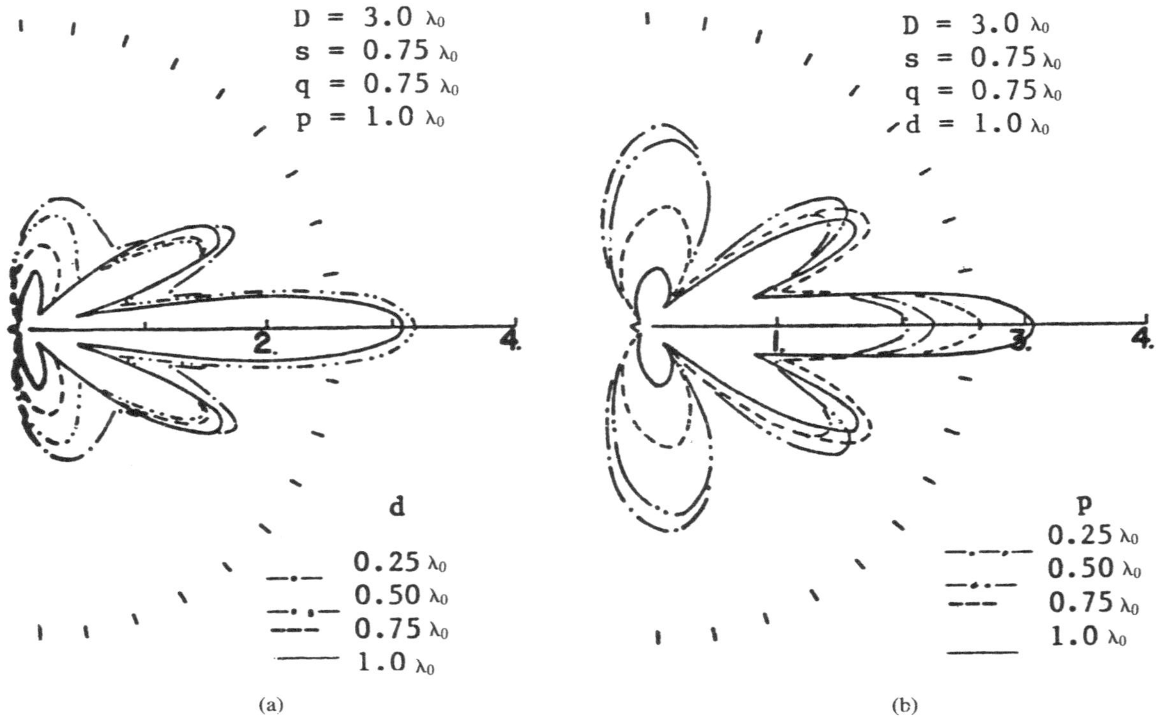

Figure 5.69(a) Radiation pattern as a function of small reflector size (from [36])
(b) Radiation pattern as a function of lip protrusion length (from [36])

The on-axis gain is approximately the same in Figure 5.69(a) and (b), but the position of the first sidelobe is more sensitive to lip protrusion than to the small reflector size. In addition, the radiation pattern becomes suppressed, and displays erratic behavior due to an improper choice of small reflector size and lip protrusion length. If the line source is placed closer to the large reflector, extremely high sidelobe levels result (Figure 5.70(a)). By placing the line source closer to the small reflector, a sharp beam with very low backlobes can be obtained, as shown in Figure 5.70(b).

In 1983, Changlu and Ximing [38] reported the *equivalent reflector method* for the analysis of an SBFA. They assumed the SBFA to have a "focal point," like a paraboloid antenna. Figure 5.71(a) and (b) shows the schematic of the SBFA radiating source and a geometry for calculating the equivalent focal point, s is assumed to be the equivalent radiating aperture of the SBFA, and the aperture fields are assumed distributed with equal amplitude E_0. The field at point F is given by:

$$E = j\frac{CE_0}{\lambda_0}\int_S \frac{e^{-jKr}}{r}\,dS \tag{5.20}$$

where

C is a constant,
K is wave number,
$r = (f^2 + R^2)^{1/2}$ is the distance from the radiating element to point F, and
$dS = Rd\phi dR$ is the radiating area element.

For a limited size of reflector diameter, (5.20) can be rewritten as:

$$\begin{aligned} E &= j\frac{CE_0}{\lambda_0}\int_0^{2\pi} d\phi \int_0^{R_0} \frac{e^{-jK}\sqrt{f^2+R^2}}{\sqrt{f^2+R^2}}\,RdR \\ &= -CE_0(e^{-jK}\sqrt{f^2+R_0^2} - e^{-jKf}) \end{aligned} \tag{5.21}$$

The field E in (5.21) will reach its maximum if

$$Kf = 2n\pi K \cdot \sqrt{f^2 + R^2} = (2n+1)\pi \quad (n = 1, 2, 3, \ldots) \tag{5.22}$$

The field E in (5.20) will give its maximum value. The shortest focal distance ($n = 1$) is $f = \lambda_0$ and $R_0 = 1.12\lambda_0$. The radiating field of an SBFA in space can be

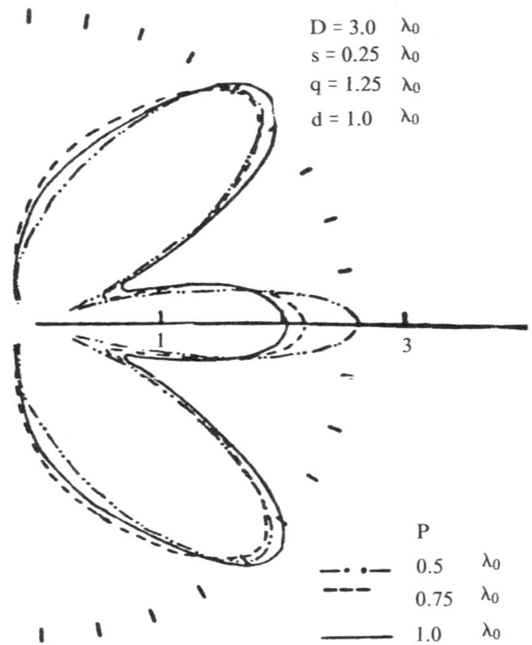

(a)

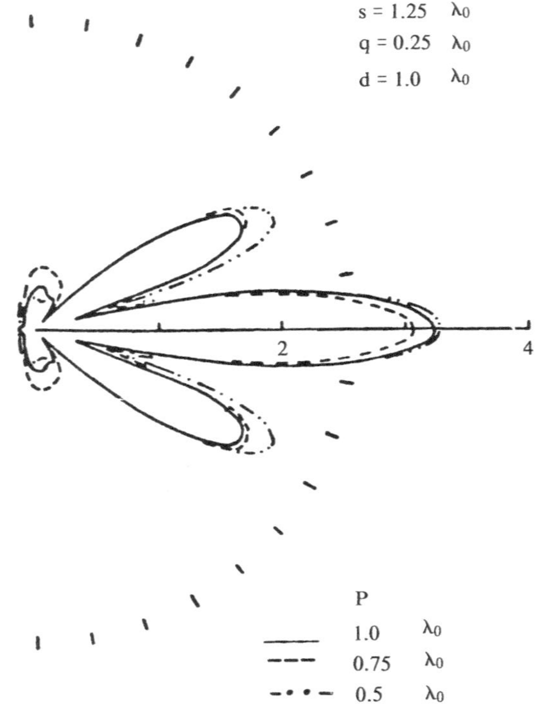

(b)

Figure 5.70(a) Line source placed closer to the large reflector (from [36])
(b) Line source placed closer to the small reflector (from [36])

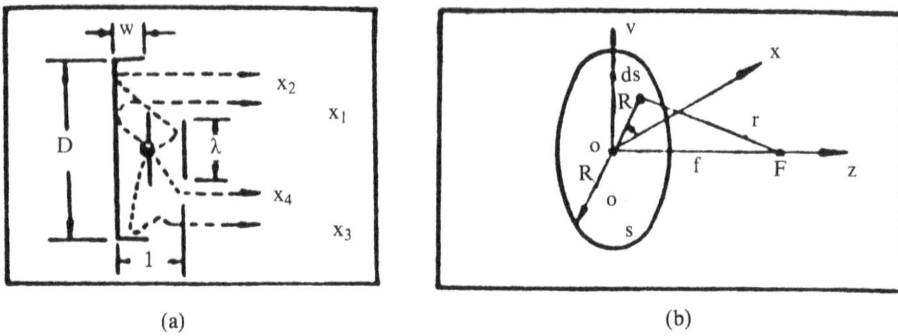

Figure 5.71(a) An SBFA radiating source (from [38])
(b) Geometry to calculate "equivalent focal point" (from [38])

considered to consist of four parts, as shown in Figure 5.71(a):

X_1 = radiating field reflected by the large reflector;
X_2 = radiating field reflected by the small reflector and then reflected by the large reflector;
X_3 = radiating field intercepted and reflected by edge ring; and
X_4 = all diffracted field through the edge of the small reflector.

In Figure 5.71(a), the geometric parameters D, d, W, l, and the distances from the feed to the large reflector and to the small reflector (the feed is placed at the equivalent focal point) are adjusted to make X_1, X_2, X_3, and X_4 in phase or with a phase difference of $2n\pi$. Figures 5.71(c) and (d) show the calculated results for optimal feed position and optimal rim width, respectively.

The gain of the SBFA, with the feed placed at the equivalent focal point, is about 2.5 dB higher than the normal SBFA. The SBFA has an optimal rim width of about $0.5\lambda_0$, and thus its gain is about 0.5 dB higher than that of the normal SBFA, with an edge ring size of about $0.25\lambda_0$. An SBFA can be analyzed as an equivalent reflector antenna.

Marougi [39] investigated the influence of the large and small reflectors used in SBFAs on the radiation impedance of a dipole feed element using near-field analysis. The two circular reflectors are divided into a number of small squares, as shown in Figure 5.72. The current density is assumed constant over the area of the square and equal to the calculated current density at the center of the square. Each square is then approximated by a current element carrying the evaluated current density. The current density induced at the center of each square due to the radiation by the dipole element is given by:

$$J = j\frac{I_0}{2\pi}\frac{h}{(Y^2 + h^2)}\{\exp(-j\beta R_1) + \exp(-j\beta R_2)$$
$$- 2\exp(-j\beta R_0)\cos\beta H\} \tag{5.23}$$

where

$\beta = 2\pi/\lambda_0$, and the geometrical parameters of (5.23) are shown in Figure 5.72.

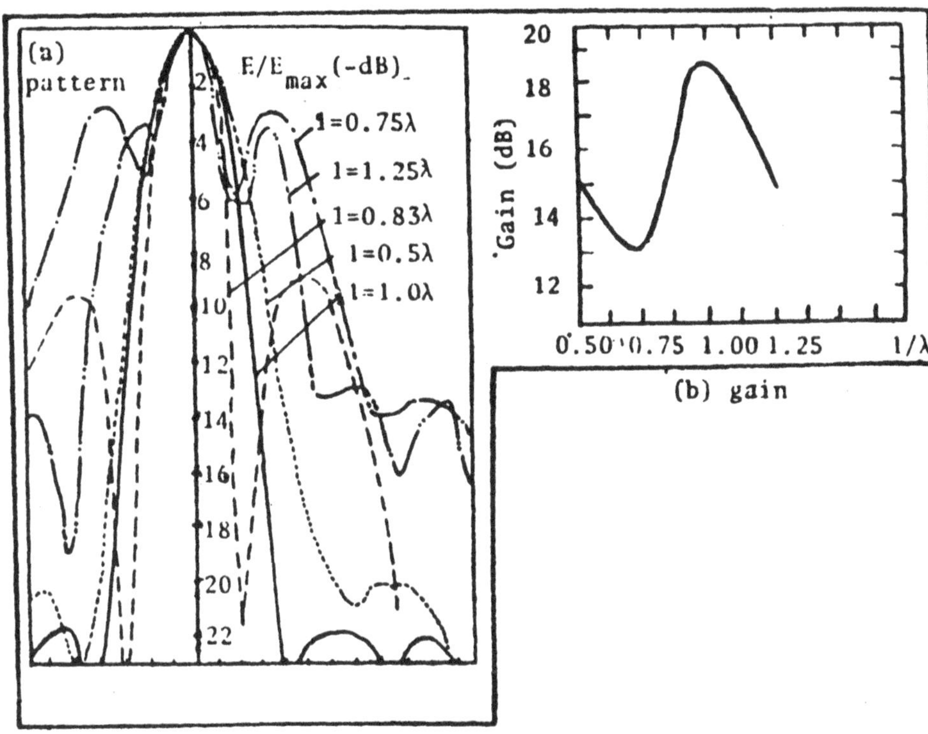

Figure 5.71(c) The effect of feed position on radiation patterns and gain (from [38])

The tangential field component produced by the current element for any point on the dipole is given by:

$$E_{tang} = \frac{jI_0}{2\pi}\frac{h}{(Y^2 + h^2)}w^2\frac{Z_0\exp(-j\beta r)}{4\pi r^2}$$

$$\times \{\exp(-j\beta R_1) + \exp(-j\beta R_2)$$
$$- 2\exp(-j\beta R_0)\cos\beta H\}\left\{\left(\frac{2}{r^2} + \frac{2}{j\beta r^3}\right)\right.$$
$$\left.\times (X - l)^2 - \left(\frac{j\beta}{r} + \frac{1}{r^2} + \frac{1}{j\beta r^3}\right)(h^2 + Y^2)\right\} \tag{5.24}$$

where w^2 is the area of the square which has been assumed as a current element.

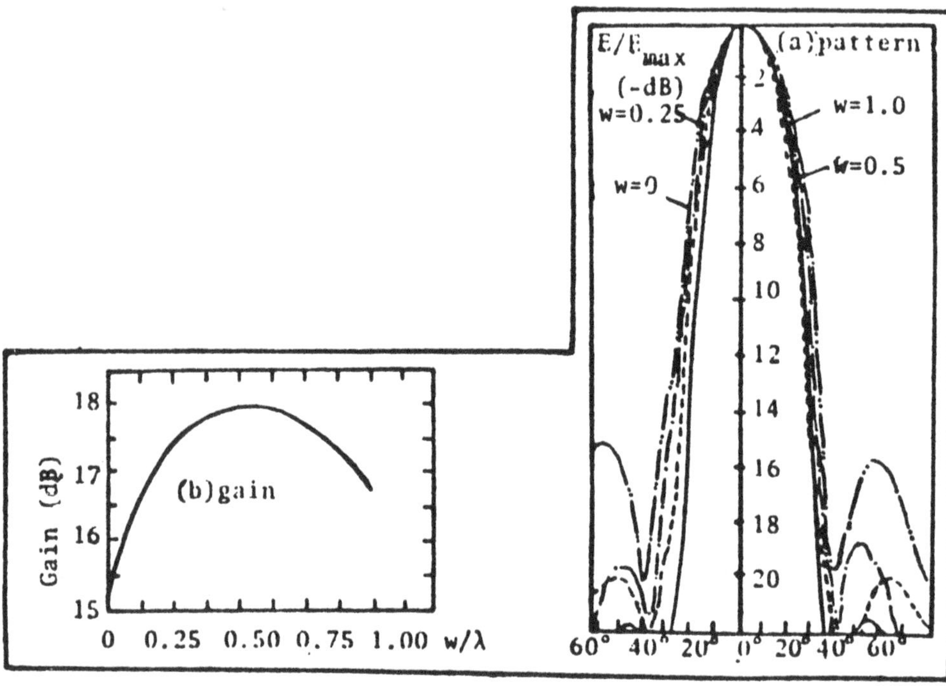

Figure 5.71(d) The effect of rim width on patterns and gain (from [38])

The resistive part ΔR and reactive part ΔX of the change dipole impedance can be evaluated by summing all the current elements representing the surface wave reflectors. Figure 5.73 (a) and (b) shows the calculated results.

At resonance ($\beta H = 90°$), most of the change in the radiation resistance of the dipole is due to the presence of a large reflector, and the small reflector contributes

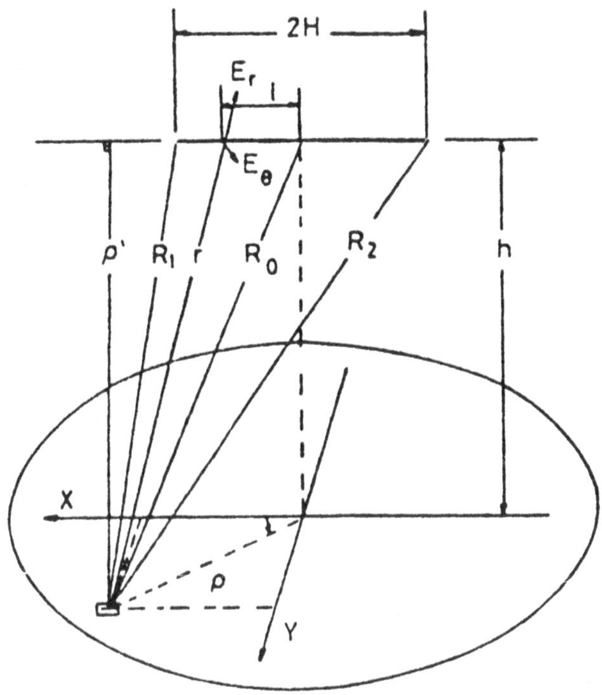

Figure 5.72 Geometrical definitions of variables on the surface wave reflector and the dipole element (from [39])

little to the antenna's input resistance. However, the change in the reactive part is equally affected by both reflectors.

Kishk *et al.* [40] used a numerical method (with different excitation types) to study the performance of an SBFA configuration. A numerical optimization performed for the antenna is described in Figure 5.74 ($D_m = 2.0\ \lambda_0$, $D_s = 0.5\ \lambda_0$, $H_r = 0.25\ \lambda_0$). In these calculations, the dipole current distribution is assumed to be sinusoidal. Figure 5.74 shows the radiation patterns of an antenna which has dimensions of $D_m = 2.2\ \lambda_0$, $D_s = 0.6\ \lambda_0$, and $H_r = 0.45\ \lambda_0$, which are optimal to achieve 1.4 dBi gain higher than Ehrenspeck's antenna. An excitation by a coaxial waveguide provides higher gain than a circular waveguide. Figure 5.75 shows geometry ($D_m = 2.64\ \lambda_0$, $D_s = 0.8\ \lambda_0$, $W_e = 0.15\ \lambda_0$, and $H_r = 0.6\ \lambda_0$) and radiation patterns of an SBFA excited by a coaxial waveguide. Table 5.16 shows a comparison of the gains and aperture efficiencies of these antennas.

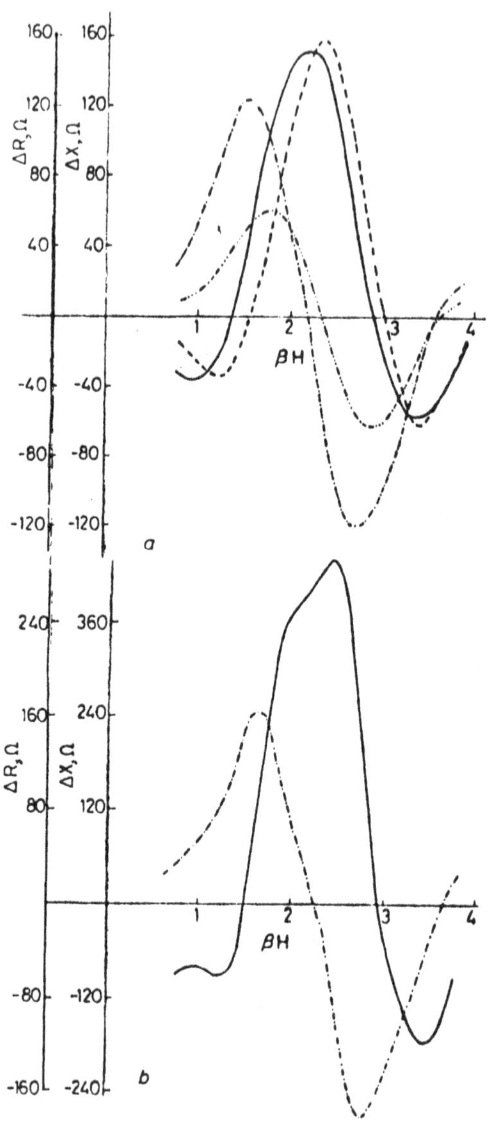

Figure 5.73 Plots of change in impedance of the antenna (from [39]): (*a*) ΔR due to large reflector ———; ΔR due to small reflector -----; ΔX due to large reflector —·—·—·—·—· ; ΔX due to small reflector —···—···—··· ; (*b*) due to presence of two reflectors, ΔR ———; and ΔX —·—·—·—·

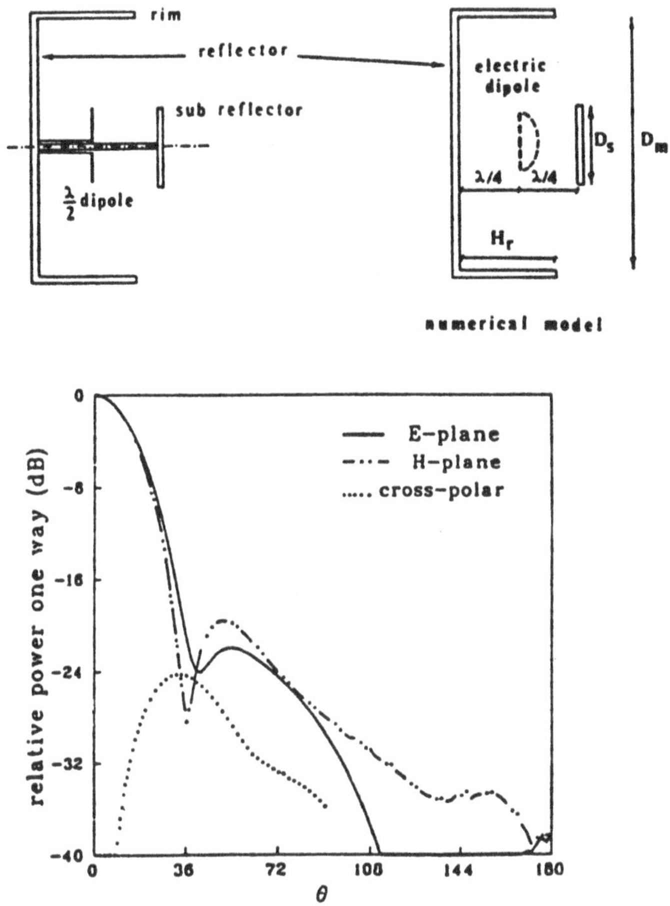

Figure 5.74 Radiation patterns of the optimized antenna fed by a dipole with different excitation type (from [40])

5.4 MICROSTRIP BACKFIRE ANTENNAS

5.4.1 Introduction

Over the past several years, the printed-circuit antenna has proven to be an efficient and effective radiator. This section discusses only *microstrip backfire antennas*

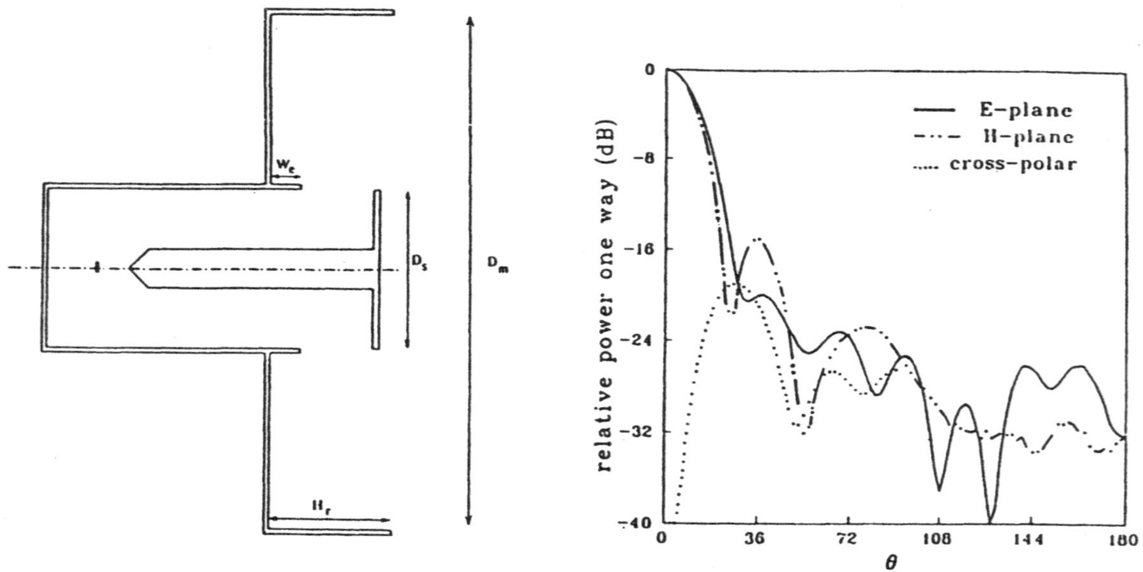

Figure 5.75 Structural dimensions and radiation patterns of a coaxial-waveguide-fed SBFA (from [40])

Table 5.16 Comparison of SBFAs

Parameters	*Ehrenspeck's Antenna*	*Optimized Dipole (Excited)*	*Coaxial Waveguide (Excited)*
Gain	15.00 dBi	16.390 dBi	18.035 dBi
Aperture efficiency	80.15%	91.17%	92.45%

(MBFAs). The MBFA configuration combines a microstrip-type element with a backfire cavity. This type of antenna was studied by Kaloi, *et al.* [41] and Van de Capelle, *et al.* [42–43]. The MBFA is an interesting antenna element, because it can provide more than three times the gain over a single microstrip element by optimizing its parameters. Therefore, it can replace four or more elements of a conventional microstrip array.

MBFAs may be useful for applications in aircraft, spacecraft, and missiles because of their excellent electrical performance, light weight, and small size. In this section, we will give a detailed description of MBFA construction. We will also discuss experimental and theoretical results, including the design parameters of several different types of MBFA.

5.4.2 Description of MBFA Construction

The MBFA consists of a microstrip element between a main reflector and a subreflector. The antenna works like a Fabray-Perot cavity, characterized by multiple reflections of electromagnetic waves between the reflectors and by the energy being radiated normal to and through the subreflector.

Figures 5.76 and 5.77 show MBFAs using typical square and circular elements, respectively. In these figures, a microstrip radiating element is mounted approximately one quarter-wavelength ($\lambda_0/4$) above a ground plane, and a subreflector is mounted parallel above the radiating element, approximately one quarter-wavelength ($\lambda_0/4$) from it. The cavity formed by the space between the main reflector and subreflector can be air or a dielectric material. Where the cavity is filled with a dielectric material, the spacing between the main reflector and the subreflector should be approximately a half-wavelength in the dielectric material.

The shape of the microstrip radiating element can be chosen based on the desired antenna characteristics. The shape of the subreflector can be different from that

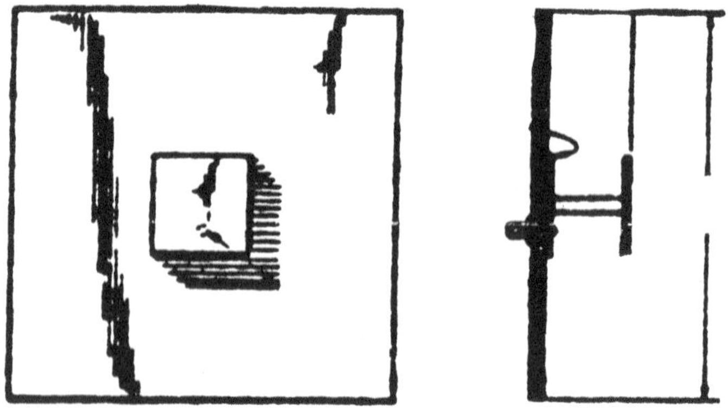

Figure 5.76 MBFA with square shape (from [41])

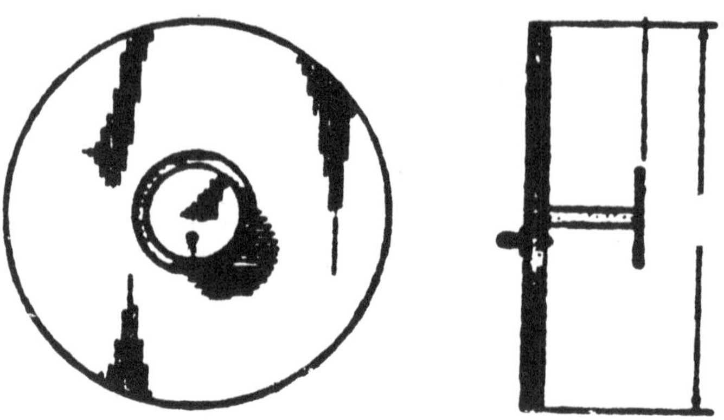

Figure 5.77 MBFA with circular shape (from [41])

of the radiating element, but in most cases it is the same. In the above figures, square and circular shapes of subreflectors are chosen, respectively, for the square and the circular radiating elements. The subreflector is supported above the radiating element by means of a central support made of a suitable dielectric material.

Figure 5.78 shows an MBFA design using a square coplanar microstrip radiating element. The main reflector is coplanar with the element. The subreflector

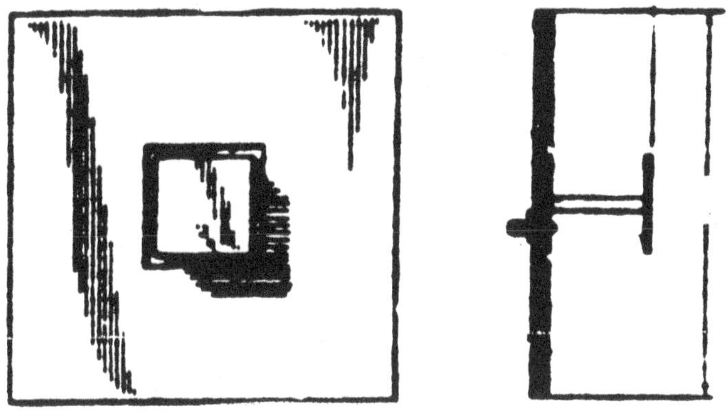

Figure 5.78 Coplanar square MBFA (from [41])

is approximately the same size as the radiating element, and the shapes of both reflectors are made square to provide symmetry with it. Figure 5.79 shows an MBFA design using a circular coplanar microstrip radiating element. The main reflector is coplanar, with the element and subreflector again approximately the same size. In this configuration, both reflectors are made circular, to provide symmetry with the microstrip radiating element.

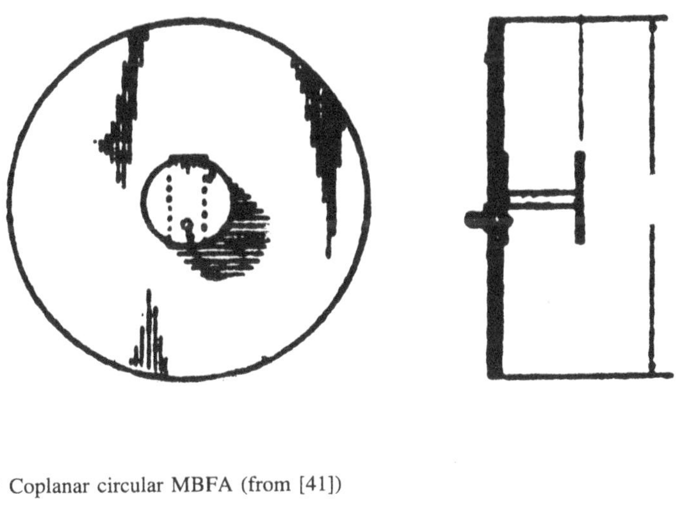

Figure 5.79 Coplanar circular MBFA (from [41])

Van de Capelle, *et al.* [42] studied MBFA with a square resonator of side W_p and a square subreflector of side W_s at X-band. The ground plane is surrounded by a vertical metal rim of height H_r. The distance between the subreflector and ground plane is H_s, and between ground plane and microstrip radiating element is H_p, as shown in Fig. 5.80.

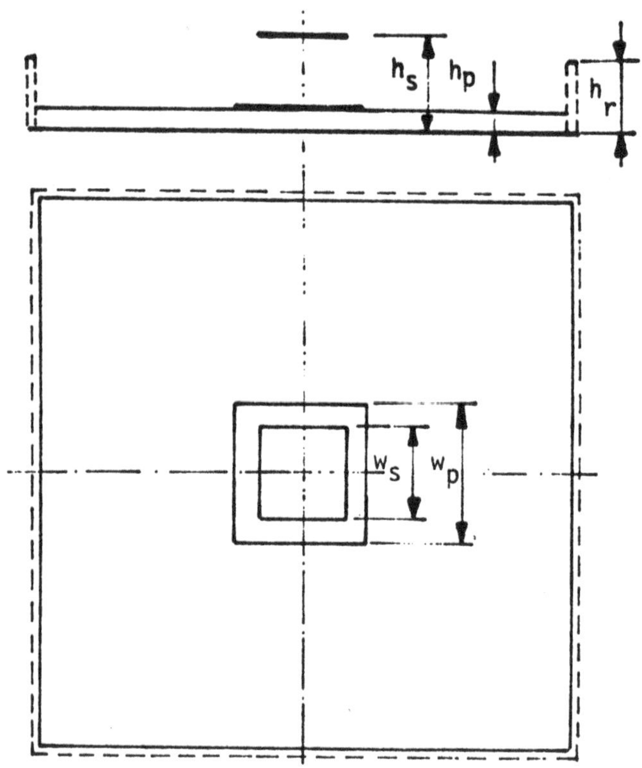

Figure 5.80 Square MBFA (from [42])

For mechanical convenience, the subreflectors are etched onto a thin dielectric substrate, with the same side dimensions as the main reflector (ground plane). This substrate is mounted with four teflon screws above the main reflector.

5.4.3 Experimental Studies

Van de Capelle, *et al.* [42] experimentally studied the square-shaped MBFA shown in Fig. 5.80. They varied the height H_s of the subreflector in steps of 0.5 mm, and recorded the reflection pattern as a function of frequency. They studied four sizes of subreflector, shown in Table 5.17, with the influence of the dielectric substrate

Table 5.17 Dimensions of Investigated Small Reflectors (from [42])

Small Reflector	*Ratio* (W_s/W_p)
A	1.00
B	0.92
C	0.85
D	0.79

taken into account during the measurement. In Figure 5.81(a), the measured minimum reflection as a function of the ratio H_s/λ_0 for the subreflectors *A*, *B*, *C*, and *D* is designated by solid lines. The minimum reflection as a function of the ratio H_s/λ_0 for the substrate without subreflector is shown in the figure by a dotted line. Figure 5.81(b) shows the 10 dB reflection bandwidth as a function of H_s/λ_0 for four different subreflectors. It is evident that the bandwidth increases linearly with the height H_s.

Figure 5.81(c) shows the *H*-plane half-power beamwidth *versus* H_s/λ_0 for the four different reflectors. This was derived from the measured *H*-plane pattern. The half-power beamwidths and the sidelobe level as functions of the size of the subreflector (W_s/W_p) are plotted in Figure 5.81(d) for a fixed value of $H_s = 0.6\ \lambda_0$. The 3 dB beamwidth decreases linearly with W_s/W_p, and the sidelobe level increases at about the same rate. Depending on the value of the acceptable sidelobe level, the minimum possible beamwidth can be derived.

Figure 5.81(e) shows the plot of sidelobe level *versus* frequency of a square MBFA for the four different subreflectors. The left sidelobe level and the right sidelobe level are shown by solid lines and dotted lines, respectively. The best level is shown around the resonant frequency of the MBFA (or at a slightly higher value). However, the lowest reflection value occurs at frequencies below the resonant frequency.

The *H*-plane patterns for the microstrip resonator antenna (with and without small reflector) are shown in Figure 5.81(f). The microstrip antenna pattern and MBFA pattern are shown by dotted lines and solid lines, respectively. Figure 5.81(g) shows the *H*-plane pattern of an MBFA with vertical rim and coplanar metal plane subreflector *A*. An *H*-plane 3 dB beamwidth of 32 degrees and sidelobe level below −20 dB are achieved, as shown in the figure.

The parameters that provide control for obtaining optimum operation conditions for the MBFA are the small reflector and large reflector diameters, the small reflector and large reflector separation, symmetry between the microstrip radiating element and the reflectors, parallelism and consistency of the reflection plane, and matching of the microstrip radiating element to the backfire cavity formed by the space between the small reflector and the large reflector. With a large reflector diameter of

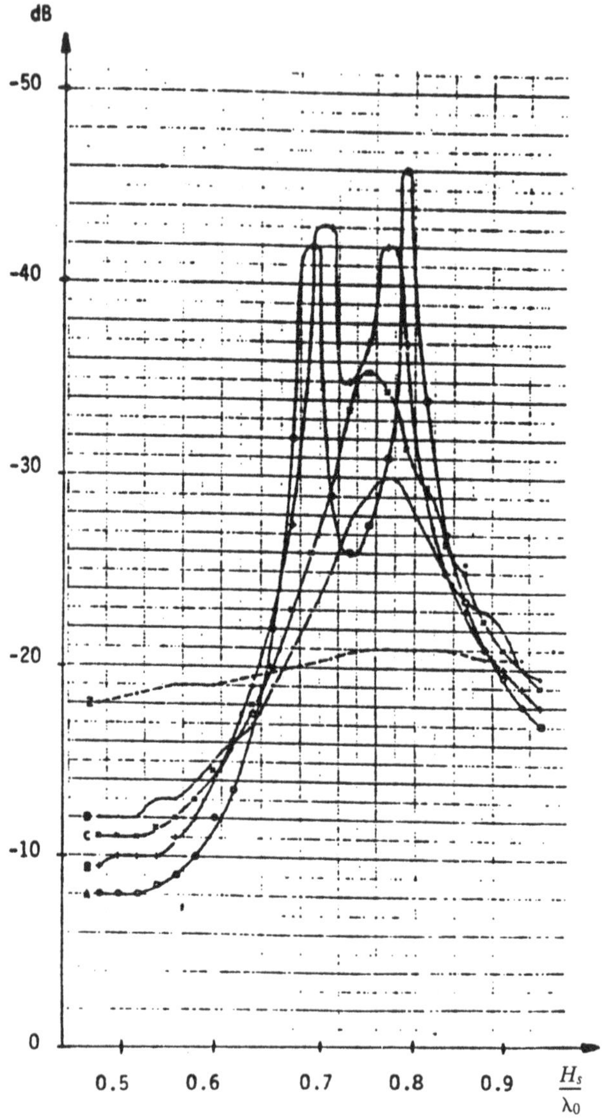

Figure 5.81(a) Minimum reflector *versus* H_s/λ_0 of a square MBFA (from [42])

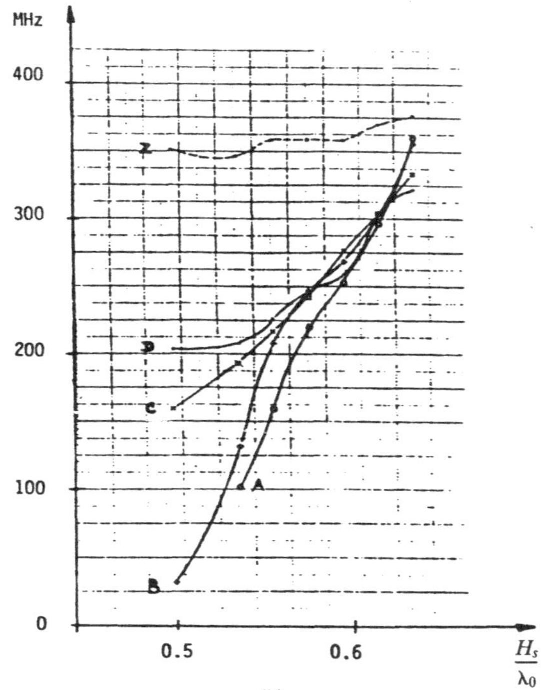

(b)

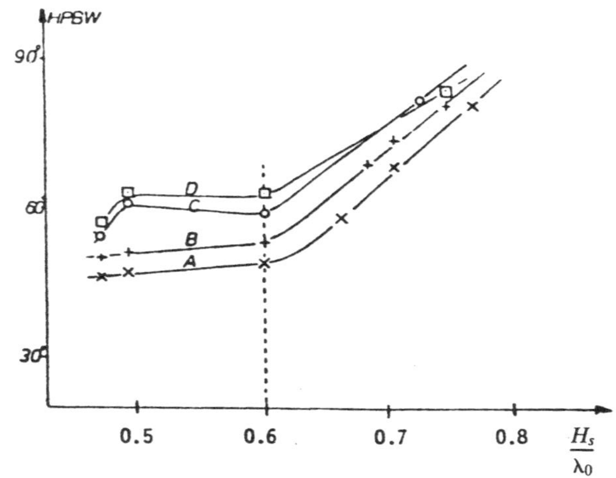

(c)

Figure 5.81(b) 10 dB bandwidth as a function of H_s/λ_0 of a square MBFA (from [42])

(c) 3 dB beamwidth *versus* H_s/λ_0 of a square MBFA (from [42])

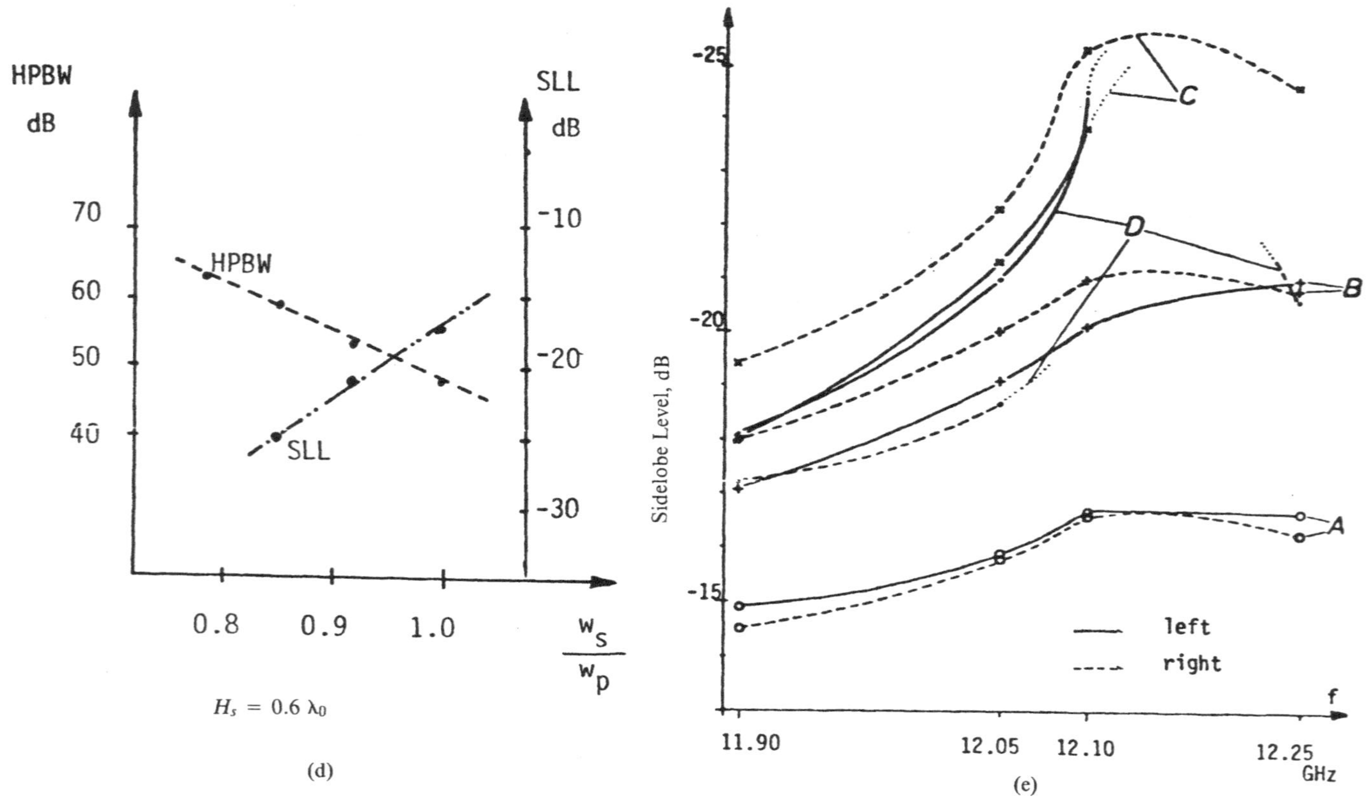

Figure 5.81(d) Dependence of sidelobe level and half-power beamwidth on small reflector size (from [42])

(e) Sidelobe level *versus* frequency plot of a square MBFA (from [42])

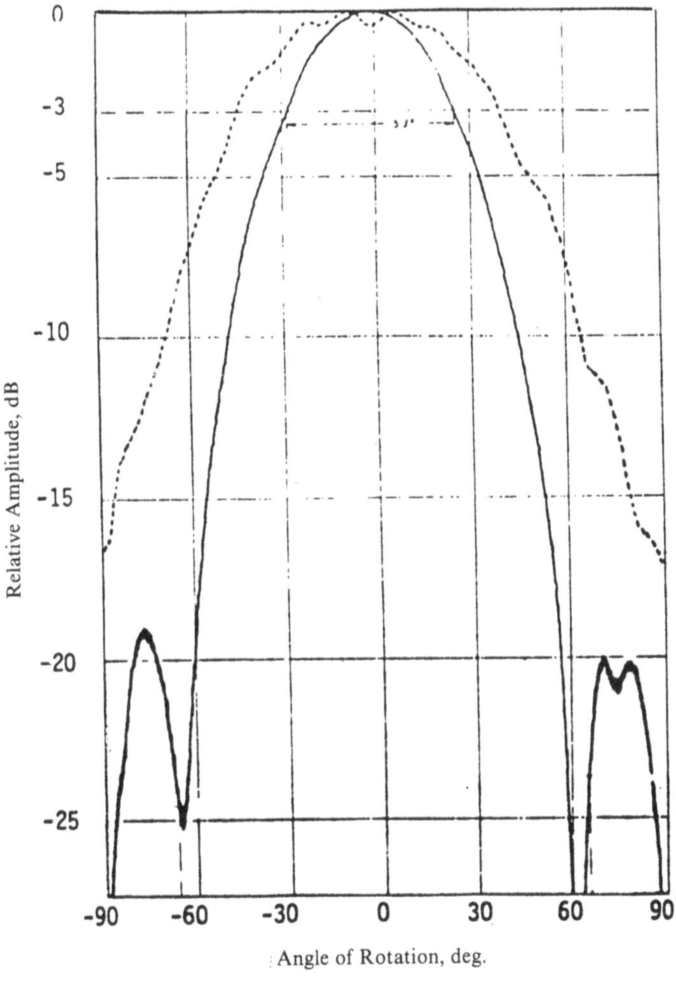

Figure 5.81(f) *H*-plane radiation patterns of a microstrip resonator antenna and an MBFA with small reflector *A* (from [42])

1.5 λ_0, a gain of approximately 10 dB is obtained, with a circular radiating element and circular coplanar reflectors. A maximum gain of approximately 12 dB has been obtained for large reflectors having greater diameters, but there is no substantial increase in gain for a large reflector with a diameter greater than 2 λ_0.

Figure 5.82(a) shows a radiation pattern of the MBFA shown in Figures 5.76 and 5.77. Figure 5.82(b) shows a radiation pattern of the antenna shown in Figures 5.78 and 5.79. The large reflector diameter for the radiation pattern shown in Fig.

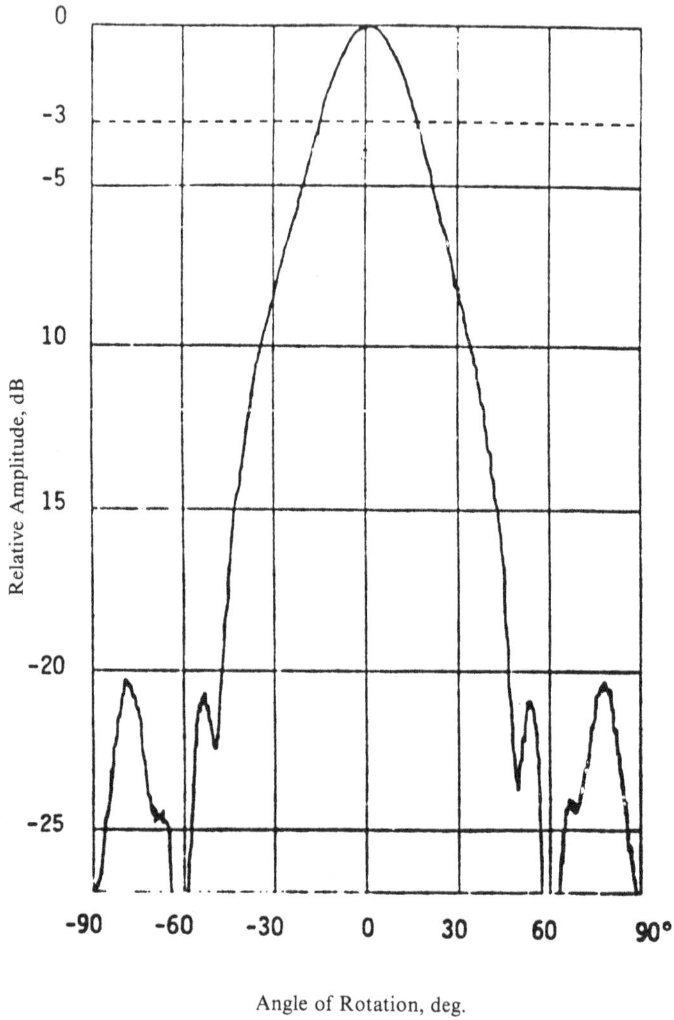

Figure 5.81(g) *H*-plane radiation pattern for an MBFA, with vertical rim and coplanar metal plane reflector *A* (from [42])

5.82(a) is 1.5 λ_0, whereas for the radiation pattern shown in Figure 5.82(b), the large reflector has a diameter of 2 λ_0.

5.4.4 Theoretical Analysis of the MSBFA

Various authors [44–48] have analyzed the microstrip antenna using the moment method. In the microstrip antenna, a dielectric layer is placed between the element

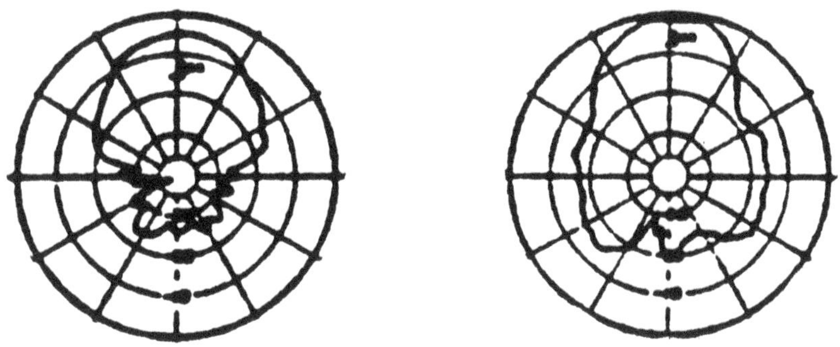

(a) (b)

Figure 5.82(a) Radiation pattern of an MBFA with 1.5 λ_0 large reflector diameter (from [41])
(b) Radiation pattern of an MBFA with 2 λ_0 large reflector diameter (from [41])

and the ground plane. Usually, the lower values of the real and imaginary parts of the permittivity are used for microstrip antennas in aerospace industries [49–50], which require low loss and light weight. In the presence of a dielectric slab, we use the dyadic Green's functions, which satisfy the boundary conditions for a unit current located in the plane of the microstrip patch.

Figure 5.83(a) and (b) shows the theoretical configuration of the microstrip antenna and the MSBFA, which were investigated by Vandenbosch *et al.* [43,52], Vandenbosch [51] and Kaoli [52a]. These authors have derived integral equations to calculate the total field. The two electric field integral equations express the conditions to be satisfied by the current distributions: the total tangential electric field must be zero both on the patch and on the small reflector. Thus,

$$\bar{n} \times \bar{E}^{inc} = \bar{n}x\left(\frac{1}{4\pi j\omega\varepsilon}\iint_{patch} (-\omega^2\mu\varepsilon\bar{J}_p + \bar{\nabla}_p'\cdot\bar{J}_p\cdot\bar{\nabla}\phi)\mathrm{d}A \right.$$
$$\left. + \frac{1}{4\pi j\omega\varepsilon}\iint_{subreflector} (-\omega^2\mu\varepsilon\bar{J}_s + \bar{\nabla}_s'\cdot\bar{J}_s\cdot\bar{\nabla}\phi)\mathrm{d}A\right) \quad (5.25)$$

where

J_p is the current distribution on the microstrip patch, and
J_s is the current distribution on the small reflector

A set of two matrix equations is derived from the transformation of the integral equations, as follows:

$$E^{inc}_{patch} = pp \cdot J_p + Sp \cdot J_S \tag{5.26}$$

$$E^{inc}_{subreflector} = pS \cdot J_p + SS \cdot J_S \tag{5.27}$$

where *pp*, *Sp*, *pS* and *SS* are matrices of elements. These matrices are dependent on the choice of expansion and testing functions. Vandenbosch, *et al.* [43] showed that a rooftop function is very efficient. Testing is done by integrating the equation along a one-dimensional interval (Figure 5.83(c)).

$$J_S = (pS \cdot pp^{-1} \cdot Sp - SS)^{-1} \cdot (pS \cdot pp^{-1} \cdot E^{inc}_{patch} - E^{inc}_{subreflector}) \tag{5.28}$$

$$J_p = pp^{-1} \cdot E^{inc}_{patch} - pp^{-1} \cdot Sp \cdot J_s \tag{5.29}$$

The radiation patterns of the antenna are calculated as a superposition of the field radiated by the independent source and the currents J_p and J_s.

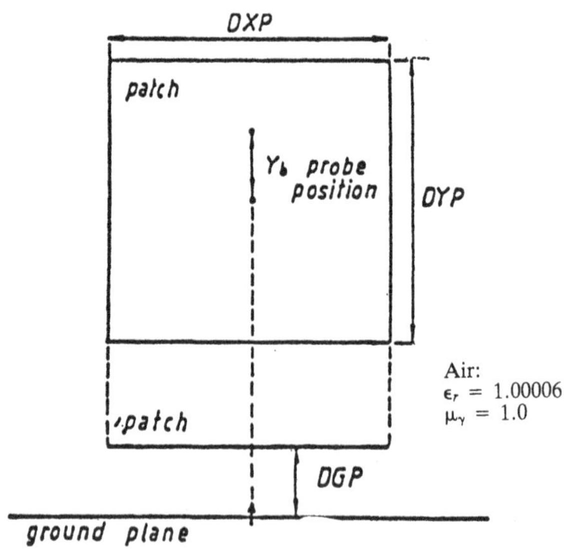

Figure 5.83(a) The microstrip antenna (from [43])

For a microstrip antenna, the small reflector is omitted and the currents J_p and J_S can be written as:

$$J_S = 0 \tag{5.30}$$

$$J_p = pp^{-1} \cdot E^{inc}_{patch} \tag{5.31}$$

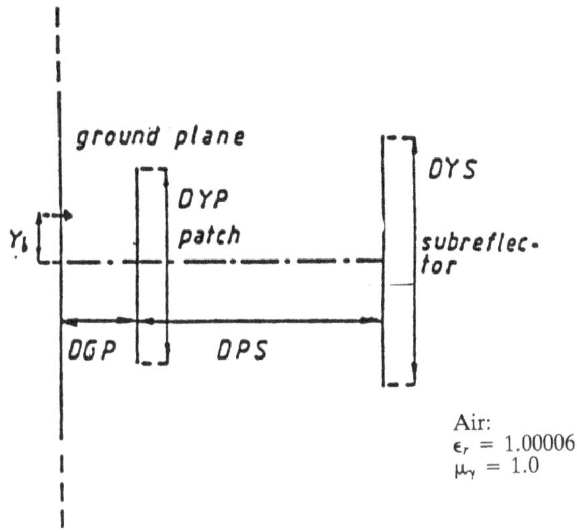

Figure 5.83(b) The MBFA (from [43])

Figure 5.84(a) and (b) shows the measured and calculated E-plane, H-plane, and cross-polarization patterns of a microstrip antenna. The antenna dimensions are shown in Figure 5.83(b) and Table 5.18. Figure 5.85(a) and (b) shows the calculated real and imaginary parts of the impedance, with variation in frequency.

Figure 5.83(b) and Table 5.19 show the dimensions of the MBFA used for the analysis.

Figure 5.86(a) and (b) shows the measured and calculated E-plane, H-plane, and cross-polarization patterns of the antenna at 1.5885 GHz. The calculated and the measured $Re(Zin)$ are 131.64 and 118.96 ohms, respectively. The imaginary part of the calculated and measured impedances are -204.74 and -176.07 ohms, respectively. The agreement between calculated and measured results is good. However, an accelerated convergence in function of the number of expansion functions can produce more accurate results.

5.5 SBFA ARRAYS

Single-element SBFAs are discussed in detail in previous sections. This type of antenna is suitable as an array element for high-gain antennas. Because it has a relatively wide radiating aperture, the number of SBFAs needed for illuminating a reflector area is much smaller than the number of dipoles or slots required for illuminating the same area. As far as gain and pattern quality are concerned, each element of an

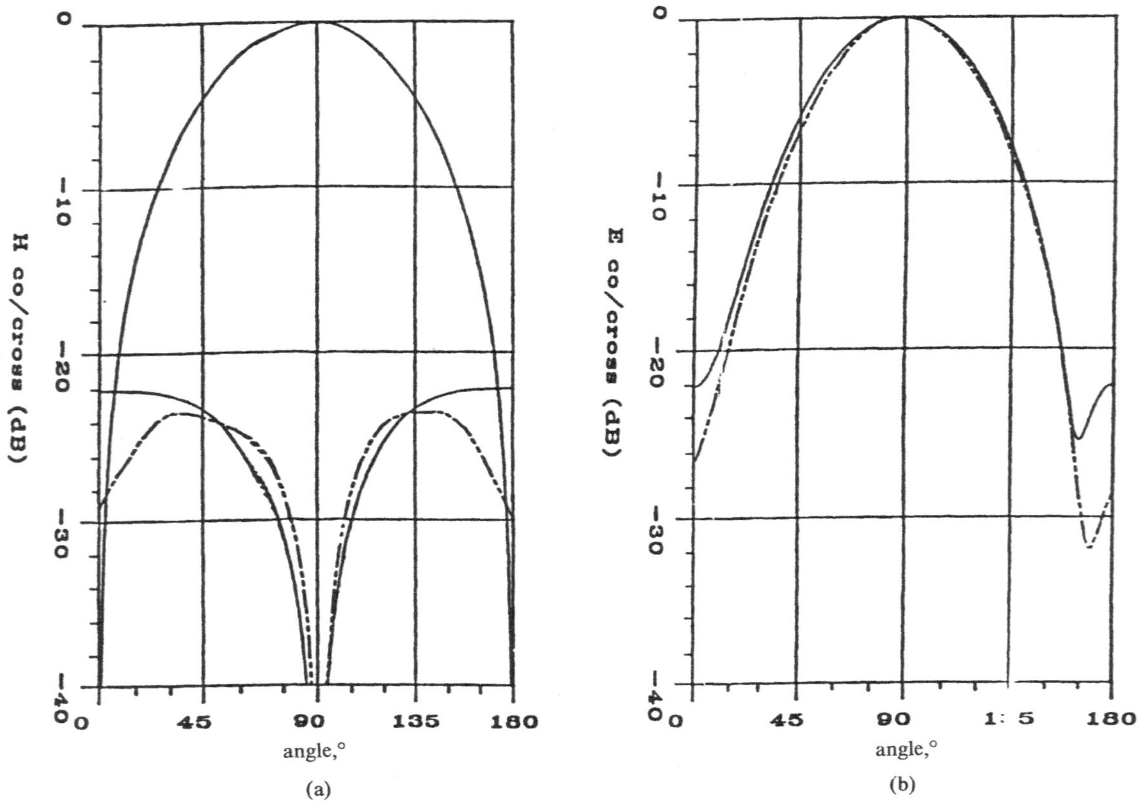

Figure 5.84(a) *H*-plane co-polar and cross-polar pattern of a microstrip antenna (from [51]): measured (1.5455 GHz) ———; calculated (1.5885 GHz) -·-·-·-·-

(b) *E*-plane co-polar pattern of a microstrip antenna (from [51]): measured (1.5455 GHz) ———; calculated (1.5885 GHz) -·-·-·-

Table 5.18 Typical Microstrip Antenna Dimensions

Parameters (See Figure 5.83(b))	*Dimensions in m*
DXP	0.0825
DYP	0.0825
X_b	0.0000
Y_b	0.0220
DGP	0.0070

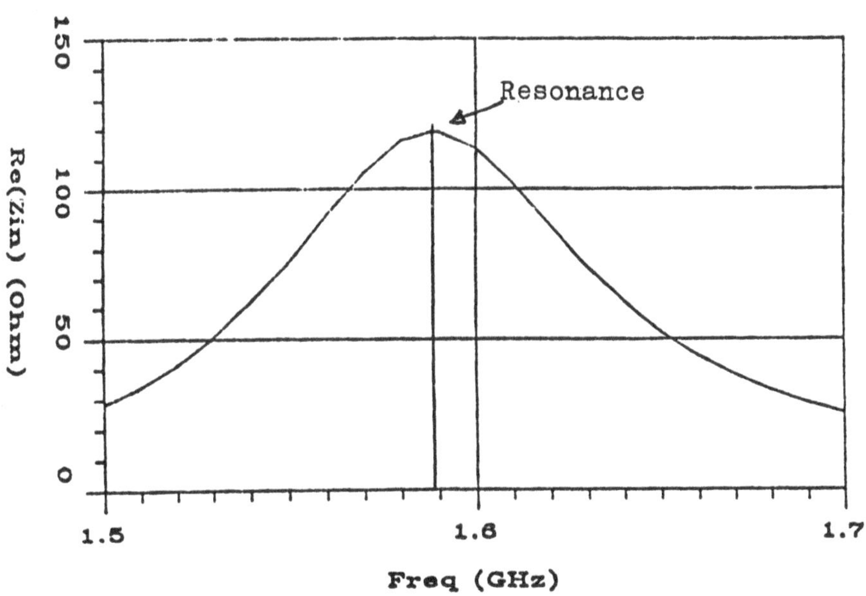

Figure 5.85(a) Calculated plot of *Re*(*Zin*) *versus* frequency of a microstrip antenna (from [51])

SBFA array is the equivalent of from four to six elements of a conventional multidipole array. In this section, we discuss the experimental and theoretical work on SBFA arrays.

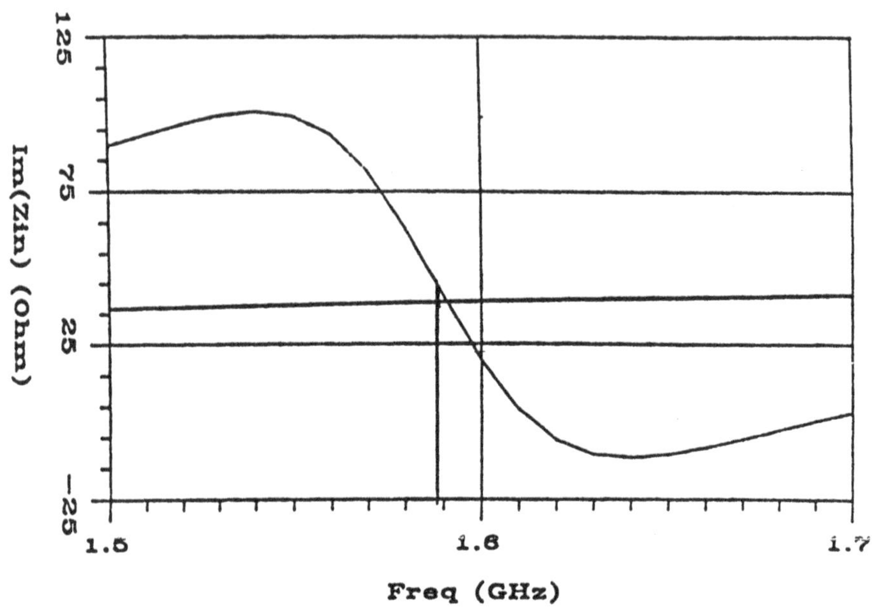

Figure 5.85(b) Calculated *Im*(*Zin*) *versus* frequency of a microstrip antenna (from [51])

Table 5.19

Parameters	*Dimensions in m*
DPS	0.0805
DGS	0.0805

5.5.1 Experimental Investigations of SBFA Arrays Fed by Crossed Dipoles

Ehrenspeck and Strom [53] have investigated a two-element SBFA as an array. Figure 5.87 shows a sketch of the array, which consists of two crossed-dipole feeds, each with a reflector disk supported by a rod, and the oval-shaped planar reflector with a rim. Table 5.20 gives the optimized dimensions and electrical characteristics of the array at 3 GHz. Figure 5.88 shows the *E*-plane and *H*-plane radiation patterns of the two-element SBFA array.

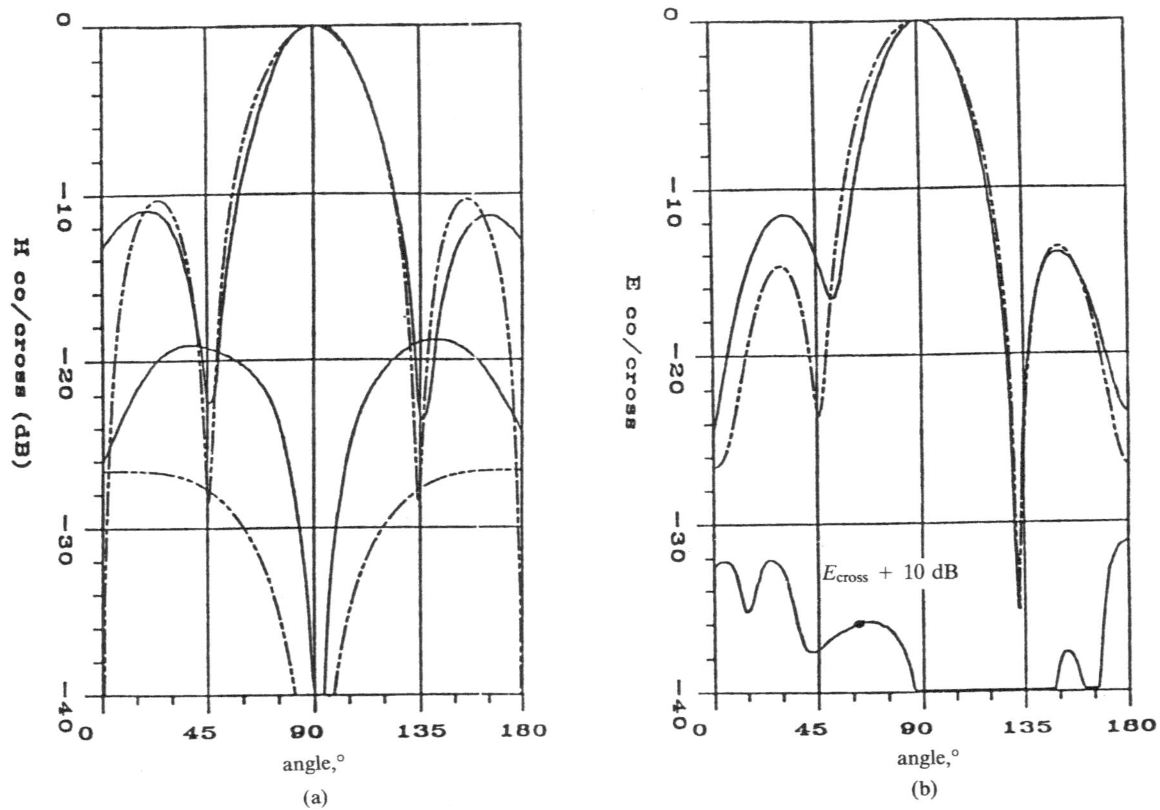

Figure 5.86(a) Measured and calculated H-plane co-polar and cross-polar patterns of an MBFA at 1.5885 GHz (from [51]): measured ———; calculated - - - - - -

(b) Measured and calculated E-plane co-polar and cross-polar patterns of an MBFA at 1.5885 GHz (from [51]): measured ———; calculated - - - - -

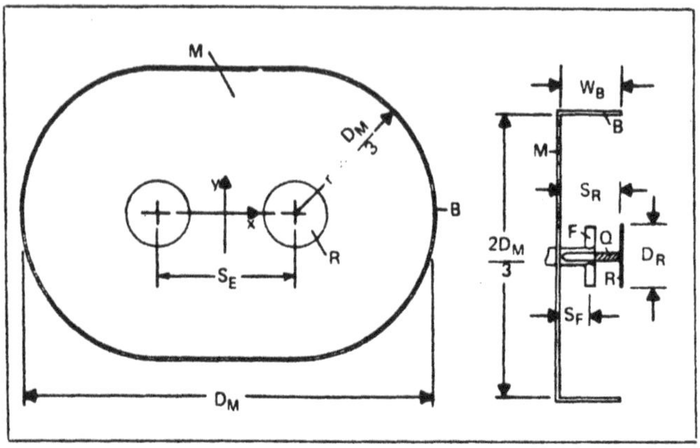

Figure 5.87 Sketch of two-element SBFA array (from [53])

Table 5.20 Dimensions and Electrical Characteristics of the Two-Element SBFA Array

$$D_M = 3.00\lambda_0$$
$$D_R = 0.50\lambda_0$$
$$S_E = 1.00\lambda_0$$
$$S_F = 0.25\lambda_0$$
$$S_R = 0.50\lambda_0$$
$$W_B = 0.50\lambda_0$$
$$M = 5.14\lambda_0^2 \text{ (area)}$$

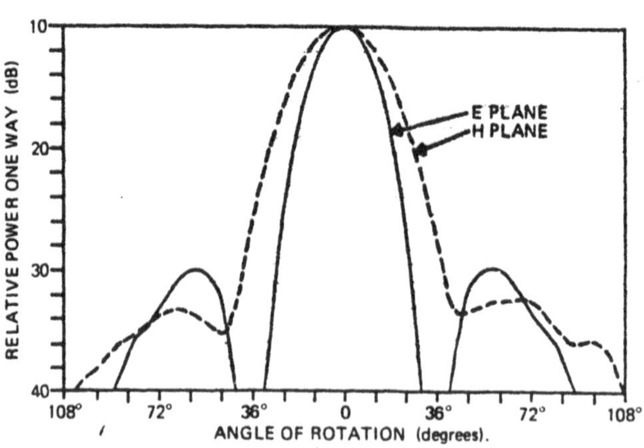

Figure 5.88 Radiation patterns of two-element SBFA array [53]. *E*-plane ———; *H*-plane ------

The sketch of a four-element SBFA array is shown in Figure 5.89. Its dimensions (in wavelengths) are shown in Table 5.21. The array consists of four separately-fed SBFA elements. The E-plane and H-plane radiation patterns, and directivity of the antenna as functions of the small reflector diameter were measured. The deduced directivity *versus* small reflector (disk) diameter, half-power beamwidth *versus* small disk diameter, and sidelobe level *versus* disk diameter are plotted in Figure 5.90(a) and (b).

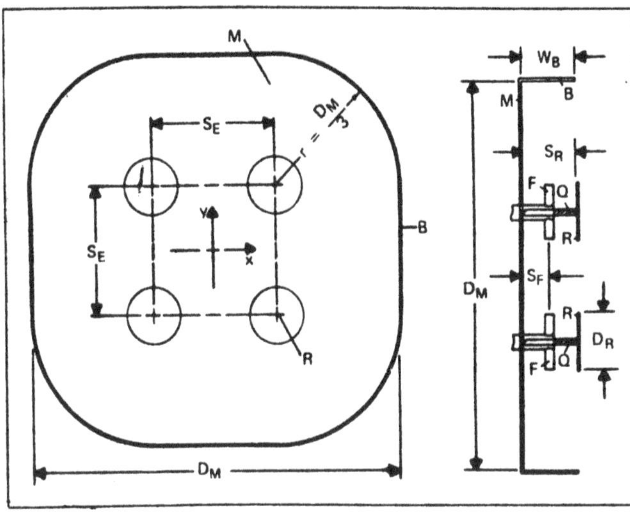

Figure 5.89 Sketch of a four-element SBFA array (from [53–55])

Table 5.21 Dimensions of the Four-Element SBFA

$$D_M = 3.00\ \lambda_0$$
$$S_M = 1.00\ \lambda_0$$
$$W_B = 0.50\ \lambda_0$$
$$S_R = 0.50\ \lambda_0$$
$$S_F = 0.25\ \lambda_0$$
$$M = 8.14\ \lambda_0^2\ \text{(area)}$$

Figure 5.91(a), (b) and (c) shows the four-element SBFA array directivity, sidelobe level, and half-power beamwidth as functions of frequency for the small reflector diameters of 0.450 λ_0, 0.475 λ_0, and 0.500 λ_0, respectively. A maximum directivity of more than 20 dB has been shown for a certain frequency f; sidelobe

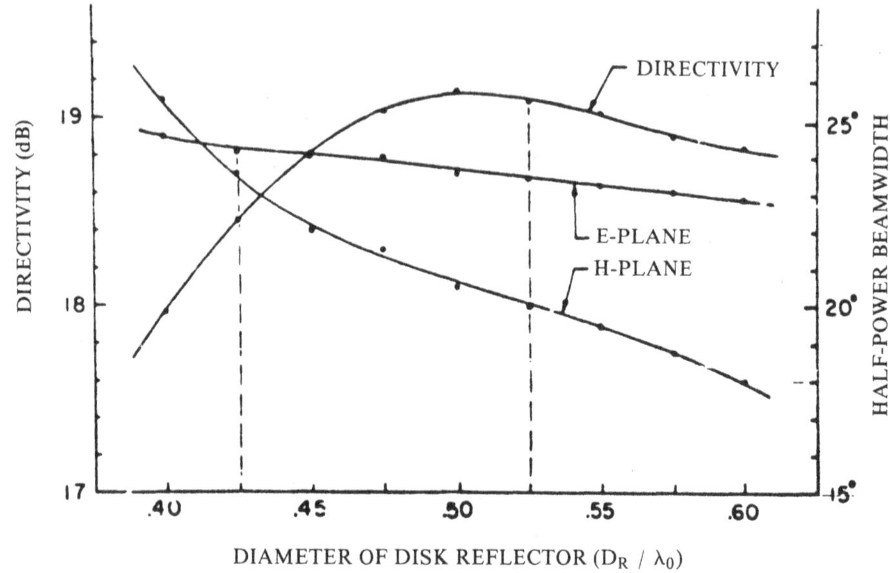

Figure 5.90(a) Directivity and half-power beamwidth of the four-element SBFA array, as functions of small reflector diameter (from [55])

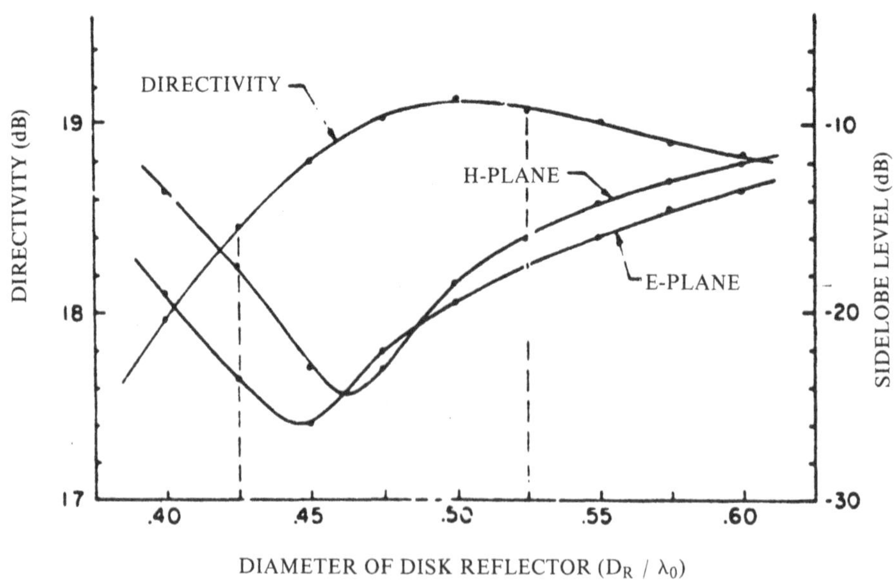

Figure 5.90(b) Directivity and sidelobe level of the four-element SBFA array, as functions of small reflector diameter (from [54])

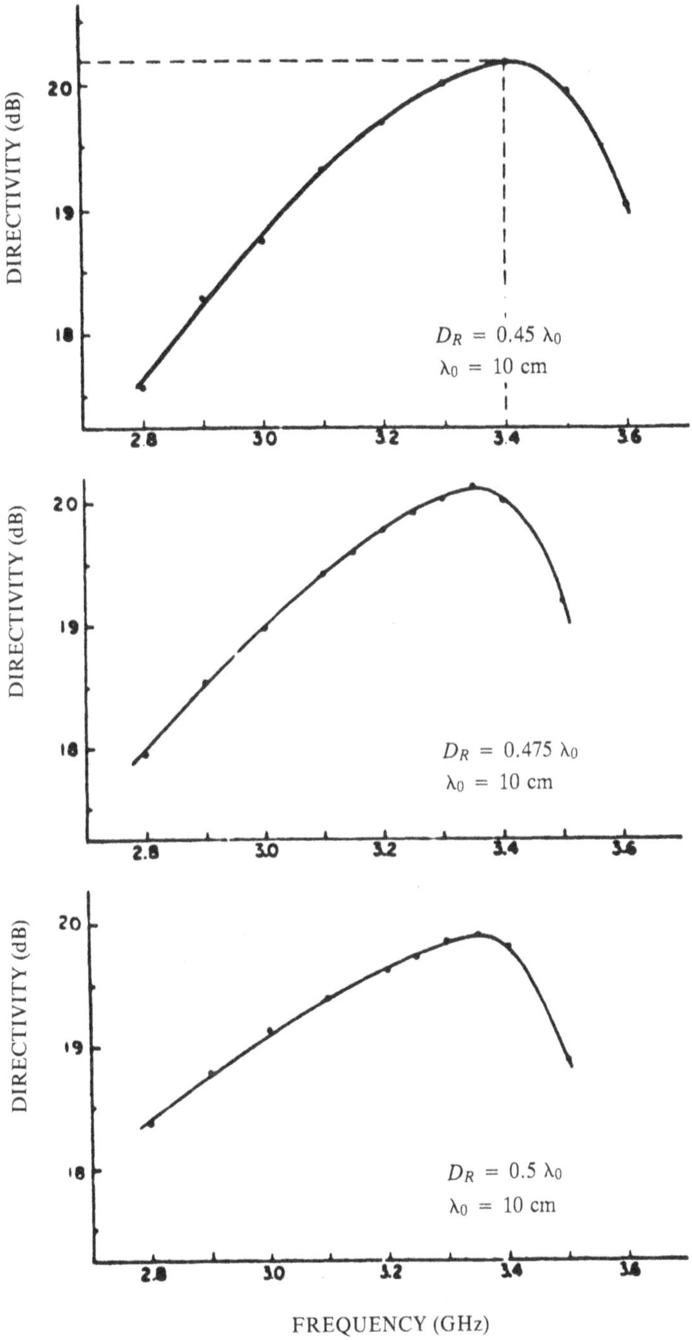

Figure 5.91(a) Directivity *versus* frequency of the four-element SBFA array for three diameters of small reflector (from [54])

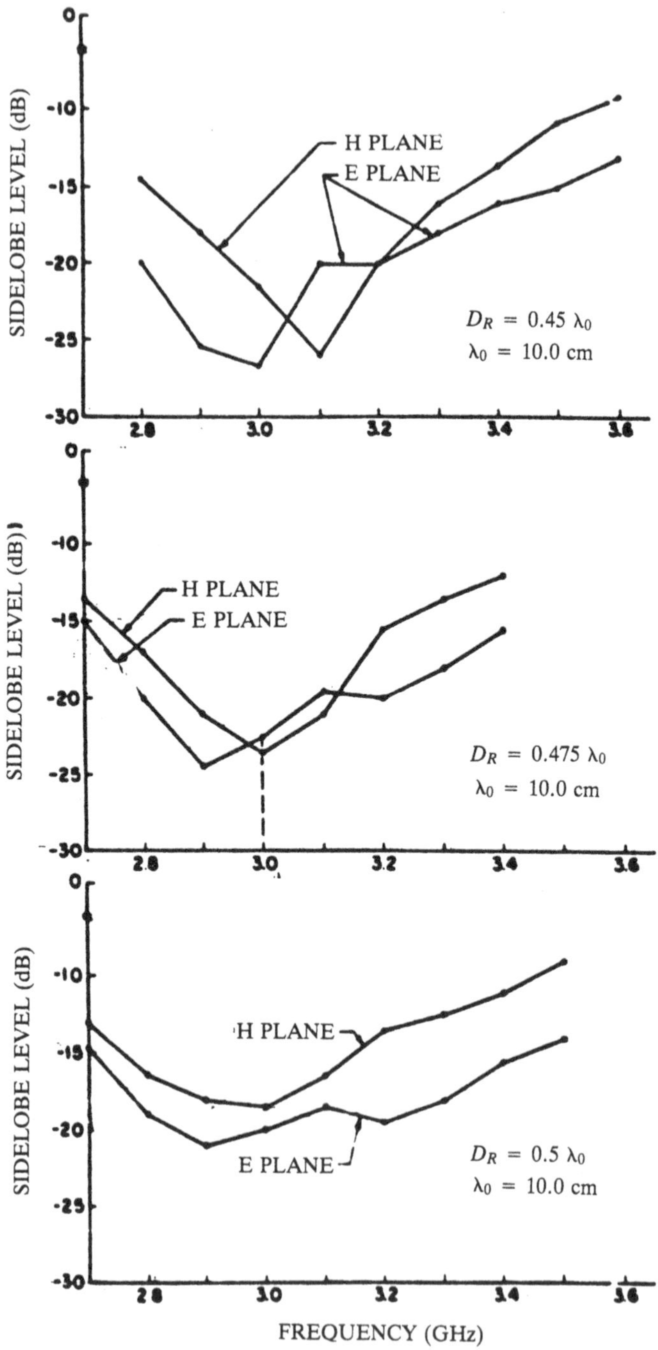

Figure 5.91(b) Sidelobe level *versus* frequency of the four-element SBFA array for three diameters of small reflector (from [54])

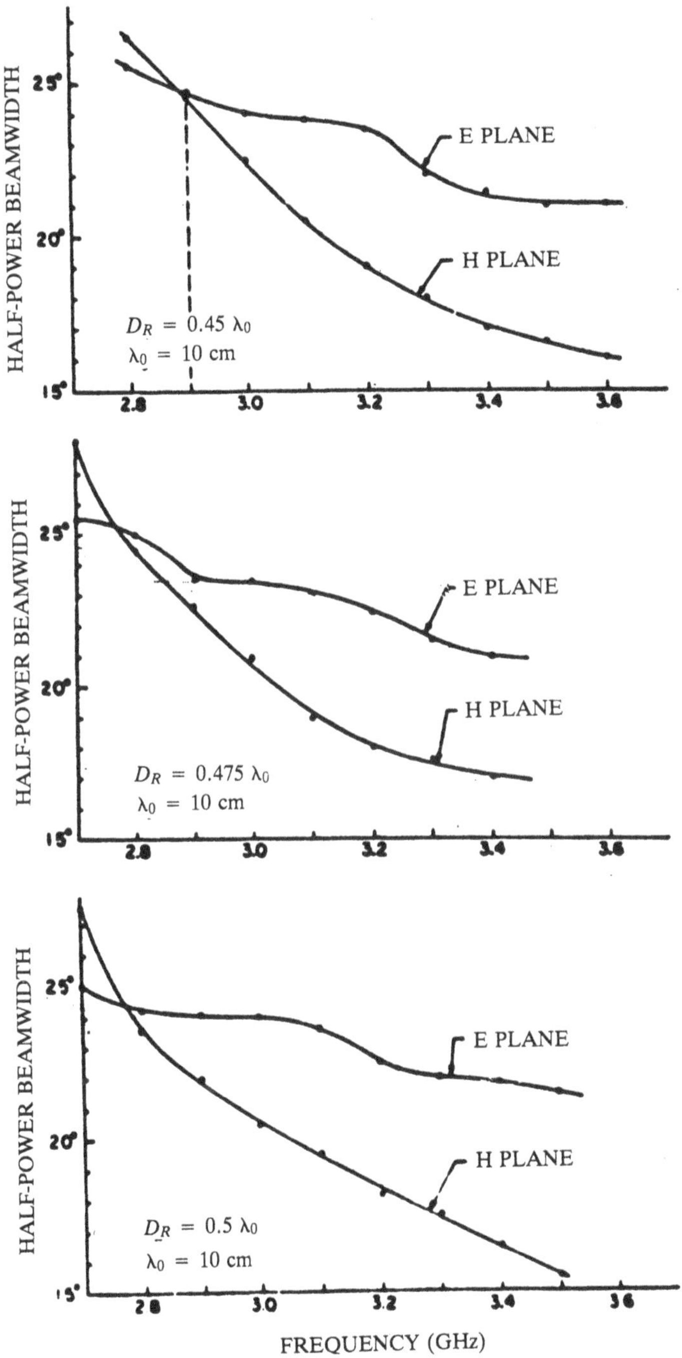

Figure 5.91(c) Half-power beamwidth *versus* frequency of the four-element SBFA array for three diameters of small reflector (from [54])

levels are more than −20 down from the main lobe for a certain frequency f; the half-power beamwidths in the E-plane and H-plane are equal at a certain frequency f. Optimal design specifications of the four-element SBFA array for any desired frequency are derived as a normalized table of parameters according to the following procedure:

Table 5.22 summarizes the optimal structural dimensions and electrical characteristics of the SBFA array for highest directivity and lowest sidelobes in the E-plane and H-plane. The optimal structural dimensions and electrical characteristics of the antenna array for equal half-power beamwidths, and equal beam shapes in E-plane and H-plane patterns, are given in Table 5.23.

Table 5.22 Optimal Dimensions and Electrical Characteristics of SBFA Arrays

For Highest Directivity		*For Lowest Directivity*	
f/f_0 =	1.13	f/f_0 =	1
D_M =	$3.40\ \lambda_0$	D_M =	$3.40\ \lambda_0$
r =	$1.13\ \lambda_0$	r =	$1.00\ \lambda_0$
D_R =	$0.475\ \lambda_0$	D_R =	$0.475\ \lambda_0$
W_B =	$0.57\ \lambda_0$	W_B =	$0.50\ \lambda_0$
S_E =	$1.13\ \lambda_0$	S_E =	$1.00\ \lambda_0$
S_R =	$0.57\ \lambda_0$	S_R =	$0.50\ \lambda_0$
S_F =	$0.28\ \lambda_0$	S_F =	$0.25\ \lambda_0$
M =	$10.4\ \lambda_0^2$ (area)	M =	$8.14\ \lambda_0^2$ (area)
Aperture Efficiency =	82%	Aperture Efficiency =	78%
First Sidelobe Level =	−14 dB	First Sidelobe Level =	−22.5 dB (E-Plane)
E-Plane 3 dB Beamwidth =	22°	=	−23.5 dB (H-Plane)
H-Plane 3 dB Beamwidth =	17°	All Other Sidelobes =	Below −30 dB
		E-Plane 3 dB Beamwidth =	23.5°
		H-Plane 3 dB Beamwidth =	21.0°

1. From Figure 5.91(a), (b), and (c), find the small reflector diameter for the directivity, sidelobe level, and half-power beamwidth of the desired values;
2. From the same figure, find the frequency f for the desired values of directivity, sidelobe level, or half-power beamwidth for an optimum (new) value of the small reflector diameter D_R (This sets the ratio f/f_0.);
3. Multiply all parameters, including the new value of the small reflector diameter D_R, by the factor f/f_0. (This normalizes all parameter values for the small reflector diameter D_R and frequency f_0.)

Figure 5.92(a) shows an eight-element SBFA array described by Ehrenspeck *et al.* [55]. The common reflector M is shaped like a four-leaf clover, because this configuration has the advantage of the smallest possible reflector area. For optimal

Table 5.23 Optimal Dimensions and Electrical Characteristics of SBFA Arrays

Equal Half-Power Beamwidths in E-Plane and H-Plane		*Equal Half-Power Beamwidths in E-Plane and H-Plane*	
f/f_0 =	0.97	f/f_0 =	0.93
D_M =	2.90 λ_0	D_M =	2.80 λ_0
r =	0.97 λ_0	r =	0.93 λ_0
D_R =	0.43 λ_0	D_R =	0.44 λ_0
W_B =	0.48 λ_0	W_B =	0.47 λ_0
S_E =	0.97 λ_0	S_E =	0.93 λ_0
S_R =	0.48 λ_0	S_R =	0.47 λ_0
S_F =	0.24 λ_0	S_F =	0.23 λ_0
M =	7.60 λ_0^2 (area)	M =	7.10 λ_0^2 (area)
Aperture Efficiency =	71%	Aperture Efficiency =	71%
E-Plane 3 dB Beamwidth =	24.5°	E-Plane 3 dB Beamwidth =	24.0°
H-Plane 3 dB Beamwidth =	24.5°	H-Plane 3 dB Beamwidth =	24.0°

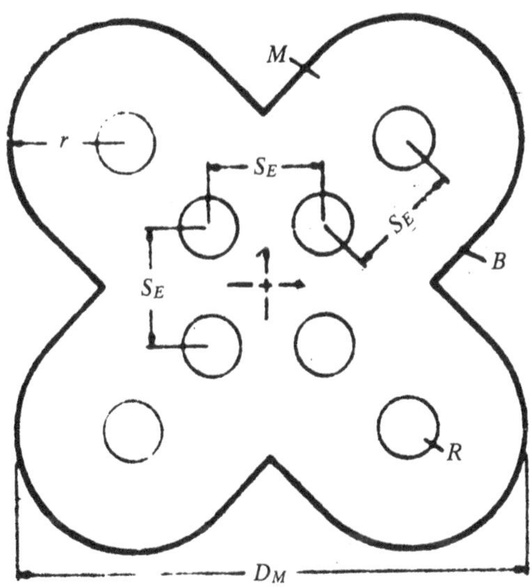

Figure 5.92(a) Sketch of eight-element SBFA at 3 GHz (from [55])

patterns, the diameters of the small reflector disks have been made slightly different in size, as shown in Table 5.24. The E-plane and H-plane patterns are respectively shown in Figure 5.92(b) and (c). Electrical characteristics of the array are shown in Table 5.24.

Table 5.24 Dimension and Performance of the Eight-Element Array at 3 GHz

Large reflector diameter (D_M)	$= 5.41\ \lambda_0$
Distance between the center of the outer element and the reflector edge (r)	$= 1.00\ \lambda_0$
Diameter of small reflector for those nearest the array center (D_R)	$= 0.40$ to $0.45\ \lambda_0$
Diameter of small reflector for those nearest to the array reflector edge (D_R)	$= 0.50\ \lambda_0$
Distance between elements (center to center) (S_E)	$= 1.00\ \lambda_0$
Area of the antenna (M)	$= 15.88\ \lambda_0^2$
Rim depth (W_B)	$= 0.50\ \lambda_0$
Distance between the large reflector and the dipole (S_F)	$= 0.25\ \lambda_0$
Distance between the large and the small reflector (S_R)	$= 0.50\ \lambda_0$
First sidelobe level in E-plane	$= -19.0$ dB
First sidelobe level in H-plane	$= -17.5$ dB
Measured directivity	$= 22.0$ dB
Calculated directivity for beamwidth values of 15.5 degrees and 15.0 degrees	$= 21.8$ dB

A 16-element SBFA array [55] is shown in Figure 5.93(a). The array dimensions are shown in Table 5.25. Figure 5.93(b) and (c) shows the experimental curves of half-power beamwidth and first sidelobe level of the array as functions of frequency. The experimental curve of directivity *versus* frequency is shown in Figure 5.93(d).

The three-term theory was applied to study the circuit and field properties of crossed-dipole SBFA arrays by King and Sandler [56]. The arrays are composed of both driven and parasitic elements. The theoretical formulation takes into account all effects of coupling between the elements, including changes in distribution of currents. Because the theory can be applied only to linear elements, the SBFA array (with its reflecting disk of finite size) is modified by the substitution, for the disk, of a set of elements that corresponds exactly to the "image array" in an infinite reflecting plane. Figure 5.94(a) and (b) shows six- and nine-element modified SBFA arrays, respectively. The admittances for the arrays in Figure 5.94(a) and (b) are in Table 5.26.

Figure 5.94(c) and (d) shows a "diskpole" antenna, which consists of a pair of crossed dipoles and a circular disk called *the director*, all mounted above a ground

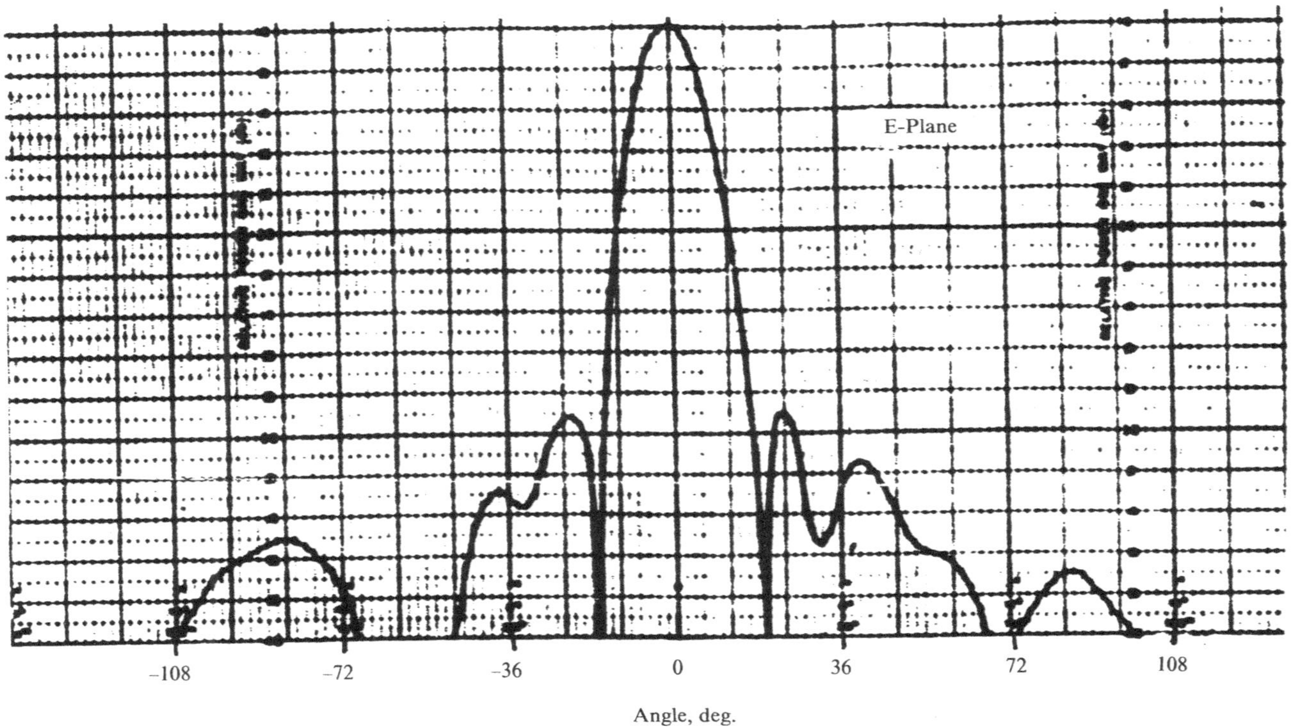

Figure 5.92(b) *E*-plane pattern of eight-element array at 3 GHz (from [55])

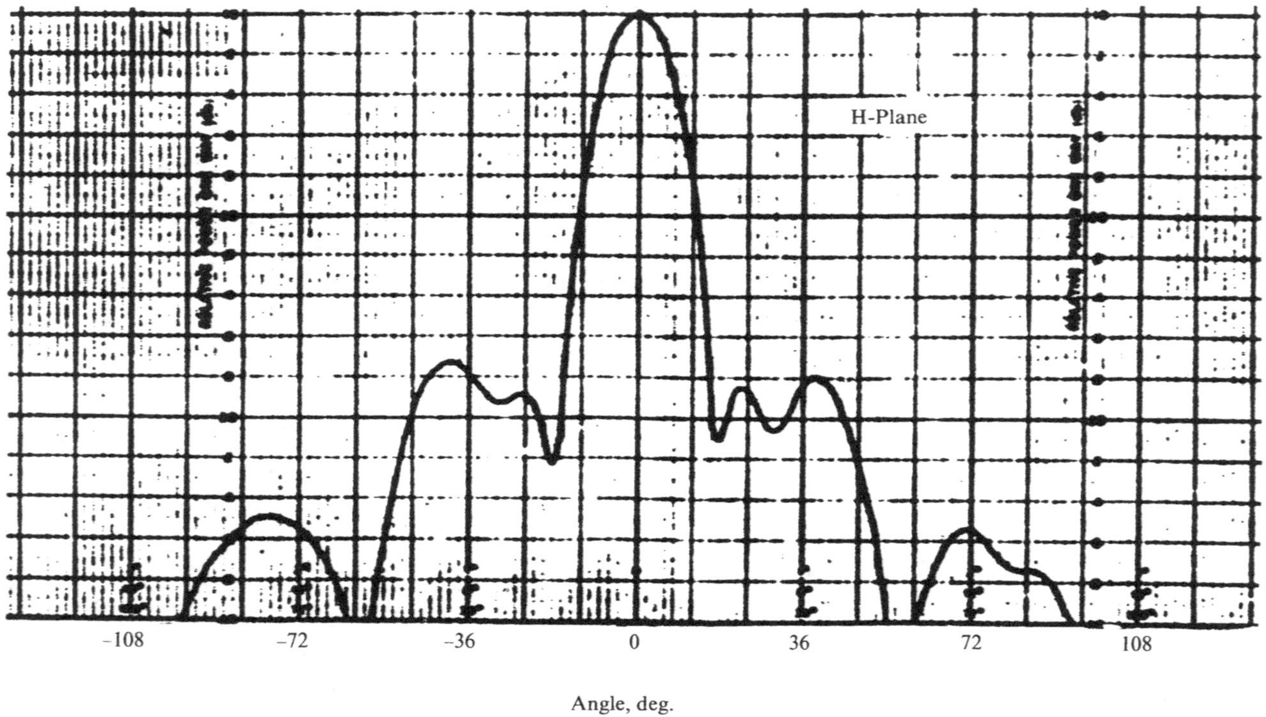

Figure 5.92(c) *H*-plane pattern of eight-element array at 3 GHz (from [55])

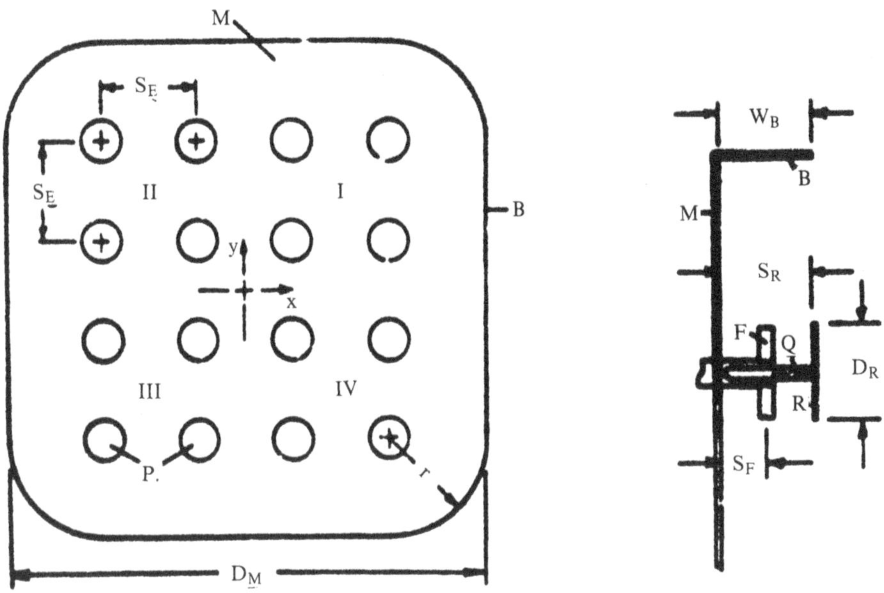

Figure 5.93(a) Sketch of a 16-element SBFA array (from [55])

Table 5.25 Array Dimensions for $f_0 = 3.0$ GHz

$$D_M = 5.63\ \lambda_0$$
$$r = 1.13\ \lambda_0$$
$$D_R = 0.50\ \lambda_0$$
$$S_R = 0.50\ \lambda_0$$
$$S_F = 0.25\ \lambda_0$$
$$W_B = 0.50\ \lambda_0$$
$$M = 30.55\ \lambda_0^2\ \text{(area)}$$

plane [57]. Dual-sense CP can be achieved using two orthogonal dipoles fed from a quadrature hybrid, as discussed in Chapter 4. The relative dimensions of the dipole and disk are adjustable so that the mutual coupling between the elements in the array may be varied. (See Figure 5.94(e).)

This antenna is easily analyzed by the present theory when the disk is replaced by an essentially equivalent crossed dipole, as shown in Figure 5.95(a) and (b).

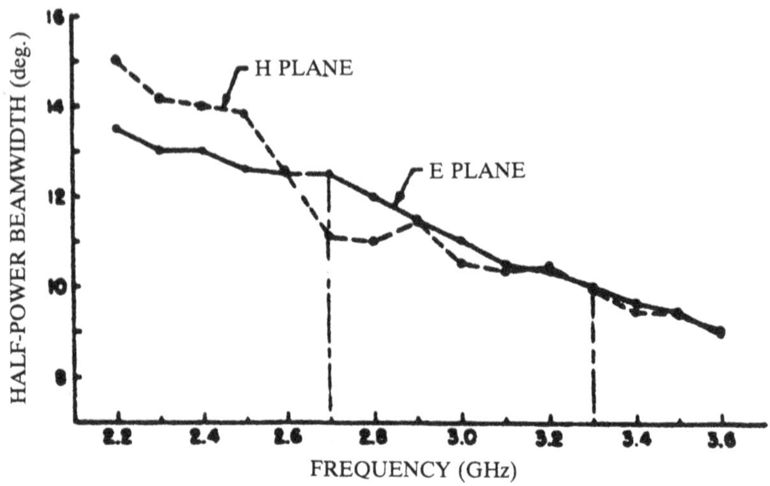

(b)

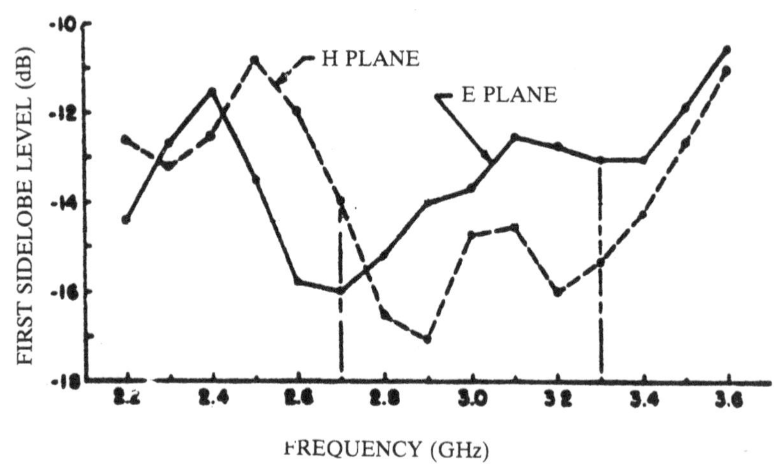

(c)

Figure 5.93(b) Measured half-power E-plane and H-plane beamwidth *versus* frequency (from [55])
(c) Measured first sidelobe level *versus* frequency of the array (from [55])

Figure 5.95(c) and (d) shows the H-plane radiation patterns as functions of the spacing and half-length of the reflector. The forward gain corresponding to reflector spacing and reflector half-length is also shown. The optimum gain is shown for a reflector

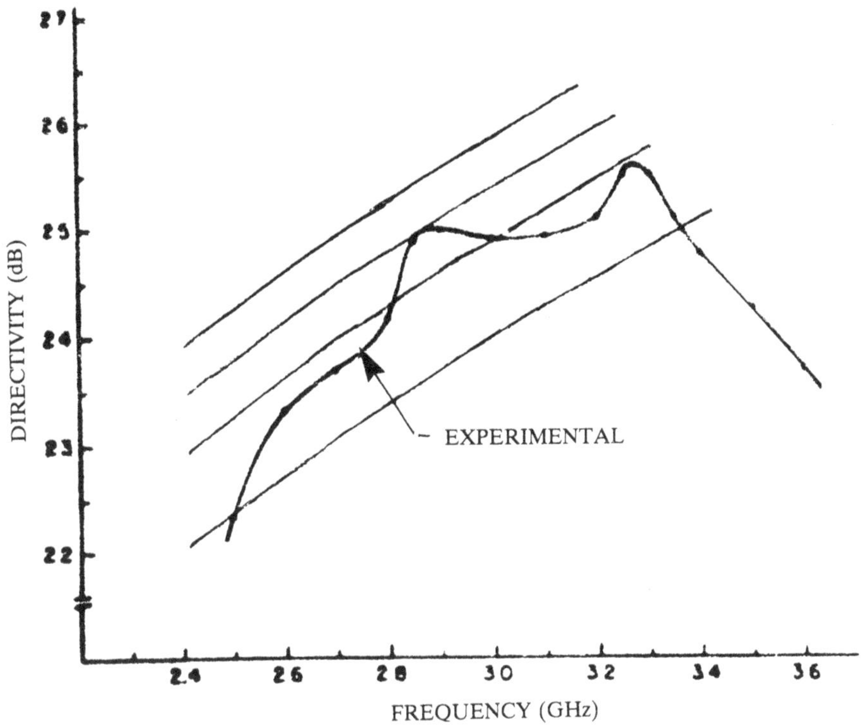

Figure 5.93(d) Directivity *versus* frequency curve for a 16-element SBFA array (from [55])

of half-wavelength $h_R/\lambda_0 = 0.24$ and a reflector spacing $b_R/\lambda_0 = 0.12$. Table 5.27 shows the driving-point admittances corresponding to Figure 5.95(c) and (d).

The size and location of the disk are the most important factors in determining the shape of the radiation patterns. Pattern shape is affected by the dipole distance above the ground plane. The aperture efficiency of a diskpole element is better than 97%. Figure 5.95(e) shows an array of 7 × 7 elements spaced at 0.87 λ_0 apart. Measured E-plane, H-plane and 45° plane patterns at 1.6 GHz are shown in Figure 5.95(f); a gain loss of less than 0.2 dB for a 10° scan is achieved. Also, grating lobe suppression in excess of 20 dB is found. The coupling coefficient, measured between the center element of the 7 × 7 array and its neighbors, is −23 dB, and coupling to the next ring of elements is −45 dB. Figure 5.95(g) shows that the VSWR of the center elements of a 7 × 7 array, for a 5% band centered at 1.6 GHz, is less than 2:1. The active reflection coefficient *versus* scan angle in the E-plane at 1.56, 1.60, and 1.64 GHz is shown in Figure 5.95(h).

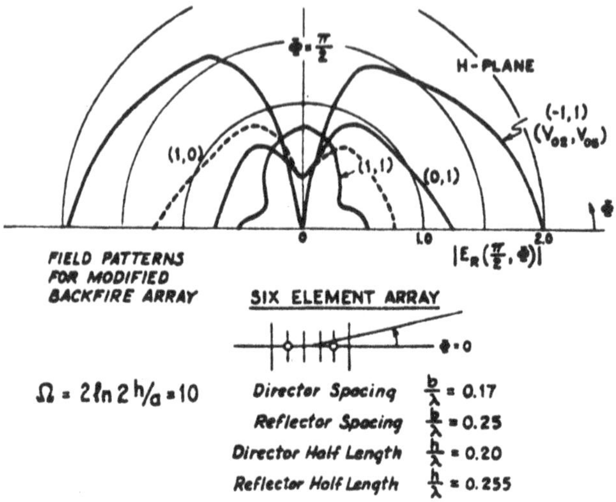

(a)

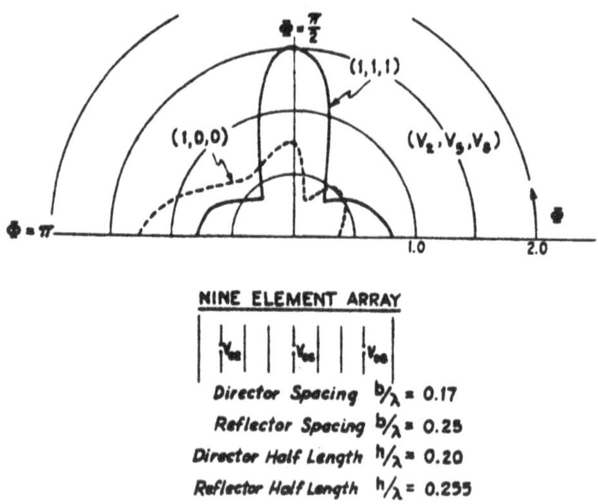

(b)

Figure 5.94(a) Six-element modified SBFA array (from [56])
(b) Nine-element modified SBFA array (from [56])

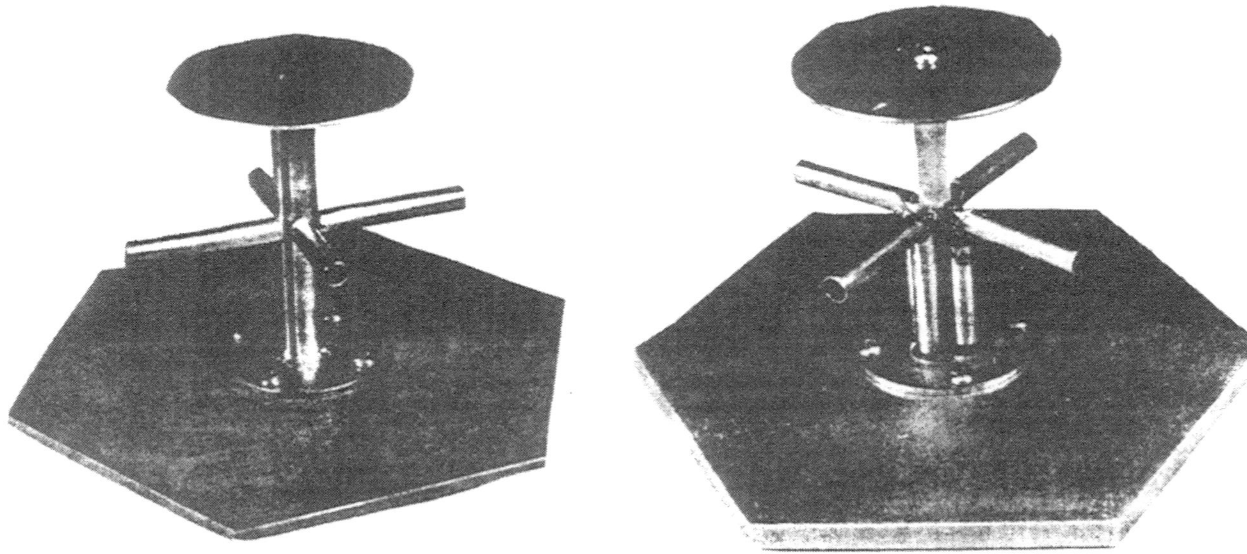

(c) (d)

Figure 5.94(c) Dipoles fed by a split-coaxial balun
(d) Dipole fed by coaxial inputs

Table 5.26 Driving-Point Admittances of Modified SBFA Arrays

Six-Element Array	
$V_{02} = 0, \quad V_{05} = 1$	$Y_{05} = 5.944 + j9.017$ mmho
$V_{02} = 1, \quad V_{05} = 1$	$Y_{02} = Y_{05} = 2.918 + j12.511$ mmho
$V_{02} = -1, \quad V_{05} = 1$	$Y_{02} = Y_{05} = 8.970 + j5.524$ mmho
Nine-Element Array	
$V_{02} = 1, \quad V_{05} = 0, \quad V_{08} = 0$	$Y_{02} = 5.195 + j10.302$ mmho
$V_{02} = 1, \quad V_{05} = 1, \quad V_{08} = 1$	$Y_{05} = Y_{02} = 4.076 + j9.751$ mmho,
	$Y_{05} = 0.732 + j14.345$ mmho

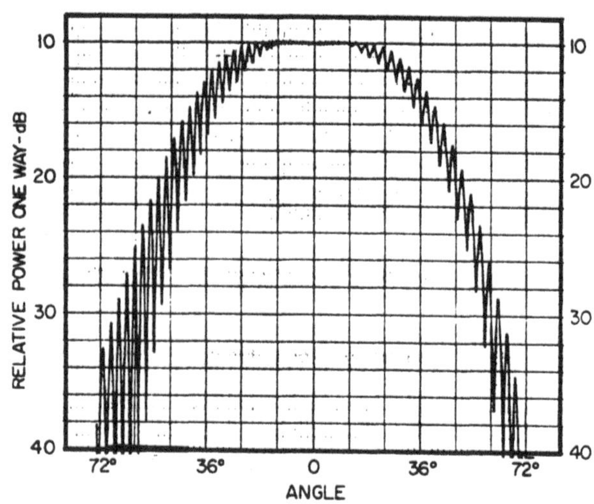

Figure 5.94(e) Axial ratio patterns of the radiating elements shown in Figure 5.94(c) and (d) (from [57])

5.5.2 Waveguide and Slotted SBFA Arrays

Large [16–18] developed a trough-shaped aperture, which is fed by a sectoral horn, as shown in Figure 5.96(a). The E-plane radiation characteristics of the array are shown in Figure 5.96(b). These characteristics are formed by the backfire aperture, which consists of the large reflector trough of height H and small reflector strip of width W, spaced apart by a distance S. The H-plane radiation characteristics are similar to those of a horn without the backfire aperture.

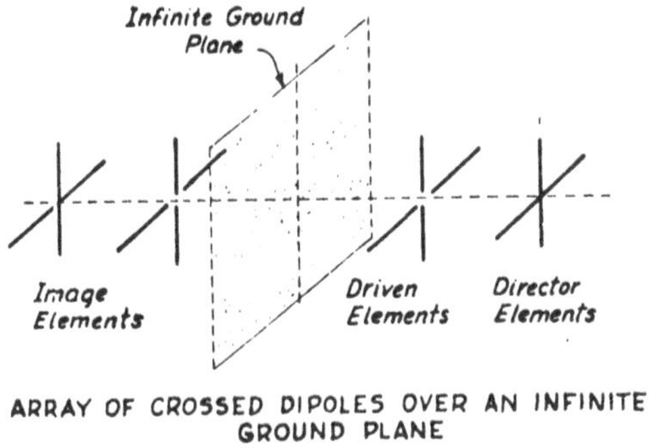

(a)

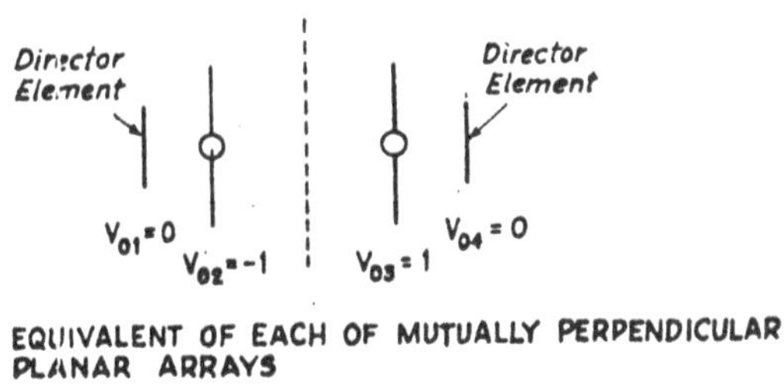

(b)

Figure 5.95 Crossed-dipole array (a) and equivalent (b) (from [56])

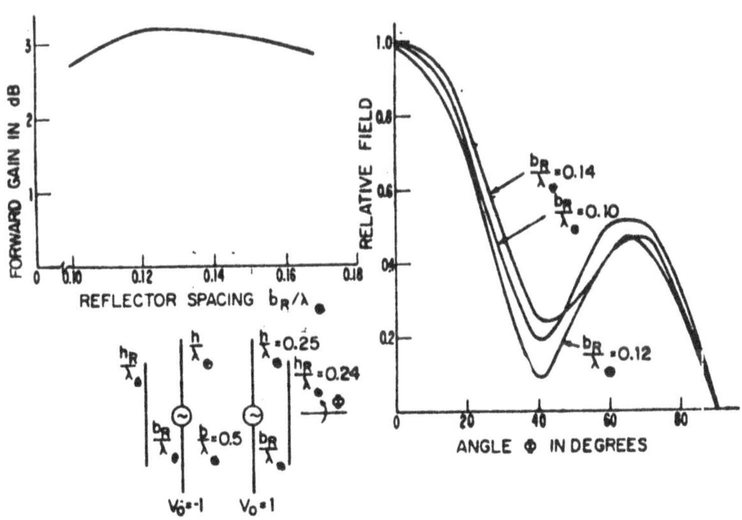

Figure 5.95(c) Forward gain and relative field of four-element array for half-length of reflectors $h_R/\lambda_0 = 0.24$ (from [56])

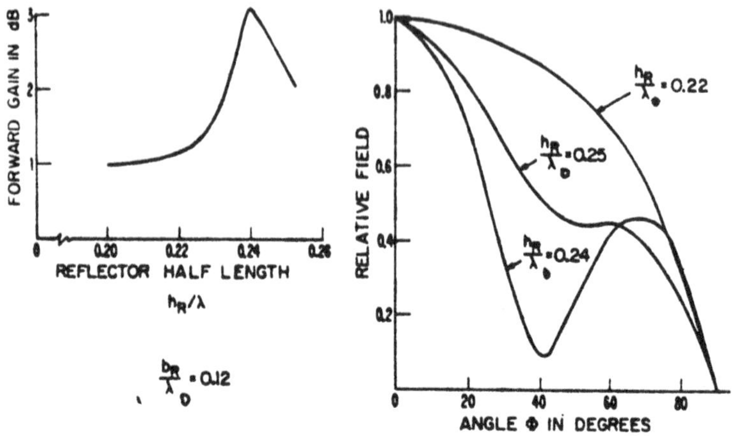

Figure 5.95(d) Forward gain and relative field of four-element array for reflector spacing $b_R/\lambda_0 = 0.12$ (from [56])

Table 5.27 Driving-Point Admittances

Case I[a]	
Reflector Spacing (b_R/λ)	Y_0 (mmho)
0.10	0.848 − *j*18.232
0.12	0.768 − *j*14.113
0.14	0.835 − *j*11.395
0.16	0.972 − *j*9.441
Case II[b]	
Reflector Length (h_R/λ)	Y_0 (mmho)
0.18	8.513 − *j*3.347
0.20	11.38 − *j*3.271
0.21	15.03 − *j*4.457
0.22	19.99 − *j*13.04
0.225	15.33 − *j*21.19
0.23	6.21 − *j*21.66
0.235	1.84 − *j*17.51
0.24	0.769 − *j*14.11
0.245	0.761 − *j*11.92
0.25	1.033 − *j*10.50

[a] $b/\lambda_0 = 0.5$, $h/\lambda_0 = 0.25$ (driven elements); $\Omega = 2 \ln (2h/a) = 10$; $h_R/\lambda_0 = 0.24$.
[b] $b/\lambda_0 = 0.5$, $h/\lambda_0 = 0.25$ (driven elements); $\Omega = 2 \ln (2h/a) = 10$; $b_R/\lambda_0 = 0.12$.

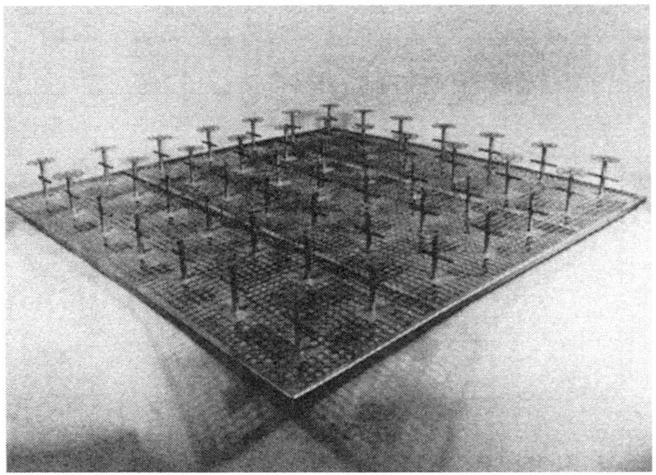

(e)

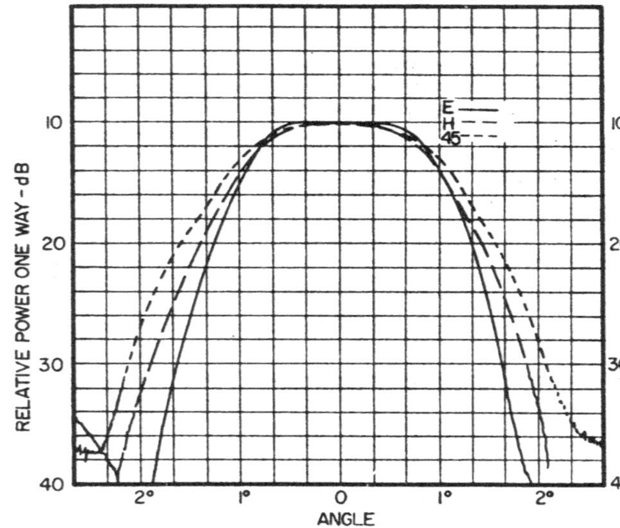

(f)

Figure 5.95(e) A 7 × 7 array constructed from diskpole elements (from [57])

(f) Measured E-plane, H-plane, and 45° plane patterns of 49-element array at 1.60 GHz (from [57])

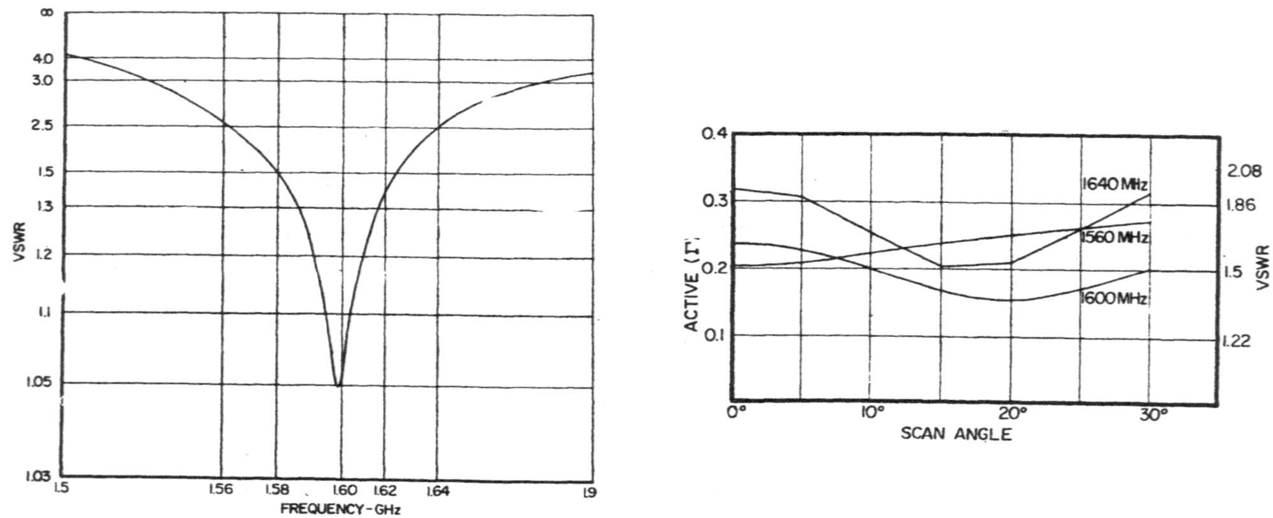

(g) (h)

Figure 5.95(g) VSWR *versus* frequency plot of linearly-polarized diskpole elements (from [57])
(h) Diskpole active reflection coefficient in the E-plane, as calculated from measured coupling coefficients (from [57])

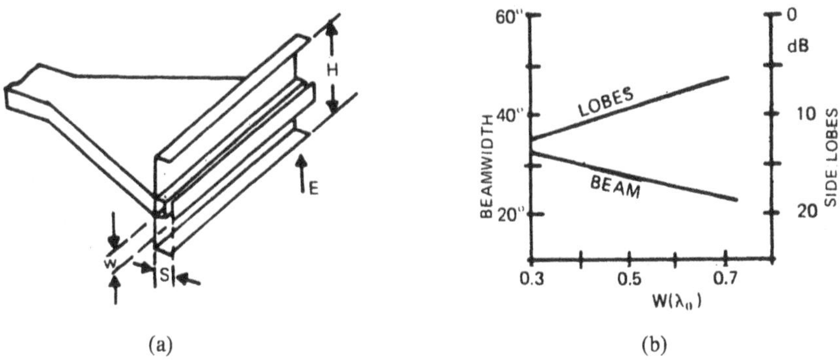

Figure 5.96(a) A sectoral-horn-fed SBFA array (from [18])
(b) E-plane radiation characteristics of the antenna (from [18])

A slotted linear array, fitted with a backfire aperture formed by a trough-shaped large reflector and small reflector strip running the whole length of the slotted aperture, is shown in Figure 5.96(c). The E-plane pattern of the antenna, measured at 9.4 GHz, is shown in Figure 5.96(d). Figure 5.96(e) and (f) also shows the measured radiation patterns for a vertically-polarized linear array (with and without a backfire plane mechanism) at 9.4 GHz. The H-plane patterns are measured for two elevation tilt angles: the array waveguide mounted so that its slotted wall is normal to the incident radiation from the transmitter (elevation tilt = 0°), and the array tilted back to 15° (elevation tilt = 15°). The H-plane patterns for the same array, fitted with a backfire aperture, show that at 0° elevation tilt, the highest sidelobes are −24 dB, (i.e., approximately 2 dB higher with the backfire aperture fitted). At 15° elevation tilt, the normal sidelobes are at a similar level to that of the array without the backfire aperture, but the grating lobes are reduced by approximately 7 dB.

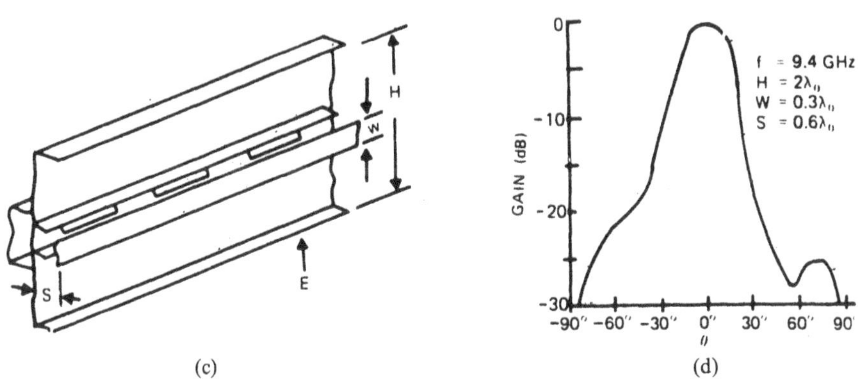

Figure 5.96(c) A slotted SBFA array (from [18])
(d) E-plane pattern of the antenna at 9.4 GHz (from [19])

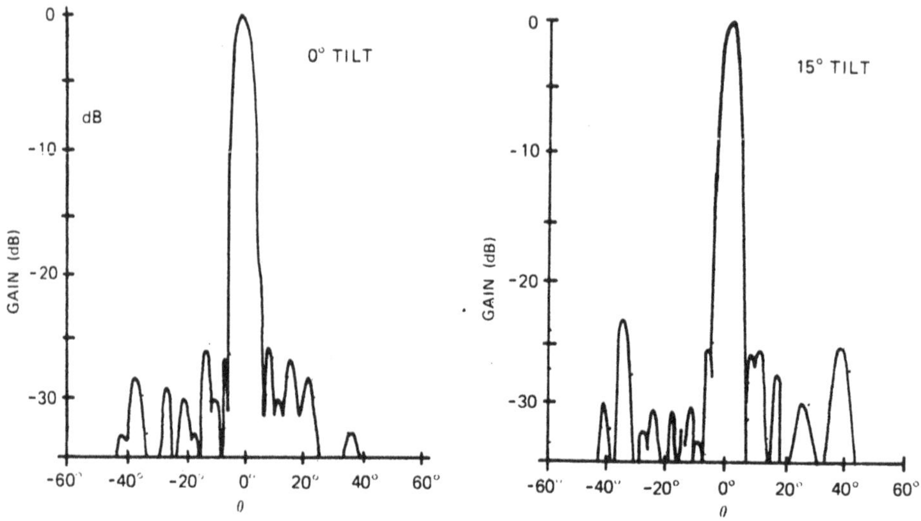

(e)

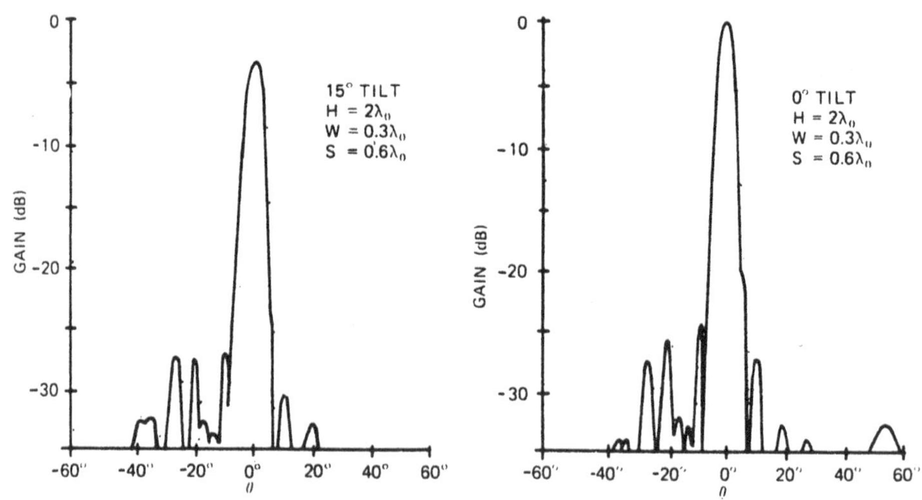

(f)

Figure 5.96(e) *H*-plane patterns of a vertically-polarized array of 24 shunt displaced slots without backfire aperture (from [18])

(f) *H*-plane patterns of a vertically-polarized array of 24 shunt displaced slots with backfire aperture (from [18])

Several X-band arrays with rectangular backfire elements were constructed and examined by Hristov and Urumov [58]. Figure 5.97(a) shows the E-plane radiation patterns of the studied SBFA array (a^+) and horn antenna (b^+) with the same cross-sectional dimensions. These antennas have approximately equal gain, but the SBFA

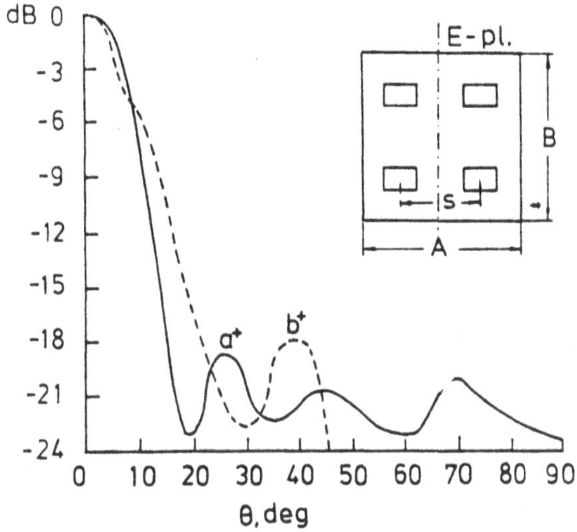

Figure 5.97(a) E-plane patterns of an SBFA array and rectangular horn of the same aperture size (from [58]): SBFA array pattern a^+; horn antenna pattern b^+

array is about one-third the length of the horn antenna. Figure 5.97(b) shows another example of a multiple SBFA array. The linear array consists of five rectangular SBFA elements with a spacing $S = 2\ \lambda_0$. The larger reflector dimension is approximately 10 by 2.4 λ_0. The E-plane pattern (a) has a beamwidth of 4°, which is exceptionally good for such an array dimension. The H-plane radiation pattern (b) is shown in Figure 5.97(b).

The application of a waveguide feed to an SBFA makes possible a high gain, compact radiator, which can be employed as a very efficient array element. Kumar [59] and Hristov [60] have developed a linear antenna array comprised of SBFA elements, which is shown in Figure 5.98(a). Figure 5.98(b) shows the E-plane radiation pattern of the array at 8.8 GHz. Table 5.28 summarizes the electrical characteristics of the array.

Two- and four-element SBFA arrays [20,61] with corrugated rims on the common large reflectors are shown in Figure 5.99(a) and (b). Figure 5.99(c) shows the E-plane and H-plane patterns of the two-element SBFA arrays at 34 and 35 GHz,

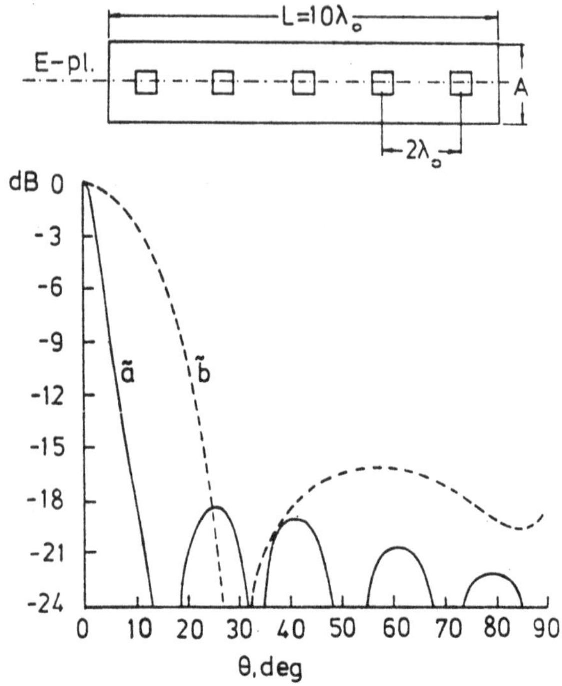

Figure 5.97(b) E-plane and H-plane radiation patterns of an SBFA array (from [58]): E-plane pattern $\tilde{a}$; H-plane pattern $\tilde{b}$

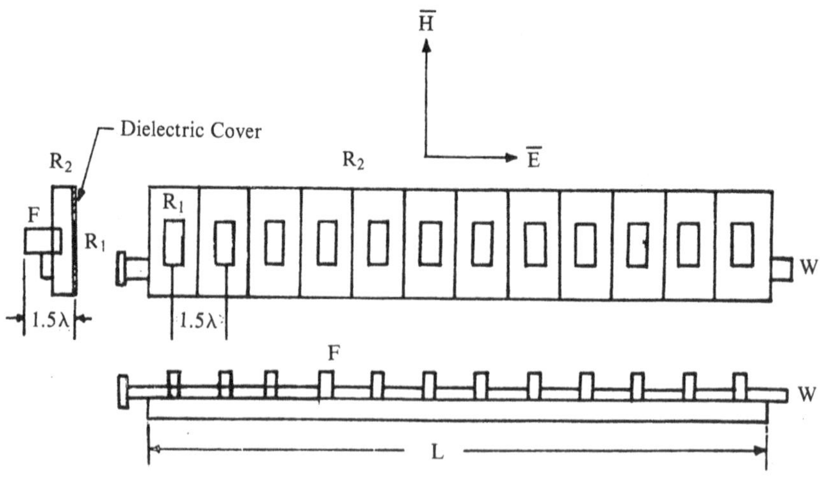

Figure 5.98(a) Twelve-element SBFA array (from [60])

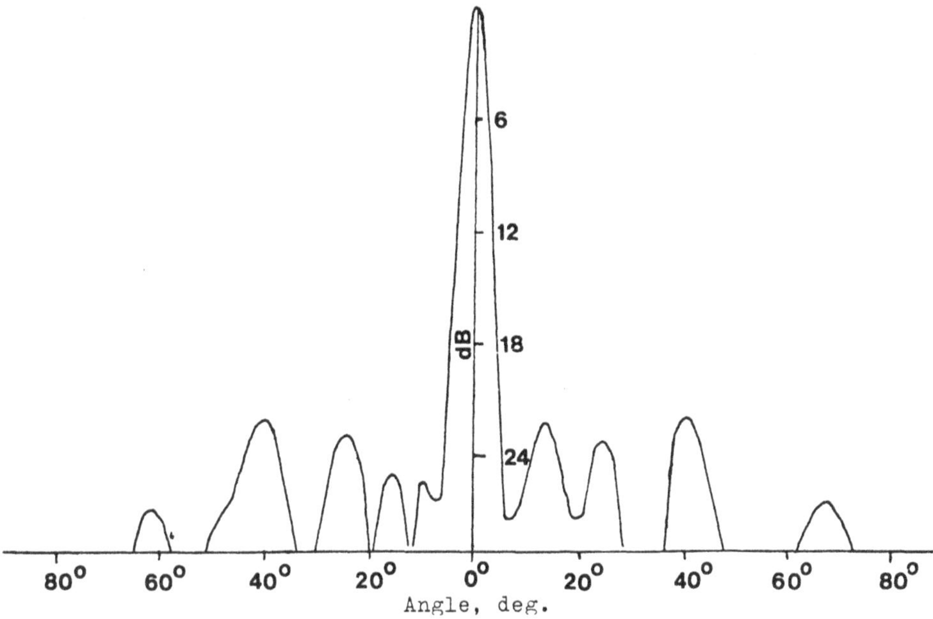

Figure 5.98(b) *E*-plane pattern of a 12-element SBFA array (from [59,60])

Table 5.28 Electrical Characteristics of the Twelve-Element SBFA Array

3 dB Beamwidths: 1) *E*-Plane	= 3.4°
2) *H*-Plane	= 22°
Highest Sidelobe Level	= −22 dB at ±40°
Far Sidelobe Level (Angle Greater than ±40°)	= −32 dB

with and without a corrugated rim on the large reflector. Figure 5.99(d) shows the *E*-plane and *H*-plane patterns of the four-element SBFA array, with and without a corrugated rim on the common large reflector. Figure 5.99(e) and (f) also shows the *E*-plane and *H*-plane patterns of two- and four-element SBFA arrays [62] at 34 and 35 GHz, with and without a dielectric lining on the rim of the common large reflector. Table 5.29 summarizes the electrical characteristics of the two-element array.

The waveguide-backed reactively-loaded aperture arrays were investigated by Harrington [63], Luzwick *et al.* [64] and Sinha *et al.* [65–67]. The reactively-loaded SBFA array may be defined as an *N*-port backfire radiating system, with one of the

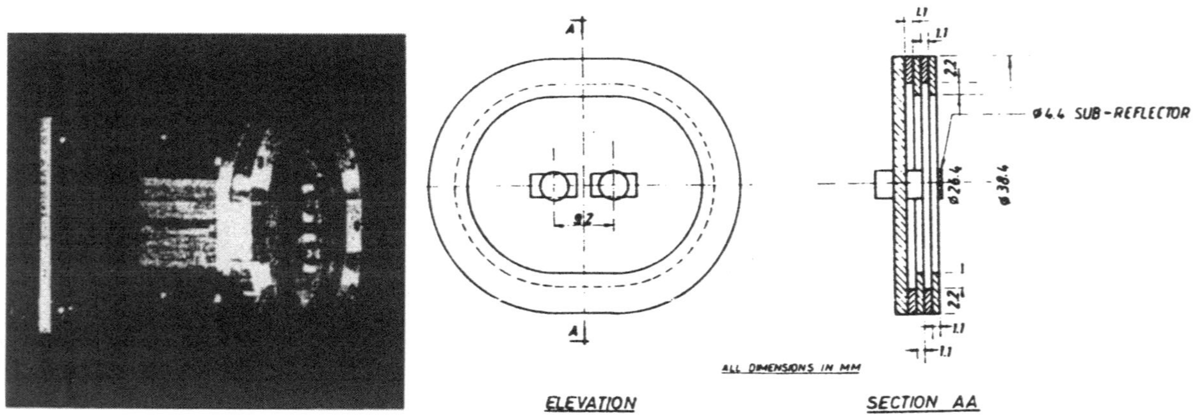

Figure 5.99(a) Two-element SBFA array, with corrugated rim on a common large reflector (from [20,61])

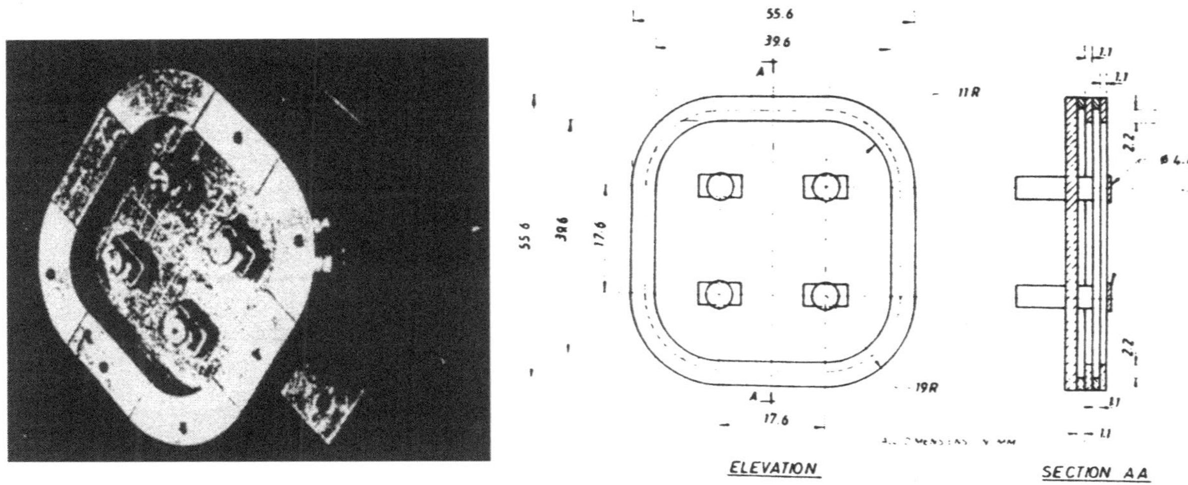

Figure 5.99(b) Four-element SBFA array, with corrugated rim on a common large reflector (from [20,61])

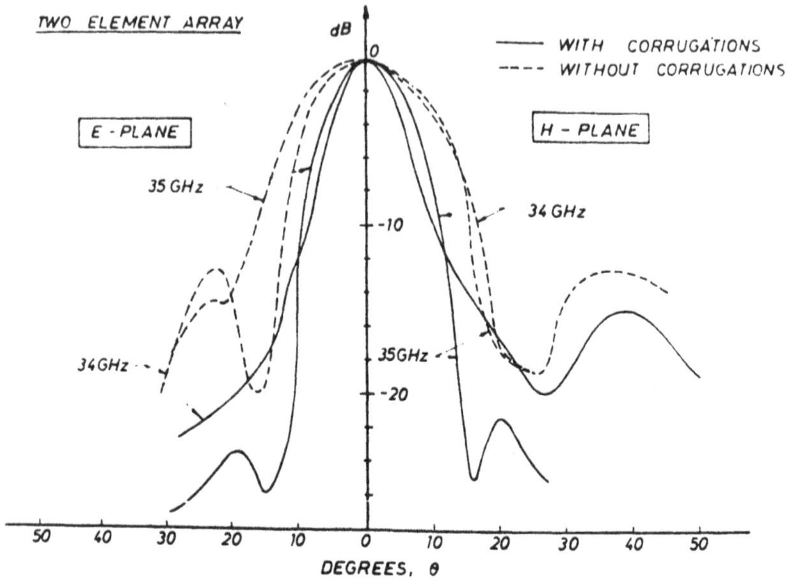

(c)

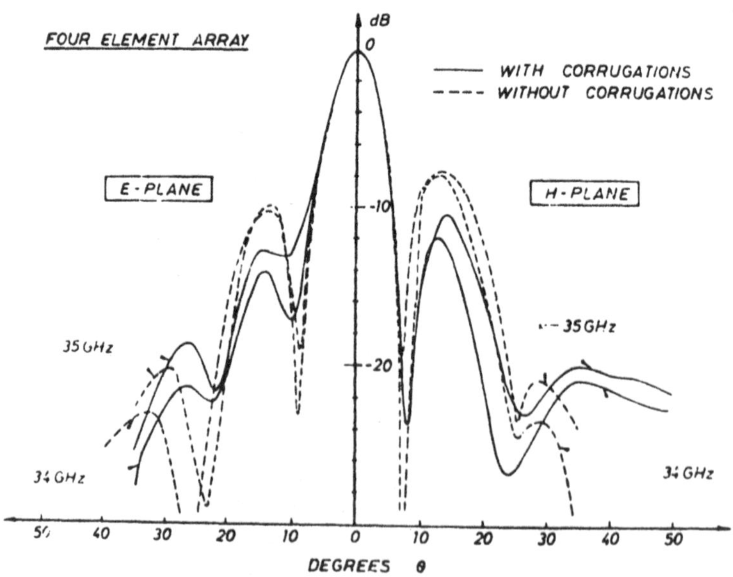

(d)

Figure 5.99(c) E-plane and H-plane patterns of a two-element SBFA array (from [20,61])

(d) E-plane and H-plane patterns of a four-element SBFA array (from [20,61])

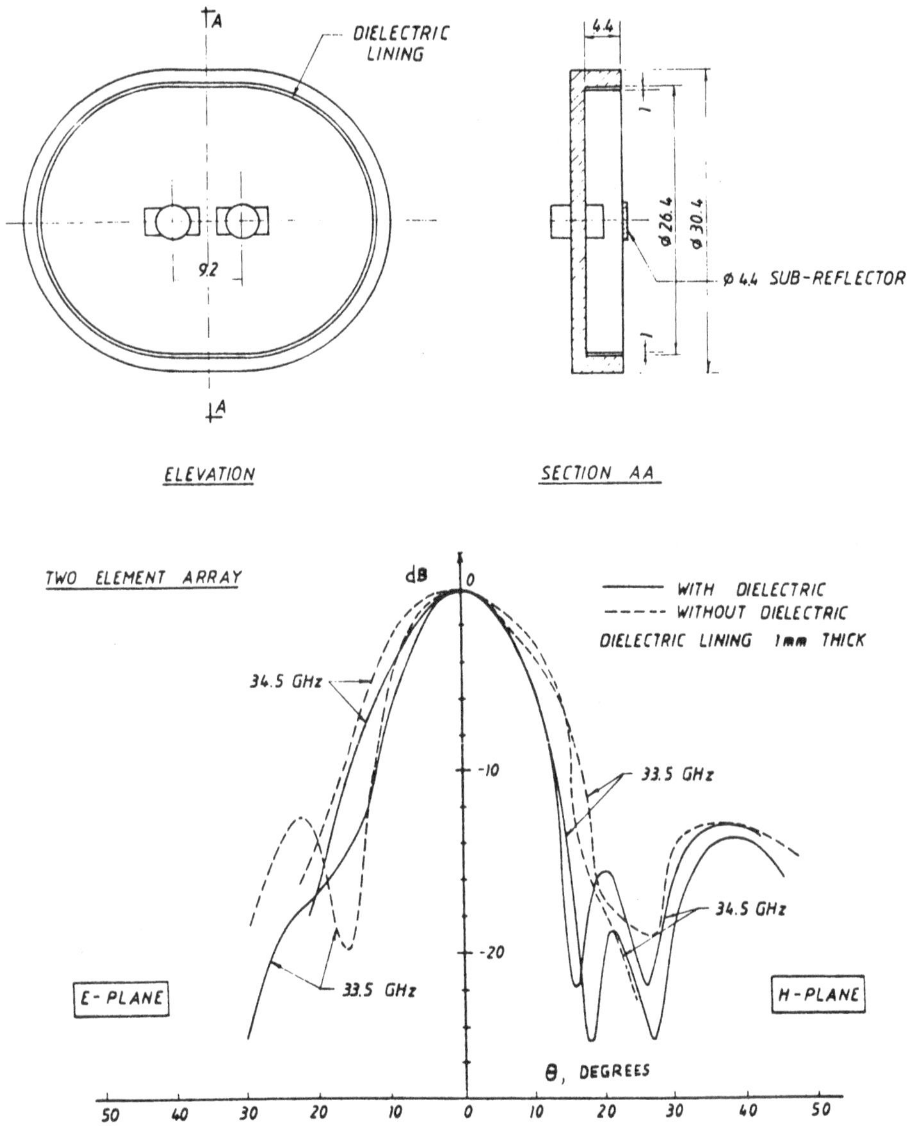

Figure 5.99(e) E-plane and H-plane patterns of a two-element SBFA array, with dielectric lining on the rim of the large reflector (from [62])

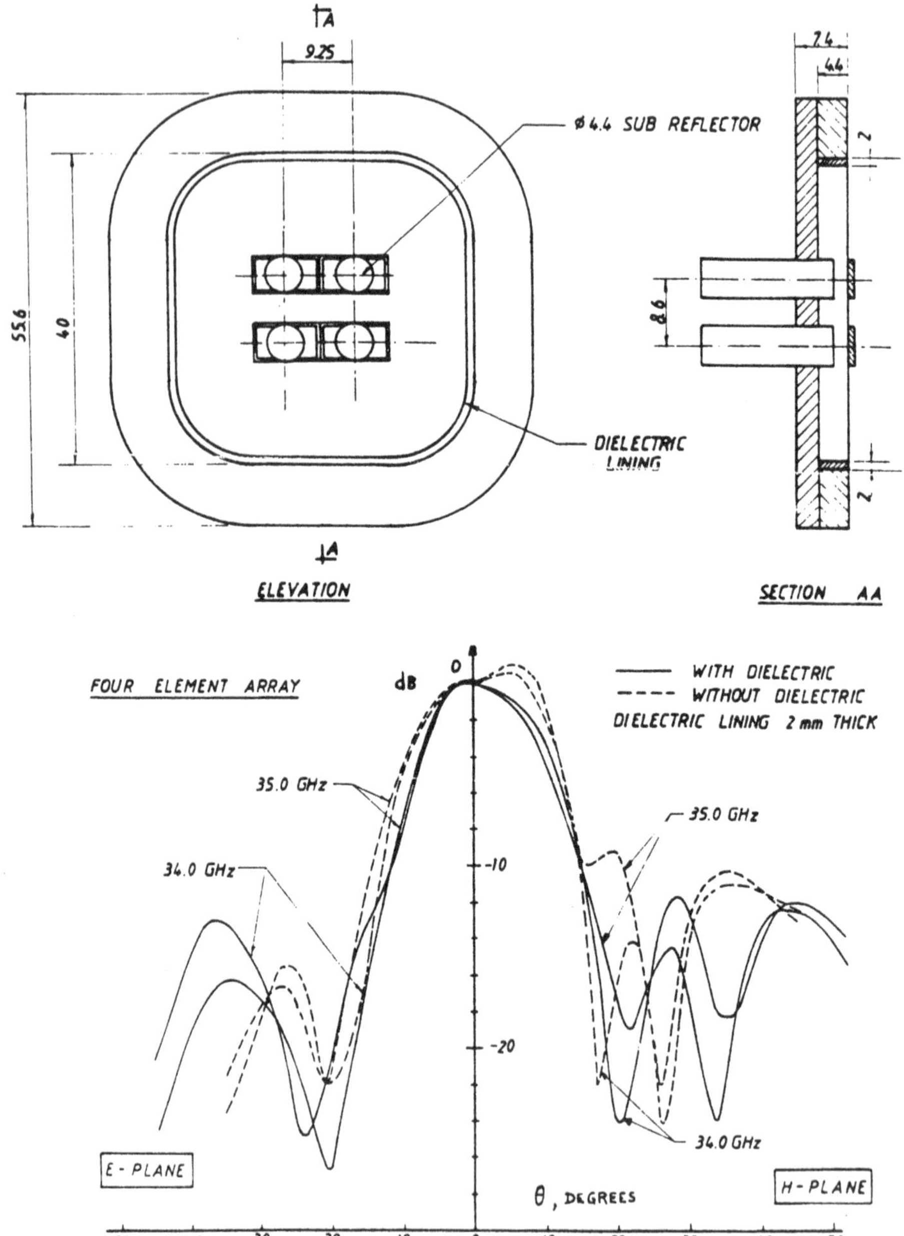

Figure 5.99(f) *E*-plane and *H*-plane patterns of a four-element SBFA array, with dielectric lining on the rim of the large reflector (from [62])

Table 5.29 Electrical Characteristics of Two- and Four-Element SBFA Array with Plane and Corrugated Rim of the Large Reflector

Electrical Parameters	*Two-Element SBFA Array*		*Four-Element SBFA Array*	
	Plane Rim	*Corrugated Rim*	*Plane Rim*	*Corrugated Rim*
3 dB H-plane pattern at 34 GHz	20°	14°	7.5°	7°
3 dB H-plane pattern at 35 GHz	20°	10°	7.5°	7°
3 dB E-plane pattern at 34 GHz	16°	8°	8°	8°
3 dB E-plane pattern at 35 GHz	20°	11°	8°	8°
E-plane sidelobe level at 34 GHz	−13 dB	−22 dB	−10 dB	−14 dB
E-plane sidelobe level at 35 GHz	−15 dB	−23 dB	−10.3 dB	−13 dB
H-plane sidelobe level at 34 GHz	−12.5 dB	−15 dB	−8.3 dB	−10.7 dB
H-plane sidelobe level at 35 GHz	−12.5 dB	−21 dB	−8 dB	−12 dB
Directive Gain	21.1 dB	25 dB	25 dB	28.6 dB

ports driven and the rest terminated in reactive loads. This type of array is an efficient directional radiator, which differs from its waveguide counterpart in that a subreflector is placed in front of the apertures at an appropriate distance. Sinha *et al.* [65] made experimental investigations on a waveguide-fed SBFA array with modified reflectors (Figure 5.100(a)), as shown in Table 5.30. An important advantage of the array is that the gain remains constant, because the main beam is steered by varying the reactive loads.

The slight decrease in gain may be due to an increase in VSWR, but this can be compensated for by matching the waveguide feed. An increase in sidelobe level and the appearance of a grating lobe take place when the beam is steered to angles of more than 20°. The curved small reflectors in this array system have these disadvantages: it is difficult to adjust the curvature in larger arrays; the E-plane pattern shows high sidelobe level; and the distance between the apex of the large and small reflectors is relatively large. To overcome these disadvantages, Sinha *et al.* [65–67] have used a rectangular small reflector, in conjunction with a half-wavelength rim around the edge of the large reflector, resulting in a smaller and simpler structure having larger power gain, smaller beamwidth, and lower sidelobe levels. Figure

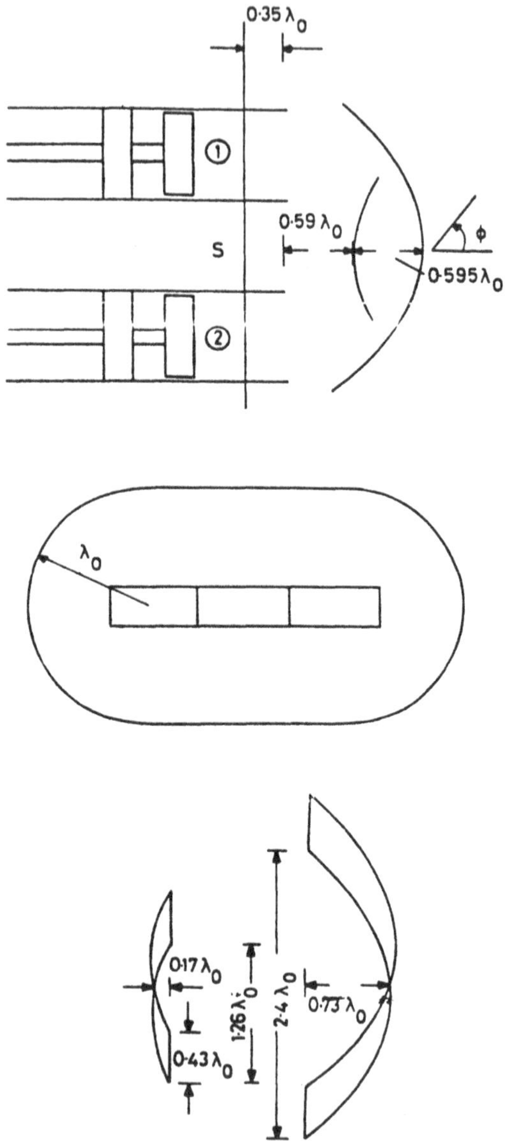

Figure 5.100(a) Sketch of a reactively-loaded waveguide-fed SBFA array, including the shape of reflectors, at 9.98 GHz (from [65])

Table 5.30 Properties of Reactively-Loaded Waveguide-Fed SBFA Array for Various Reactive Loads (from [65])

*Reactive Loads**		*Beam Position*	*3 dB Beam Width in*	*Sidelobe*		
1	*2*	ϕ	*H-Plane*	*Level,—dB*	*VSWR*	*Gain*
0.0	0.0	0°	24°	19	1.21	10.8
0.0	0.25517j	−5°	24°	19.5	1.31	10.6
−0.2417j	0.3057j	−10°	26°	18	1.33	10.4
11.4169j	0.2386j	−15°	19°	14.5	1.47	10
2.5675j	0.46018j	−20°	17°	9	1.52	9.2
1.11618j	1.2272j	−25°	17.5°	Grating lobe	1.61	—

*Calculated theoretically from the length of short-circuited transmission line

5.100(b) shows a section through a seven-element, reactance-loaded, waveguide-fed SBFA array at 9.98 GHz. Tables 5.31(a), (b) and (c) show the dimensions and the radiation characteristics (for $\phi = 0°$) of three-, five-, and seven-element arrays; Table 5.32 shows the performance characteristics of the seven-element array for different beam positions.

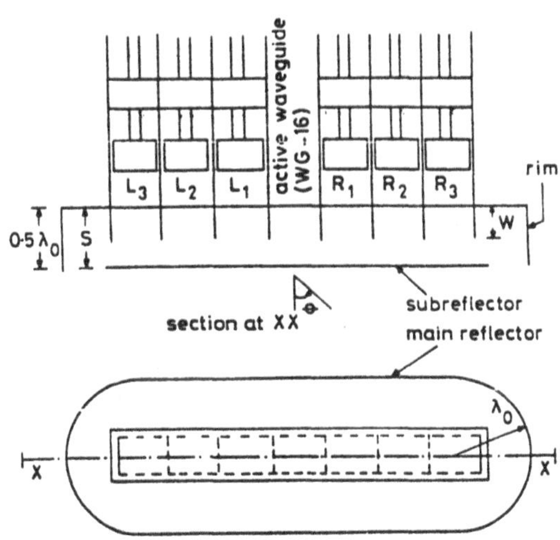

Figure 5.100(b) Section through seven-element reactance-loaded SBFA array (from [65])

Table 5.31(a) Dimensions of a Reactance-Loaded Waveguide-Fed SBFA Array (from [65])

SBFA Array	*S in λ_0*	*W in λ_0*	*Small reflector size in λ_0*
3-element	0.50	0.20	0.50 × 2.66
5-element	0.50	0.18	0.50 × 4.33
7-element	0.50	0.23	0.50 × 6.00

Table 5.31(b) Radiation Characteristics of Array for $\phi = 0°$

	3 dB Beamwidth		*Power*		*Sidelobe Level*		*Normalized Reactive Loads* (X)*					
	E-Plane	*H-Plane*	*Gain*	*VSWR*	*E-Plane*	*H-Plane*	R_1	R_2	R_3	L_1	L_2	L_3
			dB		dB	dB						
3-element array	20°	20°	13.8	3.5	12	25	−10.05	—	—	−1.64	—	—
5-element array	18°	13°	15.5	2.8	12.5	25	−2.2	−3.21	—	−2.41	−1.22	—
7-element array	16°	8°	16.0	3.2	15	25	−1.32	−7.11	0.073	−1.13	−3.01	0.22
conventional s.b.a.	23°	30°	12.0	1.4	11.5	18	—	—	—	—	—	—
3-element array with curved subreflectors[1]	15°	24°	10.8	1.21	5	19	0.0	—	—	0.0	—	—

*Calculated theoretically from the length of short-circuited waveguides

Table 5.31(c) Performance Characteristics of the Seven-Element SBFA Array for Different Beam Positions

Beam Position 0	*3 dB Beamwidth in H-Plane*	*Power Gain*	*VSWR*	*Sidelobe Level*	*Normalized Reactive Loads (X)* R_1	R_2	R_3	L_1	L_2	L_3
		dB		dB						
0°	8°	16	3.2	25	−1.32	−7.11	0.073	−1.13	−3.01	0.22
−10°	9°	15.5	1.76	16	−2.95	0.16	−6.8	−0.53	−2.67	−1.47
−15°	10°	13.6	1.2	10	4.49	−0.54	2.09	0.036	−1.42	−0.71

Table 5.32 SBFA Array Element Performance

Bandwidth	1535–1559 and 1636–1660 MHz
Peak gain	15.3 dB
Edge of coverage (±9°) gain	14.3 dB
Axial ratio	<1.5 dB
VSWR	<1.20

Harrington, *et al.* [68] and Sinha, *et al.* [69] used a moment method to analyze the reactively-loaded waveguide-fed SBFA array. An iterative procedure was applied in order to arrive at the values of reactive loads for maximum gain in a given direction. They have formulated the problem for the E-plane and H-plane reactively loaded SBFA array. A comparison of computed and measured radiation patterns indicates good agreement between theory and experiment.

Many researchers [70–81] have developed L-band and C-band SBFA arrays for advanced multibeam and multiservice mobile communications satellites. An extensive investigation of possible single-element antennas and subarrays has led to SBFA as a first choice. Molker [70] developed C-band SBFA arrays with sixteen and nineteen elements. Figure 5.101(a) and (b) shows the sixteen- and nineteen-element SBFA arrays, respectively. The radiation patterns of individual elements were measured and the radiation pattern of the complete arrays were computed based on the measurements and assumed excitations.

An L-band SBFA array was designed [70–80] to produce 24 dB minimum edge coverage gain within a fixed nineteen-beam pattern. The performance of the SBFA elements is presented in Table 5.32. To evaluate the losses in the elements, the directivity was computed by two methods, measured radiation patterns, and radiation patterns calculated by the software program ROT2 [77–79], based on the electrical field integral equation method, with triangular expansion functions and Galerkin's method [77], and compared to the gain. The losses in the elements are approximately 0.8 dB. The coupling between the elements in the array is less than −55 dB, which means it has negligible influence on the performance of the array. Figure 5.102(a and b) shows the practical realization of the eighteen SBFA elements and a nineteen beam global coverage pattern, respectively. The array has a hexagonal outer shape. All eighteen radiating SBFA elements are identical and positioned in a triangular lattice to give a structure of 11 λ_0. These arrays are made of aluminum and *carbon-fiber-reinforced plastic* (CFRP). The weight saved by using CFRP rather than

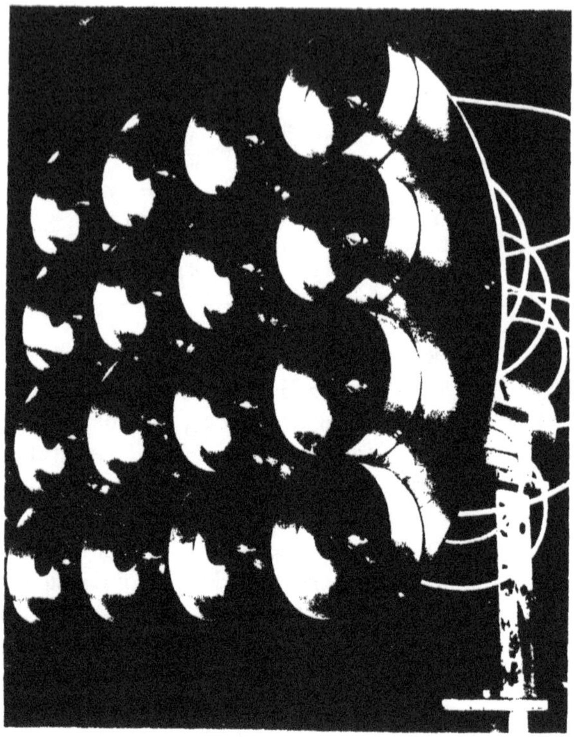

Figure 5.101(a) Sixteen-element *C*-band SBFA array (from [70])

aluminum is more than 12 kg, which is significant for space applications. A metal-plated CFRP material discussed by Kumar [82,82a] was classified as highly unreliable because it introduced *passive intermodulation* (PIM).

Brian and Talbot [83] investigated the reduction in *effective isotropic radiated power* (EIRP) of the array, due to element failure in an eighteen-element SBFA system and the array *passive intermodulation product* (PIMP) under multicarrier drive conditions. They assumed that each element of the array shown in Figure 5.102(a) is driven by a class A 3 W amplifier on transmitting, while on receiving, each element has a receiver with an individual noise figure of 2.6 dB. The *beam-forming network* (BFN) shown in Figure 5.103(a), generates the nineteen beams and consists of: a set of nineteen phase-splitting amplifiers, which generates the four quadrature signals required for each beam input; a resistive matrix with 76 × 18 nodes, which provides the beam-element interconnections with the correct amplitude and phase relationships; a set of eighteen low-input-impedance summing amplifiers, which combine the element signals; and a power-supply filter, which rejects the low-frequency ripple present on the power supply lines.

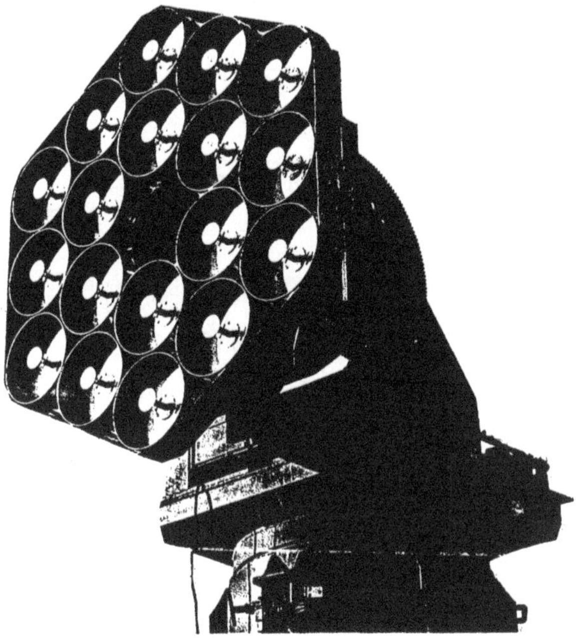

Figure 5.101(b) Nineteen-element *C*-band SBFA array (from [70])

Figure 5.102(a) Eighteen-element *L*-band array [70–76]

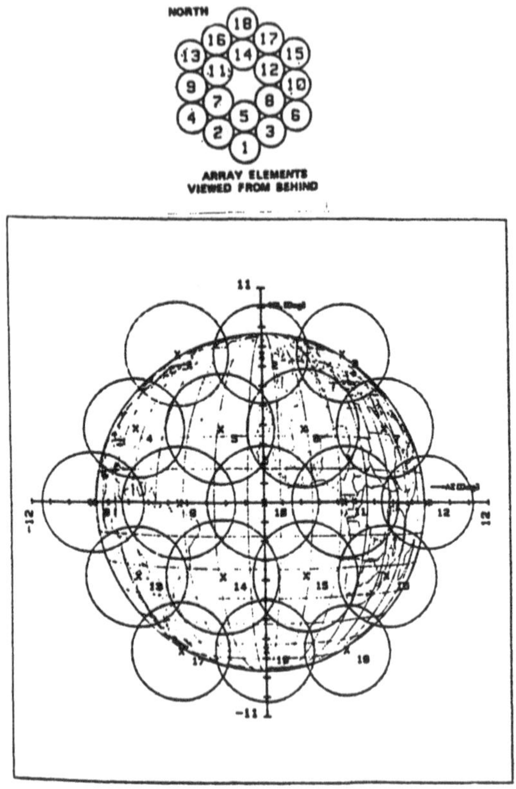

Figure 5.102(b) Array 24 dB contours [71]

Figure 5.103(b) shows the mechanical construction of the BFN manufactured by Marconi Space and Defense System (UK). Table 5.33 gives a summary of test results for BFN matrices at an intermediate frequency (IF) of around 30 MHz. Reference [71] gives full details on the BFN and associated electronics.

Special emphasis has been put on the amplitude and phase errors at the element outputs and, due to the large amount of testing which must be done, an automatic test set-up has been implemented. The error distributions for the amplitude and phase are shown in Figures 5.103(c) and 5.103(d), respectively.

Measurements of beam patterns, EIRP and G/T have been performed on the cylindrical near-field range at British Aerospace and on a conventional far-field range at the Marconi Research Labs. These results are summarized by Brian and Talbot [83] in Table 5.34.

In a phased array system, the failure of elements has minimal effect on EIRP and beam pattern, when compared to conventional global and multiple-feed system

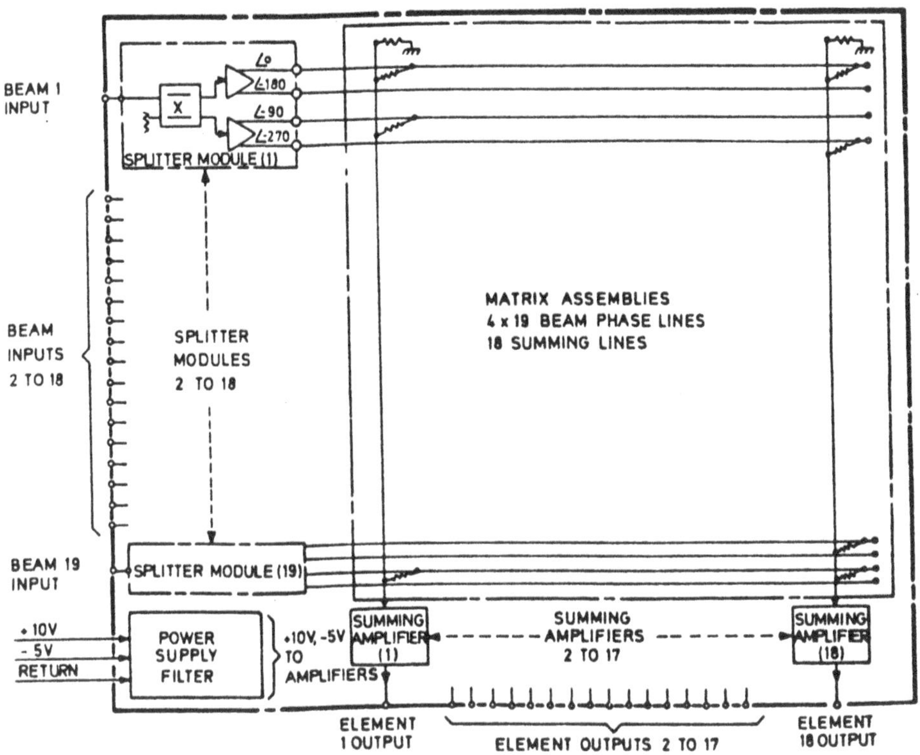

Figure 5.103(a) A block diagram of a BFN matrix (from [71])

failure as a whole. The reduction of power for the failure of elements can be calculated by the formula: reduction in power in the total beam = ((no. of working elements)/(total no. of elements))^2. The reduction in EIRP due to three element failures in an eighteen-element system is 1.6 dB. Figure 5.104(a) and (b) shows the measured EIRP contours for beams 10 and 15, with all elements functioning and with three elements disabled, respectively. In this particular example, a reduction in EIRP of 2.8 dB is noticed. However, a more random three-element failure (Figure 5.104(c)) shows a smaller reduction in EIRP.

Figure 5.104(d) shows the measured third- and fifth-order PIMP, due to three different carriers in beams 2, 5, and 6. The resultant spreading of some of the products outside the earth coverage area leads to a reduction in C/I noise received by an earth terminal.

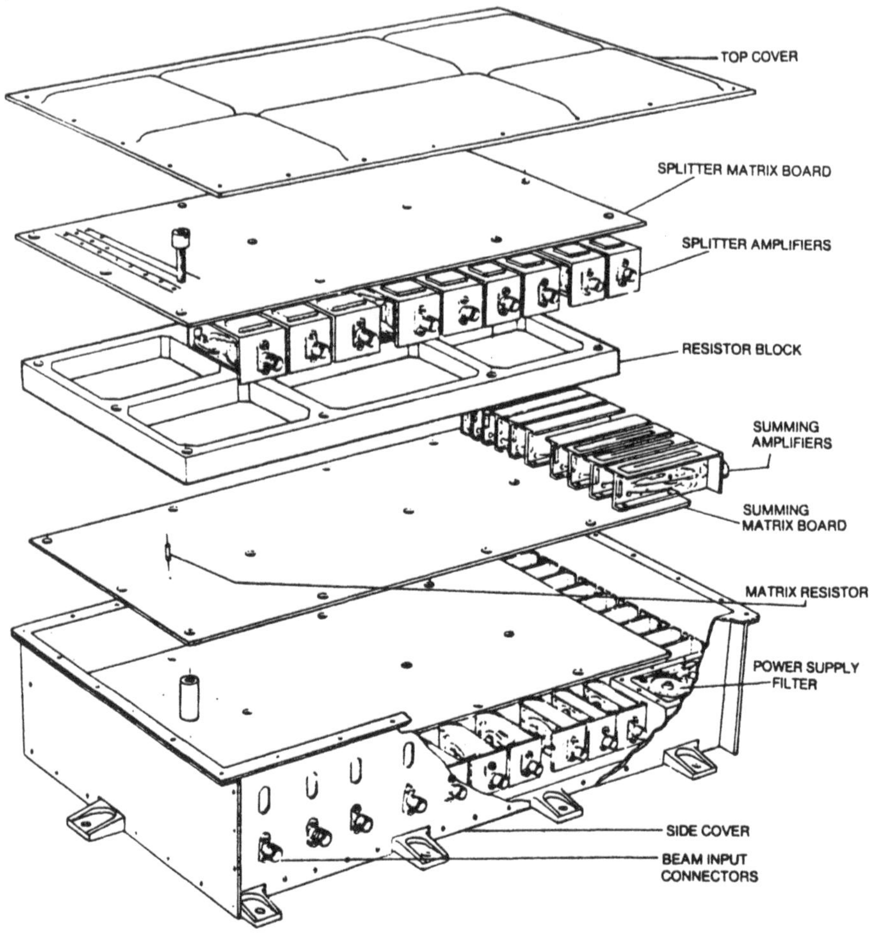

Figure 5.103(b) BFN matrix assembly (from [71])

5.6 APPLICATIONS OF THE SBFA

The SBFA is adaptable to many configurations, from a lightweight portable single element to large fixed arrays. It can be used for satellite tracking, telemetry applications, and ground-to-ground and ground-to-air communications, including those via satellites. It offers these advantages: a high gain antenna; extremely low sidelobes and backlobes; optimum performance in the gain range between 12 and 20 dB; elimination of complex parabolic surfaces; utilization of a simple feed; light weight; com-

Table 5.33 Summary of BFN Matrix Test Results (from [71])

Parameter	*Requirement*	*Unit*	*Result*
Phase error 27.5–35 MHz	±9 max	deg	see Figure 5.103(d)
Gain 27.5–35 MHz	−5.0 nominal	dB	see Figure 5.103(c)
Third-order intermodulation (three carriers. 50 kHz minimum spacing)	40 min	dBc	43 min
Power consumption	24 max	W	18.7
Mass	5.0 max	kg	3.8
Volume	200 × 350 × 80 max	mm^3	OK

pactness in size and ease of transport; tremendous tactical advantage because of low overturning moment; and low construction cost.

The performance of the SBFA was first demonstrated by AFCRL in 1962. In 1969, it was placed in production as the primary transmitting and receiving antenna for the *Tactical Satellite Communications System* (TACSATCOM). In a joint effort of the Navy, Marine Corps, and Air Force, the system was tested in a network of 13 stations located on military vehicles, ships, and submarines in the area between Bermuda and Hawaii. In this test of TACSATCOM [84–87], all 13 stations were contacted from a control station at Fort Monmouth, N.J., via the 35,200 km high TACSAT I satellite with a high voice quality. Figure 5.105(a) shows a typical SBFA, which can be opened and closed like an umbrella and even packed in a rucksack for easy transport.

Reynolds [88] described a collapsible team pack antenna, similar to Figure 5.105(a), for use as part of the ground station equipment for communication, via synchronous satellite, during tactical situations. The antenna consists of an octagonal back reflector with an octagonal lip, a set of crossed sleeve dipoles, a circular disk front reflector, matching transformers, and a 3 dB coupler for forming circular polarization. The antenna weighs less than 14 kg, including ancillary equipment. The antenna handles 40 W of average power and operates in the frequency range of 240 to 315 MHz. An SBFA was designed which can handle more than 2 kW average power and higher gain than earlier versions. Both these antennas were constructed from a heavy marquisett-weave metalized cloth.

A four-element SBFA array (Figure 5.89) can be used for tracking applications. Figure 5.105(b) and (c) also shows the *E*-plane and *H*-plane difference patterns, in which there are extremely sharp minima, all sidelobes are lower than 24 dB, and the backlobe is more than 30 dB below the main lobe. Figure 5.105(d) shows the sum patterns measured for cuts at 45° between the *X* and *Y* axes, obtained after rotating the array around its diagonal axis. Comparing these with the *E*-plane and

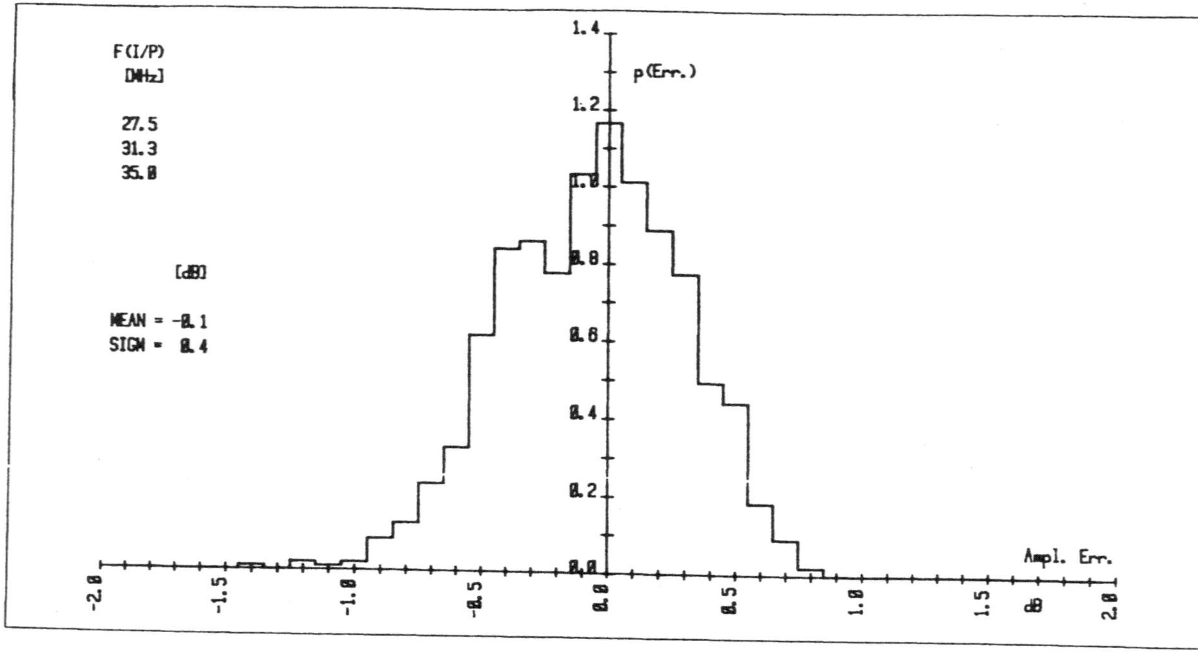

Figure 5.103(c) Amplitude error distribution at BFN output (from [71])

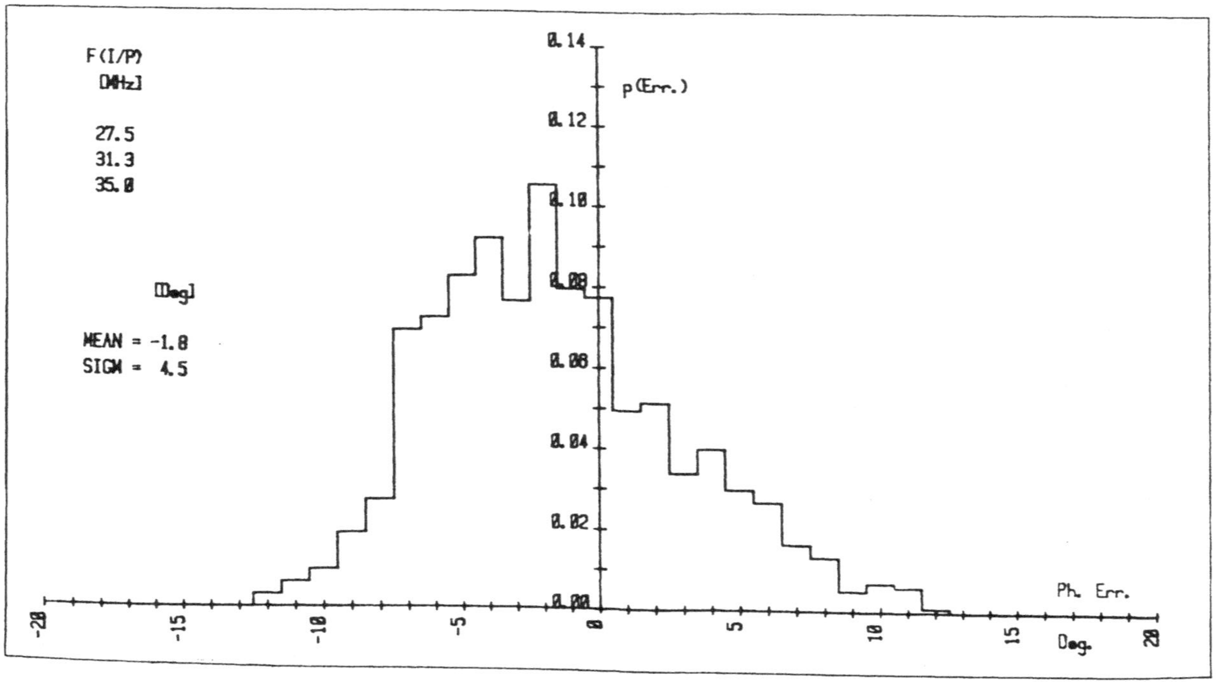

Figure 5.103(d) Phase error distribution at BFN output (from [71])

Table 5.34 Measured EIRP and G/T Results (measurement error ±0.5 dB)

Beam Location	*Beam No.*	*EIRP in* dBW	*G/T in* dB/K
Center	10	44.4	−0.6
Intermediate	12	43.5	−2.3
Edge	18	44.0	−1.3

H-plane patterns of the array (Figure 5.105(b) and (c)), some minor changes occur, which have negligible influence on the array's performance. Katulski [89–90] proposed a similar concept of an antenna tracking system for a medium-class ship terminal. The results of his experiments show that the four-element SBFA array and the amplitude comparison monopulse system may be applied in ship terminals. The system is very simply constructed and satisfactory in its technical parameters when compared with other systems. The amplitude comparison monopulse system [91] is used for the antenna system. Figure 5.105(e), (f), and (g) shows the *E*-plane and *H*-plane differences, and 45° sum patterns of an eight-element SBFA array, respectively.

Figure 5.105(h) shows a NASA 36-slot array [92], a European tracking 48-dipole array [93], and an AFCRL eight-element SBFA array [55]. In satellite tracking applications, the SBFA array is capable of the same gain, pattern quality, and reliability as the NASA 36-slot array and European 48-dipole array, as well as those composed of Yagi, disk-on-rod or helical antennas.

An eight-element SBFA array of three wavelengths (across the large common reflector) was designed for shipboard or buoy use for the relay of large amounts of undersea telemetry data to satellites. The common cavity has a sidewall depth of 0.5 λ_0 at center band, and the crossed dipoles are spaced at 1.2 λ_0. The measured gain of the array is better than 18.5 dB, and the sidelobe levels are more than 15 dB down, in both principal and diagonal planes.

Hartmann and Engelhardt [94] developed three one-element SBFAs at 40 MHz, 140 MHz, and 360 MHz in the Max Planck Institute (Lindau, Germany) for use with the ATS-6 radio beacon experiment. For all three antennas, sidelobe suppression was better than 20 dB and forward-backward ratio was better than 30 dB. The relative dispersion-phase shift between the 40 MHz, 140 MHz, and 360 MHz carriers and their + 1 MHz sidebands was measured accurately for dispersion. The SBFAs seem to be the best possible compromise of any antenna type for the specific purpose of differential group delay measurements.

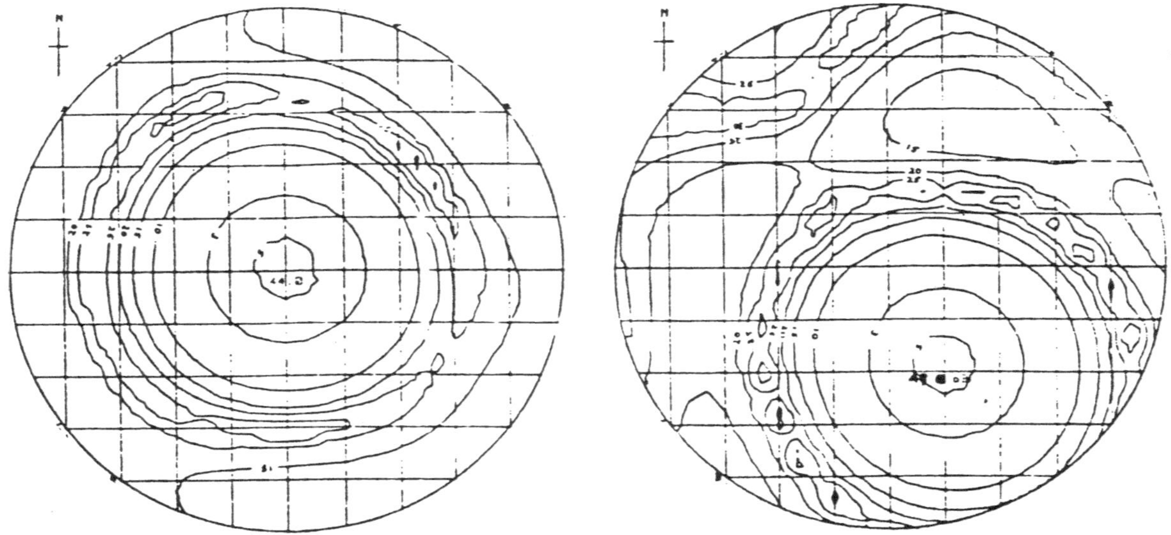

Figure 5.104(a) EIRP contour plot (with all elements working) for beams 10 and 15 (from [83])

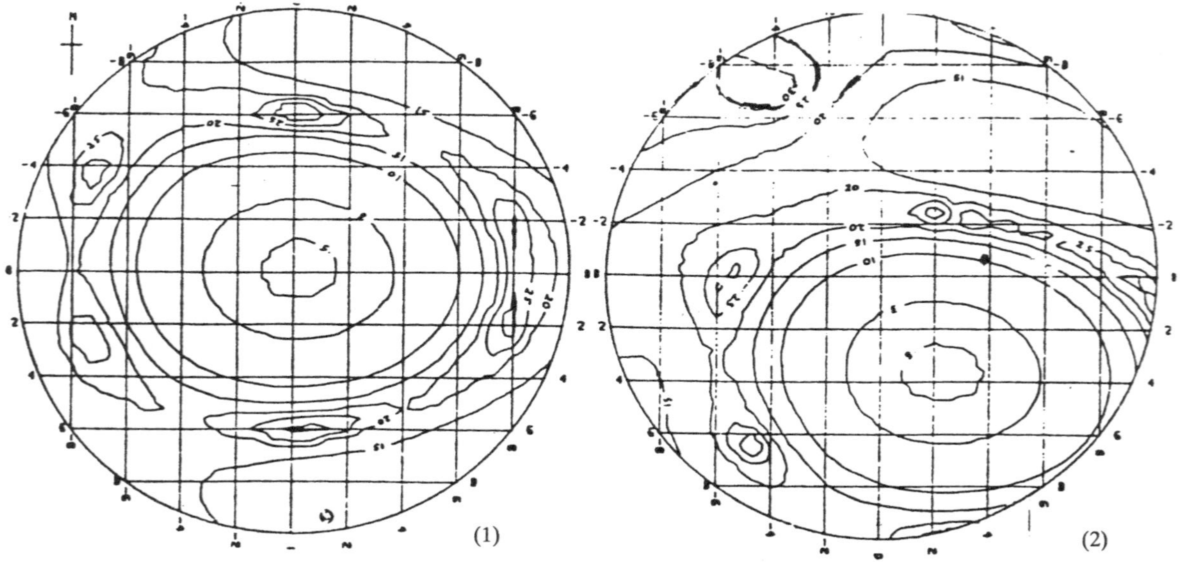

(1) (2)

Figure 5.104(b) EIRP contour plot (with elements 6, 10, and 15 disabled) for beams 10 and 15 (from [83]): (1) Transmit beam 10, EIRP 41.1 dBW; (2) Transmit beam 15, EIRP 40.3 dBW

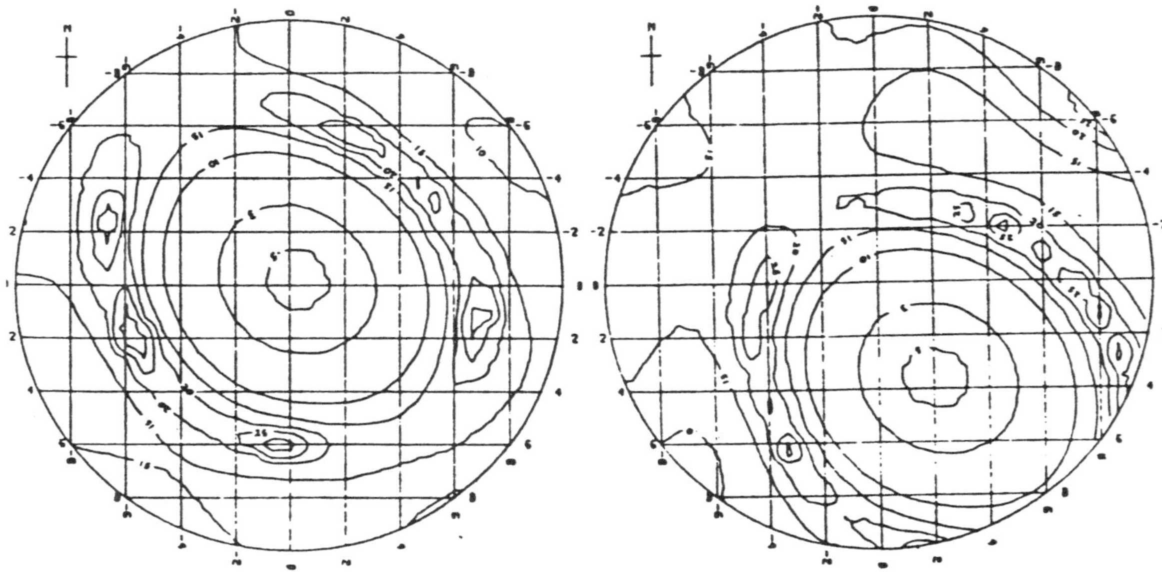

Transmit Beam 10
EIRP 41.7 dBW

Transmit Beam 15
EIRP 41.1 dBW

Figure 5.104(c) EIRP contour plot (with elements 3, 5, and 14 disabled) for beams 10 and 15 (from [83])

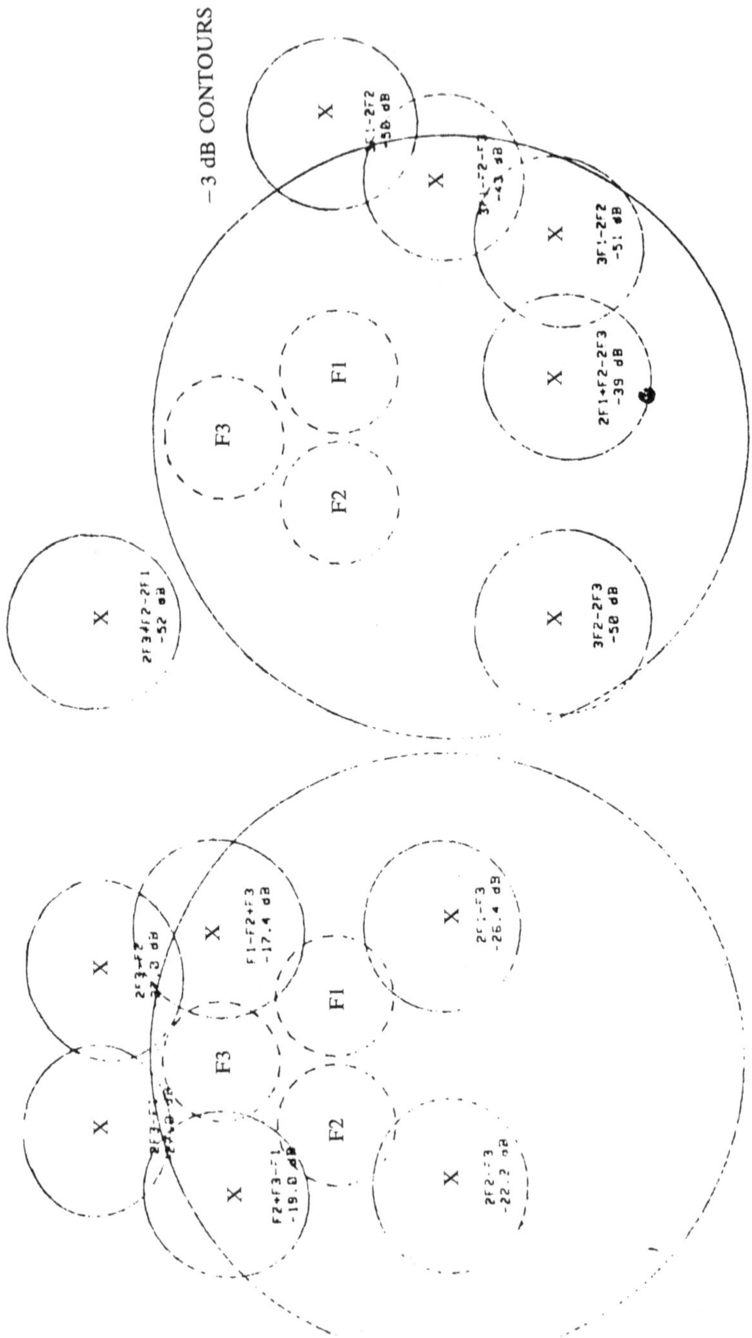

gure 5.104(d) Measured PIMP for three different carriers in beams 2, 5, and 6 (from [83])

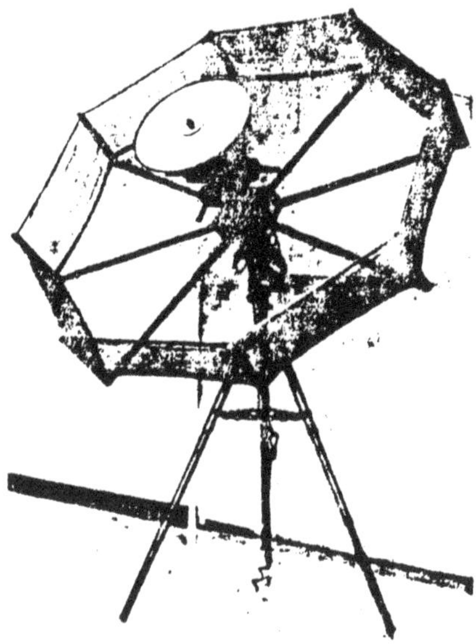

Figure 5.105(a) An umbrella-type SBFA (from [85–88])

For reduction of weight, mesh instead of solid reflectors are used. Various papers [95–97] have been published on the design of mesh reflectors. For a grid consisting of circular rods with diameters $2a$ and a grid constant d being smaller than the wavelength λ_0, the attenuation A of the reflector is given by:

$$A = 10 \log [1 + \{(2d/\lambda_0) \ln (d/(2\pi a))\}^2]^{-1} \qquad (5.32)$$

Figure 5.106(a) and (b) shows a mesh design for the SBFA, and the 10 m diameter large reflector of a 40 MHz SBFA, respectively.

Hristov and Urumov [98,99] described some backfire-aperture antenna designs that have been applied as radar for surveillance or for navigational purposes. A typical navigational marine radar antenna is required to have a radiation pattern that is narrow in the azimuthal plane (0.8°–4°) and wide in the elevation plane (15°–40°). The narrow beamwidth in azimuth is often obtained from a linear array of slot elements cut along one wall of a waveguide.

Large [18] developed a vertically-polarized slotted linear array of 24 shunt displaced slots, the radiation patterns of which are shown in Figure 5.96(e) and (f). Hristov *et al.* [98,99] proposed and patented a similar construction having a horizontal polarization, which is usually the case in marine radar applications. Figure

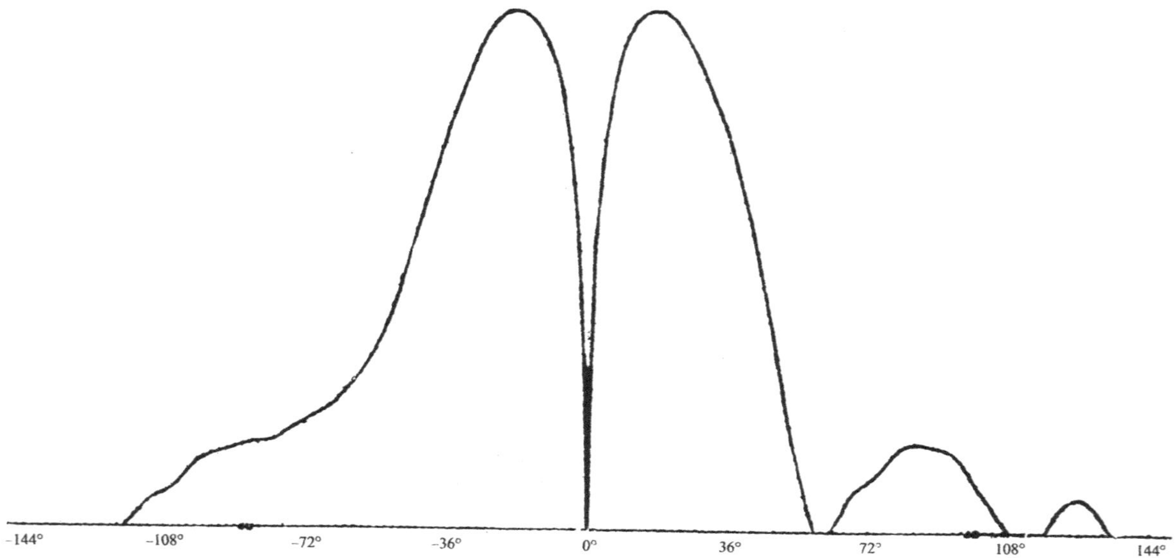

Figure 5.105(b) *E*-plane difference pattern of a four-element SBFA array at 3 GHz (from [55])

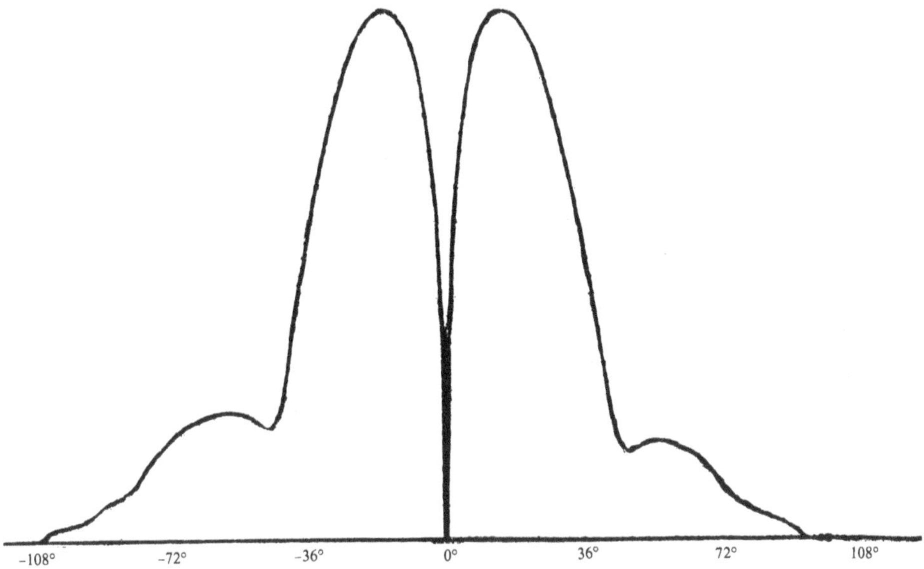

(c)

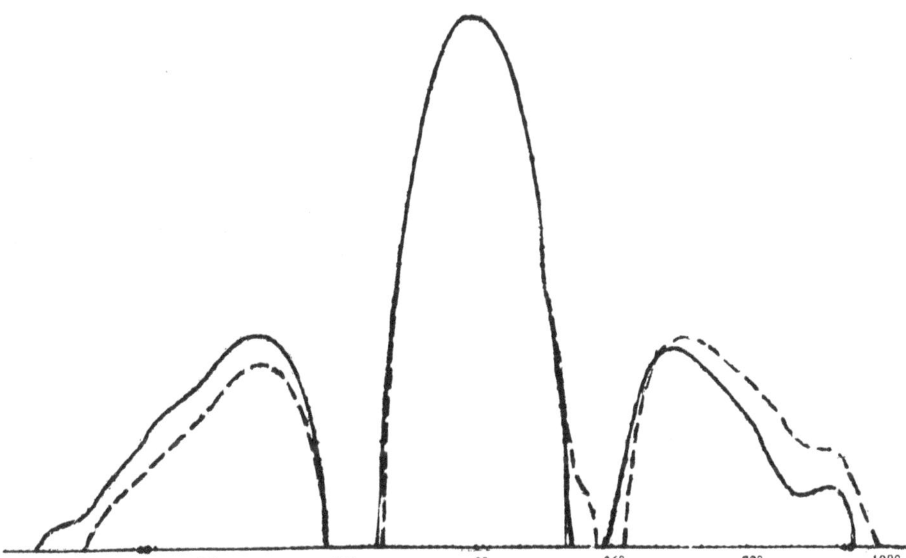

(d)

Figure 5.105(c) *H*-plane difference pattern of a four-element SBFA array at 3 GHz (from [55])
(d) 45° cuts between *E*-plane and *H*-plane (from [55])

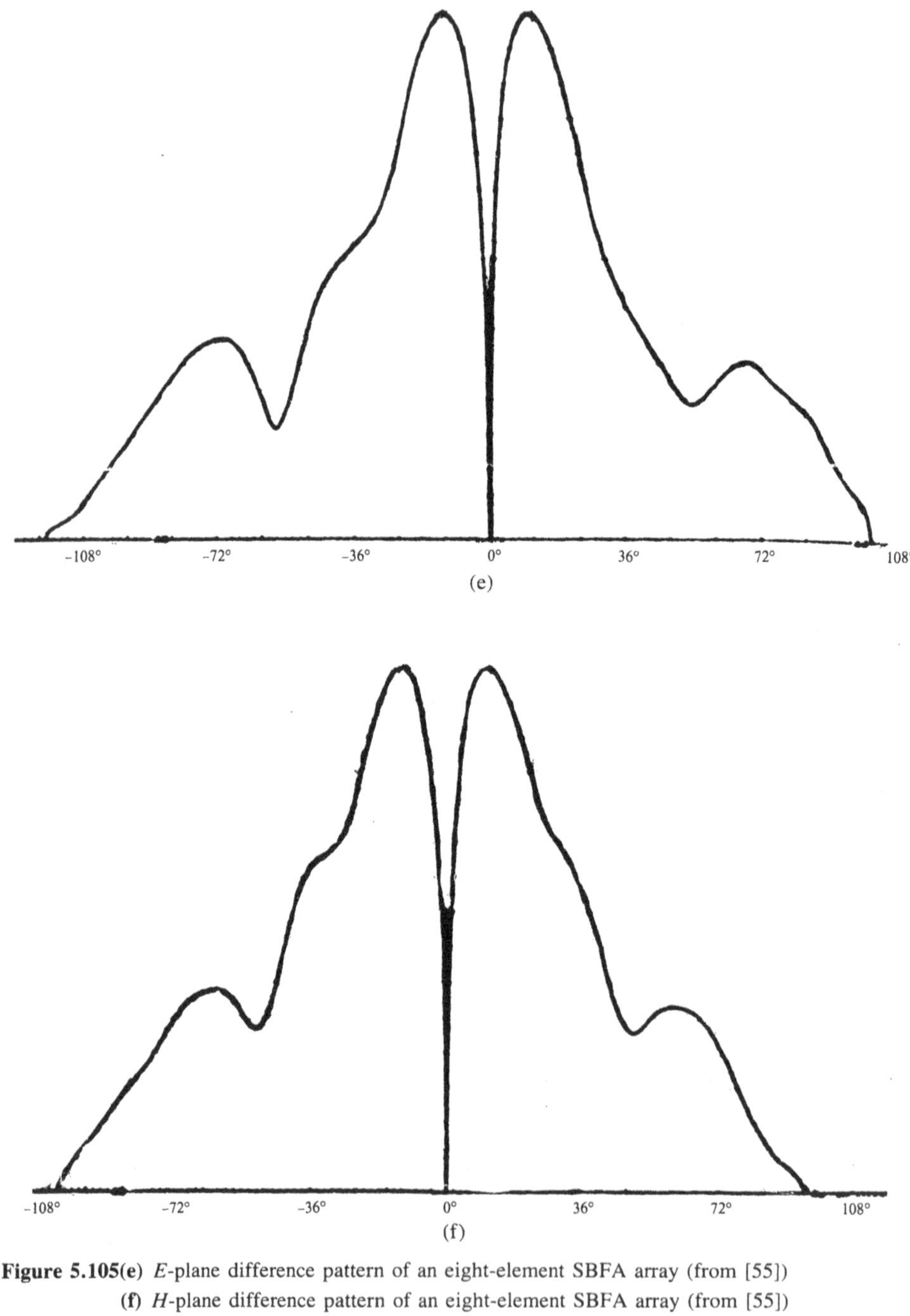

Figure 5.105(e) *E*-plane difference pattern of an eight-element SBFA array (from [55])
(f) *H*-plane difference pattern of an eight-element SBFA array (from [55])

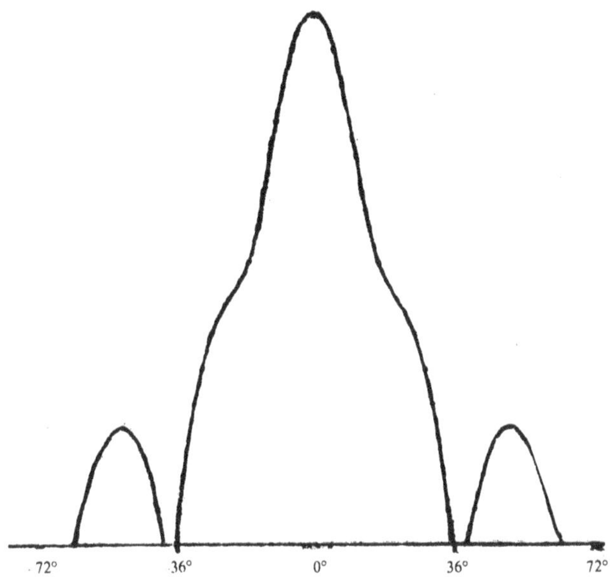

Figure 5.105(g) 45° cuts between *E*- and *H*-plane patterns of an eight-element SBFA array (from [55])

5.107 shows an *X*-band antenna model that is 2 m in length, 6 cm in height, and 4.5 cm in depth, which produces an azimuthal and an elevation beamwidth of 1.2° and 18°–30°, respectively. In Figure 5.107, (1) is a rectangular waveguide, (2) is the large reflector, (3) is the small reflector, (4) shows the array slots, and (5) shows the parallel metal strips on each side of the slots. The vertical antenna size is $H = 2\ \lambda_0$ and the strip reflector width is $W = 0.3\ \lambda_0$. These values depend on the size and position of the small reflector. The small reflector is not a continuous strip, but a multiple-reflector array [100,101], which is printed on the thin dielectric plate [101]. Table 5.35 shows the basic dimensions and characteristics of this typical widely used navigational antenna, and the new SBFA array. The SBFA array is shorter and lighter in weight compared to horn and dielectric surface wave navigational radar antennas.

Heckert [102] investigated *L*-band shipboard antennas for maritime satellite applications and concluded that an *L*-band SBFA subsystem is the best. Table 5.36 shows the summary of characteristics and performance of the antenna. Figure 5.108(a)

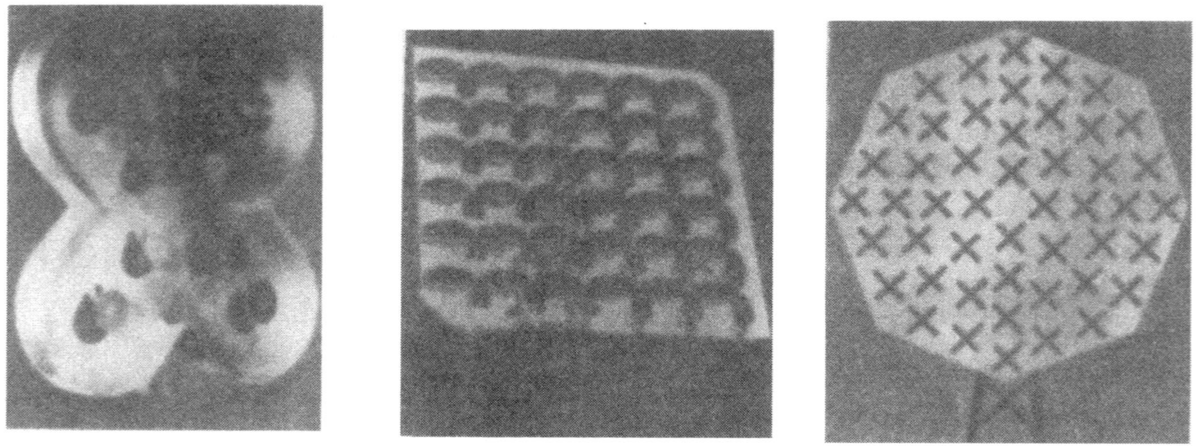

(a) (b) (c)

Figure 5.105(h) Comparison of three equal gain satellite tracking arrays: (a) NASA 36-slot array (from [92]), (b) European tracking network 48-dipole array (from [93]), and (c) AFCRL eight-element SBFA array (from [55])

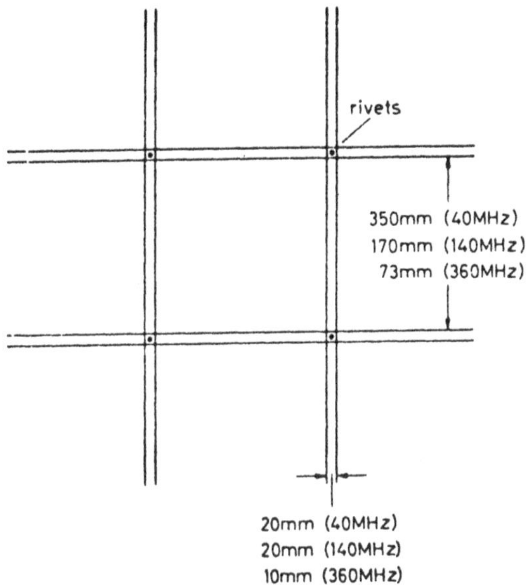

(a)

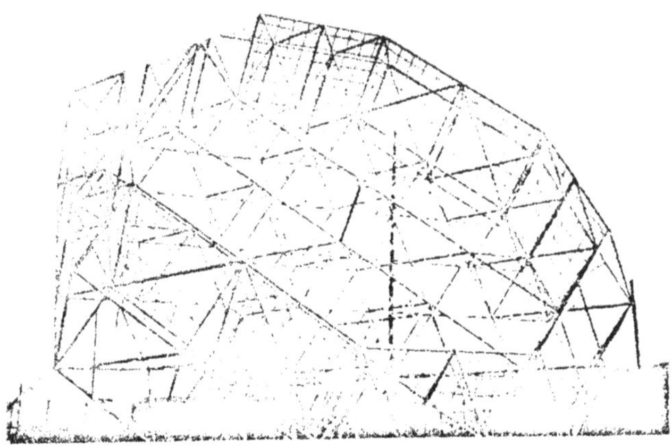

(b)

Figure 5.106(a) Mesh for SBFA reflector design (from [94]): material—aluminium plate strips 1 mm thick, fixed with rivets.

(b) 40 MHz SBFA with 10 m diameter large reflector (from [94])

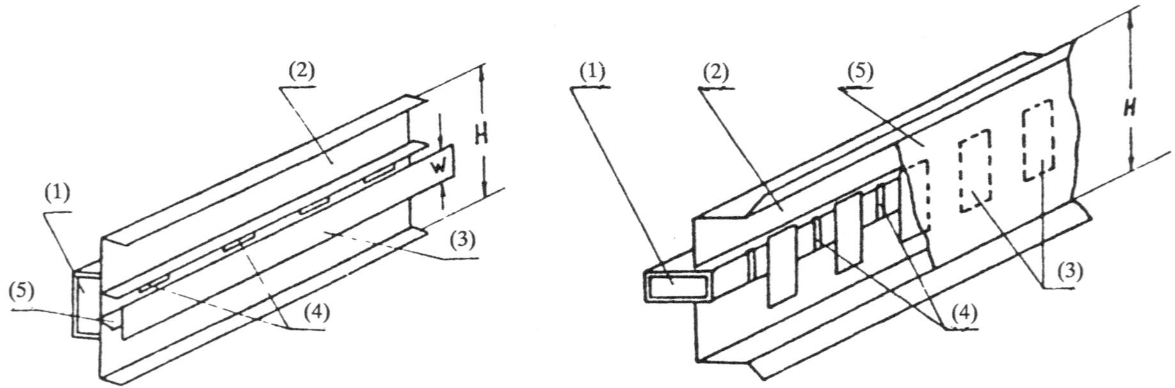

Figure 5.107 SBFA array for navigational radar application (from [99])

Table 5.35 Constructional Detail and Characteristics of Horn, Dielectric and SBFA (from [98–99])

Antennas	*Height* H/λ_0	*Depth* D/λ_0	*Elev. 3 dB beamwidth* *degrees*	*Sidelobe levels* dB	*Cross-sectional views*
With flared (horn) section	4.0	6.5	23°	−25	
With dielectric structure	1.5	7.5	23	−15	
Large's antenna	2.0	1.0	28	−22	
Hristov's antenna model	2.0	1.4	24	−20	

Table 5.36 Summary of Characteristics and Performance of a Recommended Shipboard Antenna Subsystem

Antenna Subsystem Characteristic	*Performance*
Antenna Gain, 1535–1543.5 MHz	14.0 dB peak, (13.0 dB min. operating in system)
Antenna Gain, 1636.5–1645 MHz	14.2 dB peak
Antenna Radiator Type	Short-Backfire
Antenna Feed	Monopulse, (>0.02 volts/volt-deg)
Antenna Size	39 cm (15.4 in.) diam. by 10 cm (3.9 in.) deep
Half-Power Beamwidth	~33°
Polarization	Circular
Pedestal	*X-Y* Mount

Table 5.36 (cont'd)

Antenna Subsystem Characteristic	*Performance*
Coverage:	360° Azimuth, and:
Ship size >150 m	−30° to 90° elevation
Ship size <150 m	−45° to 90° elevation
Angular Rate Capability:	
Ship size >150 m	15 deg/s, 7.5 deg/s^2
Ship size <150 m	45 deg/s, 45 deg/s^2
Wind Loads	75 knot design criteria, Survivability at 150 knots
Pointing Modes:	
1) Manual	Console operator
2) Autotracking	Single-channel Monopulse (Monoscan)
Monoscan Coupling Factor	12 dB
Position Tkg Loop (Noise) Bandwidth	2 Hz (Large Ships), 4 Hz (Small ships)
Electronics Mounted With Antenna	Diplexer, Preamplifier (3.5 dB N.F.) Solid State Transmitter (~60 W), *L*-band/70 MHz Up- and Down-Converters

and (b) shows the slave antenna pointing concept and the shipboard terminal block diagram with monoscan autotracking equipment.

A compact ship terminal antenna for the *Aeronautical-Maritime Engineering Satellite* (AMES) system was developed by the Radio Research Laboratory of Japan [103]. Table 5.37 shows its electrical characteristics and dimensions. The antenna consists of an SBFA with a large reflector of 0.4 m and gain of approximately 14 dBi.

Shiokawa, *et al.* [104] proposed a modified SBFA for the standard B ship earth station (for smaller ships and vessels) due to its small size, light weight, high efficiency, and simplicity in construction. Figure 5.109(a) shows the construction of the modified SBFA. The antenna has two small reflectors and a large reflector with a step configuration. The size of reflectors d_1, d_2, and D_1 are 0.448 λ_0, 0.436 λ_0 and 1.7 λ_0 at 1.54 GHz, respectively. Figure 5.109(b) shows the experimental values of gain and axial ratio of the antenna, with respect to the step height, *hs* when D_1 = 1.7 λ_0 and D_2 = 0.7 λ_0.

The SBFA has been used to suppress the reflected wave by polarization mismatch. Figure 5.109(c) shows that the horizontally-polarized wave is reflected perfectly, while the vertically-polarized wave is reflected with a fairly large attenuation at incident angles of less than 20°. The polarization of the reflected wave becomes elliptical, with the opposite sense of rotation with respect to the incident circular

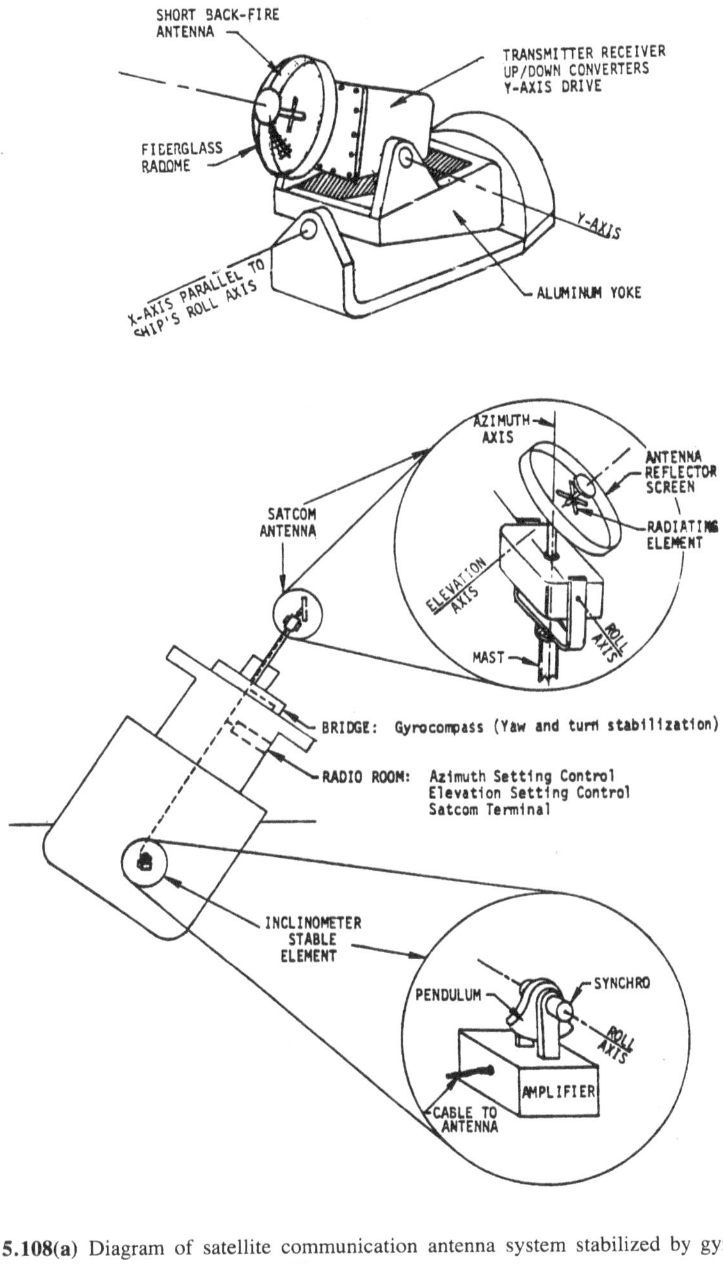

Figure 5.108(a) Diagram of satellite communication antenna system stabilized by gyrocompass and inclinometer (from [102])

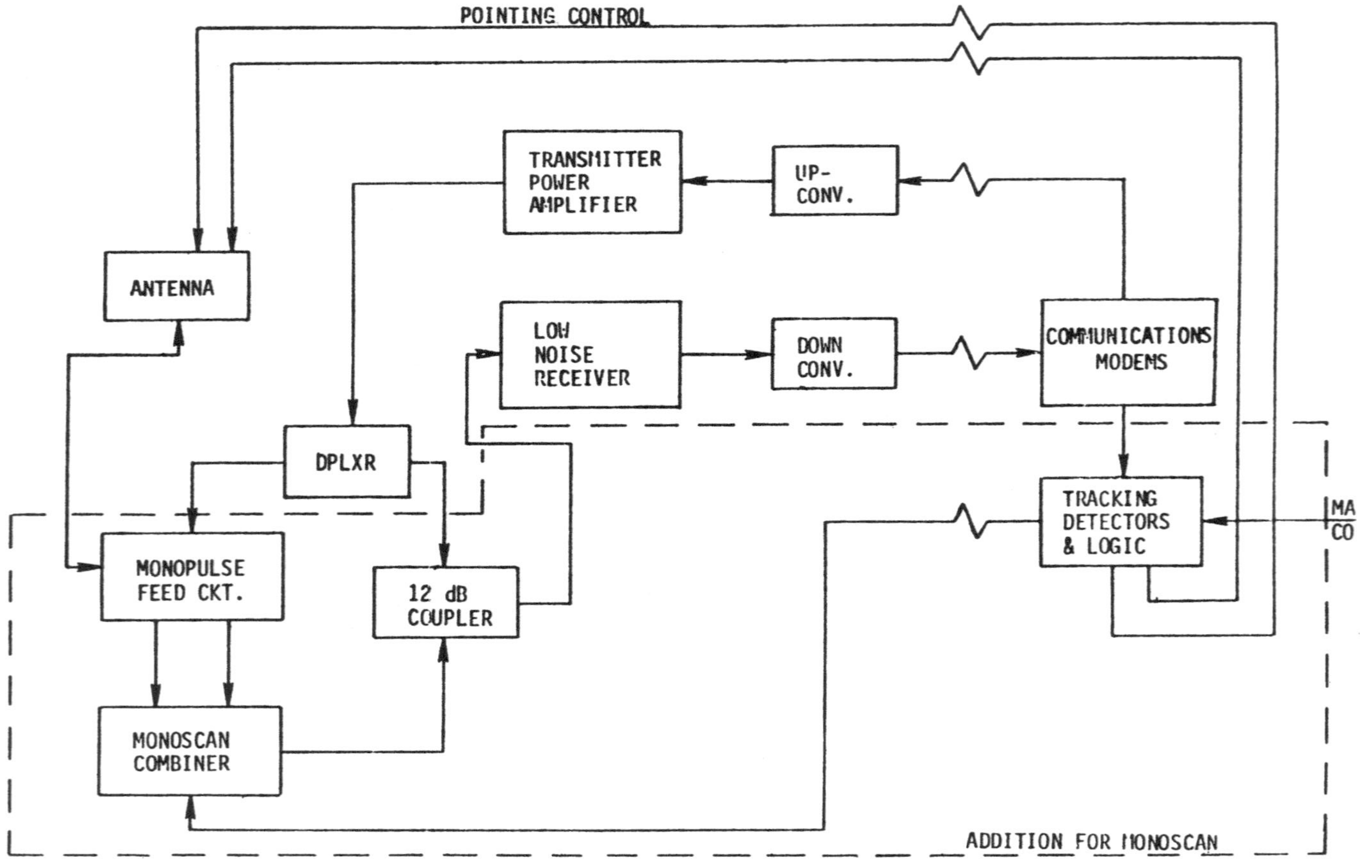

Figure 5.108(b) Shipboard terminal block diagram with monoscan autotracking (from [102])

Table 5.37 Electrical Characteristics and Dimensions of the AMES System

Antenna Gain (1.5 GHz)	14 dB
G/T (elevation angle = 5°)	−9.7 dB/K
Axial ratio	1 dB
Side lobe level	−22 dB
Half power angle	17 degrees
Noise temperature of LNA	145 degrees
Modulation method	NBFM, ΔM-PSK, PN-PSK
Weight of above-deck equipment	40 kg
Diameter of radome	64 cm
Height of radome	60 cm
Tracking accuracy	10 degrees

polarization, at elevation angles greater than about 5°, and its major axis is nearly horizontal. The reflection coefficients of sea surface are also shown in Figure 5.109(d). As a result, the fading due to sea surface scattering can be sufficiently reduced, although some polarization loss of a direct wave signal persists. Figure 5.109(e) shows an SBFA with the polarization shaping method. In this method, the cross-polar elements are set diagonally, and phase is adjusted by using a variable phase shifter. Since the axial ratio of the SBFA can be controlled arbitrarily, the orthogonal polarization to the reflected wave can be easily obtained. The fade depth corresponding to 99% of the time decreases from 4.1 dB to 1.9 dB by use of the polarization shaping method. The polarization-shaped SBFA is excellent for suppressing a large amount of fading due to sea surface scattering [105].

Martin [106] suggested using an SBFA instead of a conventional horn antenna for doppler sensing equipment. He showed that a 0.68 λ deeper rim on the main reflector not only reduces the gain by 0.25 dB, but also reduces the sidelobe level to approximately −35 dB and the bandwidth to more than 5%.

An SBFA array can be used as a feed cluster for an offset reflector antenna for a mobile satellite communication system [107,108]. The SBFA feed is designed to extend the features of the largest Cassegrain antenna to smaller diameters than are presently feasible with the Cassegrain design. Recently, Syrigos [109] described a conical scan parabolic antenna with an SBFA feed, which achieves equal E-plane and H-plane beamwidths, VSWR better than 1.4, high gain, and an overall performance which is better than Cassegrain designs.

Hristov [110] has developed two SBFAs for data receiving from satellites under a contract from the Bulgarian Central Laboratory for Space Research. These antennas are an undersized SBFA designed for use in the frequency band of 136–138 MHz, and a dual-frequency SBFA for use in the frequency bands of 136–138 MHz and 460–480 MHz. Figure 5.110 shows a sketch of an undersized SBFA used in the frequency range 136–138 MHz. The size of the large reflector is approximately 1.2

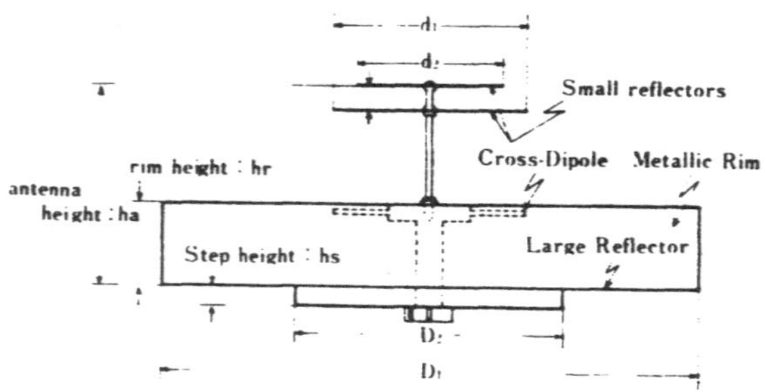

Figure 5.109(a) Construction details of a modified SBFA (from [104])

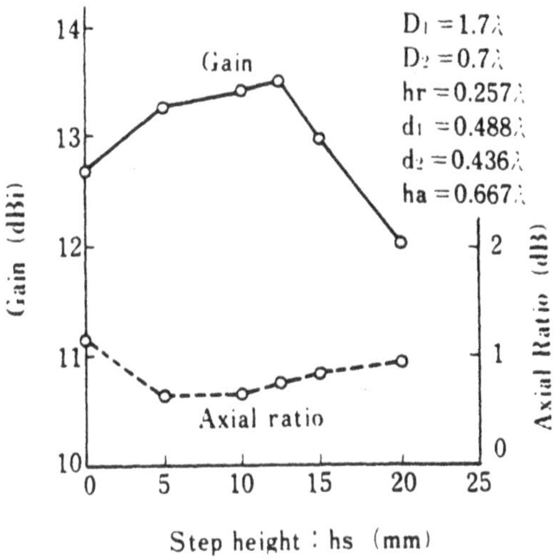

(b)

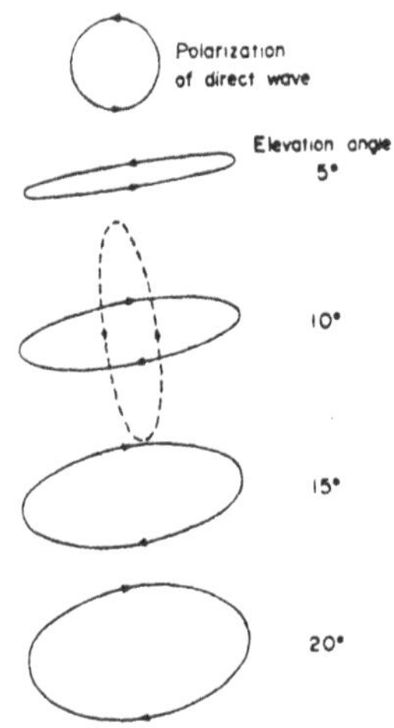

(c)

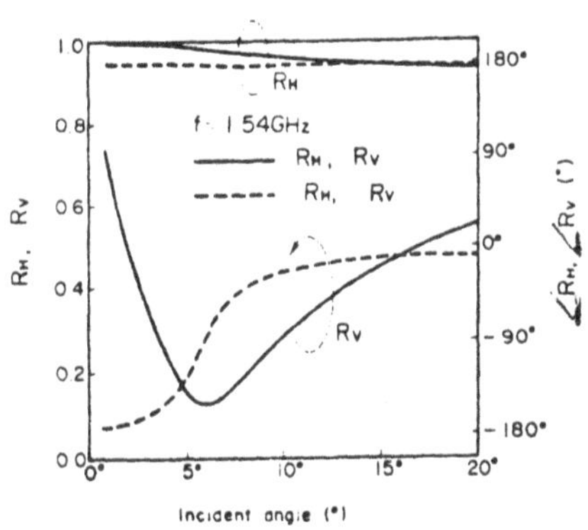

Horizontal polarization: $R_H = |R_H| \exp(j\angle R_H)$
Vertical polarization : $R_V = |R_V| \exp(j\angle R_V)$

(d)

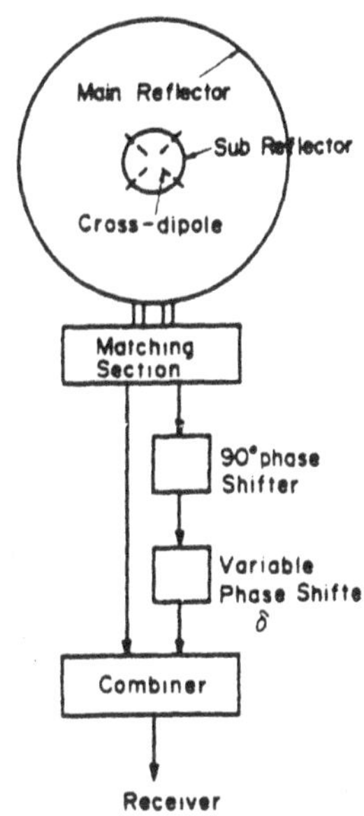

(e)

Figure 5.109(b) Characteristics of a modified SBFA *versus hs* (from [104])
(c) Polarization ellipse of a reflected wave at sea surface (from [105])
(d) Reflection coefficients of sea surface (from [105])
(e) SBFA with the polarization shaping method (from [105])

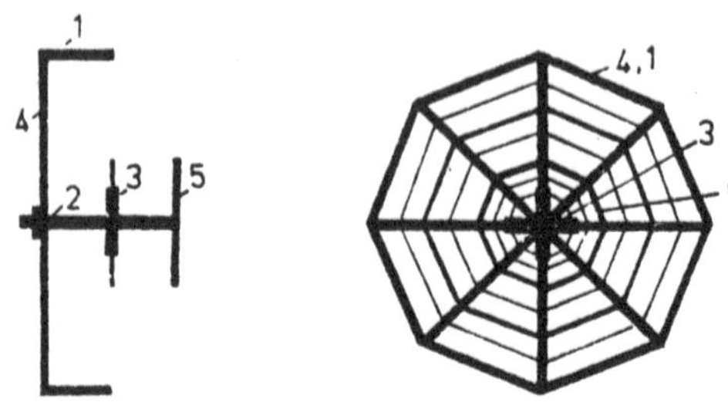

Figure 5.110 An undersized SBFA designed in the frequency range 136–138 MHz (from [110]): 1—rim of the SBFA, 2—connection between coaxial feeder and large reflector, 3—crossed dipole, 4—gridded 1.2 λ_0 diameter large reflector (octagonal), and 5—octagonal small reflector.

λ_0, with an octagonal wire-grid configuration. The antenna is excited by a pair of crossed dipoles for linear or circular polarizations.

The dual-frequency antenna system consists of two gridded octagonal SBFAs, as shown in Figure 5.111. The antenna has three gridded reflectors and two pairs of crossed dipoles. The octogonal gridded reflector (2) is a large reflector for the 460–480 MHz band and a small reflector for the 136–138 MHz band. These antennas were optimized experimentally to produce approximately 15 to 17 dB gain. This antenna was used in the Soviet-Bulgarian telemetry satellite launched in 1981 for scientific remote sensing experiments.

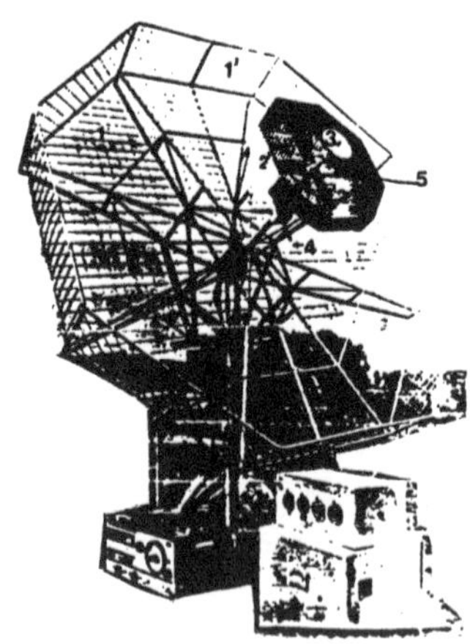

Figure 5.111 Dual-frequency SBFAs (from [110]): 1—gridded octagonal large reflector for 136–138 MHz, 1′—rim of the large reflector (136–138 MHz), 2—gridded octagonal small reflector for 136–138 MHz, which is also used as large reflector for 460–480 MHz band, 2′—rim for the reflector, 3—small reflector for 460–480 MHz band, 4—crossed dipole (136–138 MHz band), and 5—crossed dipole (460–480 MHz band)

Figure 5.112 shows the HS 601 satellite bus for the US Navy UHF follow-on satellite, which has a large four-element SBFA array for transmitting [111]. The satellite payload will have about twice the communications capacity of FLTSAT-COM, and will be manufactured with materials [112] which are not susceptible to nuclear radiation. The payload will have 21 narrow-band channels with a 5 kHz bandwidth, 17 relay channels with a 25 kHz bandwidth, and one fleet broadcast channel with a 25 kHz bandwidth.

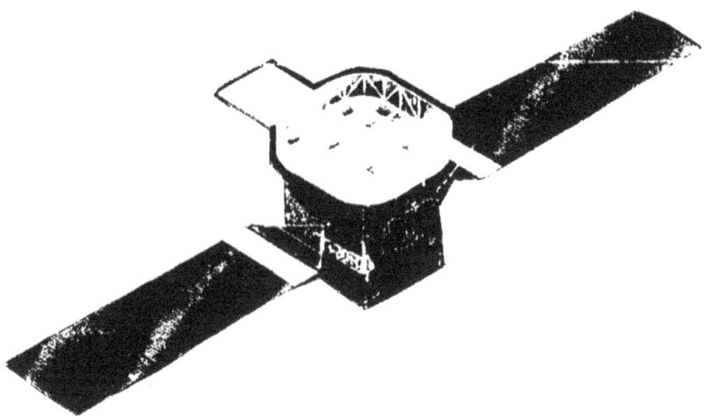

Figure 5.112 Hughes HS 601 bus for Navy UHF follow-on satellite with four-element SBFA array for transmit bands (from [111])

REFERENCES

1. Trentini, G.V., "Partially Reflecting Sheet Arrays," *IRE Trans. Antennas and Propagation*, Vol. AP-4, No. 4, October 1956, pp. 666–671.
2. Ehrenspeck, H.W., "The Short-Backfire Antenna," *Proc. IEEE*, Vol. 53, No. 4, August 1965, pp. 1138–1140.
3. Dod, L.R., "Experimental Measurements of the Short-Backfire Antenna," *Tech. Report No. X-525-66-480*, October 1966.
4. Nielsen, E.D., "Influence of Variation of Backfire Antenna Parameters II: Short-Backfire Antenna," *Report No. S182, R82*, The Technical University of Denmark, Lyngby, Denmark, October 1970.
5. Ehrenspeck, H.W., "A New Class of Medium-Size High-Efficiency Reflector Antennas," *IEEE Trans. Antennas and Propagation*, Vol. AP-22, No. 2, March 1974, pp. 329–332.
6. Ehrenspeck, H.W., and J.A. Strom, "The Short-Backfire Antennas as Element for High-Gain Arrays," *AFCRL-71-0234*, Air Force Cambridge Research Lab., Bedford, Mass, April 1971.
7. Ehrenspeck, H.W., and J.A. Strom, "Short-Backfire Antenna—A High Efficient Array Element," *Microwave Journal*, Vol. 20, No. 5, May 1977, pp. 47–49.

8. Woodward, O.M., "Short-Backfire Antenna with Sum and Error Patterns," US Patent 4240080, December 1980.
9. Kooi, P.S., M.S. Leong, and T.S. Yeo, "Dipole-Excited Short Backfire Antenna with Corrugated Rim," *Electronics Letters,* Vol. 15, No. 14, July 1979, pp. 421–423.
10. Ohmori, S, M. Shuichi, K. Kiyonobu, and Y. Hironitsu, "An Improvement in Electrical Characteristics of a Short Backfire Antenna," *IEEE Trans. Antennas and Propagation,* Vol. AP-31, No. 4, July 1983, pp. 644–646.
11. Katulski, R.J., "The Optimization of the Crossed Dipoles as a Radiating Source in the Short Backfire Antenna," *Third Int. Conf. on Antennas and Communications,* ICAP 83, IEE Conf. Pub. No. 219, IEE London, U.K., 1983, pp. 412–416.
11a. Keen, K.M., "Feeder Errors Cause Antenna Circular Polarization Deterioration," *Microwave System News,* Vol. 14, No. 5, May 1984, pp. 102–108.
12. Hristov, H.D., and G.S. Kirov, "*X*-Band Short Backfire Antenna," *Scientific Session of The Higher Institute of Mechanical and Electrical Engineering,* Varna, Bulgaria, March 1971.
13. Hristov, H.D., and D. Taylor, "Rectangular Backfire Antenna With Dielectric Surface-Wave Structure," *Electronics Letters,* Vol. 8, No. 6, March 1972, pp. 163–165.
14. Hristov, H.D., "*X*-Band Rectangular Backfire Antenna," Bulgarian Patent 18149, July 1972.
15. Large, A.C., "The Application of a Waveguide Feed to a Short Backfire Antenna," *ASWE Tech. Report No. TR-73-16,* Admiralty Surface Weapons Establishment, Portsdown, U.K., April 1973.
16. Large, A.C., "A Linear Array of Greatly Reduced Size and Weight," *ASWE Tech. Report No. TR-74-2,* Admiralty Surface Weapons Establishment, Portsdown, U.K., February 1974.
17. Large, A.C., "Improvements in or Relating to Short Backfire Antennae," British Patent 1420207, January 1976.
18. Large, A.C., "Short Backfire Antennas With Waveguide and Linear Fields," *Microwave Journal,* Vol. 19, No. 8, August 1976, pp. 49–52.
19. Leong, M.S., and P.S. Kooi, "Rectangular-Waveguide-Excited Short Backfire Antenna With Corrugated Rim," *Electronics Letters,* Vol. 15, No. 17, August 1979, pp. 533–535.
20. Kooi, P.S., and M.S. Leong, "*Q*-Band Short Backfire Antenna Arrays," *Proceedings 10th European Microwave Conference,* Poland, September 1980.
21. Emerson & Cuming Technical Data on Eccosil 4122.
22. Kumar, A., "A Hemispherical Coverage Antenna for Spacecrafts," *Electronics Letters,* Vol. 18, No. 10, May 1988, pp. 631–633.
23. Leong, M.S., and P.S. Kooi, "Performance of a Rectangular-Waveguide-Excited Short Backfire Antenna Employing a Conical Main Reflector," *Electronics Letters,* Vol. 20, No. 18, August 1984.
24. Leong, M.S., P.S. Kooi, and K.N. Wong, "Radiation Characteristics of *Q*-Band Short Backfire Antennas using a Conical Main Reflector," *Proc. Int. Symposium on Antennas and Electromagnetic Theory,* Beijing, China, 1985, pp. 718–723.
25. Hristov, H.D., and A. Kumar, "*X*-Band Waveguide-Fed Long Backfire Antenna With Dielectric Surface Wave Structure," Research Report Queen Mary College, London, U.K., 1977.
26. Chen, K.M., D.P. Nyquist, and J.L. Lin, "Radiation Field of The Short Backfire Antenna," *IEEE Trans. Antennas and Propagation,* Vol. AP-16, No. 5, September 1968, pp. 596–597.
27. Loh, S.C., and W.S. Leung, "The Radiation Field of the Short Backfire Antenna," *The Radio and Electronic Engineer,* Vol. 39, No. 4, April 1970, pp. 198–200.
28. Nielsen, E.D., and K. Pontoppidan, "Backfire Antenna with Dipole Elements," *IEEE Trans. Antennas and Propagation,* Vol. AP-18, No. 3, May 1970, pp. 367–374.
29. Nielsen, E.D., and K. Pontoppidan, "Backfire Antennas" *Report S183, R69,* Technical University of Denmark, Lyngby, Denmark, January 1969.
30. Nielsen, E.D., "Influence of Variation of Backfire Antenna Parameters, 1: Yagi Backfire Antenna," *Report No. 2, Contract F 61052 67 C 0056, S183, R80,* Technical University of Denmark, Lyngby, Denmark, September 1970.

31. Nielsen, E.D., Private Communication.
32. Patel, D., "Short Backfire Antenna Far-Field Radiation Patterns," *Int. J. Electronics,* Vol. 30, No. 2, February 1971, pp. 185–187.
33. Hsieh, T.Z., D.P. Nyquist, and K.M. Chen, "The Short-Backfire Antenna: A Numerical-Physical Optics Study of its Characteristics," USAF Contract F19628-70-C-0072, 1970.
34. Kumar, A., "Theoretical Analysis of a Long Backfire Antenna Containing a Dielectric Structure," *IEE J. on MOA,* Vol. 2, No. 3, May 1978, pp. 91–95.
35. Kumar, A., "The Short Backfire Antenna: A Theoretical Approach," *IEEE Conference on IECI,* Toronto, Canada, 1985.
36. Gibson, G.A., S. Govind, and L.L. Tsai, "Analysis of the Short-Backfire Antenna," *IEEE APS Symposium,* Massachusetts, IEEE Cat No. 76CH1121-3AP, October 1976, pp. 229–232.
37. Harrington, R.F., *Field Computation by Moment Methods,* Macmillan, New York, 1968.
38. Changlu, L., and S. Ximing, "Short Backfire Antenna Research," *IEEE APS Symposium,* IEEE Cat No. 83CH1860-6, May 1983, pp. 138–141.
39. Marougi, S.D., "Effect of Disc Reflector on Radiation Impedance of Short-Backfire Antenna," *Electronics Letters,* Vol. 18, No. 4, February 1982, pp. 154–156.
40. Kishk, A.A., and L. Shafal, "Gain Optimization of Short Backfire Antenna with Different Excitation Types," *IEEE APS Symposium,* IEEE Cat No. 86CH2325-9, June 1986, pp. 923–926.
41. Kaloi, C., D. Hatfield, and P. Simon, "Microstrip Backfire Antenna," *IEEE APS Symposium,* IEEE Cat No. 81CH1672-5, June 1981, pp. 343–346.
42. Van de Capelle, A., E. Beyne, and L. Schiltz, "Experimental Study of the Microstrip Backfire Antenna," *L'Onde Electrique,* Vol. 65, No. 2, March–April 1985, pp. 88–92.
43. Vandenbosch, G., and A. Van de Capelle, "Analysis of the Microstrip Backfire Antenna," *Proc. 5th Int. Conf. on Antennas and Propagation,* ICAP 87, IEE Conf. Pub. No. 274, IEE London, U.K., March–April 1987, pp. 292–295.
44. Newman, E.H., and P. Tulyathan, "Analysis of Microstrip Antennas using Moment Methods," *IEEE Trans. Antennas and Propagation,* Vol. AP-29, No. 1, January 1981, pp. 47–53.
45. Bailey, M.C., and M.D. Deshpande, "Integral Equation Formulation of Microstrip Antenna," *IEEE Trans. Antennas and Propagation,* Vol. AP-30, No. 4, July 1982, pp. 651–656.
46. Pozar, D.M., "Input Impedance and Mutual Coupling of Rectangular Microstrip Antennas," *IEEE Trans. Antennas and Propagation,* Vol. AP-30, No. 6, November 1982, pp. 1191–1196.
47. Pozar, D.M., "Improved Computational Efficiency for the Moment Method Solution of Printed Dipoles and Patches," *Electromagnetics,* Vol. 3, 1982, pp. 299–309.
48. Newman, E.H., J.H. Richmond, and B.W. Kwan, "Mutual Impedance Computation between Microstrip Antennas," *IEEE Trans. Microwave Theory and Tech.,* Vol. MTT-31, No. 11, November 1983, pp. 941–945.
49. Kumar, A., "A Review on Microstrip Dielectric Materials for Space Application," *Report No. 88-ECC-02,* Electromagnetic Canada Consultants, Montreal, Canada, September 1988.
50. Kumar, A., "Effect of Nuclear Radiation on Microstrip Satellite Antennas," *Proc. JINA'88,* Nice, France, November 1988, Paper No. 3.16.
51. Vandenbosch, G., Private Communication, 1988.
52. Vandenbosch, G., and A. Van de Capelle, "Improved Analysis of the Microstrip Backfire Antenna," *Proc. JINA'88,* Nice, France, November 1988, Paper No. 3.1.
52a. Kaloi, C, Private Communication.
53. Ehrenspeck, H.W., and J.A. Strom, "Short-Backfire Arrays," *Report No. AFCRL-71-0368,* Air Force Cambridge Research Lab., May 1971.
54. Ehrenspeck, H.W., and J.A. Strom, "The Four-Element SBF Array: Variation of Parameters for Optimization of Performance," *Report No. AFCRL-71-0569,* Air Force Cambridge Research Lab., 1971.

55. Ehrenspeck, H.W., and J.A. Strom, "A Sixteen Element SBF Array," *Report No. AFCRL-72-0500,* Air Force Cambridge Research Lab., August 1972.
56. King, R.W.P., and S.S. Sandler, "A Modified Backfire Antenna," *IEEE Trans. Antennas and Propagation,* Vol. AP-22, No. 2, March 1974, pp. 349–352.
57. Phelan, H.R., and J.L. Harrison, "Phased-Array Cost Cuts By New Radiating Elements," *Microwaves,* Vol. 13, No. 1, January 1973, pp. 40–43.
58. Hristov, H.D., and J.R. Urumov, "*X*-Band Backfire Antenna Arrays," *5th European Microwave Conference,* Hamburg, Germany, September 1975, pp. 112–115.
59. Kumar, A., "Waveguide Fed Short Backfire Antenna Array," *IEEE National Telecommunication Conference,* Houston, Texas, 1980.
60. Hristov, H.D., "Short Backfire Antenna Arrays," *Proc. 5th Int. Symposium Radioelectronics-74,* Varna, Bulgaria, Vol. 3, June 1974.
61. Kooi, P.S., Private Communication.
62. Leong, M.S., and P.S. Kooi, "*Q*-Band Dielectric Loaded Short Backfire Antenna Arrays," *2nd Int. Conf. on Antennas and Propagation,* IEE Conf. Pub. No. 195, IEE, London, U.K., April 1981, pp. 13–16.
63. Harrington, R.F., "Reactively Controlled Directive Array," *IEEE Trans. Antenna Propagation,* Vol. AP-26, 1978, pp. 390–395.
64. Luzwick, J., and R.F. Harrington, "A Study of a Reactively Loaded Finite Planar Rectangular Waveguide Antenna Array," *Report No. 9, Contract No. N 0014-76-C-0225,* Naval Research, February 1979.
65. Sinha, S.N., and R.P. Agrawal, "Reactance-Controlled Short Backfire Array," *Electronics Letters,* Vol. 16, No. 7, March 1980, pp. 253–254.
66. Sinha, S.N., and R.P. Agrawal, "Experimental Investigations on Reactance-Loaded Waveguide Arrays," *Int. J. Electronics,* Vol. 51, No. 5, May 1981, pp. 683–691.
67. Sinha, S.N., and R.P. Agrawal, "Reactance-Controlled Short Backfire Array: New Results," *Electronics Letters,* Vol. 16, No. 20, September 1980, pp. 764–765.
68. Harrington, R.F., and J.R. Mautz, "A Generalized Network Formulation for Aperture Problems," *IEEE Trans. Antennas and Propagation,* Vol. AP-24, No. 11, November 1986, pp. 870–873.
69. Sinha, S.N., D.K. Mehra, and R.P. Agrawal, "Analysis of Reactively Loaded Waveguide Fed Short Backfire Arrays," *IEEE Montech'86 Conference on Antennas and Communications,* IEEE Cat. No. TH0156-0, Montreal, Quebec, Canada, September-October 1986, pp. 271–274.
70. Molker, A., "High-Efficient Phased Array Antenna for Advanced Multibeam; Multiservice Mobile Communication Satellite," *Third Int. Conf. on Satellite Systems for Mobile Communications and Navigation,* IEE Conf. Pub. No. 222, IEE London, U.K., June 1988, pp. 75–77.
71. Coirault, R., and W. Kreodte, "Multibeam Generation at *L*-Band: A Phased Array Approach," *ESA Journal,* Vol. 4, 1980, pp. 319–353.
72. Aasted, J., and A. Roederer, "A Multibeam Array for Communication With Mobile," *Proc. 7th AIAA Communications Satellite System Conference,* 1977, pp. 400–406.
73. Mica, G., "Technology Development Programme for Communication Satellite Payload in ESA," *Proc. 8th AIAA Communication Satellites System Conference,* Orlando, Florida, 1978.
74. Coirault, R., "L-Band Phased Array Transmitter for Communication Satellites," *Proc. 10th AIAA Communication Satellites System Conference,* 1980, pp. 484–489.
75. Rudge, A.W., "Current Trends in Antenna Technology and Prospects for the Next Decade," *IEEE APS News Letters,* Vol. 25, No. 6, December 1983, pp. 5–12.
76. Brian, M.T., Z.F. Voyner, and B. Howe, "Measurement of the MAM III Antenna System Using the Near Field Facility at BAe Stevenage," *Proc. European Microwave Conference,* Helsinki, 1982.
77. Ligthart, L.P., "A New Numerical Approach to Planar Phased Arrays," *Proc. 5th European Microwave Conference,* 1975, pp. 453–457.

78. Mautz, J.R., and R.F. Harrington, "Radiation and Scattering from Bodies of Revolution," *Appl. Sci. Res.*, June 1969, pp. 405–435.
79. Andersson, A., P. Bengtsson, and A. Molker, "Short Backfire Antenna Arrays for Space Communications," *IEEE APS Symposium,* Stanford, California, June 1977, pp. 194–197.
80. Kumar, A., "Short Backfire Antenna Arrays," *IEEE APS/URSI Symposium Digest,* 1984, pp. 20.
81. Ehrenspeck, H.W., "Backfire—Antennen," *NTZ,* Vol. 22, No. 5, 1969, pp. 286–289 (in German).
82. Kumar, A., "Passive IM Products Threaten High Power Satcom Systems," *Microwaves & RF,* Vol. 26, No. 13, December 1987, pp. 98–103.
82a. Kumar, A., "Microwave Holographic Imaging of Passive IM Product Generation on Antennas," *Research Report 88-AK-04,* Electromagnetic Canada Consultants, Montreal, Quebec, Canada, October 1988.
83. Brian, M.T., and A.C. Talbot, "Some Characteristics of an Experimental *L*-Band Multibeam Phased Array Payload," *Proc. IEE Conference on Satellite Systems for Mobile Communications and Navigation,* London, June 1983, pp. 78–82.
84. TACSATCOM Tested, *The Hansconian* (Bedford, Mass.), 10 Marz 1969, pp. 10 (in German).
85. Ehrenspeck, H.W., "Short-Backfire Antennenals UHF Fernsehantennen," *Funk-Technic,* Vol. 16, 1971, pp. 598–600 (in German).
86. Ehrenspeck, H.W., *et al., AFCRL Progress Report,* 1969–1972, pp. 251–253.
87. Dod, L.R., "Backfire Antennas for Telemetry and Tracking Applications," *Proc. NTC'69,* Washington, D.C., 1969, pp. 77–84.
88. Reynold, B.C., "A Discussion of the Short Backfire Antenna and its Application in Satellite Communication and Tracking Systems," *IEEE Int. Conv. Digest,* New York, March 1970, pp. 384–385.
89. Katulski, R.J., "The Ship Terminal Antenna of the Maritime Satellite Radiocommunication System," *Proc. Seventh Int. Coll. Microwave Communication,* Budapest, Hungary, 1982, pp. 315–318.
90. Katulski, R.J., "The Medium System for a Medium Class, Ship Terminal of the Satellite Communications," *Proc. Satellite Communications Conf.*, Ottawa, Canada, 1983, pp. 16.4.1–4.
91. Skolnik, M.I., *Introduction to Radar Systems,* McGraw-Hill, New York, 1962.
92. Lantz, P.A., "Handbook of NASA/GSFC Tracking, Data-Acquisition and Communication Antennas," *Tech. Report X-525-64-222,* Goddard Space Flight Center, Greenbelt, Md, 1964.
93. Oettl, H., and L. Thomanek, "Monopuls-Antenne der Bodenstation fur Satellitenfunk der Deutschen Versuchsanstalt fur Luft und Raumfahrte V," *NTZ-Nachrtech. Z.*, Vol. 21, No. 10, 1968, pp. 631–634.
94. Hartmann, G.K., and W. Engelhardt, "Dispersion Measurements of One-Element Short Backfire (SBF) Antenna," *IEEE Trans. Antennas and Propagation,* Vol. AP-23, No. 3, March 1975, pp. 289–293.
95. Larsen, T., "A Survey of the Theory of Wire Grids," *Proc. IRE Trans. Microwave Theory and Tech.*, Vol. MTT-10, No. 3, May 1962, pp. 191–201.
96. Yampolskii, V.G., *Radiotekhnika* (Moscow), Vol. 10, No. 9, 1955, pp. 39–48.
97. Yampolskii, V.G., *Radiotekhnika* (Moscow), Vol. 11, No. 11, 1956, pp. 33–37.
98. Hristov, H.D., and J.R. Urumov, "Backfire-Aperture Radar Antennas," *Proc. Seventh Int. Coll. Microwave Communication,* Budapest, Hungary, September 1982, pp. 299–302.
99. Hristov, H.D., and J.R. Urumov, "Multielement Backfire Antenna for Centimeter-Waves," Bulgarian Patent No. 24 034, October 1976.
100. Ehrenspeck, H.W., "High-Frequency Waveguide Feed in Combination with Backfire Antenna," U.S. Patent 3774223, November 1972.
101. Hristov, H.D., and J.R. Urumov, "Multielement Backfire Antenna for Centimeter Wave," Bulgarian Patent 24034, October 1976.

102. Heckert, G.P., "Investigation of *L*-Band Shipboard Antennas for Maritime Satellite Applications," *NASA Report No. NASA-CR-126032*, February 1972.
103. Miura, S., "Aeronautical-Maritime Engineering Satellite," *Proc. IEEE Tokyo Section*, No. 20, 1981, pp. 12–15.
104. Shiokawa, T., Y. Karasawa, and H. Yokoi, "Compact Antenna Systems for INMARSAT Standard-B Ship Earth Stations," *Proc. IEE* ICAP '81, IEE Conf. Pub. No. 195, IEE London, U.K., April 1981, pp. 303–307.
105. Shiokawa, T., and Y. Karasawa, "Shipborne Antennas Suppressing Multipath Fading in Maritime Satellite Communication," *IEEE APS Symposium*, Albuquerque, New Mexico, May 1982, pp. 390–393.
106. Martin, A.G., "Short Backfire Antenna for Doppler Sensing," *Microwave Journal*, Vol. 24, No. 10, October 1981, pp. 93–96.
107. Kumar, A., "SBFA Feed for an Off-set Reflector Antenna," *Research Report 82-AK-05*, Engineering Education and Training Consultant, Conventry, U.K., July, 1982.
108. Horstein, M., "Requirements for Mobile Communications Satellite System," *Technical Report TRW*, California, Contract No. NAS3-23257, November 1983.
109. Syrigos, H., "Backfire Feed Antenna Beats Cassegrain Design," *MSN & Communication Technology*, Vol. 13, No. 10, October 1983, pp. 164–167.
110. Hristov, H.D., private communication.
111. Dornheim, M.A., "Navy Selects Hughes HS601 As Its UHF Follow-on Satellite," *Aviation Week & Space Technology*, August 8, 1988, p. 22.
112. Kumar, A., "Effect of Nuclear Radiation on Microstrip Antennas," *Proc. JINA*, Nice, France, November 1988, pp. 278–281.

Chapter 6
Long Backfire Antennas

The early work on Yagi long backfire antennas proved that when a gain of 15–30 dBi is desired, they have a number of advantages when compared to conventional types of antennas. The ordinary endfire for such a gain becomes very long. Dipole and slot broadside arrays, besides requiring complicated feed systems, have limited bandwidth, and are expensive when compared to backfire antennas. Although parabolic reflector antennas are usually suitable when higher gain is required, the long backfire antenna combines the structural advantages of a single endfire model with the high gain of a reflector antenna. The use of a simple open-ended waveguide instead of a dipole feed to excite a continuous surface wave structure (a dielectric rod, for example) allows construction of very effective and simple long backfire antennas for microwave frequencies. A very promising way to improve further the backfire antenna is to replace the uniform surface wave structure with a nonuniform (modulated) one.

For millimeter-wave band, a long backfire antenna with an insular image guide feed that is compatible with dielectric integrated circuit technology has been proposed. Thus, depending on the surface wave structures and feed configurations, many varieties of long backfire antennas covering the entire radio-frequency band have been developed and examined. This chapter summarizes the main theoretical and experimental results of research on long backfire antennas. Many practical antenna designs are described in detail, and their basic electromagnetic parameters are given.

6.1. THE UNIFORM LONG BACKFIRE ANTENNA: AN APPROXIMATE APPROACH TO ITS DESIGN

Long backfire antenna operation, as explained by Ehrenspeck in his first publications on the subject, [1–10] is based on the idea of doubling the physical length of the classical endfire antenna by means of a large plane reflector placed at the open end of the endfire structure, perpendicular to its axis. In other words, long backfire antenna performance can be analyzed by viewing the antenna as a modified version of an ordinary endfire slow-wave antenna, consisting of a reflector R_1, primary source F, and a surface wave line S (Figure 6.1(a)). The surface wave structure can take many forms, but a Yagi-dipole assembly, using corrugated (disk-on-rod), or dielectric rods is most popular. According to the *Hansen-Woodyard* (HW) condition, the maximum gain is proportional to the length of the surface wave structure, provided that the wave phase velocity is adjusted to its optimum value for any specific antenna length.

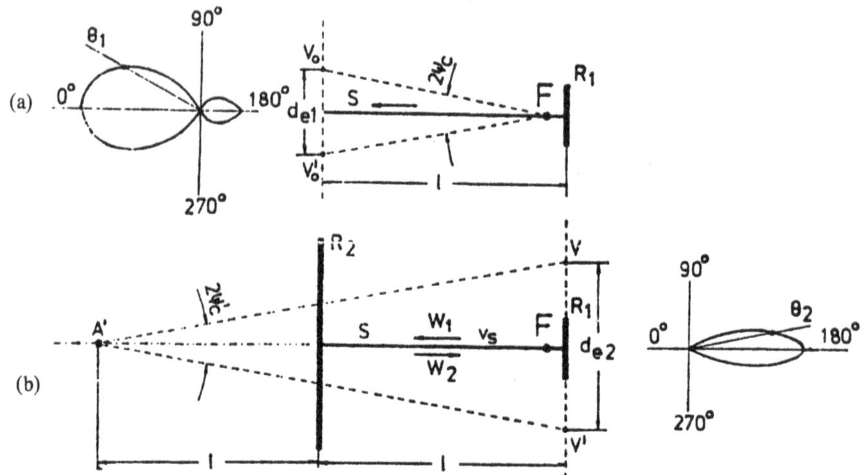

Figure 6.1 (a) Ordinary long endfire antenna; (b) principle of the long backfire antenna

The backfire antenna, as initially proposed, consists of a surface wave structure placed between two planar reflectors of different sizes (Figure 6.1(b)). The large reflector R_2 acts as a mirror that turns the surface wave backward, thus forcing the wave to traverse the physical length at least twice. As a result, the backfire antenna acts as an endfire antenna of nearly double its length, with a surface wave adjusted to the optimum directivity for such an endfire system. This implies that the backfire design has a much larger aperture (with effective diameter) for a given size, and a much more directive pattern. Properly tuned, a backfire antenna will give about four times (6 dBi) more gain than an endfire component of equal length. Experimental and theoretical study of the backfire radiation mechanism has shown that optimum antenna performance depends on obtaining an extremely high VSWR over the entire antenna length. This condition can be fulfilled only when the distance between R_1 and R_2 is nearly a multiple of a half-wavelength.

Thus, the basic long backfire antenna is essentially a cavity-resonator component with electromagnetic characteristics comparable to those of a laser cavity. Although the primary source is localized in the single feed dipole instead of being

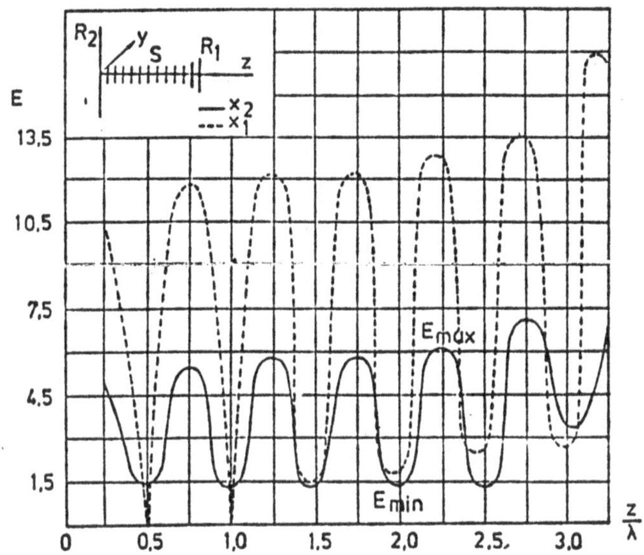

Figure 6.2 Field amplitude distribution along the z-axis, for $x = 0.25\lambda$ (dashed line) and $x = 0.90\lambda$ (solid line) (from [11–13])

distributed over the cavity, its energy distribution is similar to that of the laser. The dominant mode is simply a standing wave surrounding the surface wave structure. The smaller reflector acts like the partially transparent laser endmirror, with its size corresponding to the degree of transparency. The cavity properties are illustrated in Figure 6.2, which shows the longitudinal amplitude distribution of the electric field in the long backfire antenna with a disc-on-rod surface wave structure. The distance between the two reflectors is 4λ. The graphs are drawn for two values of the coordinate x (0.25λ and 0.90λ).

The near-field distribution of an antenna is very important for understanding the radiation mechanism and making a correct model of its operation. Figure 6.3 shows the relative amplitude distribution in the virtual aperture plane (*VA*) in terms of the normalized coordinate x/λ ($y = 0$, $z = 0$). The pattern shape may be explained by taking account of the "shade effect" of the smaller reflector R_1. Amplitude distributions are measured for two positions of the feed element: $\lambda/4$ from the large reflector (left) and $\lambda/4$ from the small reflector (right). The electromagnetic cavity properties of the backfire antenna can be studied in the amplitude-phase contour plots of Figure 6.4. These are H-plane plots of the near field, measured for a backfire antenna 2.0λ long (Figure 6.4(b)) and for an endfire antenna of equal length (Figure 6.4(a)), both tuned for maximum gain. In the figures e, f, g, h, i, and k designate lines of constant phase plotted in intervals of 360°. A, B, C, and D are lines of

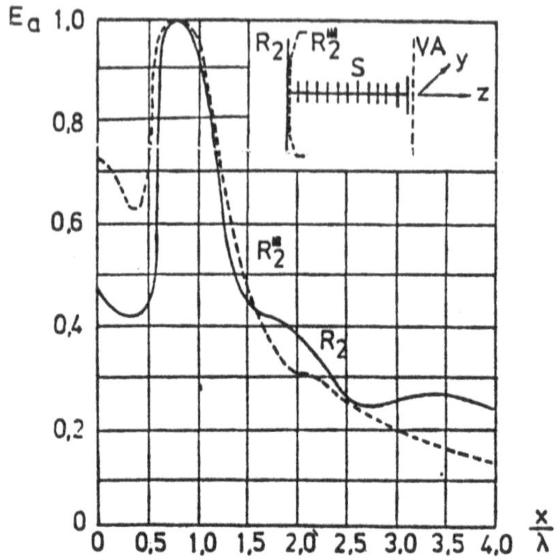

Figure 6.3 Relative amplitude distribution in the virtual aperture, with R_2 as the plane disk reflector and R_2''' as the plane-conical reflector with rim (from [12])

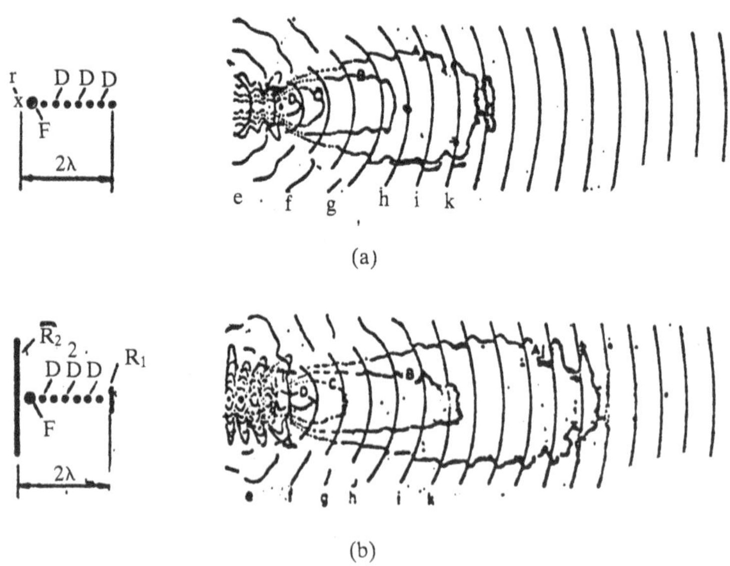

Figure 6.4 Amplitude-phase plots of (a), Yagi endfire and (b) Yagi backfire antennas with equal lengths (2.0λ) (from [10])

constant amplitude level, decreasing in steps of 5 dB from the axis. From these figures, obtained from near-field measurements, we see that, in general, the long backfire antenna can be considered as an open cavity-type antenna. The equivalent radiating aperture is about twice as wide as the aperture of the corresponding endfire antenna. If we assume nearly constant phase distribution in the aperture plane, then we may conclude that the gain of the long backfire increases by about 6 dB over that of a conventional endfire of the same length [10].

Following the experimental results and conclusions obtained by Ehrenspeck [10], Zucker [14–15] has suggested a qualitative approach to long backfire antenna design, specified later by Hristov [12,16,17] and Kumar [18–19]. Figure 6.5 shows a simplified mirror-image model of a long backfire antenna. According to the original HW condition for a very long continuous linear array of length 1, a maximum directivity is obtained when

$$u_0' = (\beta - \beta_s).z_0 = -1.47 \tag{6.1}$$

the traveling wave current distribution $G(z)$ along the axis is assumed to be

$$G(z) = e^{j\beta_s z} \tag{6.2}$$

where β is a free-space phase constant and β_s is a surface wave phase constant. Bach [4] proved that the above condition can be practically applied for both continuous and discrete uniform linear endfire antennas (except for very short arrays) in the modified form:

$$u'(l/\lambda) = (\beta - \beta_S)Z_0 \tag{6.3}$$

where the function $u'(l/\lambda)$ for a continuous surface wave structure is plotted in Figure 6.6. In Figure 6.6, HW is equal to -1.47. If we set $u = -u'$, [12] (6.3) could be written as:

$$u = \frac{\pi\, l_e}{\lambda}(h - 1) \tag{6.4}$$

or

$$h = 1 + \frac{\lambda}{\pi l_e}\, u\left(\frac{l}{\lambda}\right) \tag{6.5}$$

where $h = \lambda/\lambda_s$ is a coefficient of the surface wave shortening. A similar expression for h is obtained experimentally for an optimally tuned Yagi antenna:

$$\tilde{h} = 1 + \frac{\lambda}{ql} \tag{6.6}$$

where $q(l/\lambda)$ can be found from Figure 6.7.

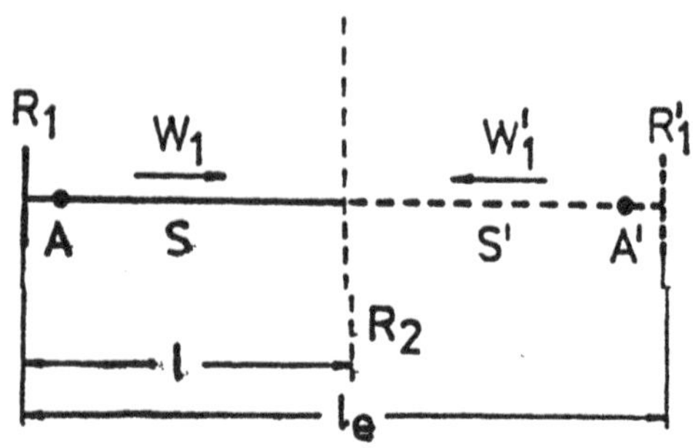

Figure 6.5 Simplified model of a long backfire antenna

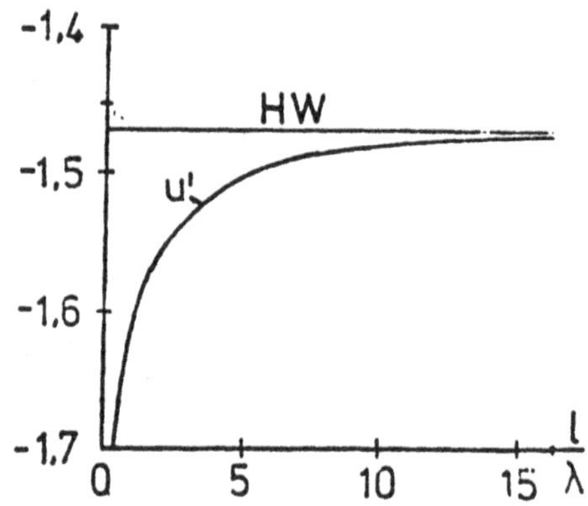

Figure 6.6 Optimum value of u' as a function of l/λ (from [20])

If for any length $l_e = l$ the coefficients h and $\tilde{h}$ are equal, then from (6.5) and (6.6) we obtain

$$q = \frac{\pi}{u} \tag{6.7}$$

In accordance with Figure 6.6, the function $u(l/\lambda)$ could be approximated by the following expression:

$$u = 0.065\frac{\lambda}{l} + 1.47 \quad (6.8)$$

After substituting (6.8) into (6.5), we obtain:

$$h = 1 + 0.234\frac{\lambda}{l} + 0.021\left(\frac{\lambda}{l}\right)^2 \quad (6.9)$$

or (6.10) for very long antennas,

$$h \approx 1 + 0.23\frac{\lambda}{l} \quad (6.10)$$

As in any cavity, the length l of a long backfire antenna must satisfy a relation of the form:

$$L = \frac{n\lambda_s}{2} + c_r, \quad n = 1, 2, 3 \ldots \quad (6.11)$$

where c_r corrects for the radiation from the cavity. Experiments with backfires show that

$$l = n\frac{\lambda}{2} \quad (6.12)$$

Equation (6.12) shows that the set of optimized backfire antenna lengths is discrete.

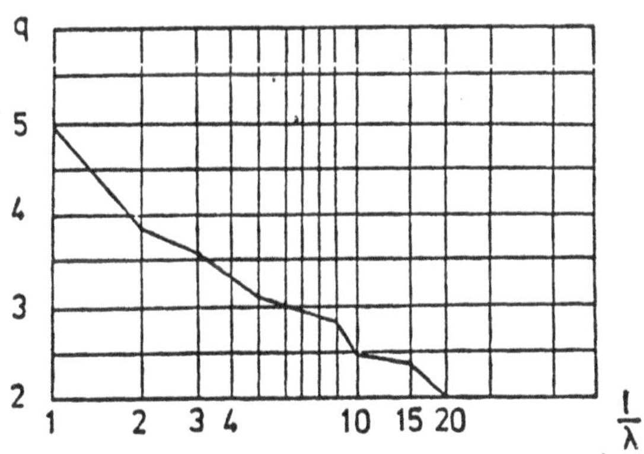

Figure 6.7 Plot of the function of l/λ (from [12])

In accordance with the experimental study of the endfire near field, it is accepted that, in the sector of the so-called "critical angle" ψ_c (Figure 6.9), the surface wave is plane. From this figure it is evident that the surface wave structure transforms the primary spherical wave into a plane surface wave [12,21]. The critical angle depends on the surface-wave phase velocity, and is given approximately by

$$\tan \psi_c = \sqrt{\left(\frac{\lambda}{\lambda_S}\right)^2 - 1} = \sqrt{h^2 - 1} \tag{6.13}$$

If the large reflector R_2 (SW reflector) is a plane metal disk, its diameter is obtained as follows: the diameter of the circular region A–B on the large reflector R_2 is related to the critical angle ψ_c (Figure 6.8) by

$$d_2' = 2l \tan \psi_C \tag{6.14}$$

From (6.10), (6.13), and (6.14) we have

$$d_2' = 2\lambda \sqrt{\frac{u_1}{\pi}\left(\frac{l}{\lambda} + \frac{u_1}{4\pi}\right)} \tag{6.15}$$

Because $u_1/4\pi \ll l/\lambda$, the second member of the expression under the radical sign can be neglected, then from [12], we have:

$$\frac{d_2}{\lambda} = 2\sqrt{\frac{ul}{\pi\lambda}} \tag{6.16}$$

or, according to (6.8), the diameter d_2' is given approximately by:

$$\frac{d_2}{\lambda} = 1.37\sqrt{\frac{l}{\lambda}} \tag{6.17}$$

The optimum surface wave reflector diameter d_2 depends on the admissible phase shift of the field on the periphery of the disk reflector. If the maximum phase shift is 90°, it is not difficult to obtain the simple relationship for the normalized disk diameter d_2 [12]:

$$\frac{d_2}{\lambda} \approx 2\sqrt{\frac{l}{\lambda}} \tag{6.18}$$

where the length l is defined by the desired gain and the resonant properties of the long backfire antenna. Some possible main reflector configurations are shown in Figure 6.10: a simple disk reflector (R_2); a disk reflector with a rim on its periphery (R_2'); a plane reflector with a step correction (R_2''); and a plane reflector with a linearly corrected plane-conical reflector (R_2''').

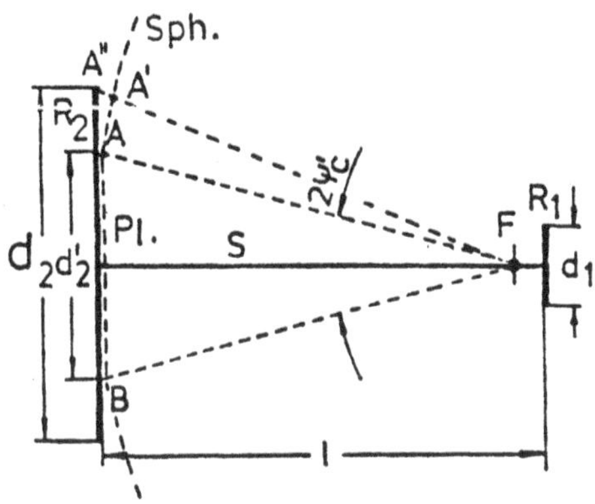

Figure 6.8 Long BFA with plane disk reflectors (from [14])

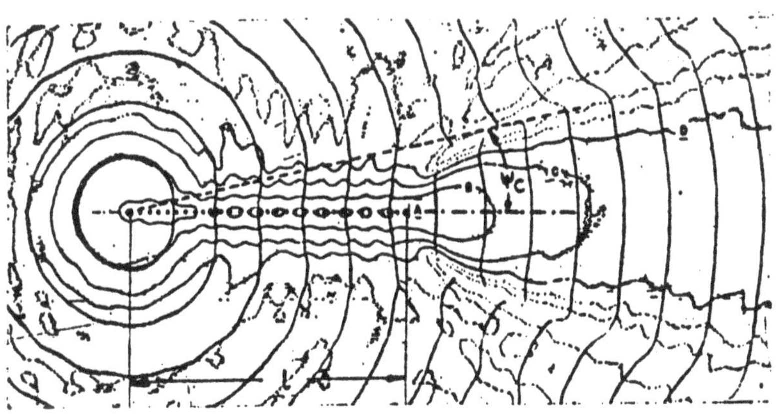

Figure 6.9 Amplitude-phase plots around a 6λ long Yagi endfire antenna without reflector (from [21])

The external diameter of R_2'' can be found from the relationship:

$$d_{2k} = (1.4 - 1.5)d_2 \tag{6.19}$$

which is in good agreement with the experiment. The plane-conical reflector R_2''' could be designed according to the following relationships:

$$d_{20} = (0.9 - 1.0)d_2 \tag{6.20}$$

$$W_1 = W_2 = 0.25\lambda \tag{6.21}$$

We have done experiments comparing the stepped reflector R_2'' and the plane-conical reflector R_2''' and have found that the latter gave better antenna performance. The small reflector diameter d_1 is also very important for field amplitude-phase distribution in the antenna aperture and for the excitation efficiency of the slow surface wave. We get maximum directivity with an optimum value of d_1.

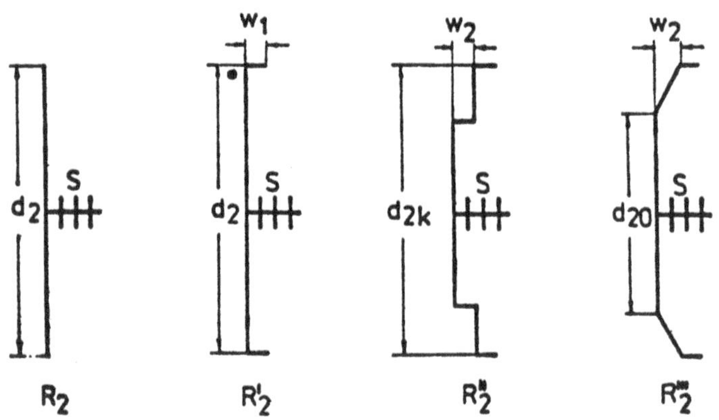

Figure 6.10 Some possible main reflector configurations (from [12])

The positioning of the feed dipole along the antenna is also of great importance to input and radiation properties. Two antenna designs are shown in Figure 6.11, in which only the dipole feed position differs. The experiments were carried out for the frequency $f = 2.3$ GHz and three typical antenna lengths: 2λ, 3λ, and 4λ. The large reflector is a metal disk with a rim having dimensions according to (6.18) and (6.21).

The small reflector of the antenna design in Figure 6.11 (a) has optimum diameters of 0.8λ, 1.0λ, and 1.2λ for antenna lengths 2λ, 3λ, 4λ, respectively (i.e., $d_{1opt} = 0.29d_2$). Antenna directivity for the optimized dimensions can be calculated by the approximate formula from [22–23]:

$$D_{max} = 13.5 + 10 \log \frac{l}{\lambda} \quad \text{dBi} \tag{6.22}$$

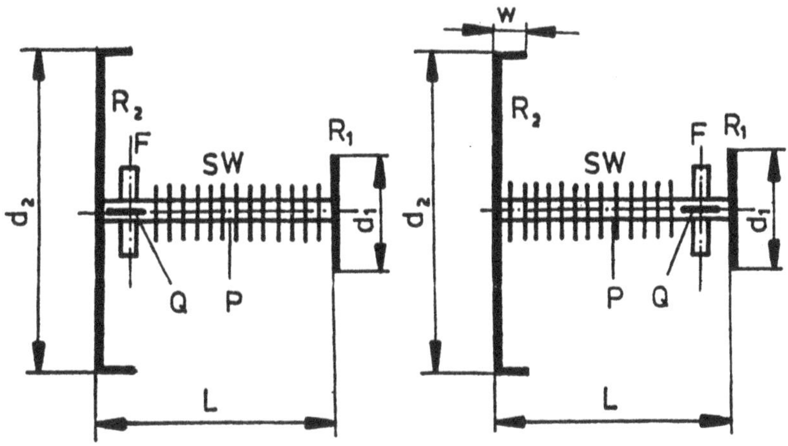

(a) (b)

Figure 6.11 Two long backfire antenna designs with different feed positions: (a) $\lambda/4$ from R_2; (b) $\lambda/4$ from R_1 (from [22,23])

In the second case (Figure 6.11(b)), the small reflector diameter d_1 has optimum values of 0.45λ, 0.55λ, and 0.70λ for antenna lengths 2λ, 3λ, and 4λ, respectively (i.e., $d_{1opt} = 0.17d_2$), and maximum directivity is given approximately by

$$D_{max} = 17.5 + 10 \log \frac{1}{\lambda} \quad \text{dBi} \tag{6.23}$$

It is interesting to compare the directive properties of optimally tuned endfire and long backfire antennas by using the following formulas [12]:

(a) directivity of an endfire antenna, tuned according to the HW condition:

$$D_{max} = 10 \log \left(10 \frac{l}{\lambda}\right) \text{dBi} \tag{6.24}$$

(b) directivity of a long backfire antenna with large disk reflector R_2:

$$D_{max} = 10 \log \left(40 \frac{l}{\lambda}\right) \text{dBi}$$

(c) directivity of a long backfire antenna with rimmed disk reflector R_2':

$$D_{max} = 10 \log \left(50 \frac{l}{\lambda} \right) \text{ dBi} \tag{6.25}$$

(The formulas in (6.23) and (6.25) refer to the same antenna design, and give equal values for maximum directivity.)

(d) directivity of a long backfire with plane stepped reflector R_2'':

$$D_{max} = 10 \log \left(60 \frac{l}{\lambda} \right) \text{ dBi} \tag{6.26}$$

(e) directivity of long backfire antenna with plane-conical reflector R_2''':

$$D_{max} = 10 \log \left(70 \frac{l}{\lambda} \right) \text{ dBi} \tag{6.27}$$

As an example we will calculate directivity using the above formulas for an antenna length of $4.0\lambda_0$. We obtain 16.02 dBi, 22.04 dBi, 23.01 dBi, 23.80 dBi, and 24.47 dBi for the cases (a),(b),(c),(d), and (e), respectively.

We can define an aperture efficiency of the long backfire as

$$e = \frac{A_{eff}}{A_e} \tag{6.28}$$

where A_{eff} is the effective area of the equivalent virtual aperture, A_e is the geometrical area of the equivalent aperture, and e is the aperture efficiency.

The effective area is related to the maximum directivity by

$$A_{eff} = \frac{\lambda^2}{4\pi} D_{max} \tag{6.29}$$

Using (6.27), the expression for the long backfire with plane-conical main reflector for D_{max} and assuming $d_e = d_{2k}$, we calculate the aperture efficiency as $e = 0.77$. This value, compared with other types of aperture antennas, is too big.

The usual parabolic antenna with dipole feed has an aperture efficiency of 0.5–0.7. For small antenna diameters, which are on the order of backfire main reflector diameters, we use $e = 0.5$. If we let d_p and l_p designate the parabolic antenna diameter and length respectively, the following relationships are valid:

$$d_p = 1.2\, d_{2k} \tag{6.30}$$

$$\frac{l_{p\cdot}}{\lambda} = 1.3 \sqrt{\frac{l}{\lambda}} \tag{6.31}$$

where d_{2k} and l relate to the long backfire antenna with a plane-conical main reflector.

6.2 THE YAGI BACKFIRE ANTENNA

6.2.1. Theoretical Analysis

The first attempt at a theoretcial investigation of Yagi backfire antennas (long backfire antennas with a dipole surface wave structure) may be the work performed by Davies [24] in 1967. In his calculations, however, he uses measured values for the mutual impedances between the dipole elements of the antenna. His method is, therefore, not directly applicable for designing backfire antennas by the optimization of their parameters. Another theoretical approach was proposed by Hristov in 1973 for a Yagi backfire antenna with two disk reflectors [12]. The rigorous diffraction theory for electromagnetic systems consisting of two disks and a dipole prirmary source, as developed by Ivanov [25] and James [26], was applied to find its radiation field. Then, using the superposition principle, the total field of the assembly of two coaxial disks, large and small, and a row of N dipoles arranged and excited as a Yagi structure, was obtained (Figure 6.12). The disk centers $0_{\pm}$ are also origins of the local coordinate systems, rectangular ($x_{\pm}, y_{\pm}, z_{\pm}$) and spheroidal ($u_{\pm}, v_{\pm}, \psi_{\pm}$); $P_1, P_2, P_3, \ldots, P_i, \ldots, P_N$ are electric moments of the dipole elements ($\bar{P}_i = \bar{P}_{ix}$); $a_{\pm}$ are disk radii; M is a field point and l is an antenna length.

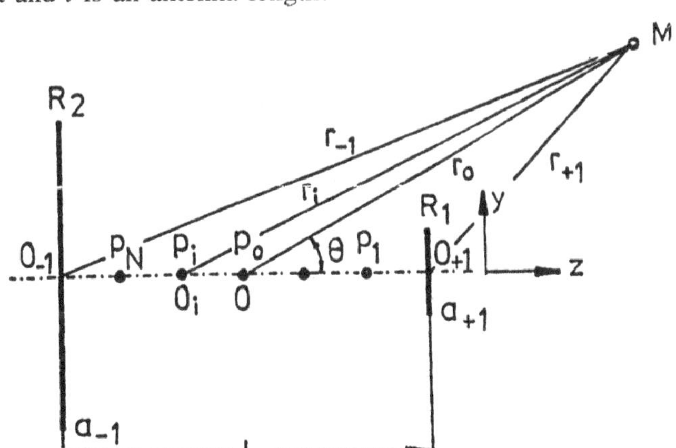

Figure 6.12 Geometry of Yagi backfire antenna with disk reflectors (from [12])

The total radiation electric field is given by

$$E_v = \frac{120\pi\beta^2}{r_0} \sin\varphi \cos\theta \sum_{i=1}^{N} p_i e^{j\beta r_i} e^{j\psi_i} W_{1i}(\theta) \tag{6.32}$$

$$E\varphi = -\frac{120\pi\beta^2}{r_0} \sin\varphi \sum_{i=1}^{N} p_i e^{j\beta r_i} e^{j\psi_i} W_{2i}(\theta) \tag{6.33}$$

where P_i and ψ_i are an amplitude and a phase of the ith dipole moment.

$$\begin{aligned} W_{1i}(\theta) = 1 - 2e^{\pm j\beta d_i \cos\theta} \Bigg[& \sum_{n=1,3\dots} nj^n A_n^{-1} R_{0n}^{(1)\prime}(-jc_{+1}, j\theta) \\ & \times S_{0n}(-jc_{+1}, \cos\theta) + e^{-j\beta l \cos\theta} \sum_{n=1,3,\dots} nj^n \\ & \times A_n^{-1} R_{0n}^{(1)\prime}(-jc_{-1}, j0) S_{0n}(-jc_{-1}, \cos\theta) \Bigg] \end{aligned} \tag{6.34}$$

$$\begin{aligned} W_{2i}(\theta) = 1 - 2e^{\pm j\beta d_i \cos\theta} \Bigg[& \sum_{n=1,3,\dots} nj^n A_n^{+1} R_{0n}^{(1)\prime}(-jc_{+1}, j0) \\ & \times S_{0n}(-jc_{+1}, j0) + e^{-j\beta l \cos\theta} \sum_{n=1,3\dots} nj^n A_n^{-1} R_{0n}^{(1)\prime}(-jc_{-1}, j0) \\ & \times S_{0n}(-jc_{-1}, \cos\theta) - tg\theta \sum_{n=1,3,\dots} \\ & \times j^n B_n^{+1} R_{1n}^{(1)}(-jc_{+1}, j0) S_{1n}(-jc_{+1}, \cos\theta) \\ & - e^{-j\beta l \cos\theta} \times tg\theta \sum_{n=1,3,\dots} j^n B_n^{-1} R_{1n}^{(1)}(-jc_{-1}, j0) S_{1n}(-jc_{-1}, \cos\theta) \Bigg] \end{aligned} \tag{6.35}$$

$R_{1n}^{(1)}(-jc_{+1}, j0)$ is the spheroidal radial function of the first kind, first order, and nth degree;

$R_{0n}^{(1)\prime}(-jc_{+1}, j0)$ is the first derivative of the spheroidal radial function of the first kind, zero order, and nth degree;

$S_{0n}(-jc_{+1}, \cos\theta)$ and $S_{1n}(-jc_{+1}, \cos\theta)$ are the spheroidal angular functions of zero and first order, respectively;

$A_n^{\pm 1}, B_n^{\pm 1}, C_{\pm 1}$ are constants obtained from the boundary conditions on the disks.

The diffraction method described, although rigorous in principle, is not suitable for practical use. There are some rough assumptions in the proposed model, which introduce inaccuracies in the final results. Furthermore, the computation of spheroidal functions is a difficult task, even on a large computer.

Figure 6.13 shows H-plane radiation patterns of a 4.0λ long Yagi backfire antenna having two disk reflectors (R_2 is 4.0λ and R_1 is 0.7λ in diameter).. The plots are: an experimental pattern for the corresponding endfire antenna as comparison; experimental, for $f = 3.7$ GHz; theoretical, according to (6.32) to (6.35).

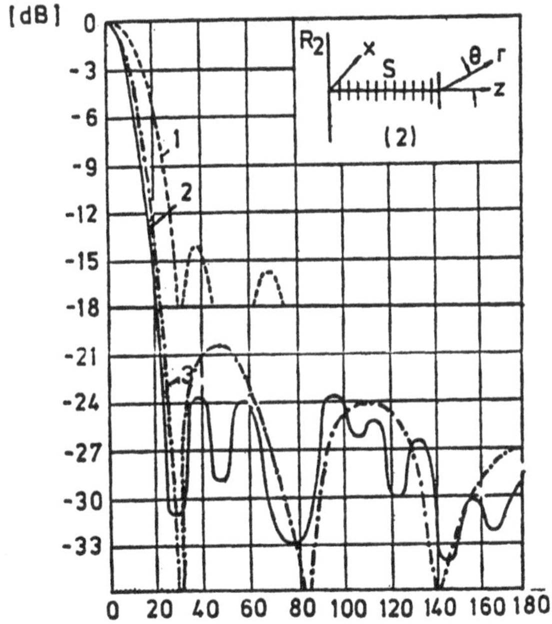

Figure 6.13 H-plane radiation patterns: (1) endfire antenna (dash); (2) backfire antenna, experimental (solid); (3) backfire antenna, theoretical (dot-dash) (from [12])

The most comprehensive theoretical and numerical study of Yagi backfire antenna was by Nielsen [27] in his Ph.D. thesis, using a method proposed by himself and Pontoppidan in 1970 [28–29]. We will briefly describe here this method and some of the numerical results, directly following the indicated publications.

A photograph of a model Yagi backfire antenna, which is a copy of the antenna designed by Ehrenspeck, is shown in Figure 6.14 and its geometry is given in Figure

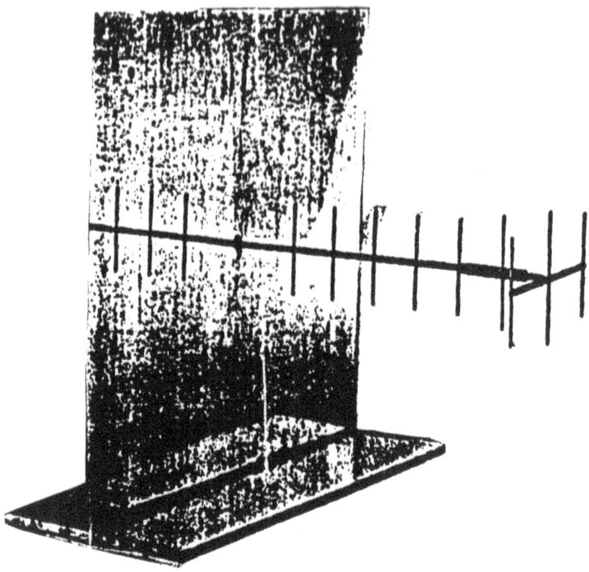

Figure 6.14 Yagi backfire model photograph (from [27–29])

6.15. The feed is placed in the origin of a rectangular coordinate system. The dipole elements are numbered from 1 to $N + M$, where M is the number of the dipole elements, which form the small reflector, and N is the number of the Yagi structure elements.

In order to determine the field radiated from the antenna, it is necessary to determine all current sources on the separate parts of the antenna structure. The image theory is used for this, with the main or surface wave reflector considered (in the first instance) as an infinite, perfect plane mirror. In this way, an antenna system composed of both real dipoles and image dipoles is obtained. Mutual coupling between all elements is determined using the induced EMF method for calculation of the mutual and the self-impedances, assuming sinusoidal current distribution along the resonant half-wave dipoles. Formulas for the mutual impedances as developed by Baker and La Grone [29a] are used for numerical computations, assuming infinitely thin dipoles. However, when self-impedances are to be calculated, the diameter of the dipole elements is a finite quantity, and so another method is applied — the induced EMF method as proposed by Jordan [30].

After determining all coupling impedances, a complex matrix equation for the induced currents in the dipole is set up:

$$[[Z_{mm}] - [Z_{mm'}]][I_m] = [V_m] \tag{6.36}$$

where the prime denotes image dipoles. The submatrices Z are of the type:

$$[Z_{mm'}] = \begin{bmatrix} z_{11'} & z_{12'} & \dots & z_{1N'} \\ z_{21'} & z_{22'} & \dots & z_{2N'} \\ \vdots & \vdots & & \vdots \\ z_{N1'} & z_{N2'} & & z_{NN'} \end{bmatrix} \tag{6.37}$$

where Z_{pq} is the mutual impedance between dipole p and the image of the dipole q, and N is the total number of dipoles. Because all dipoles in the Yagi structure, except for the feed dipole, are short-circuited, the excitation voltage vector is

$$[V_m] = [1, 0, 0, \dots 0] \tag{6.38}$$

with element number 1 as the feed dipole. The unknown induced dipole currents are

$$[I_m] = [I_1, I_2, \dots, I_p, \dots, I_{N+M}] \tag{6.39}$$

By solving the complex matrix equation, the above complex currents $[I_m]$ are determined.

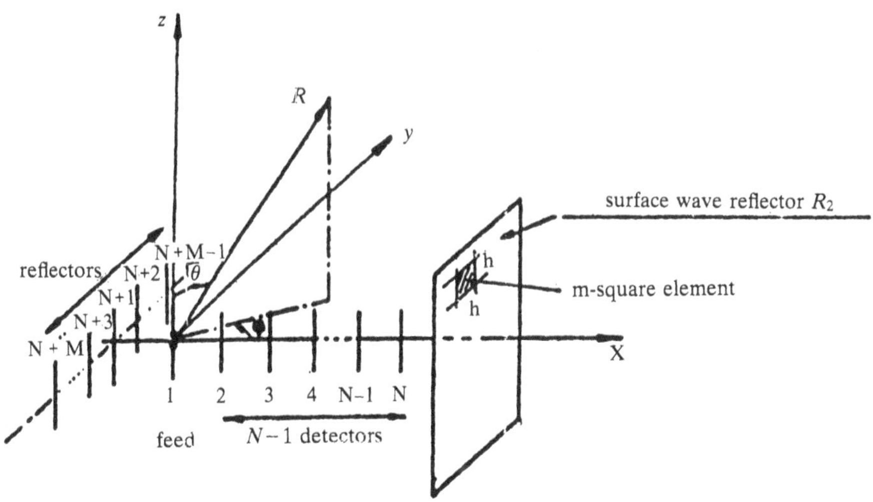

Figure 6.15 Geometry of an antenna system (from [28])

The total field is found by superposition of the fields radiated from physically existing sources, the dipole currents and the surface current distribution on the surface wave reflector. The currents in the real dipoles are assumed to be the same as those found in simple image theory. The surface currents on the finite rectangular reflector are approximated by the currents which would exist if the plate were a part of an infinite conducting plane. Due to the approximations involved, the surface currents do not vanish at the edges of the plate; this effect, as well as the edge diffraction phenomena in general, is not taken into account. Once all currents are found, the total far field from the Yagi backfire antenna is obtained [27] as:

$$\bar{E} = \left[-\frac{jW}{2\pi \sin\theta} \cdot \sum_{n=1}^{N} \frac{I_n}{R} e^{j\beta(R - \bar{R}^0 \cdot \bar{r}_n)} \cdot \frac{\cos(\beta l_n \cos\theta) - \cos(\beta l_n)}{\sin(\beta l_n)} - \frac{j\beta W h^2}{4\pi} \sin\theta \cdot \sum_{m=1}^{M} \frac{J_m}{R} e^{j\beta(R - \bar{R}^0 \cdot \bar{r}_m)} \right] \cdot \bar{\theta}^0 \tag{6.40}$$

where M is the total number of squares representing the surface wave reflector, $W = 120\pi$ is the free wave impedance, n is the number of dipole elements with length $2l_n$, $\bar{R}$ is the radius vector from the origin of the rectangular coordinate system to the point in the far region, $\bar{R}^0$ is the unit vector, m is the number of squares, and $\bar{r}_n$ and $\bar{r}_m$ are distance vectors from the origin to the nth dipole and the mth square, respectively (Figure 6.15). Figure 6.16 shows the computed surface current distribution on a square large (surface wave) reflector with sides up to 10 wavelengths. The solid and dotted lines represent curves of constant current amplitude and phase, respectively, one quarter-wavelength from the whole reflector.

The actual size of the reflector in the examined Yagi backfire antenna is two wavelengths (Figure 6.14). The antenna consists of nine parallel dipoles, three of which act as parasitic small reflectors. The spacing is 0.2λ between directors and 0.3λ between the small reflector dipole elements. The distance from the last director to the square surface wave reflector is 0.3λ. The lengths of the dipoles are: feed length—0.450λ; director length—0.381λ; center reflector length—0.514λ; outside reflector length—0.412λ. The radius of all dipoles is 0.00635λ. The antenna model has been designed for a frequency of 1.5 GHz.

Computed and measured radiation patterns of the antenna model, both in E-plane and H-plane, are plotted for comparison in Figure 6.17 and Figure 6.18, respectively. The measurements were made in the Radio Anechoic Chamber of the Technical University of Denmark. Very good agreement between measured and computed results was obtained. The agreement is equally good in both principal planes.

Using the theoretical method described, many other numerical investigations of the influence of the antenna parameters on the radiation and input characteristics of these antennas were performed. The results proved the validity of the theory and numerical methods involved. The numerical investigations carried out by Nielsen

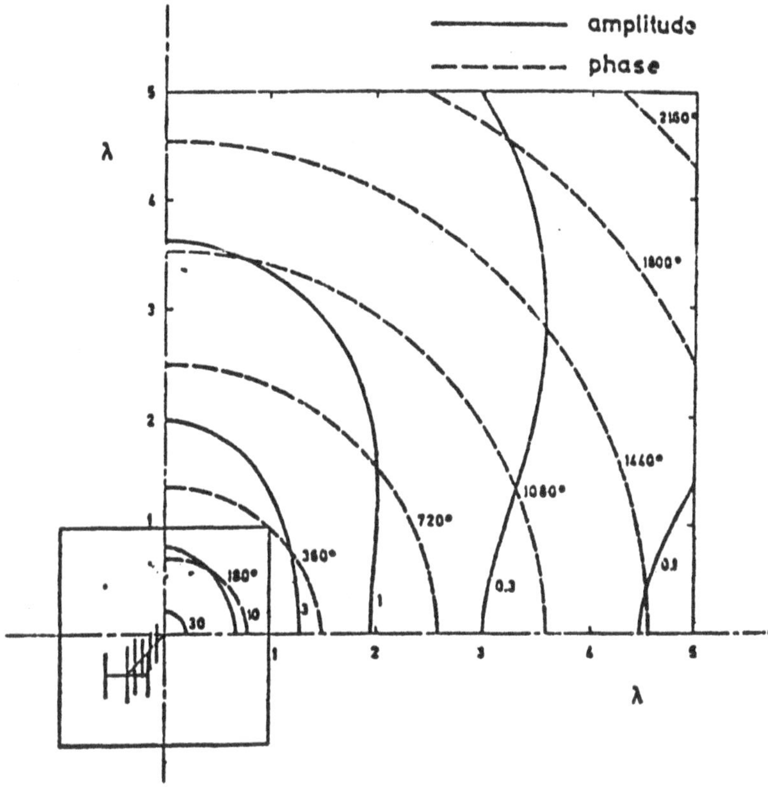

Figure 6.16 Surface current distribution on a square reflector plate (from [27–29])

[29] led to the development of a package of computer programs, written in the FORTRAN IV language.

6.2.2 Other Yagi Backfire Antenna Designs: Experimental Results

At the AFCRL in the early 1960s, Ehrenspeck constructed and experimentally examined many Yagi backfire designs to determine the optimum antenna dimensions, method of excitation, and shape of the SW reflector. For a length of 2.0λ, the Yagi backfire with a surface wave disk reflector of 1.0λ in diameter has a gain of about 12 dBi, a 3 dB beamwidth of 25°, first sidelobe level of −12 dB, and backlobe level of −7 dB. After replacing the SW reflector with a larger one (d_2 = 2.0λ), the antenna

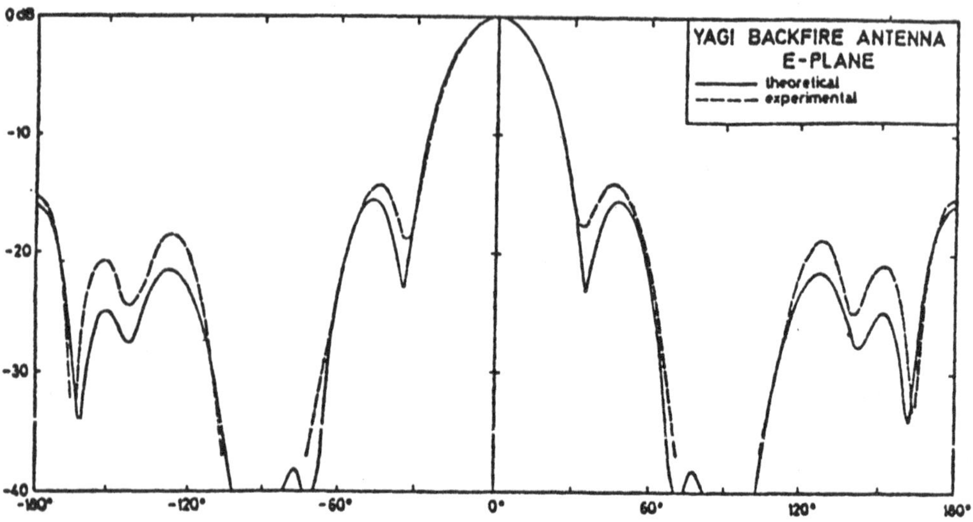

Figure 6.17 *E*-plane nine-element Yagi backfire antenna radiation patterns (from [29])

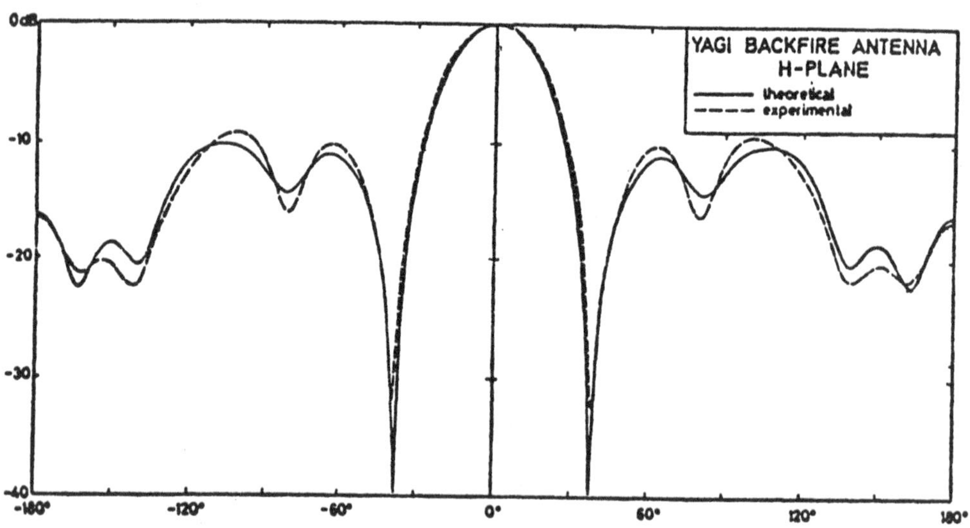

Figure 6.18 *H*-plane nine-element Yagi backfire antenna radiation patterns (from [29])

gain increased 2 dB; such a backfire antenna has a 3 dB beamwidth of 20°, first sidelobe level of −13 dB, and backlobe level of −18 dB.

After optimizing the feed efficiency, wave phase velocity on the Yagi dipole structure, and disk reflector diameters, the antenna gain was increased several times.

For example, an optimally tuned Yagi backfire with a SW reflector 4.0λ in diameter and 2.5λ in length has a gain of 21 dBi and 3 dB beamwidth 16°. Figure 6.19 shows the gain as a function of the surface wave reflector diameter, measured from a Yagi backfire antenna 4.0λ long.

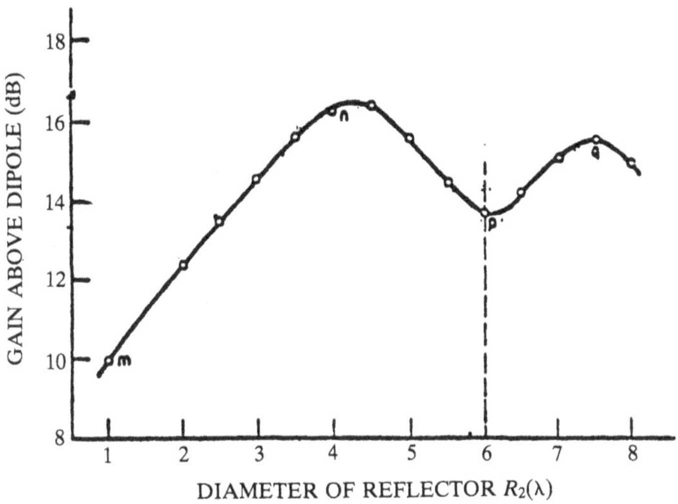

Figure 6.19 Yagi backfire gain as a function of SW reflector diameter (from [10])

After a first continuous increase in gain proportional to the increase in diameter of R_2 (from point *m* to point *n*), a maximum of 16.5 dB above dipole is reached for a reflector diameter of 4.25; the gain then drops to 13.8 dB for a diameter of 6.0 (at point *p*), and later rises to a second maximum of 15.5 dB for a diameter of 7.5λ (at point *q*), *et cetera*. It is not difficult to explain such behavior [10]: the backfire radiation pattern is the vector sum of the field E_1 (radiated from the antenna's virtually equivalent aperture), the field E_2 (radiated directly from the feed *F*), and other high-order fields due to diffraction effects. When all these fields are in phase, the antenna gain has a maximum value.

In a scientific report published by Strom and Ehrenspeck in 1963 [31], the continued experimental investigation of Yagi backfire antennas was described in great detail. The investigation was carried out on four different backfire antennas, designed for frequencies of 220, 600, 3000, and 9080 MHz, respectively. A drawing of *S*-band Yagi backfire antenna with stepped SW reflector is shown in Figure 6.20(a).

This antenna has a length of 4.0λ, the total diameter of the SW stepped reflector d_{2k} of 6.0λ, and small reflector diameter d_1 of 1.0λ. The dipole feed is located near the small reflector. For these dimensions, the antenna has the following electric parameters:

3 dB beamwidth in E-plane = 13°;

3 dB beamwidth in H-plane = 10.5°;

maximum sidelobe levels (in both planes) are under −20 dB;

backlobe level = −30 dB;

gain = 23.5 dBi.

This Yagi backfire antenna is compared with the telemetry array of Yagi endfire elements (Figure 6.20(b)). The latter has 16 nine-dipole Yagis (128 dipoles altogether). These two antennas have almost the same gains, but the Yagi backfire possesses many advantages: compactness, small size, a simple feed system, and reduced cost. Furthermore, the Yagi backfire antenna has lower sidelobes (under −20 dB for the backfire and about −11 dB for the telemetry array). This is shown in Figure 6.20(c). The radiation patterns of the Yagi backfire antenna indicate a wider pattern bandwidth than anticipated. The approximated gain, calculated from the radiation patterns as a function of frequency, is shown in Figure 6.21.

(a) (b)

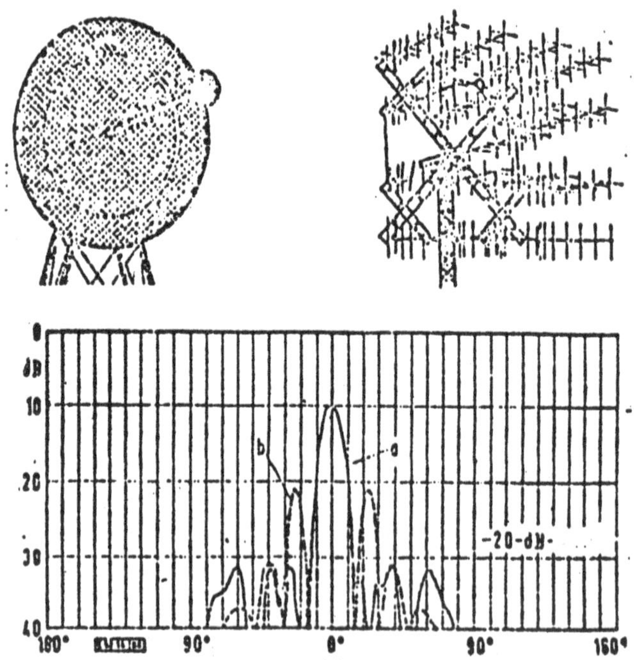

(c)

Figure 6.20 Yagi backfire antenna with stepped large SW reflector (a); compared with a telemetry array of 16 endfire Yagi antennas (b); sidelobe levels for Yagi backfire antenna (c) (from [2])

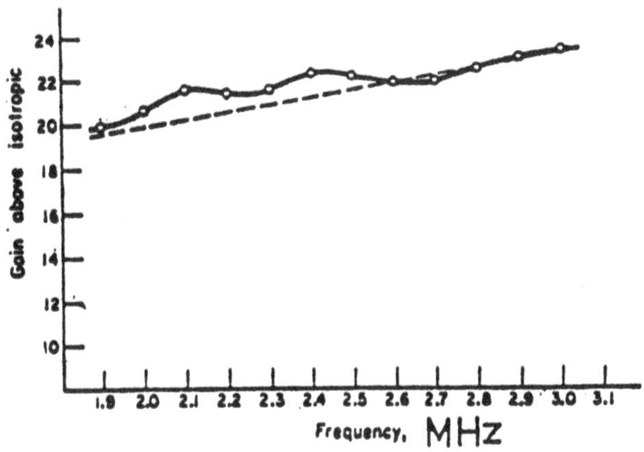

Figure 6.21 Gain of a 4.0λ long backfire antenna as a function of frequency. The dashed plot is an equivalent-area paraboloidal antenna, with 60% efficiency (from [10])

In 1967 Longva [31a] reported experiments carried out in connection with the installation of a 136.5 MHz Yagi backfire antenna for the NTNF satellite telemetry station in Tromso, Norway. The antenna had a gain of 17.5 dBi and a sidelobe level of 13 dB below the main lobe. The standing wave ratio was 1.3, and linear or circular polarization was optional.

6.3. BACKFIRE ANTENNAS WITH DIPOLE-FED SURFACE WAVE STRUCTURE

6.3.1. Antenna Radiation Mechanism and Analysis

A special feature of the backfire antenna is the multiple partial reflection from the small reflector. The surface wave structure transforms the spherical wave, radiated by the primary source, into a plane surface wave. This transformation, however, is not complete (i.e., a part of the primary field is radiated directly as a space spherical wave). The radiation antenna mechanism is illustrated by Figure 6.22. As a first approximation, the antenna far field is given as the sum of the radiation fields of two fictitious antennas: a backfire antenna with SW structure S and without primary source (b), and a backfire antenna with a primary source F and without SW structure (c). This is given as:

$$E = E_s + E_d \tag{6.41}$$

where E_s is the radiation field of the equivalent antenna aperture, and E_d is the radiation field of the primary source (direct space field).

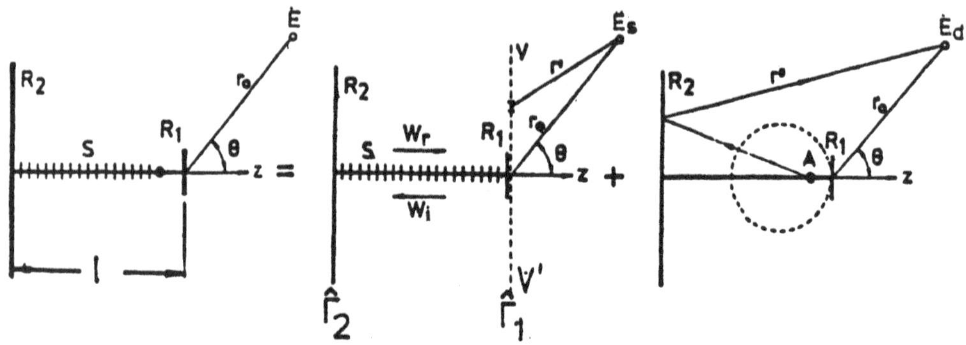

Figure 6.22 Radiation antenna mechanism (from [12])

Due to the multiple reflection phenomenon, the electromagnetic field E_{sw} in the equivalent virtual aperture $V - V'$ can be represented as a superposition of inhomogeneous (surface wave) plane waves having the same transverse amplitude distribution, but different amplitudes and phases. In a similar way [12], the far field E_s is given as:

$$\begin{aligned}\frac{E_s}{\Gamma_2} &= E_{s1} + E_{s2} + \ldots + E_{sk} = AE_{sm}F_s(\phi, \theta)e^{-2\alpha_s l}e^{j\phi l} \\ &+ AE_{sm}\Gamma_1 F_s(\phi, \theta)e^{-4\alpha_s l}e^{j2\phi l} + \ldots \\ &+ AE_{sm}\Gamma_1^{k-1}F_s(\phi, \theta) \times e^{-2k\alpha_s l}e^{jk\phi l}.\end{aligned} \tag{6.42}$$

The radiation direct field E_d can be written as:

$$E_d = E_{dm}F_d(\phi, 0) = BE_{sm}F_d(\phi, \theta) \tag{6.43}$$

Then, the total far field is obtained:

$$\begin{aligned}E &= \Gamma_2 AE_{sm}F_s(\phi, 0)[e^{-D}e^{j\phi_l} + \Gamma_1 e^{j\phi_1}e^{-2D}e^{j2\phi l} + \ldots \\ &+ \Gamma_1^{k-1}e^{j(k-1)\phi_1}e^{-kD}e^{jk\phi_l}] \\ &+ BE_{sm}F_d(\phi, \theta)\end{aligned} \tag{6.44}$$

where A and B are amplitude constants,

$F_s(\phi, \theta)$ is the normalized radiation pattern of the equivalent aperture,

$F_d(\phi, \theta)$ is the normalized radiation pattern of the primary (dipole) source,

α_s is the attenuation constant of the surface wave,

$D = 2\alpha_s l$ is the surface wave attenuation after double traveling on the SW structure,

$\Gamma_{1,2} = \Gamma_{1,2}e^{j\phi_{1,2}}$ are complex reflection coefficients of the small and large reflectors, respectively,

$\phi = 2\pi/\lambda\ 2l(h - \cos\theta)$ is the phase difference between E_s and E_d,

and $h = \lambda/\lambda_s$ is the coefficient of the surface wave "shortening."

Having $\phi = \phi_1 + \phi_L$, and $\hat{G} = -D + j\phi$, and keeping in mind that, theoretically, there is an infinite number of partial reflections from the small reflector for the complex amplitude of the total antenna far field, the following expression can be obtained [12]:

$$E = -AE_{sm}\left[\left(\frac{1}{\Gamma_1}F_s(\phi, \theta)\frac{MP - NQ}{P^2 + Q^2} + \frac{B}{A}F_d(\phi, \theta)\right) + \frac{1}{\Gamma_1}F_s(\phi, \theta) \times \frac{NP + QM}{P^2 + Q^2}\right], \tag{6.45}$$

where, for the reflection coefficient from the large reflector $\hat{\Gamma}_2$ we assume that

$$\hat{\Gamma}_2 = -1,$$

$$M = \cos\phi_1(1 - \Gamma_1 C \cos V_c + \Gamma_1 S \cos V_s) + \Gamma_1 \sin\phi_1(S \sin V_s - C \sin V_c);$$

$$N = \sin\phi_1(-1 + \Gamma_1 C \cos V_c - \Gamma_1 S \cos V_s) + \Gamma_1 \cos\phi_1(S \sin V_s - C \sin V_c);$$

$$P = 1 - 2\Gamma_1 C \cos V_c + \Gamma_1^2;$$

$$Q = 2\Gamma_1 C \sin V_c:$$

$$V_c = -\text{arctg}(thD\ tg\phi);$$

$$V_s = -\text{arctg}(cthD\ tg\phi);$$

$$C = \sqrt{sh^2D + \cos^2\phi};$$

$$S = \sqrt{sh^2D + \sin^2\phi}.$$

From (6.45), we see that to calculate the antenna pattern, it is necessary that we know the following quantities: radiation pattern of the equivalent aperture F_s; radiation pattern of the primary source F_d; antenna length l; reflection coefficient of the small reflector $\hat{\Gamma}_1$; SW propagation constant; amplitude relation A/B (i.e., the excitation efficiency of the surface wave structure). The aperture radiation field E_s is a basic part of the total far field, and its determination has a primary significance in the theoretical analysis. For that purpose, we must first find out the equivalent aperture field distribution. Neglecting the diffraction field components which are of secondary importance, we assume that the aperture field is obtained only by the SW transverse field components. For a backfire antenna with a metal corrugated rod structure which has a fundamental surface mode T_{11}, the aperture is divided into two main regions [12]: (a) region 1, for $(0 \leq \rho \leq a_1)$,

$$E_{ax} \approx E_{ay} \approx H_0 E_0 \tag{6.46}$$

(b) region 2, for $(a_1 \leq \rho \leq \infty)$,

$$\begin{aligned} E_{ax} &= \frac{E_0}{2}\left[\frac{K_1(v)}{v} - K_1'(v)\right]\sin 2\psi \\ E_{ay} &= -E_0\left[\frac{K_1(v)}{v}\cos^2\psi + K_1'(v)\sin^2\psi\right] \end{aligned} \tag{6.47}$$

where $v = \beta_e \rho = 2\pi/\lambda \sqrt{h^2 - 1}\,\rho$.

Region I is small compared with region II and, according to the experiments, the field on it is accepted as constant. The aperture blockage due to the small reflector is treated by a selection of an appropriate value for the constant H_0 $(0 \leq H_0 \leq 1)$. The far field can be obtained using the Huygens-Kirchoff scalar integral expression:

$$E_s = j\frac{e^{-j\beta r_0}}{2\lambda r_0}(1 + \cos\theta)\int_\rho\int_\psi E_a(\rho, \psi)e^{j\beta\rho\sin\psi\sin\theta}\rho\, d\rho\, d\psi \tag{6.48}$$

After putting (6.46) and (6.47) into (6.48) and solving the relevant integrals, we obtain the following radiation field components:

$$E_s^E(\theta) = C_0 f_E(\theta) \tag{6.49}$$

$$E_s^H(\theta) = C_0 f_H(\theta) \tag{6.50}$$

where $f_E(\theta)$ and $f_H(\theta)$ are radiation patterns of the equivalent antenna aperture in the E-plane and H-plane, respectively, and are given as:

$$\begin{aligned} f_E(\theta) = F_0(\theta)\Bigg[H_0 \frac{J_1(u_1)}{u_1} &+ \frac{1}{(u_1^2 + v_1^2)[K_2(v_1) - K_0(v_1)]} \\ &\times [-u_1 J_1(u_1) K_0(v_1) - v_1 J_2(u_1) K_3(v_1) \\ &+ v_1 J_0(u_1) K_1(v_1) + u_1 J_3(u_1) K_2(v_1)]\Bigg] \end{aligned} \tag{6.51}$$

$$\begin{aligned} f_H(\theta) = F_0(\theta)\Bigg[H_0 \frac{J_1(u_1)}{u_1} &+ \frac{1}{(u_1^2 + v_1^2)[K_2(v_1) - K_0(v_1)]} \\ &\times [-u_1 J_1(u_1) K_0(v_1) + v_1 J_0(u_1) K_1(v_1) \\ &- u_1 J_3(u_1) K_2(v_1) + v_1 J_2(u_1) K_3(v_1)]\Bigg] \end{aligned} \tag{6.52}$$

Here the Bessel functions of the first kind, nth order, are represented by $J_n(u)$, and the modified Bessel functions of the second kind, nth order, are represented by $K_n(v)$. Also,

$$C_0 = j \frac{e^{-j\beta r_0}}{2\lambda r_0} E_0;$$

$$F_0 = 1 + \cos\theta;$$

$$u_1 = \beta a_1 \sin\theta;$$

$$v_1 = \beta e^{a_1}$$

$$\beta = \frac{2\pi}{\lambda};$$

$$\beta_e = \sqrt{\beta_s^2 - \beta^2};$$

$$h = \frac{\beta}{\beta_s} = \frac{\lambda_s}{\lambda} < 1$$

Then we find these formulas for the antenna radiation parameters:
(a) 3 dB beamwidth of the main pattern lobe,

$$2\theta^0_{3dB} = 2 \text{ arc sin}\left[0.122\,\frac{\lambda}{a_1}\sqrt{\frac{\dfrac{K_1(v_1)}{K_0(v_1)}}{\dfrac{2}{v_1^2}\dfrac{K_1(v_1)}{K_0(v_1)} + 1}}\right] \tag{6.53}$$

(b) directivity, due to aperture radiation only,

$$D = 160\left(\frac{a_1}{\lambda}\right)^2 D_0\left(\frac{a_1}{\lambda}\right) \tag{6.54}$$

where

$$D_0\left(\frac{a_1}{\lambda}\right) = \frac{\left[0.5\,H_0 + \dfrac{m}{v_1}\right]^2}{H_0^2 + m^{2-1}}; \quad m = \frac{K_1(v_1)}{K_0(v_1)}$$

(c) aperture efficiency,

$$g = \frac{A_{\text{eff}}}{A_e} = 4\left(\frac{a_1}{a_2}\right)^2 D_0\left(\frac{a_1}{\lambda}\right) \tag{6.55}$$

(d) reflection coefficient modulus of the small reflector,

$$\Gamma_1 = \sqrt{-\left(\frac{a_1}{a_0}\right)^2 \frac{Q_0(a_1) + Q_2(a_1)}{Q_0(a_0) + Q_2(a_0)}}$$

$$Q_0 = K_0^2(k_e a_i) - K_1^2(k_e a_i);$$

$$Q_2 = K_2^2(k_e a_i)[1 + 4k_e^{-2} a_i^{-2}] - K_2'^2(k_e a_i) \quad (i = 0, 1) \tag{6.56}$$

In these formulas, a_2 is the radius of the main reflector, a_1 is the radius of the small reflector, and a_0 is the radius of the corrugated rod.

In Figure 6.23 a family of plots, $\Gamma_1(a_1/a_0)$, is drawn for $h = c/v_s = 1.02$, where

Γ_1^c is the modulus of the reflection coefficient for a corrugated (disk-on-rod) structure

Γ_1^g is the modulus of the reflection coefficient for a Goubeau-line rod, and

Γ_1^d is the modulus of the reflection coefficient for a cylindrical dielectric rod.

From the numerous experiments on long backfire antennas with continuous uniform SW structure (corrugated rod, Goubeau-line structure, dielectric rod, *et cetera*), fed by a dipole near the small reflector, the following empirical expressions are obtained [12]:

(a) if the maximum sidelobes are less than −15 dB, the antenna directivity is

$$D = 10 \log \frac{33800}{(2\theta_{3dB}^{oE})(2\theta_{3dB}^{oH})}, \text{ dBi} \tag{6.57}$$

or, for the radiation pattern with an axial symmetry,

$$D = 10 \log \frac{33800}{2\theta_{3dB}^{o2}}, \text{ dBi} \tag{6.58}$$

(b) the 3 dB beamwidth as a function of the relative antenna length ($1/\lambda$) is

$$2\theta_{3dB}^{o} = 22 \sqrt{\frac{\lambda}{l}} \tag{6.59}$$

These formulas are valid for the optimized LBFA, with stepped or plane-conical surface wave reflector (R_2'' or R_2''').

6.3.2 Some Antenna Designs

A portable long backfire antenna for reception of satellite telemetry signals was designed at AFCRL by Strom in 1969 [32]. Because of the relatively high frequency (f = 2045 MHz) and ease of construction, a dielectric rod slow-wave structure was used instead of the dipole structure of previous models.

Figure 6.24 is a photograph of this antenna, and its longitudinal cross section is shown in Figure 6.25. The SW reflector has a stepped configuration with the following dimensions: d_{2k} = 881 mm, d_{20} = 587 mm, w_1 = w_2 = 36 mm. The dielectric surface wave structure *SW* is made of a *Plexiglas* tube (ε_r = 2.5), 45 mm in diameter, fitted on a 16 mm diameter aluminium tube. The surface wave along such a structure is slowed by the coefficient h = 1.02. The antenna length is l = 587 mm (4.0λ). The small reflector is a disk with a diameter d_1 = 147 mm (1.0λ). The outer ends of the feed dipole are loaded with small disks, 16 mm in diameter, and the dipole length is 54 mm. Each dipole half is fed separately with microcoaxial

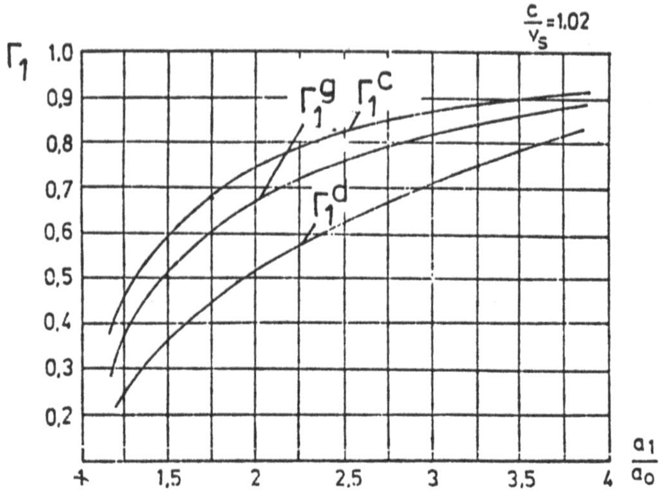

Figure 6.23 Reflection coefficients for corrugated (Γ_1^c), Goubeau-line (Γ_1^g), and dielectric rod (Γ_1^d) structures as functions of a_1/a_0, for $h = c/v_s = 1.02$ [12]

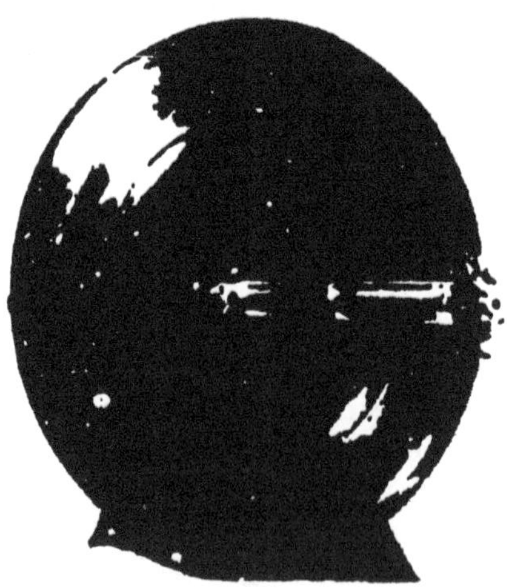

Figure 6.24 Photograph of a backfire antenna with a dielectric rod structure (from [32])

cable through the aluminum tube. The outputs of the cables are paired and phased for proper balance and matching (Figure 6.26).

In the antenna described above, both linear and circular polarizations are optional by use of cross-polarized feed dipoles. A plot of VSWR *versus* frequency is

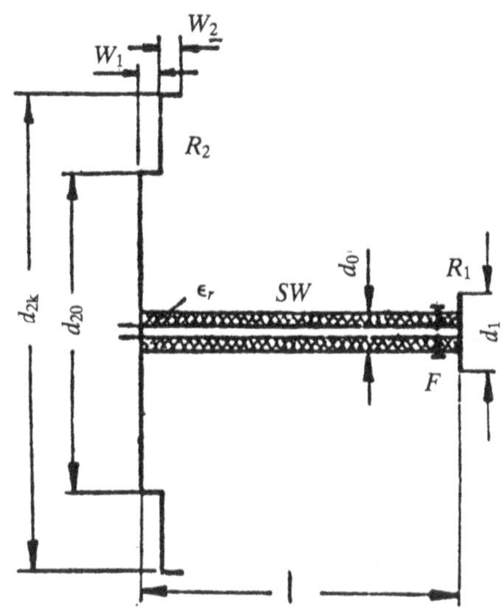

Figure 6.25 Sketch of the antenna in longitudinal cross section (from [32])

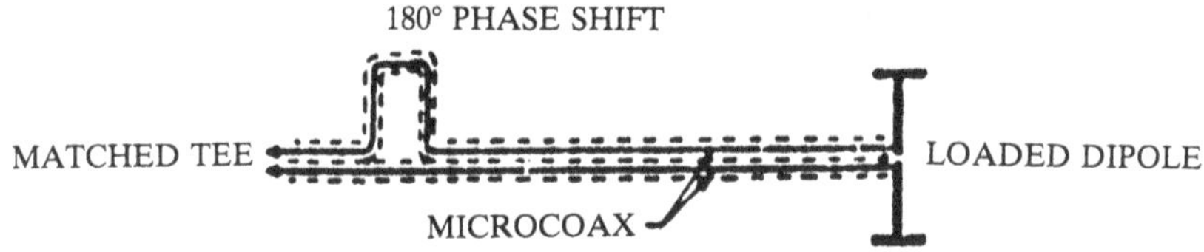

Figure 6.26 Sketch of feed circuitry (from [32])

shown in Figure 6.27. At a center frequency VSWR = 1.08, and it is less than 2.0 in a 5.6% frequency range. Figure 6.28 shows an E-plane radiation pattern at the center frequency f_0 = 2045 MHz. A similarly excellent pattern is measured in the H-plane. The main 3 dB beamwidths are 12.5° and 10° in the E-plane and H-plane, respectively. The radiation patterns in a 10% frequency range have very acceptable quality. The antenna directivity was measured with the use of a pattern integrator, resulting in a value of 24.4 dBi.

Similar backfire antenna designs were developed and examined in detail in 1971 by Hristov [13]. A photograph of a backfire model with a disk-on-rod metal SW structure and plane-conical reflector is shown in Figure 6.29, and its dimensions for the optimum frequency f_0 = 3.7 GHz are given in the longitudinal cross section in Figure 6.30. This antenna has a slotted-dipole-feed, airspaced coaxial line assembly, the output of which is a 50 ohm connector at the rear of the larger reflector.

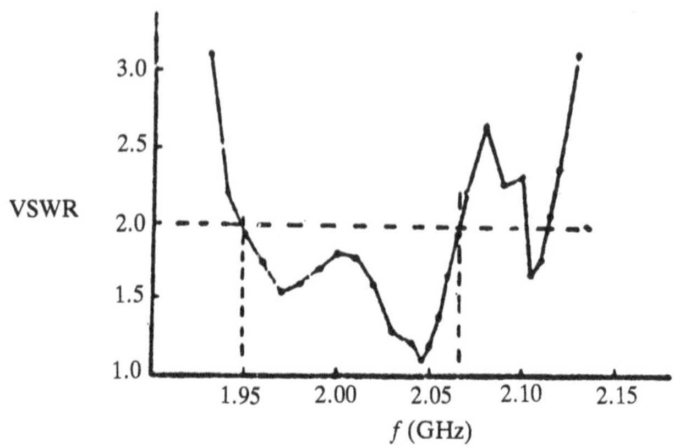

Figure 6.27 VSWR *versus* frequency (from [32])

The measured VSWR *versus* frequency is plotted on Figure 6.31 for two cases: (a) BFA with disk reflector R_2; (b) BFA with plane-conical reflector with an outer rim (R_2'''). The bandwidth properties in the second case are extremely good: VSWR varies only from 1.5 to 1.19 over a 10% frequency range.

Table 6.1 summarizes some of the main parameters of metal disk-on-rod and dielectric-rod backfire antennas, which are 4.0λ in length and tuned for their optimum frequencies.

The radiation characteristics of backfire helical and zig-zag antennas in *S*-band were studied by Loh and Jacobsen [34]. The helical backfire antenna was about 2.0λ long and had a surface wave reflector with a rim 4.0λ in diameter. The small disk reflector, 1.0λ in diameter, gave the best results. The radiation characteristics of helical antennas operating under backfire condititons showed marked improvement over those operating under endfire conditions. The examined antenna model operated satisfactorily over the frequency range 1.4:1.0, according to these criteria: (a) 3 dB beamwidth is less than 17°; (b) the sidelobe levels are down at least 5 dB; (c) the directivity is more than 10 dBi. Figure 6.32 shows the directivity computed from the measured radiation patterns *versus* frequency.

Later, similar results for a 2.0λ long backfire with a conical helix as a SW structure were obtained [33]. The replacing of the cylindrical helix with a conical one (Figure 6.33) gave a small decrease in gain (about 1.3 dB), but better frequency bandwidth was observed. An optimum axial polarization ratio of 2.04 dB was measured for the ordinary conical endfire helix, while for backfire conditions the axial polarization ratio was considerably improved, and reached a value of 1.2 dB.

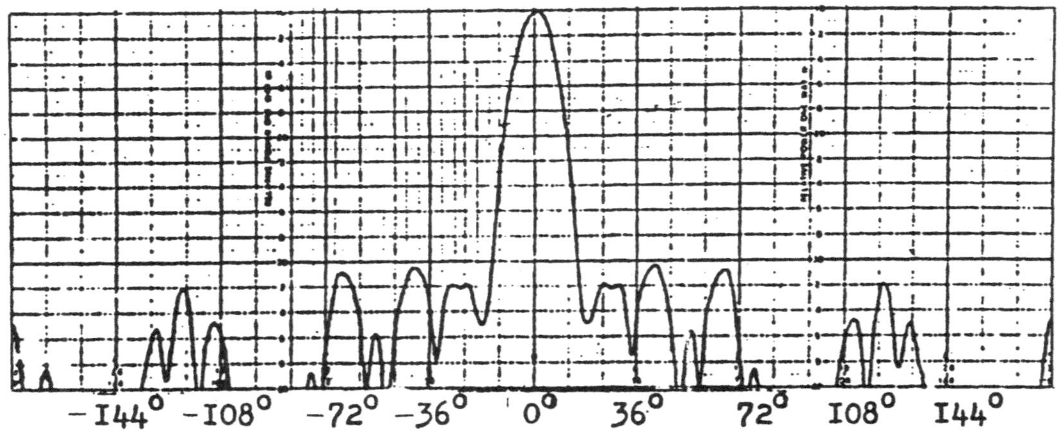

Figure 6.28 E-plane radiation pattern at f_0 = 2045 MHz (from [32])

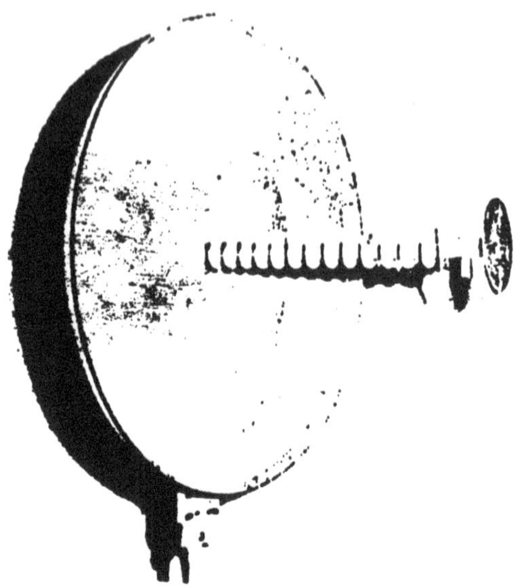

Figure 6.29 Photograph of an LBFA with disk-on-rod SW structure (from [12])

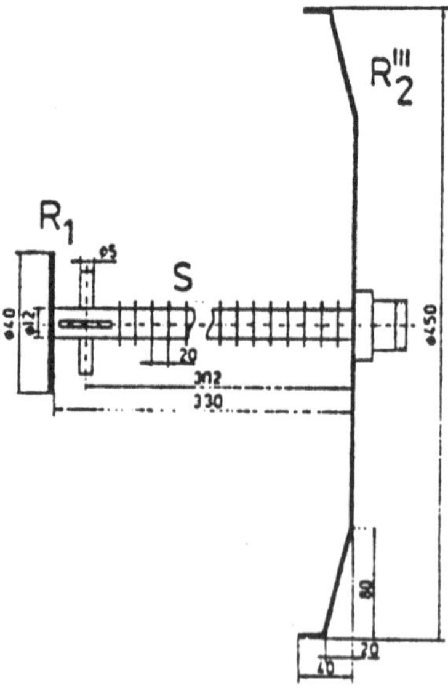

Figure 6.30 Longitudinal cross section of an LBFA with disk-on-rod structure view (from [12])

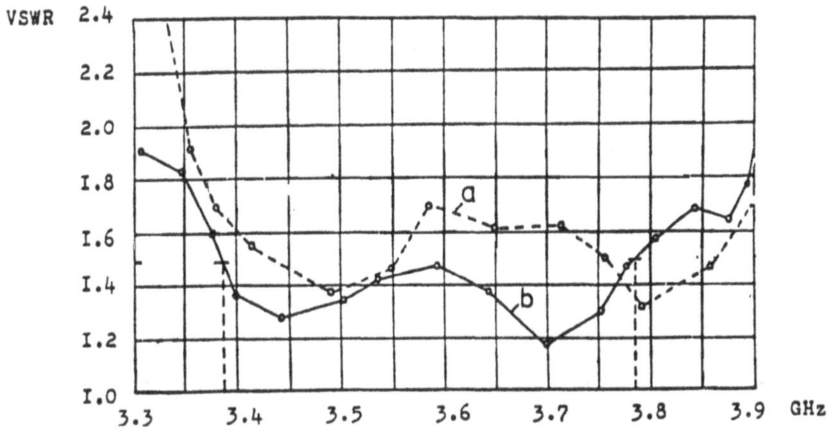

Figure 6.31 Measured VSWR *versus* frequency: (a) for antenna with disk reflector; (b) for antenna with plane-conical reflector

Table 6.1

Antennas	*Beamwidths (Degrees)*		*Max. Side Lobes (dB)*		*Back Lobes (dB)*		*Gain*
	H-plane	*E-plane*	*H-plane*	*E-plane*	*H-plane*	*E-plane*	*(dBi)*
metal disk-on-rod BFA with disk refl. R_2 (from [12])	16	13	−20.5	−19.5	−23	−22	21.8
metal disk-on-rod BFA with disk refl. R_2' (from [12])	15	12	−21.5	−21	−26	−24	22.5
metal disk-on-rod BFA with plane-conical refl. R_2''' (from [12])	11	10	−25	−23	−33	−30	24.3
dielectric-rod BFA with stepped refl. R_2'' (from [12])	12	10	−20	−21	−25	−35	24.0

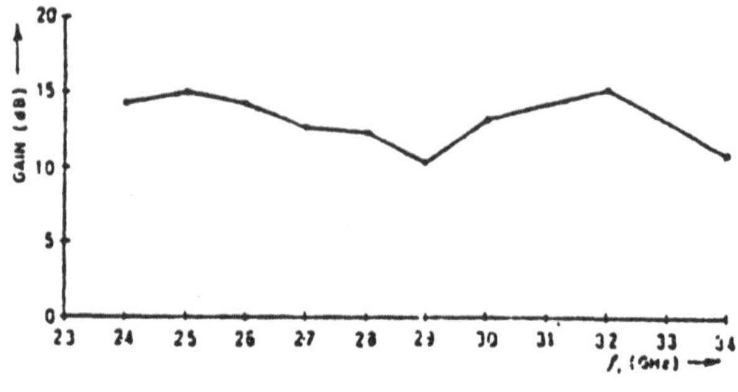

Figure 6.32 Directivity for the 2.0λ long backfire helical antenna (from [33])

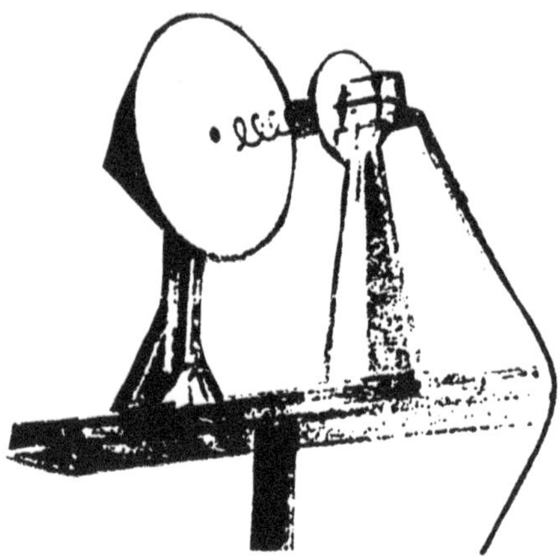

Figure 6.33 A backfire antenna with a conical helix as a SW structure (from [33])

6.4 BACKFIRE ANTENNAS WITH WAVEGUIDE-FED DIELECTRIC ROD STRUCTURE

6.4.1 Rectangular Backfire Antennas with Waveguide-Fed Dielectric Bar

For centimeter and millimeter microwaves, waveguide feeds are used instead of dipole feeds because of the physical and mechanical limitations. Furthermore, the waveguide feeds have the following several advantages:

> construction simplicity and reduction in size; high power operation; low feed system loss; and improved cross-polar performance and impedance bandwidth.

As described initially, the backfire antenna with waveguide-fed dielectric SW structure [34] (Figure 6.34) consists of a large reflector R_2', with $A_2 = B_2 = 78$ mm, and a rim width $W = 22$ mm, and a small reflector R_1 of dimensions $A_1 = B_1 = 16$ mm, placed on the end of a partially tapered dielectric (Teflon) bar as an SW structure. The antenna length is 62 mm. The launcher, which is part of the waveguide, is $Z = 15$ mm in length. The dielectric strips (DS) give a significant improvement in the radiation pattern and are of special value in minimizing the sidelobes. Included here are two versions of the small reflector (R_1 and R_1'). The use of a shaped reflector (R_1') improves the input VSWR of the antenna.

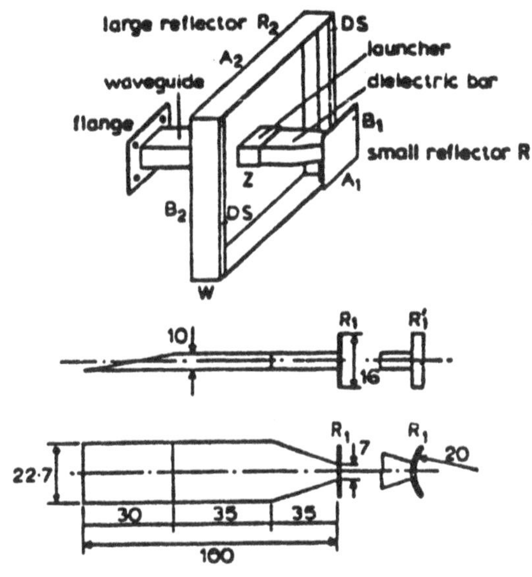

Figure 6.34 10 GHz antenna model of rectangular backfire antenna with waveguide-fed dielectric SW bar (from [34])

In the experimental tests, the cross section of the backfire antenna (i.e., the dimension of the large reflector), was deliberately made identical with that of the pyramidal horn antenna. The H-plane (Figure 6.35) and E-plane (Figure 6.36) radiation patterns of the backfire, horn, and dielectric rod antenna are shown for a frequency $f = 9.57$ GHz ($\lambda = 3.14$ cm). Table 6.2 summarizes some of the main parameters measured for these three antennas. The backfire antenna is considerably shorter than the two reference antennas. Although the radiation patterns and the directivity coefficient are remarkable for such dimensions, the loss and matching efficiency needed further improvement. Hristov and Uromov [35] achieved this by changing the antenna dimensions slightly to obtain a gain of 16.4 dB (i.e., a value

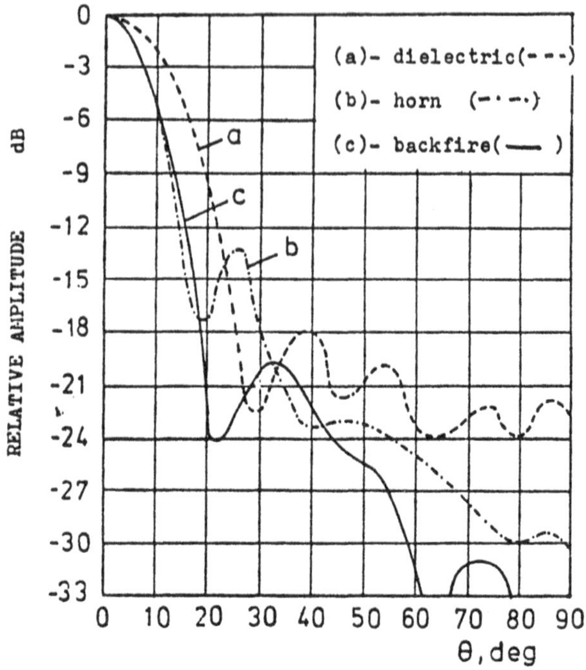

Figure 6.35 *H*-plane radiation patterns of (a) ordinary dielectric, (b) horn, and (c) backfire antennas (from [34])

close to that of the horn antenna gain (17.2 dB)). Figure 6.37 shows longitudinal and face views of three antennas: (a) the reference horn antenna with length 4.6λ = 3.2 cm; (b) the rectangular backfire antenna modification, 1.4λ long; (c) the round short backfire antenna with open-end waveguide feed having a length of 0.6λ. These are all drawn to the same scale. The dimensions of the cross section are: $A_2 = B_2 = A_0 = B_0 = 2.4\lambda$; $d_2' = 2.0\lambda$; $A_1 \times B_1 = 0.4\lambda \times 0.6\lambda$; $A_1' \times B_1' = 0.4\lambda \times 0.9\lambda$. We may easily see from Figure 6.37 that the backfire antennas (b) and (c) are 3.3 and 7.7 times shorter than the reference horn (a), respectively.

6.4.2 Circular Backfire Antennas With Waveguide-Fed Dielectric Rod

The rectangular backfire configuration is characterized by an axial asymmetry in its radiation pattern. To overcome this, a circular backfire antenna was developed [36–38]. It consists of two round reflectors and a cylindrical dielectic rod fed by an open-ended circular waveguide. The dielectric rod is made of perspex, tapered for better matching between the fundamental mode (TE_{11}) of the metal waveguide and the surface wave.

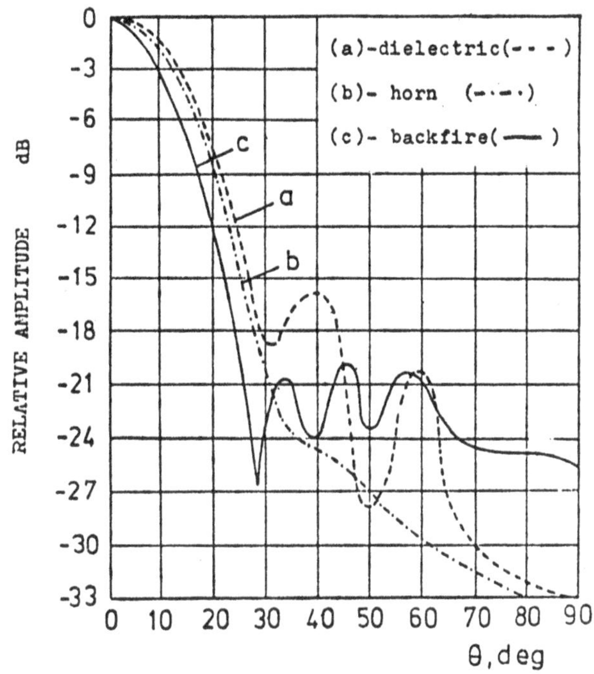

Figure 6.36 *E*-plane radiation patterns of (a) ordinary dielectric, (b) horn, and (c) backfire antennas (from [34])

Table 6.2

Antennas	*Beamwidth H-plane deg*	*Beamwidth E-plane deg*	*Maximum Sidelobe H-plane dB*	*Maximum Sidelobe E-plane dB*	*Gain dB*	*Length mm*
backfire antenna	20	16	−20	−19.8	15.2	62
dielectric antenna	26.2	24.6	−16	−18	14.8	135
horn antenna	24	15.4	—	−13	17.2	152

A model was constructed for *X*-band operation with a number of large reflectors (Figure 6.38). All antenna varieties had lengths between 142 mm and 150 mm and the measurements were made in the frequency range of 8.6 to 10.6 GHz. The main purpose was to optimize the large reflector configuration experimentally, with respect to pattern quality and bandwidth. The antenna parameters are summarized and

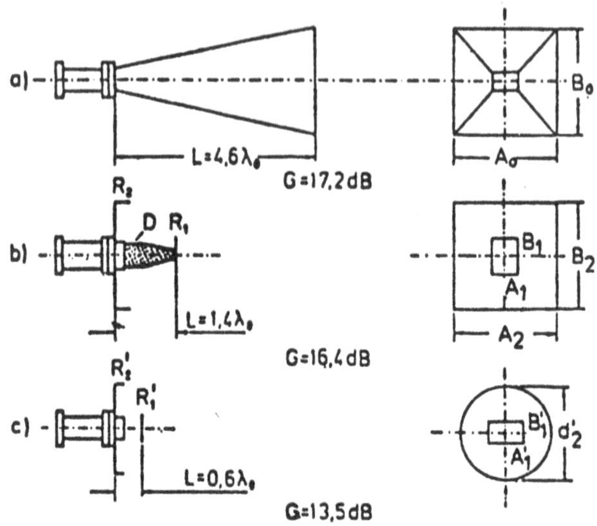

Figure 6.37 Longitudinal and face views of (a) reference horn, (b) rectangular backfire modification, and (c) short backfire antenna (from [33])

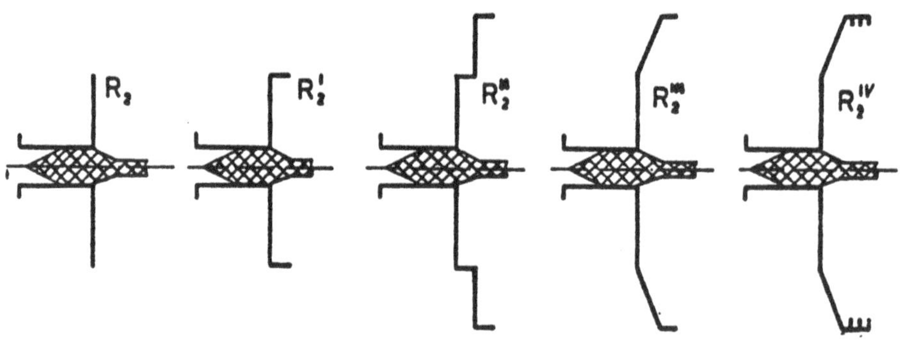

(a) (b) (c) (d) (e)

Figure 6.38 Large reflector varieties [39]: (a) R_2, a simple disk reflector; (b) R_2' a disk reflector with rim; (c) R_2'' a stepped reflector; (d) R_2''' a plane-conical reflector; (e) R_2^{IV} a plane-conical reflector with corrugated rim

compared in Table 6.3. As expected, the use of plane-conical reflectors R_2''' and R_2^{IV} gave the best antenna performance. The radiation patterns (copolar and cross-polar) of the backfire antenna with the large reflector R_2''', for an optimum frequency $f =$ 10.16 GHz, are shown in Figure 6.39. The 3 dB beamwidth in the two principal

Table 6.3 (from [39])

Reflector	*Reflector Diameter (mm)*	*3 dB Beamwidth (degrees)*	*Max. Sidelobe Levels (dB)*	*Gain (dB)*
R_2	130	19	−16	19.5
R_2'	130	18	−18	20.0
R_2''	190	13	−19	23.0
R_2'''	220	11	−22	24.2
R_2^{IV}	220	11	−24	24.4

planes (*E*-plane and *H*-plane) is less than −21 dB. For comparison in the same figure, copolar radiation patterns of a reference pyramidal horn with a gain of 22 dBi are given. These parameters do not vary much in the frequency band 9.95–10.45 GHz, and the peak cross-polar level within the bounds of the main beam is less than −29 dB.

When the antenna model studied was tested with the small disk reflector *R*, the input frequency bandwidth was too narrow (VSWR less than 2 over 1% of the frequency band and about 1.2 at the optimum frequency). This was due to the greater amount of power reflected back into the waveguide feed from the small reflector. To reduce this, a shaped reflector is used. The VSWR for the antenna with the shaped small reflector is less than 1.8 over about 7% of the frequency band and 1.04 at the optimum frequency.

A comparison was made between the long backfire antenna and a 22 dBi standard pyramidal horn, which has approximately the same aperture area but is two and a half times longer. We conclude that the long backfire antenna containing a plane-conical reflector with rim, waveguide-fed dielectric SW rod, and a small reflector has high directivity, a symmetrical radiation pattern, and low cross polarization. The investigated backfire antenna with waveguide-fed dielectric rod can be modified with a small strip reflector to operate in two discrete frequency ranges, for two orthogonal polarizations, respectively [35–36, 40] (Figure 6.40). The basic antenna configuration and dimensions are kept the same.

The circular waveguide diameter is 22.9 mm, and the outer circular part of the dielectric structure has a diameter of 10 mm. For a dielectric constant of 2.6, the slow-wave coefficient is $h = c/v_s = 1.02$. In the radiation mode, for example, the antenna functions as follows (Figure 6.41): for two orthogonal, linearly polarized electromagnetic waves ($\bar{E}_1$ and $\bar{E}_2$) in the circular metal waveguide, two different radiation mechanisms of the antenna are obtained: backfire and endfire. Figure 6.41(a)

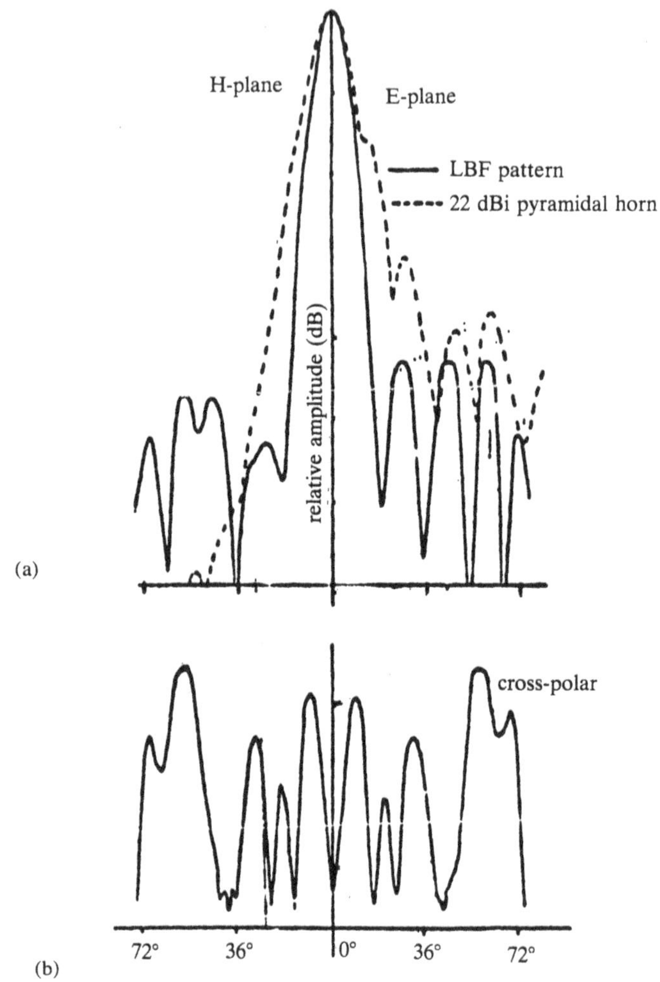

Figure 6.39 Radiation patterns at 10.16 GHz (from [37]): (a) *H*-plane and *E*-plane copolar radiation patterns of a long backfire antenna (solid) and a 22 dBi pyramidal horn (dashed); (b) cross-polar pattern of a long backfire antenna

is a face view of a backfire antenna with a solid metal small reflector, and Figure 6.41(b) is that of one with a small strip reflector. *H*-plane and *E*-plane radiation patterns are shown in Figure 6.42. Pattern 1 corresponds to the backfire radiation mode and pattern 2 corresponds to the endfire radiation mode. The 3 dB beamwidth for the backfire case, tuned for 10.2 GHz, is 11 degrees in both planes with the sidelobes in the *E*-plane and *H*-plane less than −20 dB and −23 dB, respectively.

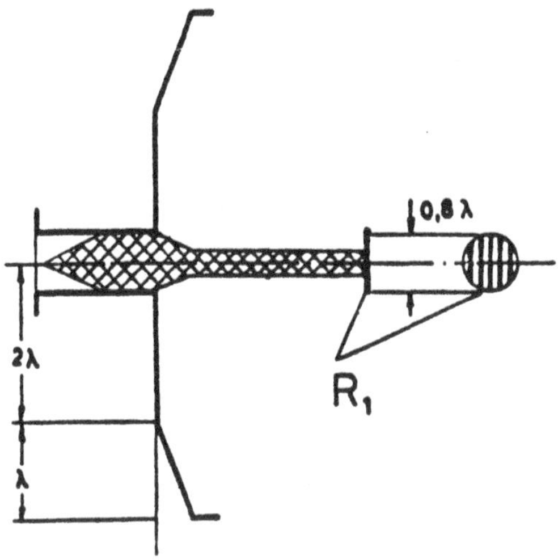

Figure 6.40 Backfire antenna with waveguide-fed dielectric rod and small strip reflector (from [40])

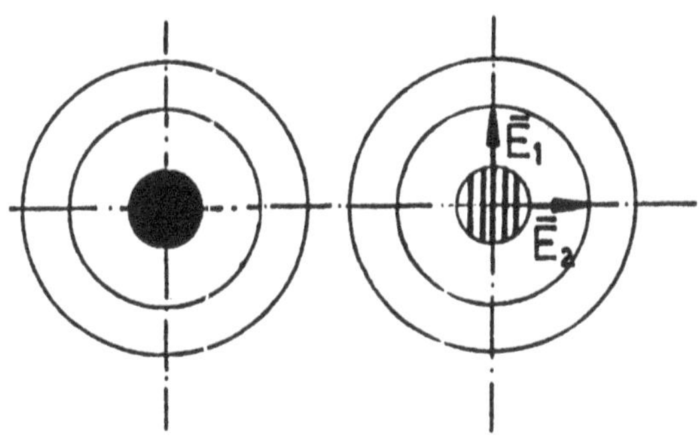

Figure 6.41 (a) Face view of the backfire antenna with small solid reflector; (b) Face view of the same antenna with a small strip reflector (from [40])

The 3 dB beamwidth for the endfire case for the same frequency is 28 degrees in both planes, and the sidelobes in the *E*-plane and *H*-plane are less than −12 dB and −13 dB, respectively. Similar patterns for the latter case were obtained at another frequency range, 8.6 to 9.0 GHz, which is outside the range for the backfire case, 8.8 to 10.2 GHz. Measured VSWR is less than 1.8 in both discrete frequency ranges, and 1.08 at 8.8 GHz and 1.04 at 10.2 GHz.

The cross-polarization antenna performance for both frequency ranges is very good. For the endfire mode at the frequency range of 8.6–9.0 GHz, the peak cross-

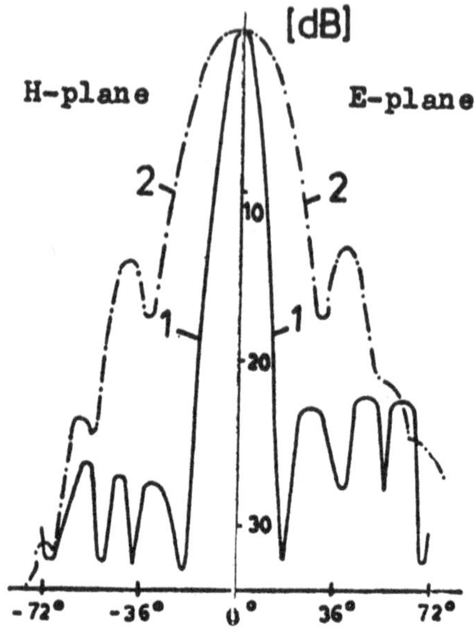

Figure 6.42 *H*-plane and *E*-plane radiation patterns for two radiation modes (from [40]): (1) backfire; (2) endfire

polar levels are less than −27 dB and, for the backfire mode at the frequency range 8.8 − 10 GHz, are less than −30 dB over the main lobe.

6.4.3 Long Backfire Antennas with Insular Image Guide Feed for Dual-Frequency Operation

Another modification of the waveguide-fed backfire with dielectric rod SW structure was proposed by Simons and Shahani [41] (Figure 6.43). The feed F consists of an insular image guide that passes through a square hole of side $2C$ in the circular reflector R. With $c > a$, a square annular gap of width $N = (c - a)$ is formed around the dielectric strip. A mode launcher has been used between the Gunn oscillator and the insular image guide feeding the long backfire antenna. The metal small reflector S (subreflector) is firmly fixed to the end of the dielectric guide. The experimental model operates in the backfire mode from 9.7 to 9.9 GHz, and in the endfire mode from 8.9 to 9.1 GHz. The measured antenna parameters for the backfire and endfire operation are given in Table 6.4.

To ascertain that the antenna truly operates in the endfire, the gain was measued (when set to 9.0 GHz) with, and without, both the main reflector and subreflector; it was found to be approximately equal. The examined antenna design, with an insular image guide, is compatible with dielectric integrated circuit technology for

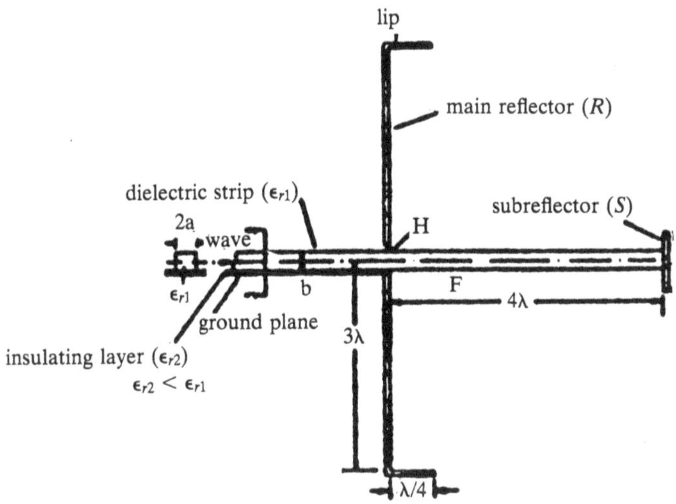

Figure 6.43 Long backfire antenna with insular image guide feed (from [41])

Table 6.4 (from [41])

Antenna Parameters	*Backfire*	*Endfire*
3 dB beamwidth (degrees)		
E-plane	12	13
H-plane	11	16
Sidelobe level (dB)		
E-plane	−16	−14
H-plane	−16	−16
Gain (dBi)	20	17
Cross polarization (dB)	−22	−20.5
Center frequency f (GHz)	9.8	9.0
VSWR at f	1.2	1.2
3 dB gain bandwidth (MHz)	400	400
Bandwidth where VSWR is less than 2.0, (MHz)	200	200

millimeter waves. It is a simple and efficient radiator suitable for operation in two discrete frequency ranges.

6.4.4 Radiation Theory of Circular Backfire Antennas with Waveguide-Fed Circular Dielectric Rod

The antenna radiation mechanism is similar to that of a backfire antenna with a dipole-fed SW rod structure, which was described in detail in Section 6.3.1. As we see in Figure 6.44, the theoretical model in the present analysis is defined more accurately. The total far field is a superposition of several field components [38–39, 42–49]:

$$\dot{E} = \dot{E}_s + \dot{E}_{F_d} + \dot{E}_{R_1}^{(d)} + \dot{E}_{R_2}^{(d)} + \dot{E}_{R_2}^{(r)} \tag{6.60}$$

where E_s is a field, radiated by the equivalent antenna aperture (cross section A-A', Figure 6.44(a));

E_{F_d} is a direct spherical wave, radiated by the feed aperture (Figure 6.44(a));

$E_{R_1}^{(d)}$ is a field component due to diffraction by the small reflector (Figure 6.44(b));

$E_{R_2}^{(d)}$ is a field component due to diffraction by the large reflector (Figure 6.44(c));

$E_{R_2}^{(r)}$ is a field component due to the reflection from the large reflector (Figure 6.44(d)).

In long backfire antenna designs, the large reflector diameter is on the order of two to six wavelengths. Moreover, the amplitude of the slow surface wave decreases in a radial direction, so we can assume that the large reflector perfectly reflects the incident surface wave (i.e., $\dot{\Gamma}_2 = e^{j\pi} = -1$). The small reflector diameter is on the order of 0.5 to 1.0 wavelengths, and it partially reflects the surface wave. The complex reflection coefficient of this reflector can be defined [12] as

$$\dot{\Gamma}_1 = \frac{\dot{E}_{sw}^-(0)}{\dot{E}_{sw}^+(0)} = \frac{|\dot{E}_{sw}^-(0)|\, e^{j}}{|\dot{E}_{sw}^+(0)|\, e^{j0}} = \sqrt{\frac{P_{sw}^-}{P_{sw}^+}} = |\dot{\Gamma}_1|\, e^{j\phi_1} \tag{6.61}$$

where $E_{sw}^-(0)$ and P_{sw}^- are the electric field intensity and the average power of the surface wave reflected from the small reflector (within the limits $0 \le r \le a_1$), and $E_{sw}^+(0)$ and $P_{sw}^+(0)$ are the electric field intensity and the average power of the incident surface wave in a cross section A-A' (within the limits $0 \le r \le \infty$).

For a weakly delayed surface wave ($h = c/v_s = 1.02 - 1.04$) in a circular dielectric rod, the reflector coefficient modulus is

$$|\dot{\Gamma}_1| = \sqrt{1 - \left(\frac{a_1}{a_0}\right)^2 \cdot \frac{Q_0(a_1)}{Q_0(a_0)}} \tag{6.62}$$

where

$$Q_0(a_i) = K_0^2(k_e a_i) - K_1^2(k_e a_i), \quad i = 0, 1, 2 \ldots$$

The reflector coefficient argument ϕ_1 has a value of less than 180°, and may be accepted as on the order of 120° to 150°. Bearing in mind the multiple reflection phenomenon by the reflectors, the electromagnetic field in the equivalent antenna aperture can be considered as an infinite sum of plane surface waves:

$$\dot{E}_{sw}(\rho, \psi, 0) = \sum_{n=1}^{\infty} \dot{E}_{sw_n}(\rho, \psi, 0) \tag{6.63}$$

By analogy, the far field can be obtained as an infinite sum of partial spherical far fields:

$$\dot{E}_s(\theta, \phi, r_0) = \sum_{n=1}^{\infty} \dot{E}_{s_n}(\theta, \phi, r_0) \tag{6.64}$$

where

$$E_{s_n} = E_0 \Gamma_1^{n-1} \Gamma_2^{n-1} e^{-(2n-1)\gamma_s L} F_s(\theta, \phi) \frac{e^{-jpr_0}}{r_0}$$

and $\gamma_s = a_s + j\beta_s$ is a complex propagation constant of the surface wave. If we set

$$E_s(\theta, \phi, r_0) = E_0 F_s(\theta, \phi) \frac{e^{-j\beta r_0}}{r_0}$$

then the far field, radiated by the equivalent antenna aperture, can be written as:

$$\hat{E}_s(\theta, \phi, r_0) = E_s(\theta, \phi, r_0) S_n \tag{6.65}$$

where

$$\begin{aligned} S_n &= \sum_{n=1}^{\infty} |\dot{\Gamma}_1|^{n-1} |\dot{\Gamma}_2|^{n-1} e^{j(\psi r_1 + \psi r_2)} e^{-(2n-1)\gamma_s L} \\ &\approx \sum_{n=1}^{\infty} |\dot{\Gamma}_1|^{n-1} e^{-(2n-1)\alpha_s L} e^{-j(2n-1)\beta_s L} \end{aligned} \tag{6.66}$$

In the same way (except for E_F), the remaining far field, components are represented by

$$\hat{E}_{R_1}^{(d)} = \dot{E}_{R_1}^{(d)}(\theta, \phi, r_1^d, r_2^d)S_n \tag{6.67}$$

$$\hat{E}_{R_2}^{(d)} = \dot{E}_{R_2}^{(d)}(\theta, \phi, r_3^d, r_4^d)S_n \tag{6.68}$$

$$\hat{E}_{R_2}^{(r)} = \dot{E}_{R_2}^{(r)}(\theta, \phi, r_a^r, r_b^r)S_n \tag{6.69}$$

where $r_1^d, r_2^d, r_3^d, r_4^d, r_a^r, r_b^r$ can be expressed by r_0 (Figure 6.44).

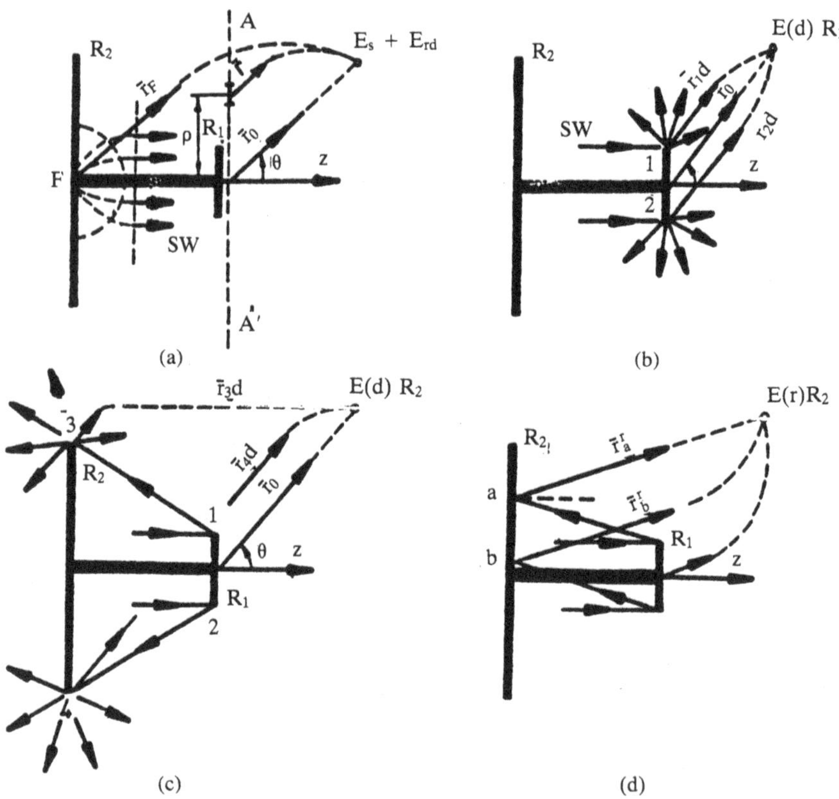

Figure 6.44 Radiation antenna mechanism (from [38–39, 43])

Then, after giving an account of the resonant antenna properties, the total far field can be written as

$$\dot{E}(\theta, \phi, r_0) = \hat{E}_s(\theta, \phi, r_0) + \hat{E}_{R_1}^{(d)}(\theta, \phi, r_0) + \hat{E}_{R_2}^{(d)}(\theta, \phi, r_0) + \hat{E}_{R_2}^{(r)}(\theta, \phi, r_0) + E_F(\theta, \phi, r_0) \tag{6.70}$$

If $v_s \approx c$, the surface wave in the dielectric rod has nearly axial symmetry and the equivalent aperture field, as a first approximation [44–45], is
(a) for region 1 ($0 \leq \rho \leq a_0$),

$$E_{sw}^{(1)}(\rho) = -AJ_0(\beta_i \rho) \tag{6.71}$$

(b) for region 2 ($a_0 \leq \rho \leq a_1$)

$$E_{sw}^{(2)}(\rho) = -AG_0 K_0(\beta_i \rho) \tag{6.72}$$

(c) for region 3 ($a_1 \leq \rho \leq \infty$),

$$E_{sw}^{(3)}(\rho) = AG_0 K_0(\beta_i \rho) \tag{6.73}$$

where $\beta_i = \sqrt{\varepsilon_r \beta^2 - \beta_S^2}$ is the transversal wavenumber of the surface wave into the dielectric rod, $\beta_e = \sqrt{\beta_S^2 - \beta^2}$ is the transversal wavenumber of the surface wave out of the dielectric rod, $\beta_S = f(\varepsilon_r, 2a_0/\lambda)$ is the surface wave phase constant of dielectric rod structure, and G_0 is a boundary amplitude coefficient, given by

$$G_0 \approx \frac{\beta_i J_1(\beta_i a_0)}{\beta_e K_1(\beta_e a_0)} \tag{6.74}$$

The far-field intensity is found from the scalar radiation integral:

$$E_s = j\frac{e^{-j\beta r_0}}{2\lambda r_0}(1 + \cos\theta)\left[\int_0^{2\pi}\int_0^{a_0} E_{sw}^{(1)}(\rho)e^{j\phi}\rho\,\mathrm{d}\rho\,\mathrm{d}\psi + \int_0^{2\pi}\int_0^{a_1} E_{sw}^{(2)}(\rho) \cdot e^{j\phi}\rho\,\mathrm{d}\rho\,\mathrm{d}\psi + \int_0^{2\pi}\int_0^{\infty} E_{sw}^{(3)}(\rho)e^{j\phi}\rho\,\mathrm{d}\rho\,\mathrm{d}\psi\right]; \quad \text{for } \phi = \beta\rho\sin\theta\sin\psi \tag{6.75}$$

After integrating and setting $a_0 = \tilde{a}_0\lambda$, $a_1 = \tilde{a}_1\lambda$, $r_0 = \tilde{r}_0\lambda$, and also $u_0 = 2\pi\tilde{a}_0 \sin\theta$; $v_{i0} = 2\pi\tilde{a}_0\sqrt{\varepsilon_r - h_S^2}$; $u_1 = 2\pi\tilde{a}_1 \sin\theta$; $v_{i1} = 2\pi\tilde{a}_1\sqrt{\varepsilon_r - h_S^2}$; $v_0 = 2\pi\tilde{a}_0\sqrt{h_S^2 - 1}$; $h_s = \lambda/\lambda_s > 1$; $v_1 = 2\pi\tilde{a}_1\sqrt{h_S^2 - 1}$; for the aperture radiation far field we obtain

$$E_s(\theta, \tilde{r}_0) = j\pi A F_0(\theta) f_1(\theta)\frac{e^{-j2\pi\tilde{r}_0}}{\tilde{r}_0} \tag{6.76}$$

where $\tilde{a}_0$, $\tilde{a}_1$, $\tilde{r}_0$ are normalized linear dimensions, and

$$F_0(\theta) = 1 + \cos\theta$$

$$f_1(\theta) = \tilde{I}_1(\theta) + \tilde{I}_2'(\theta) + \tilde{I}_2''(\theta) + \tilde{I}_3(\theta)$$

$$\tilde{I}_1(\theta) = -\frac{\tilde{a}_0^2}{u_0^2 + v_{i0}^2}\,[u_0 J_1(u_0) J_0(v_{i0}) - v_{i0} J_0(u_0) J_0(v_1)]$$

$$\tilde{I}_2'(\theta) = -G_0 \frac{\tilde{a}_1^2}{u_1^2 + v_1^2}\,[u_1 K_0(v_1) J_1(u_1) - v_{i1} K_1(v_{i1}) J_0(u_1)]$$

$$\tilde{I}_2''(\theta) = -G_0 \frac{\tilde{a}_0^2}{u_0^2 + v_0^2}\,[v_{i0} K_1(v_{i0}) J_0(u_0) - u_0 K_0(v_0) J_1(u_1)]$$

$$\tilde{I}_3(\theta) = G_0 \frac{\tilde{a}_1^2}{u_1^2 + v_1^2}\,[v_1 K_1(v_1) J_0(u_1) - u_1 K_0(v_1) J_1(u_1)]$$

Now, we find the field component $E_{R_1}^{(d)}$ due to the diffraction of the surface wave by the small reflector periphery. In the present diffraction problem, the weakly delayed surface wave falls normally on the metal disk (Figure 6.45). Using the *geometrical theory of diffraction* (GTD) [26], we obtain an approximate solution [39, 43]. The geometry of the diffraction problem is illustrated in Figure 6.46. The total diffraction field is the sum of the partial diffraction fields from points 1 and 2:

$$E_{R_1}^{(d)} = -AG_0 \sqrt{\frac{-j\tilde{a}_1}{\sin\theta}}\,\frac{K_0(v_1)}{4\pi}\left[\left(\sec\frac{\theta}{2} + \csc\frac{\theta}{2}\right) \cdot e^{j2\pi\tilde{a}_1\sin\theta} + j\left(\sec\frac{\theta}{2} - \csc\frac{\theta}{2}\right) e^{-j2\pi\tilde{a}_1\sin\theta}\right] \cdot \frac{e^{-j2\pi\tilde{r}_0}}{\tilde{r}_0} \tag{6.77}$$

As a reflection field, $R_{R_2}^{(r)}$ is considered the field due to the reflection of the primary diffraction field $E_{R_1}^{(d)}$ from the large reflector R_2, is assumed as infinite in extent. A small reflector image R_1' is introduced (Figure 6.46). The total reflection field $E_{R_2}^{(r)}$ is given by

$$E_{R_2}^{(r)}(\theta, r_0) = AG_0 \frac{1}{4\pi} \sqrt{\frac{-j\tilde{a}_1}{\sin\theta}}\,K_0(v_1)\left(\sec\frac{\theta''}{2} + \csc\frac{\theta''}{2}\right) \cdot e^{j2\pi\tilde{a}_1\cos\theta''} e^{-j4\pi\tilde{l}\cos\theta} \frac{e^{-j2\pi\tilde{r}_0}}{\tilde{r}_0}, \quad (\tilde{l}\lambda = l) \tag{6.78}$$

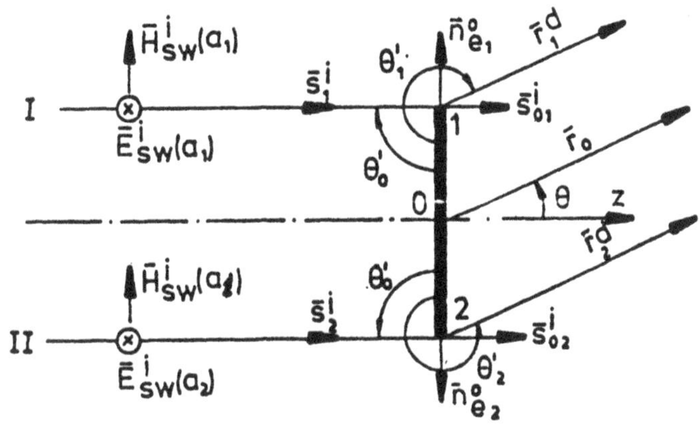

Figure 6.45 Transverse amplitude distribution of the surface wave, falling on the axially fixed small disk reflector (from [38–39, 43])

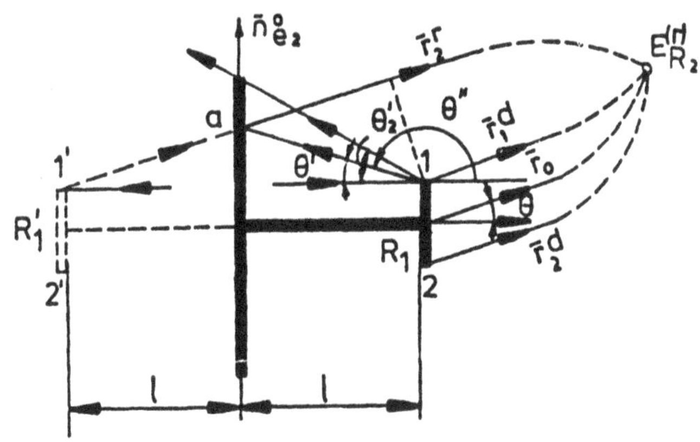

Figure 6.46 Reflection from the large reflector: ray tracing arrangement (from [39, 42])

To compute the far sidelobes and backlobes of the radiation pattern, we must consider the diffraction effects from the large reflector edge. The ray tracing picture of this diffraction case is shown in Figure 6.47. The primary diffraction rays 1 and 3 (Figure 6.47(a) and 2 and 4 (Figure 6.47(b)) give rise to secondary diffraction fields.

The total secondary diffraction field from the large reflector periphery [39, 43] is

$$E_{R_2}^{(d)} = E_3^{(d)} + E_4^{(d)} \tag{6.79}$$

where

$$\begin{aligned} E_3^{(d)} = &-\frac{AG_0}{4\pi}\sqrt{\frac{-j\tilde{a}_1}{\sin\theta}}\,K_0(\nu_1)\left(\sec\frac{\theta_{1-3}}{2} + \csc\frac{\theta_{1-3}}{2}\right) \\ &\cdot e^{j2\pi\tilde{a}_1\cos\theta_{1-3}}\,\frac{e^{-j2\pi\tilde{\rho}_{1-3}}}{\tilde{\rho}_{1-3}}\left[-\frac{1}{4\pi\sqrt{j}}\left(\sec\frac{\pi-\theta}{2}\right.\right. \\ &\left.\left.+\csc\frac{\pi-\theta}{2}\right)\right]\sqrt{\tilde{\rho}_s^{(3)}}\;e^{-j2\pi[\tilde{\rho}_{1-3}\cos(\theta+\theta_1'')-\tilde{a}_1\sin\theta]}\cdot\frac{e^{-j2\pi\tilde{r}_0}}{\tilde{r}_0} \end{aligned} \tag{6.80}$$

$$\begin{aligned} E_4^{(d)} = &\frac{AG_0}{4\pi}\sqrt{\frac{-j\tilde{a}_1}{4\pi}}\,K_0(\nu_1)\left(\sec\frac{\theta_{2-4}}{2} - \csc\frac{\theta_{2-4}}{2}\right) \\ &\cdot e^{-j2\pi\tilde{a}_1\cos\theta_{2-4}}\,\frac{e^{-j2\pi\tilde{\rho}_{2-4}}}{\tilde{\rho}_{2-4}}\left[-\frac{1}{4\pi\sqrt{j}}\left(\sec\frac{\pi+\theta}{2}\right.\right. \\ &\left.\left.-\csc\frac{\pi+\theta}{2}\right)\right]\sqrt{\tilde{\rho}_s^{(4)}}\;e^{-j2\pi[\tilde{\rho}_{2-4}\cos(\theta-\theta_1'')+\tilde{a}_1\sin\theta]}\cdot\frac{e^{-j2\pi\tilde{r}_0}}{\tilde{r}_0} \end{aligned} \tag{6.81}$$

Expressions for the secondary diffraction fields from the large disk reflector with rim (R_2'), stepped reflector with rim (R_2'') and plane-conical reflector with rim (R_2''') can also be obtained using this method. The direct radiation far-field component E_F from the feed aperture is estimated experimentally, proposing a simplified form of the expression [43]:

$$E_{Fd} = jB\,[\cos(0.75 - \theta)]^n\,e^{-j\beta l\cos\theta} \tag{6.82}$$

where B is an amplitude factor. We find experimentally that the excitation efficiency B/A is on the order of 0.05 to 0.10. The number of resonant waves between the antenna reflectors is n; in practice, it is sufficient to let $n = 2$ to 5. Theoretical results for an LBFA with waveguide-fed dielectric rod and large disk reflector with rim are compared with its experimental radiation patterns [43] in Figure 6.48. Figure 6.48 shows that the agreement between theoretical and experimental results is very good in both planes. The proposed theory may be extended to design the antenna configuration for maximum directivity.

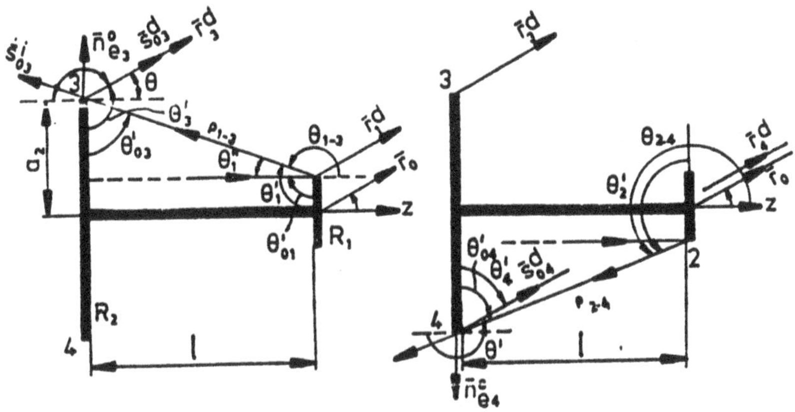

(a) (b)

Figure 6.47 Ray tracing pictures of secondary diffractions from edge points 3(a) and 4(b) (from [39, 43])

6.4.5 Cross-Polarization Characteristics of Backfire Antenna with Waveguide-Fed Circular Dielectric Rod

A plane reflection surface does not introduce cross-polar distortions in the reflected electromagnetic waves. Therefore, if the diffraction effects from the antenna plane reflectors and feed are neglected, the polarization properties of the LBFA are determined only by the polarization structure of the surface electromagnetic wave. In the previous section, the far-field solution was given in a scalar form. The expressions for the electromagnetic field components do not contain any information about cross-polar antenna characteristics. To find the cross-polar far field, it is necessary to start from the exact expressions for the two orthogonal components of the hybrid surface wave HE_{11} in the dielectric rod [44–45], which gives us the following expressions:

(a) inside the dielectric rod:

$$E_x^{(i)} = \frac{1}{2} A_0[J_0(\beta_i r)(1 + G_0) + J_2(\beta_i r)(1 - G_0)\cos 2\psi] \qquad (6.83)$$

$$E_y^{(i)} = \frac{1}{2} A_0 J_2(\beta_i r)(1 - G_0)\sin 2\psi$$

and (b) outside the dielectric rod:

$$E_x^{(e)} = \frac{1}{2} A_0 F_0 [K_0(\beta_e r)(1 + G_0) + K_2(\beta_e r)(1 - G_0)\cos 2\psi]$$
$$E_y^{(e)} = \frac{1}{2} A_0 F_0 K_2(\beta_e r)(1 - G_0)\sin 2\psi \tag{6.84}$$

where E_x is a main polarization field intensity, and E_y is cross-polarization field intensity. The HE_{11} near-field structure is illustrated in Figure 6.49.

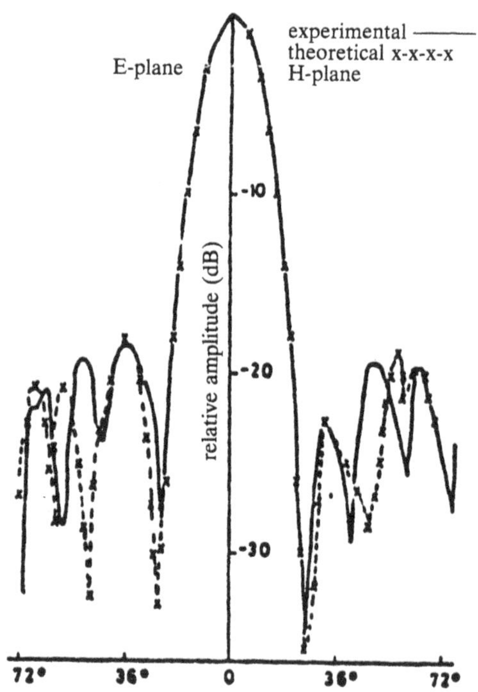

Figure 6.48 *E*- and *H*-plane radiation patterns: experimental (solid line), and theoretical (dash-and-x-line), for antenna length of 142 mm, and reflector diameter of 130 mm (f = 10.16 GHz) (from [43])

The maximum cross-polarization field component is obtained for $\psi = 45°$ and $r = a_0$. Therefore, for $r \leq a_0$ (in the dielectric medium):

$$E^{(i)}_{y\,\max} = E^{(i)}_y(a_0, 45°) = \frac{1}{2} A_0(1 - G_0) J_2(\beta_i a_0) \tag{6.85}$$

and for $r \geq a_0$ (in the air medium):

$$E^{(e)}_{y\,\max} = E^{(e)}_y(a_0, 45°) = \frac{1}{2} A_0 F_0(1 - G_0) K_2(\beta_e a_0) \tag{6.86}$$

A parameter I_{cn}, called cross-polarization coupling, is defined for the near-field region as:

$$I_c[\text{dB}] = 20 \lg \left| \frac{E_y(r, \psi)}{E_x(r, \psi)} \right| \tag{6.87}$$

The cross-polarization coupling depends on λ, ε_r, and a_0, so

$$I_c = f\left(\varepsilon_r, \frac{a_0}{\lambda_0}\right) \tag{6.88}$$

The computations show that for h in the limits 1.02 to 1.10, I_{cn} does not depend much on the dielectric constant ε_r, when $2a_0$ is 0.3 to 0.4 of the free space wavelength [46, 47]. For example, for $h = 1.03$, $I_{cn} = 35$ dB for $\varepsilon_r = 2.5$, and $I_{cn} = 37$ dB for $\varepsilon_r = 5.0$. Similarly, a parameter far-field cross-polarization coupling I_{cf} is defined, using the so-called third Ludwig definition [47a]:

$$I_c(\theta) = 20 \lg \left| \frac{E_c(\theta)}{E_m(\theta)} \right|, \quad (\phi = \text{const.}) \tag{6.89}$$

where

$$\left| \begin{aligned} E_m &= E_\theta \cos\phi - E_\phi \sin\phi \\ E_c &= E_\theta \sin\phi - E_\phi \cos\phi \end{aligned} \right. \tag{6.90}$$

with E_m as the main (copolar) far-field intensity, and E_c as the cross-polarization far-field intensity. E_θ and E_ϕ are far-field components in a spherical coordinate system, given by [3, 48]

$$\left| \begin{aligned} E_\theta &= -(N_x \cos\phi + N_y \sin\phi)(1 + h\cos\theta) \\ E_\phi &= (N_x \sin\phi - N_y \cos\phi)(h + \cos\theta) \end{aligned} \right. \tag{6.91}$$

where

$$\begin{vmatrix} N_x = N'_x - N''_x \\ N_y = N'_y - N''_y \end{vmatrix} \tag{6.92}$$

and

$$N'_x = \int_0^{2\pi}\int_{a_0}^{\infty} E_x e^{j\beta_0 r \sin\theta\cos(\phi-\psi)} r\mathrm{d}r\mathrm{d}\psi \tag{6.93}$$

$$N''_x = \int_0^{2\pi}\int_{a_0}^{a_1} E_x e^{j\beta_0 r \sin\theta\cos(\phi-\psi)} r\mathrm{d}r\mathrm{d}\psi \tag{6.94}$$

$$N'_y = \int_0^{2\pi}\int_{a_0}^{\infty} E_y e^{j\beta_0 r \sin\theta\cos(\phi-\psi)} r\mathrm{d}r\mathrm{d}\psi \tag{6.95}$$

$$N''_y = \int_0^{2\pi}\int_{a_0}^{a_1} E_y e^{j\beta_0 r \sin\theta\cos(\phi-\psi)} r\mathrm{d}r\mathrm{d}\psi \tag{6.96}$$

In the integral expressions (6.93) through (6.96), E_x and E_y must be substituted from (6.83) and (6.84). The main- or co-polarization radiation pattern $F_m(\theta)$ and cross-polarization pattern $F_c(\theta)$ may be computed as follows:

$$F_m(\theta) = 20\ \lg\left|\frac{E_m(\theta)}{E_m(\theta = 0°)}\right|, \quad (\phi = \text{const.}) \tag{6.97}$$

$$F_c(\theta) = 20\ \lg\left|\frac{E_c(\theta)}{E_c(\theta = 0°)}\right|, \quad (\phi = \text{const.}) \tag{6.98}$$

Figure 6.50 shows plots of the co-polarization ($\phi = 0°$ in E-plane) and cross-polarization ($\phi = 45°$) radiation patterns of a backfire antenna with a waveguide-fed dielectric rod [46, 47]. The antenna consists of a simple large disk reflector 180 mm in diameter, a small reflector 26.5 mm in diameter, and a waveguide-fed dielectric rod with $\varepsilon_r = 2.6$ and diameter $2a_0 = 10$ mm. The distance between the two reflectors is 140 mm. For $f = 10$ GHz, the SW delay coefficient h has a value of 1.02.

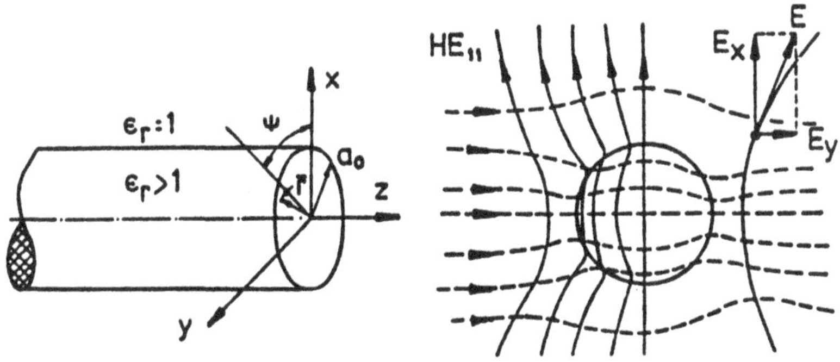

(a) (b)

Figure 6.49 (a) Circular dielectric rod; (b) HE_{11} field intensity structure

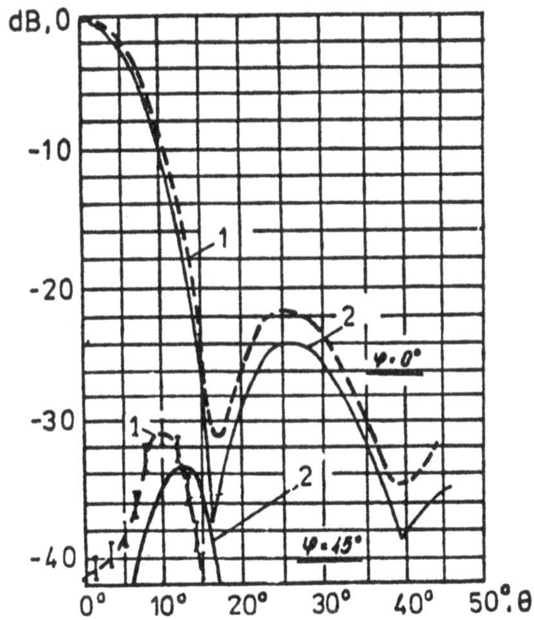

Figure 6.50 Experimental (1) and theoretical (2) radiation patterns on the copolar plane (θ = 0°) and the cross-polar plane (θ = 45°) (from [46])

6.5 BACKFIRE ANTENNAS WITH NONUNIFORM (MODULATED) SURFACE WAVE STRUCTURE

6.5.1 Modulated Antenna Surface Wave Structure: Introduction and Synthesis Method

The infinitely-long slow-wave structure supports a unidirectionally traveling wave along its axis. If such a waveguide is made open-ended at some cross section, a deformation of the wave structure occurs. In this case, the forward endfire radiation is accompanied back into the structure by a partially-reflected surface wave. The uniform endfire antenna has a simple design but, it lacks efficiency because there is a discrepancy between the requirements for maximum antenna directivity and SW feed efficiency and reflection of the SW from the open end leads to poorer matching and radiation pattern quality.

These unwanted effects are overcome to a great extent by using nonuniform (modulated) structures in the SW antennas. The SW structure is nonuniform if its transverse dimensions vary along the axis. If this variation is smooth and follows a pattern, the nonuniform SW antenna is also known as a modulated antenna. A classic example of a periodically-modulated SW endfire antenna is Simon's "cigar" antenna, which essentially is a modulated disk-on-rod antenna. It has 1 to 2.0 dB greater gain than the uniform endfire antenna of the same length.

In the electromagnetic analysis of the modulated SW antenna, the radiation pattern can be found by a given modulation [48–49] law. However, such a procedure does not specify the shape of the SW structure; this is the object of the antenna synthesis theory. The synthesis procedure of the modulated SW antenna requires that two electromagnetic problems be solved consecutively and independently: The polarization structure and amplitude-phase distribution must be obtained by a known radiation pattern in the far region. This is a classic problem of the antenna synthesis theory; after getting the near field, the relevant electrical and geometrical parameters of the antenna impedance surface must be found. This problem is solved by the synthesis method proposed by Bolljahn [49a] in 1961 and developed later by Tereshin *et cetera* [50–52].

One of the basic requirements of the synthesis method implies a pure reactive impedance of the antenna surface; therefore, there is no electromagnetic power flow through it. This means that the method is not applicable for a synthesis of an arbitrary SW structure. In practice, only two types of continuous SW structures have reactive surface impedances: dielectric metal and corrugated metal structures. The synthesized modulated SW antenna has this radiation mechanism: the primary electromagnetic field excited from the feed source is transformed by the reactive impedance surface into a secondary field, which forms the desired radiation pattern. The SW propagation on the synthesized impedance structure is characterized by changing the amplitude-phase modulation along the antenna length.

For simplicity, we will suppose that in the volume containing the synthesized structure, there is no primary source, and only the impedance surface is considered, which satisfies the electromagnetic boundary conditions. The near field around the antenna SW structure can be given as a spectrum of orthogonal slow waves. We assume that each νth wave of the spectrum is propagating in direction z_ν of a rectangular coordinate system $0\ x_\nu y_\nu z_\nu$ (Figure 6.51), and the field intensity vectors are written as an infinite series:

$$\bar{E} = \sum_{\nu=0}^{\infty} \bar{A}_\nu e - p_\nu x_\nu\, e^{-j\beta_\nu z_\nu} \tag{6.99}$$

$$\bar{H} = \sum_{\nu=0}^{\infty} \bar{B}_\nu\, e - p_\nu x_\nu\, e^{-j\beta_\nu z_\nu} \tag{6.100}$$

where

$\bar{A}_\nu$ and $\bar{B}_\nu$ are complex amplitude vectors,

β_ν is the longitudinal phase constant of the surface wave,

$p_\nu = \sqrt{\beta_\nu^2 - \beta^2}$ is the transverse wavenumber of the surface wave,

and

$$\begin{aligned} z_\nu &= z \cos \theta_\nu - x \sin \theta_\nu \\ x_\nu &= z \sin \theta_\nu + x \cos 0_\nu \end{aligned} \tag{6.101}$$

Each spectral component $\bar{E}_\nu$, $\bar{H}_\nu$ is traveling under an angle θ_ν toward the axis z of the common coordinate system $0\ x\ y\ z$ with a phase velocity $v_\nu = \omega/\beta_\nu$. The waves for which $\theta_\nu > 0$ and $v_\nu \rightarrow c$ are practically radiated completely at the end of the SW structure. By varying θ_ν, it is possible to vary the equivalent amplitude distribution. These waves are called *weakly delayed* or *secondary waves*. However, waves traveling under a negative angle ($\theta_\nu < 0$) and having a large velocity delay are strongly coupled with the impedance structure. These are called *primary waves*. The slow-wave spectrum of the SW structure consists of a fundamental wave mode and higher wave modes. Because of their much smaller amplitudes, the higher wave modes are usually neglected in the synthesis procedure. In accordance with the radiation mechanism of the modulated SW antenna just described, we assume that, at each cross section of the generalized coordinate system $0\ u\ v\ w$, where w = constant, the electromagnetic field is represented as the sum of the primary and the secondary waves:

$$\bar{E} = \bar{E}_1 + \bar{E}_2$$
$$\bar{H} = \bar{H}_1 + \bar{H}_2 \tag{6.102}$$

where $\bar{E}_{1,2}$ and $\bar{H}_{1,2}$ are primary (and secondary) wave vectors, with scalar components $E_{1,2u}$, $E_{1,2v}$, $E_{1,2w}$; and $H_{1,2u}$, $H_{1,2v}$, $H_{1,2w}$.

The power flow of the primary SW along the slow-wave structure 1 is

$$P_{1w} = Re\left[\int_u \int_v \dot{\pi}_{1w} u_0 v_0 dudv\right] \tag{6.103}$$

where

$$\dot{\pi}_{1w} = \frac{1}{2}(\dot{E}_{1u}\dot{H}^*_{1v} - \dot{E}_{1v}\dot{H}^*_{1u}) \tag{6.104}$$

is a longitudinal component of the Poynting vector, and u_0 and v_0 are the metric coefficients of the coordinates u and v. Similarly, the secondary wave power flow is written as:

$$P_{2w} = Re\left[\int_u \int_v \dot{\pi}_{2w} u_0 v_0 dudv\right] \tag{6.105}$$

where:

$$\dot{\pi}_{2w} = \frac{1}{2}(\dot{E}_{2u}\dot{H}^*_{2v} - \dot{E}_{2v}\dot{H}^*_{2u}) \tag{6.106}$$

The synthesis problems of modulated impedance antennas are solved with some finite accuracy because the primary wave amplitude at the antenna radiation end has a small finite value.

The accuracy of solution of such synthesis problems is characterized by the following coefficients [52]:

1. The efficiency coefficient of the primary wave is

$$q_1 = \frac{P_{1w}(1)}{P_{1w}(0)} = \frac{Re\left[\int\int \pi_{1w} u_0 v_0 dudv\right]_{w=1}}{Re\left[\int_u\int_v \pi_{1w} u_0 v_0 dv\right]_{w=0}} \tag{6.107}$$

where $P_{1w}(0)$ and $P_{1w}(1)$ are the average powers of the primary wave at the beginning and at the end of the antenna structure, respectively. The coefficient q_1 characterizes the relative power of the primary wave at the antenna structure end. Its value must be small, and is usually set within the limits 0.1 to 0.01.

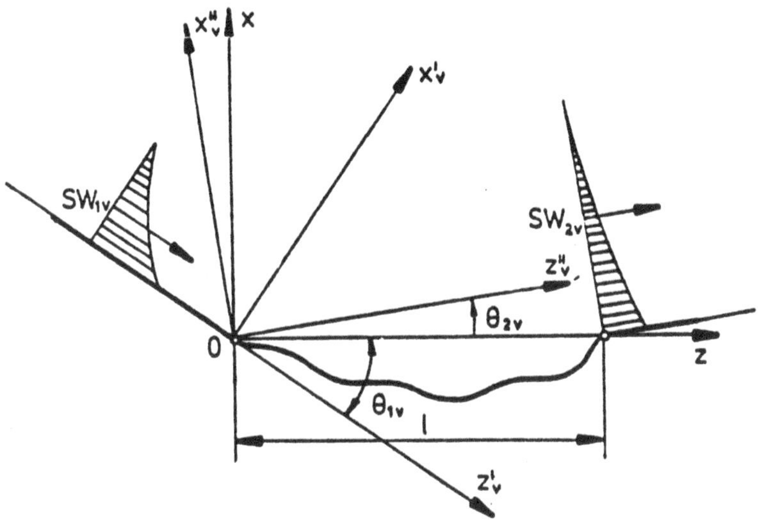

Figure 6.51 Primary and secondary wave tracing of modulated SW structure

2. The rise coefficient of the secondary wave q_2 is

$$q_2 = \frac{P_{2w}(0)}{P_{2w}(1)} = \frac{Re\left[\iint \pi_{2w} u_0 v_0 \, du \, dv\right]_{w=0}}{Re\left[\int_u \int_v \pi_{2w} u_0 v_0 \, du \, dv\right]_{w=1}} \tag{6.108}$$

This coefficient characterizes the initial power level of the secondary wave and determines the sidelobe levels of the radiation pattern. Experience shows, that for $q_2 = 0.5$ to 0.7, a maximum directivity is obtained.

The relation between the amplitude coefficients of the primary and secondary waves and the transverse antenna structure dimensions at the cross sections $w = 0$ and $w = 1$, is found by the expression of the energy balance:

$$Re\int_u\int_v \dot{\pi}_w u_0 v_0 \mathrm{d}u\mathrm{d}v_{1w=0} = Re\int_u\int_v \dot{\pi}_w u_0 v_0 \mathrm{d}u\mathrm{d}v_{1w=1} \tag{6.109}$$

where

$$\dot{\pi}_w = \frac{1}{2}(\dot{E}_u\dot{H}_v^* - \dot{E}_v\dot{H}_u^*) \tag{6.110}$$

In solving synthesis problems for the modulated impedance antenna, the slow-wave structure profile and the surface impedance change along the axis are determined first as we keep in mind that the total surrounding field is known in advance. The surface impedance is found by

$$Re\,\bar{\pi}\cdot\bar{\mathrm{n}} = 0 \tag{6.111}$$

The components of $Re\bar{\pi}$ in the coordinate system 0 u v w are given as:

$$Re\,\dot{\pi}_u = \frac{1}{2}(\dot{E}_v\dot{H}_w^* - \dot{E}_w\dot{H}_v^*) \tag{6.112}$$

$$Re\,\dot{\pi}_v = \frac{1}{2}(\dot{E}_w\dot{H}_u^* - \dot{E}_u\dot{H}_w^*) \tag{6.113}$$

$$Re\,\dot{\pi}_w = \frac{1}{2}(\dot{E}_u\dot{H}_v^* - \dot{E}_v\dot{H}_u^*) \tag{6.114}$$

The normal vector on the surface slow-wave structure is $\bar{n}$. The surface equation can be written

$$\phi(u, v, w) = 0 \tag{6.115}$$

and then

$$\bar{n} = \text{grad}\,\phi = \frac{1}{u_0}\frac{\partial\phi}{\partial u_0}\bar{u}^0 + \frac{1}{v_0}\frac{\partial\phi}{\partial v}\bar{v}^0 + \frac{1}{w_0}\frac{\partial\phi}{\partial w}\bar{w}^0 \tag{6.116}$$

where $\bar{u}^0$, $\bar{v}^0$, and $\bar{w}^0$ are unit vectors. After putting (6.112) through (6.114) and (6.115) into (6.111) we obtain:

$$\frac{1}{u_0} Re\, \dot{\pi}_u \cdot \frac{\partial \phi}{\partial_u} + \frac{1}{v_0} Re\, \dot{\pi}_v \cdot \frac{\partial \phi}{\partial v} + \frac{1}{w_0} Re\, \dot{\pi}_w \cdot \frac{\partial \phi}{\partial w} = 0 \tag{6.117}$$

The solution of the differential equation (6.117) is the surface function $\phi(u, v, w)$. The surface impedance of the corrugated SW structure with an asymmetrical (dipole type) wave HE_{11} (Figure 6.52) is given by

$$\begin{aligned} Z_{SE} &= \left[\frac{E_w}{H_v}\right]_{w=1} \\ &= j\frac{T-\Delta}{T} Z_0 \frac{M_1(\beta a, \beta b)}{J_0(\beta a)Y_1(\beta b) - Y_0(\beta a)J_1(\beta b) - M_1(\beta a, \beta b)/\beta a} \end{aligned} \tag{6.118}$$

where

$$M_1(\beta a, \beta b) = J_1(\beta a)\, Y_1(\beta b) - J_1(\beta b)\, Y_1(\beta a)$$

$$Z_{SH} = -\left[\frac{E_v}{H_w}\right]_{w=1} \approx 0 \tag{6.119}$$

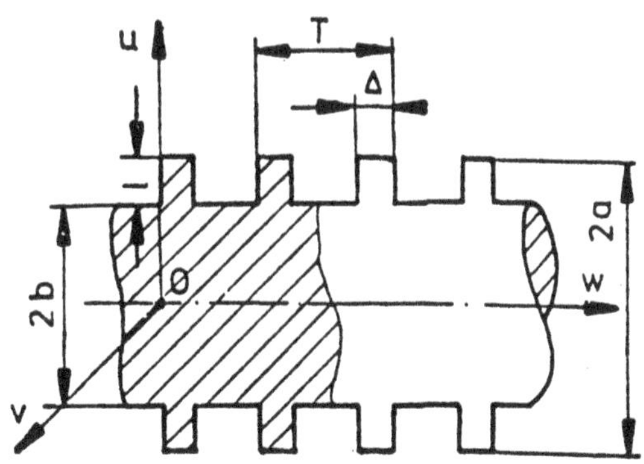

Figure 6.52 Corrugated SW structure

6.5.2 Backfire Antennas with Two-Dimensional Modulated SW Structure

For simplicity we assume that the slow wave travels along the antenna structure only twice: forward and backward [52–53]. Hence, the antenna length is utilized twice for a transformation of the primary slow wave into a secondary slow wave. The electromagnetic process is illustrated in Figure 6.53.

The primary plane slow wave SW_{1i} originates from the feed source. Traveling as an incident wave toward the large reflector ($z = 0$), the primary wave gradually transfers its energy to the weakly delayed secondary wave SW_{1i}. After reflection from the large reflector, these two waves travel backward, and their energy exchange continues (i.e., SW_{1r} transfers its energy to SW_{2r}). The incident primary plane wave in a rectangular coordinate system $0\ x'\ y'\ z'$ has components given by [23]:

$$E_{1ix'} = C'_{1i}(\beta_1/\omega\varepsilon)\, e^{-p_1 x'} e^{-j\beta_1 z'} \tag{6.120}$$

$$E_{1iz'} = jC'_{1i}(p_1/\omega\varepsilon)\, e^{-p_1 x'} e^{-j\beta_1 z'} \tag{6.121}$$

$$H_{1iy'} = C'_{1i} e^{-p_1 x'} e^{-j\beta_1 z'} \tag{6.122}$$

and those of the reflected primary wave in a coordinate system $0\ x'''\ y'''\ z'''$ are

$$E_{1rx'''} = -C'_{1r}(\beta_1/\omega\varepsilon) e^{-p_1 x'''} e^{j\beta_1 z'''} \tag{6.123}$$

$$E_{1rz'''} = jC'_{1r}(p_1/\omega\varepsilon)\, e^{-p_1 x'''} e^{j\beta_1 z'''} \tag{6.124}$$

$$H_{1ry'''} = C'_{1r} e^{-p_1 x'''} e^{j\beta_1 z'''} \tag{6.125}$$

By analogy, the incident and reflected secondary wave components are

$$E_{2ix''} = \dot{C}'_{2i}(\beta_2/\omega\varepsilon)\, e^{-p_2 x''} e^{-j\beta_2 z''} \tag{6.126}$$

$$E_{2iz''} = j\dot{C}'_{2i}(p_2/\omega\varepsilon)\, e^{-p_2 x''} e^{-j\beta_2 z''} \tag{6.127}$$

$$H_{2iy''} = \dot{C}_{2i} e^{-p_2 x''} e^{-j\beta_2 z''} \tag{6.128}$$

$$E_{2rx^{\mathrm{IV}}} = -C'_{2r}(\beta_2/\omega\varepsilon)\, e^{-p_2 x^{\mathrm{IV}}} e^{j\beta_2 z^{\mathrm{IV}}} \tag{6.129}$$

$$E_{2rz^{\mathrm{IV}}} = j\dot{C}'_{2r}(p_2/\omega\varepsilon)\, e^{-p_2 x^{\mathrm{IV}}} e^{j\beta_2 z^{\mathrm{IV}}} \tag{6.130}$$

$$H_{2ry^{\mathrm{IV}}} = \dot{C}'_{2r} e^{-p_2 x^{\mathrm{IV}}} e^{j\beta_2 z^{\mathrm{IV}}} \tag{6.131}$$

We assume that the amplitude constants C_{1i}, C_{1r}, C_{2i}, and C_{2r} are real. After translating the above electromagnetic field components into the common coordinate system $0\ x\ y\ z$, it is easy to obtain the total primary wave incident and reflected components:

$$E_{1ix} = -jC'_{1i}\frac{\dot{m}_1}{\omega\varepsilon}e^{-\dot{n}_1 x}e^{-\dot{m}_1 z} \tag{6.132}$$

$$E_{1iz} = jC'_{1i}\frac{\dot{n}_1}{\omega\varepsilon}e^{-\dot{n}_1 x}e^{-\dot{m}_1 z} \tag{6.133}$$

$$H_{1iy} = C'_{1i}e^{-\dot{n}_1 x}e^{-\dot{m}_1 z} \tag{6.134}$$

$$E_{1rx} = jC'_{1r}\frac{\dot{m}_1}{\omega\varepsilon}e^{-\dot{n}_1 x}e^{\dot{m}_1 z} \tag{6.135}$$

$$E_{1rz} = jC'_{1r}\frac{\dot{n}_1}{\omega\varepsilon}e^{-\dot{n}_1 x}e^{\dot{m}_1 z} \tag{6.136}$$

$$H_{1ry} = C'_{1r}e^{-\dot{n}_1 x}e^{\dot{m}_1 z} \tag{6.137}$$

and the total secondary wave incident and reflected components:

$$E_{2ix} = jC'_{2i}\frac{\dot{m}_2}{\omega\varepsilon}e^{-\dot{n}_2 x}e^{\dot{m}_2 z} \tag{6.138}$$

$$E_{2iz} = jC'_{2i}\frac{\dot{n}_2}{\omega\varepsilon}e^{-\dot{n}_2 x}e^{\dot{m}_2 z} \tag{6.139}$$

$$H_{2iy} = C'_{2i}e^{-\dot{n}_2 x}e^{\dot{m}_2 z} \tag{6.140}$$

$$E_{2rx} = -jC'_{2r}\frac{\dot{m}_2}{\omega\varepsilon}e^{-\dot{n}_2 x}e^{-\dot{m}_2 z} \tag{6.141}$$

$$E_{2rz} = jC'_{2r}\frac{\dot{n}_2}{\omega\varepsilon}e^{-\dot{n}_2 x}e^{-\dot{m}_2 z} \tag{6.142}$$

$$H_{2ry} = C'_{2r}e^{-\dot{n}_2 x}e^{-\dot{m}_2 z} \tag{6.143}$$

where

$$\dot{n}_1 = p_1 \cos\theta_1 - j\beta_1 \sin\theta_1 \tag{6.144}$$

$$\dot{m}_1 = p_1 \sin\theta_1 + j\beta_1 \cos\theta_1 \tag{6.145}$$

$$\dot{n}_2 = p_2 \cos \theta_2 + j\beta_2 \sin \theta_2 \tag{6.146}$$

$$\dot{m}_2 = p_2 \sin \theta_2 - j\beta_2 \cos \theta_2 \tag{6.147}$$

The total efficiency coefficient of the primary wave is given in this case as

$$q_1 = q_{1i} q_{1r} = \frac{P_{1i}(0)}{P_{1i}(-1)} \cdot \frac{P_{1r}(-1)}{P_{1r}(0)} = e^{-4p_1 1 \sin \theta_1} \tag{6.148}$$

and the total rise coefficient of the secondary wave is

$$q_2 = q_{2i} q_{2r} = \frac{P_{2i}(-1)}{P_{2i}(0)} \cdot \frac{P_{2r}(0)}{P_{2r}(-1)} = e^{-4p_2 1 \sin \theta_2} \tag{6.149}$$

The energy balance equation (6.109) for a backfire mode can be written as

$$Re \int_0^\infty \pi_z \, dx\Big|_{z=-1} = 0 \tag{6.150}$$

and, after applying the boundary electromagnetic conditions on the large reflector surface ($z = 0$) we find

$$C'_{1i} = C'_{1r} = C'_1 = C_1/2; \quad C'_{2i} = C'_{2r} = C'_2 = C_2/2 \tag{6.151}$$

Using the expressions (6.132) to (6.143), the total standing wave around the slow-wave structure with components E_x, E_z, and H_y may be obtained. To find the ratio between the amplitude coefficients of the secondary and primary surface waves, the total field components must be placed in the energy balance equation (6.150) [23, 53]. The surface profile of the two-dimensional modulated slow-wave structure is described by

$$\frac{dx}{Re\, \pi_x} = \frac{dz}{Re\, \pi_z} \tag{6.152}$$

where $\pi_x = -\frac{1}{2} E_z H_y^*$ and $\pi_z = \frac{1}{2} E_x H_y^*$, or

$$x' = \frac{dx}{dz} = \frac{Re\, \pi_x}{Re\, \pi_z} \tag{6.153}$$

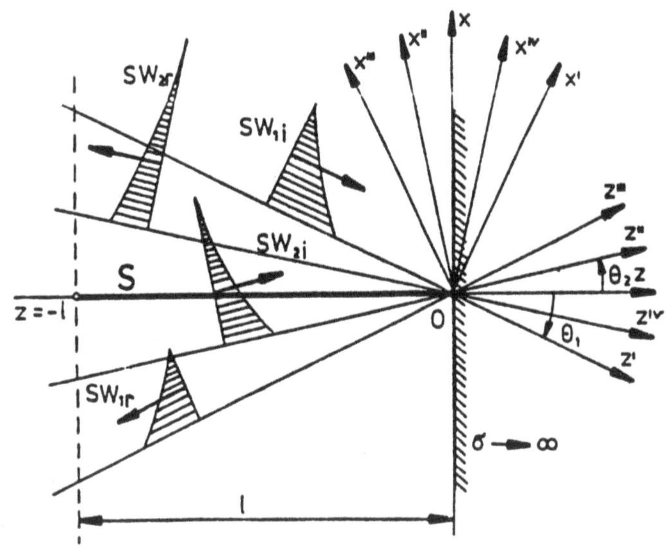

Figure 6.53 Incident and reflected wave tracing arrangement of the LBFA with two-dimensional modulated SW structure (from [23])

Finally we obtain:

$$x'(z) = \frac{\dfrac{\partial A_r^h}{\partial x} A_i^h - \dfrac{\partial A_i^h}{\partial x} A_r^h}{\dfrac{\partial A_r^h}{\partial z} A_i^h - \dfrac{\partial A_i^h}{\partial z} A_r^h} \tag{6.154}$$

where $A_r^h = Re\ H_y$ and $A_i^h = Im\ H_y$. To design the slow-wave structure, we must know the longitudinal surface impedance distribution as well as the profile function. The normalized surface impedance can be determined by the following expression:

$$\frac{X(z)}{Z_0} = \frac{1}{\beta} \cdot \frac{\left(\dfrac{\partial A_r^h}{\partial z} A_r^h + \dfrac{\partial A_i^h}{\partial z} A_i^h\right) x'(z) - \left(\dfrac{\partial A_r^h}{\partial x} A_r^h + \dfrac{\partial A_i^h}{\partial x} A_i^h\right)}{[(A_r^h)^2 + (A_i^h)^2]\sqrt{[x'(z)]^2 + 1}} \tag{6.155}$$

In Figure 6.54, the calculated profile and surface impedance functions are plotted for different values of the primary wave delay coefficient ($h_1 = \beta_1/\beta = 1.02$, 1.06, and 1.10). The modulated SW structure has a length $l = 2.125\ \lambda$. As a first

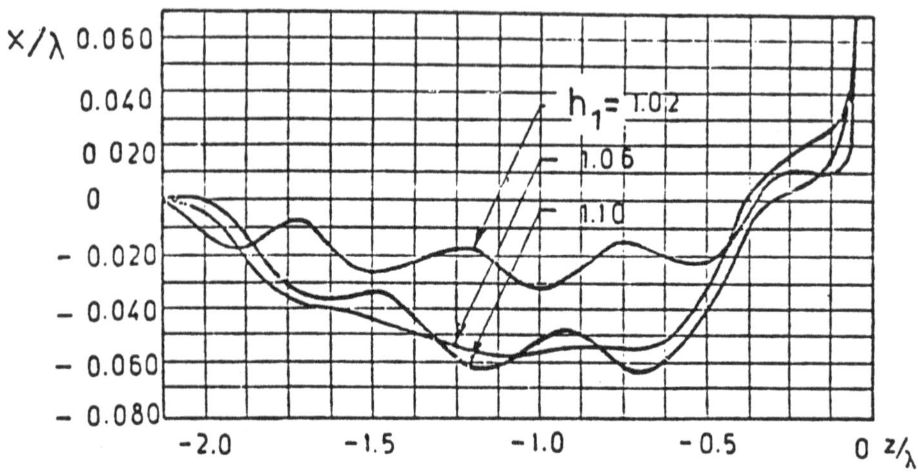

a) Profile Function Curves

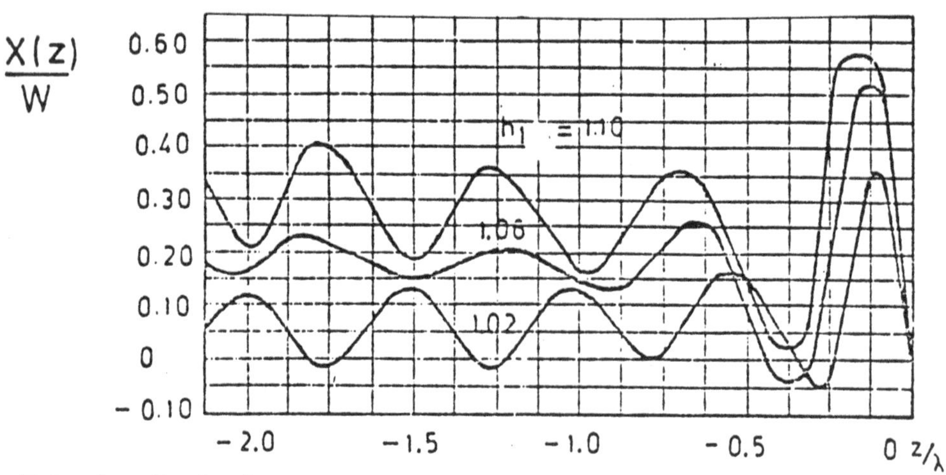

b) Impedance Function Curves

Figure 6.54 Influence of the primary wave delay coefficient on profile and impedance functions (θ_1 = 20°, θ_2 = 3°, $h_2 = \beta_2/\beta$ = 1.01) (from [23])

approximation, the antenna far field can be computed using the scalar radiation integral. The one-dimensional aperture field distribution (Figure 6.55) [23, 39] is assumed as

(a) Profile function curves

(b) Impedance function curves

$$\dot{E}_{xa} = -j\frac{1}{\omega\varepsilon}[\dot{m}_1 C_1 \exp[\mp\dot{n}_1(x_a - x_1)]\, sh[\dot{m}_1(z_a + l_A)]$$
$$+ \dot{m}_2 C_2 \exp[\mp\dot{n}_2(x_a - x_1)] sh[\dot{m}_2(z_a + l_A)]] \qquad (6.156)$$

and $E_{xa} = 0$ for

$$-\frac{D_1}{2} \leq x_a \leq +\frac{D_1}{2}$$

When the basic dimensions are known, the antenna's electrical characteristics depend mainly on the modulated structure parameters θ_1, θ_2, β_1. Ater computing the antenna far field [39], a multiparametric optimization is used to obtain maximum directivity [39] (i.e., $D(\theta_{1opt}, \theta_{2opt}, \beta_{1opt}, \beta_{2opt}) = D_{max}$. For example, for $l_A/\lambda = 4\ 0$, $l/\lambda = 3$ 125, $d_1/\lambda = 0.7$, and $d_2/\lambda = 5.5$, a maximum antenna directivity $D_{max} = 25.8$ dB is computed when $\theta_{1opt} = 13°$, $\theta_{2opt} = 0.85°$, $\beta_1/\beta = 1.045$, and $\beta_2/\beta = 1.04$.

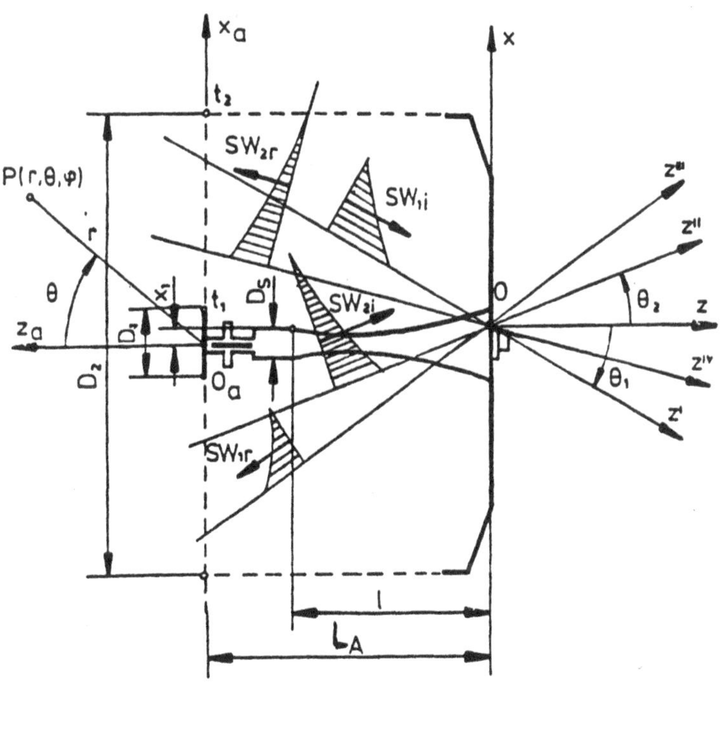

Figure 6.55 Radiation of a backfire antenna with a two-dimensional dipole-fed modulated structure (from [23])

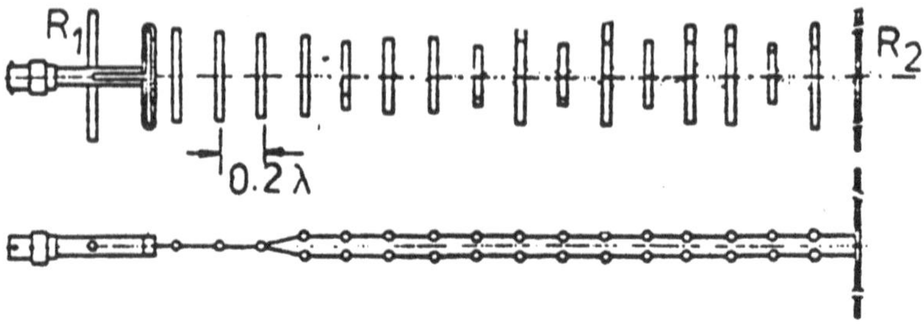

Figure 6.56 LBFA with modulated dipole structure (from [53])

These calculations show that for an arbitrary choice of the primary and secondary wave parameters, the directivity is much less than D_{max}. For instance, if $\theta_1 = 26°$, $\theta_2 = 3°$, $\beta_1/\beta = 1.08$, and $\beta_2/\beta = 1.01$, then $D = 19.7$ dB (i.e., about four times smaller than the maximum directivity obtained by computer optimization).

Keeping in mind the general character of the SW description in an arbitrary slow-wave guide, Tereshin and Kuznetzov [53] proposed, as a first practical example, a modulated dipole (Yagi) structure to be used in the LBFA. The change of the phase velocity along the antenna axis, corresponding to the surface impedance law, is realized in the Yagi structure by a variation of the dipole length. This antenna design is shown in Figure 6.56. The modulated structure consists of two rows of dipoles and is fed by a folded half-wave dipole. The antenna's E-plane and H-plane normalized power radiation patterns, plotted in a polar coordinate system, are shown in Figure 6.57. For an antenna length of about 3λ, the directivity is 17.5 dBi. In comparison, a Yagi antenna with the same directivity is 4.5 times longer.

6.5.3 Backfire Antennas with Modulated Corrugated Surface Wave Structure

The cylindrical corrugated guide is a three-dimensional surface wave structure. Using the synthesis method described in Section 6.5.1, we obtain a three-dimensional variation in SW structure size. Such a structure is very difficult to make in practice, so an approximate synthesis approach is used [46]:

1. The cylindrical corrugated structure is considered only in one longitudinal cross section (i.e., as a two-dimensional structure).
2. An independence of the electromagnetic field from the one transverse coordinate is assumed (y in synthesis procedure, ψ in realization).

These assumptions actually reduce the three-dimensional synthesis procedure followed in the previous section to one of two dimensions. After rotation of the computed profile curve about the z-axis, a modulated cylindrical structure is obtained.

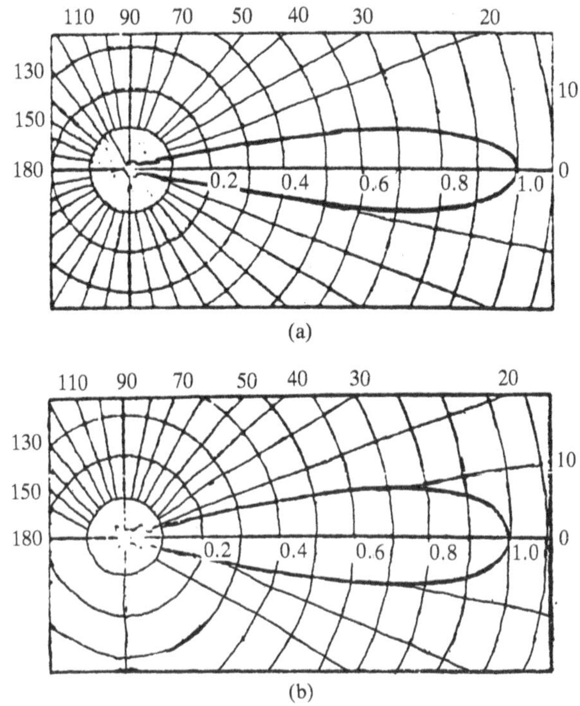

Figure 6.57 *E*-plane (a) and *H*-plane (b) radiation patterns of an LBFA with modulated dipole slow-wave structure (from [53])

Using this approximate synthesis procedure, many backfire antenna designs with corrugated nonuniform SW structures were made and tested [11, 22, 23, 39]. A comparison between the radiation parameters of long backfire antennas with different corrugated rods, modulated and uniform, is given in Table 6.5 [39], for $l_A/\lambda = 4.0$ and $d_2/\lambda = 5.5$. Figure 6.58 shows a comparison of the frequency dependence plots of the directivity for three different antenna lengths. Backfire antennas with modulated corrugated rods (Figure 6.59(a)) and uniform corrugated (disk-on-rod) structures (Figure 6.59(b)), having the same gains, respectively, are drawn in the same scale and compared for three different lengths. It is obvious that the backfire antenna with the properly modulated SW structure is smaller in size, volume, and weight compared to the antenna with a uniform structure.

An undersized BF radiometer antenna with modulated corrugated structure was designed [39] for an optimum frequency $f_0 = 2.0$ GHz (Figure 6.60). It has a gain of 25.4 dB and gain frequency bandwidth of about 10%. In this bandwidth, VSWR is less than 1.5. From Table 6.6 we conclude that for almost equal gains, the parabolic antennas have bigger apertures and volumes compared to the backfire radi-

Table 6.5

Antenna Parameters	*BFA with Modulated Structure*	*BFA with Uniform Structure*
Directivity (dB)	25.6	23.8
Beamwidth (degrees)		
a) in *E*-plane	7.0	9.5
b) in *H*-plane	7.0	8.0
Maximum sidelobe level (dB)		
a) in *E*-plane	−24.0	−21.0
b) in *H*-plane	−22.0	−20.0
Backlobe level (dB)	−30.0	−26.0

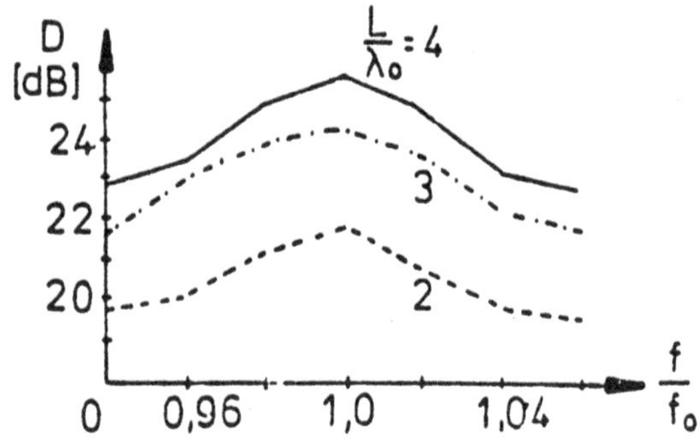

(a)

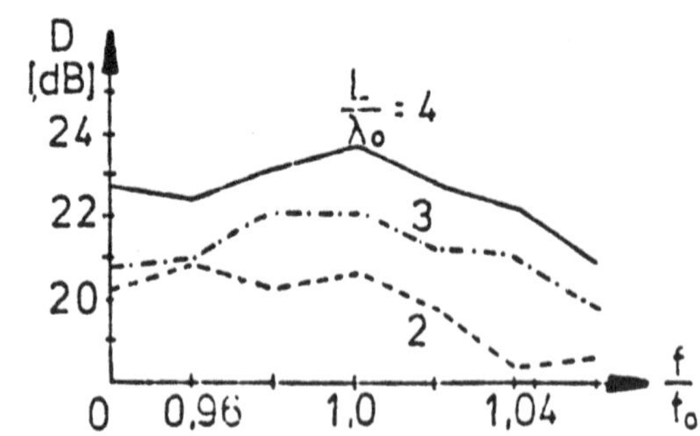

(b)

Figure 6.58 Directivity *versus* relative frequency: (a) for a BFA with modulated corrugated rod; and (b) for a BFA with uniform corrugated rod (from [39])

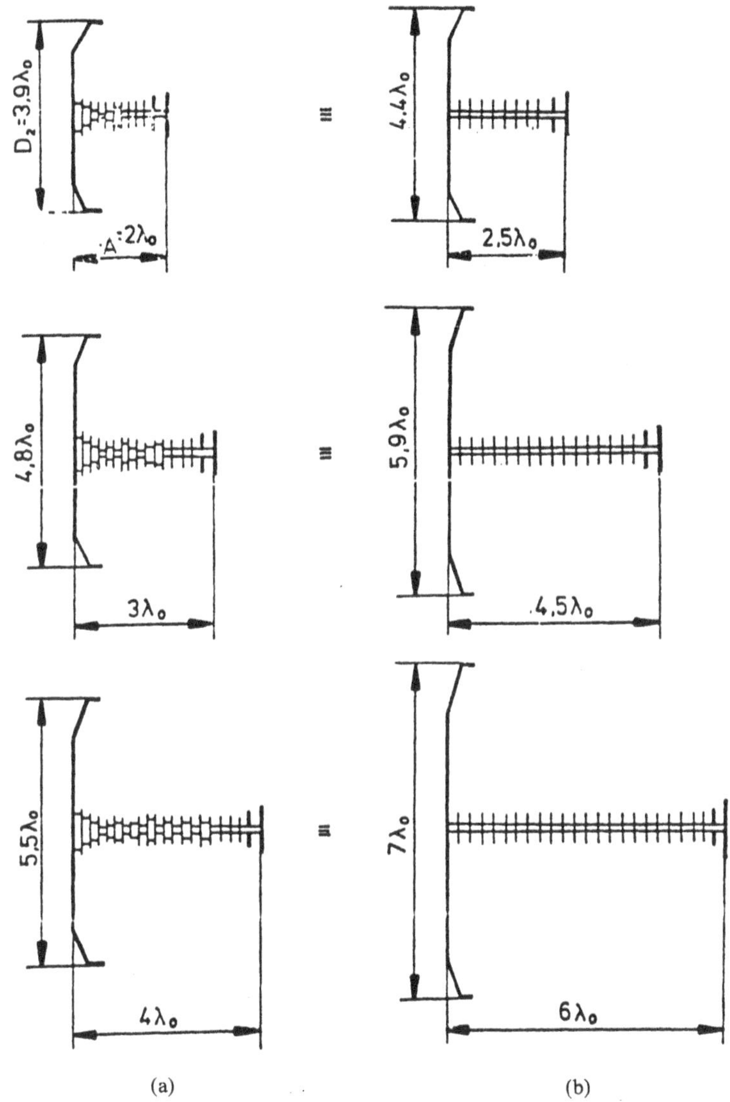

Figure 6.59 Comparison between a BFA with modulated (a) and uniform (b) corrugated structures, drawn (in the same scale), for three different lengths (from [23])

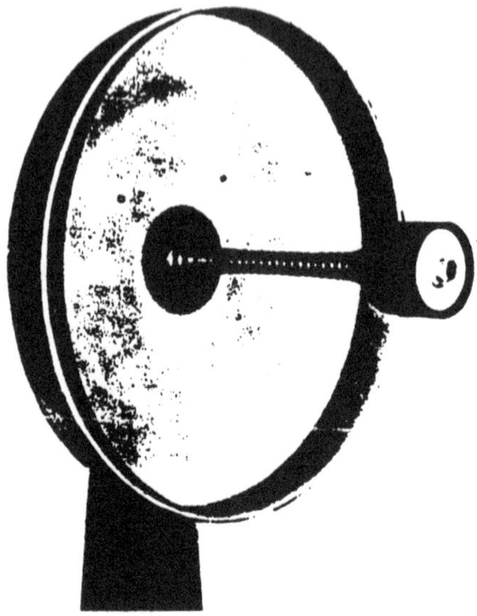

Figure 6.60 Backfire radiometer antenna photo (from [39])

Table 6.6 (From [39])

Antennas	G (dB)	D_A (m)	L_A (m)	V_A (m^3)
Radiometer backfire antenna	25.4	0.80	0.56	0.28
Parabolic antenna C/S Antennas Ltd.	25.0	1.20	0.46	0.52
Parabolic antenna 60004-21 Andrew Ant.	25.8	1.20	0.52	0.59
Optimum conical horn antenna	25.0	0.78	2.30	1.10

ometer antenna, and the latter is about four times shorter in length and smaller in volume compared to the optimum conical horn.

REFERENCES

1. Ehrenspeck, H.W., "A Family of Novel Antennas: the Backfire Antennas," *Proc. OAR Research Application Conference,* March 1967, pp. 80–104.
2. Ehrenspeck, H.W., "'Backfire' Antennen," *NTZ,* 22, No. 5, 1969, S. 286–289.

3. Ehrenspeck, H.W., *Die "Backfire"—eine Antenne mit Fassung,* NTG-URSI Tagung: Antennen and Electromagnetishe Felder, Darmstadt, 1967.
4. Ehrenspeck, H.W., "Die Backfire-antenne, Ein Neuer Langsstrahlertyp mit Hoher Richtwirkung," *Nachrichtentechn.* Fahber, 23, 1961, pp. 93–99.
5. Ehrenspeck, H.W., "High-Gain UHF Backfire Antenna for Communications, Telemetry and Radio Astronomy," *AFCRL Report 67-0568,* Bedford, MA, October 1967.
6. Ehrenspeck, H.W., "Reflection Antenna Employing Multiple Director Elements and Multiple Reflection of Energy to Effect Increased Gain," US Patent No. 3 122 745, February 1964 (filed May 1959).
7. Ehrenspeck, H.W., "The Backfire Antenna, a New Type of Directional Line Source," *AFCRL Report 61-0722,* Bedford, MA, August 1961.
8. Ehrenspeck, H.W., "The Backfire Antenna, a New Type of Directional Line Source," *Proc. IRE,* 48, No. 1, 1960, pp. 109–110.
9. Ehrenspeck, H.W., "The Backfire Antenna: New Results," *Proc. IRE,* 53, June 1965, pp. 639–641.
10. Ehenspeck, H.W., "The Backfire—a High Gain Cavity Antenna," *Int. Series of Monographs in Electromagnetic Waves, Vol II Electromagnetic Wave Theory,* Proc. of Symp., Delft, The Netherlands, Part 2, Pergamon Press, 1967, pp. 739–49.
11. Hristov, H.D., and D. Taylor, "Some Experiences with Backfire Antennas," *Discussion Meeting on Novel Types of Antennas, IEE,* Section "Antennas and Propagation," Savoy Place, London, January 1972.
12. Hristov, H.D., "Study of Backfire Antennas with Surface Wave Structures," Ph.D. Thesis, Sofia, 1973. Available in Central Inst. for Scientific and Technical Information (Tz.I.N.T.I.) (in Bulgarian).
13. Hristov, H.D., "Some Experiences with Resonator-type Antennas," *Conf. on Antennas and Communications,* IEEE Montech 86, Montreal, 1986.
14. Zucker, F.J., "The Backfire Antenna: Qualitative Approach to Its Design?" *Proc. IEEE,* 53, July 1965, pp. 746–747.
15. Zucker, F.J., and J.A. Strom, "Experimental Resolution of Surface Wave Antenna Radiation into Feed and Terminal Patterns," *IEEE Trans. Antennas and Propagation,* 18, No. 3, 1970, pp. 420–422.
16. Hristov, H.D., "Backfire Antenna," Bulgarian Patent No. 18 149, 1972.
17. Hristov, H.D., "Backfire Antennas," *Int. Symp. on Electromagnetic Compatibility of Earth and Space Communications,* Vol. I, 23–30 November, Varna, Bulgaria, pp. 83–89 (in Russian).
18. Kumar, A., "Backfire Antennas for Satellite Applications," presented at *1977 Conf. on Elect. Eng.,* Higher Inst. of Electronics, Har-Far, Malta, 1977.
19. Kumar, A., "Backfire Antennas Aim at Direct Broadcast T.V.," *Microwaves,* Vol. 17, No. 4, April 1978, pp. 106--110.
20. Bach, H., "Applicability of Hansen-Woodyard Condition," *Proc. IEE,* Vol. 119, No. 1, January 1972, pp. 38–40.
21. Jasik, H., (ed), *Antenna Engineering Handbook,* McGraw-Hill, 1961.
22. Kirov, G.S., and H.D. Hristov, "Backfire Antenna with Modulated Disk-on-rod Surface Wave Structure," Bulgarian Patent No. 23 418, 1976.
23. Kirov, G.S., "Modulated Surface Wave Antennas," Ph.D. Thesis, Sofia, 1979, Available in Central Inst. for Scientific and Technical Information (Tz.I.N.T.I.), (in Bulgarian).
24. Davies, S., "Element Vector Diagrams for Yagi-Uda Antennas," *Australian Telecommunication Research,* Vol. 1, November 1967, pp. 59–75.
25. Ivanov, E.A., *Diffraction of Electromagnetic Waves on Two Bodies,* Moscow, Nauka I Teknika Publ. House, 1968 (in Russian).

26. James, G.L., *Geometrical Theory of Diffraction of Electromagnetic Waves,* Peter Peregrinus, 1976.

27. Nielsen, E.D., Backfire Antennas with Dipole Elements, Ph.D. Thesis, Technical Univ. of Denmark, August 1974.

28. Nielsen, E.D., "Influence of Variation of Backfire Antenna Parameters," P.I. Yagi Backfire Antenna, *Sc. Report No. 2, Contract F 61 052 67, C 0056, S 183, R.80,* Laboratory of Electromagnetic Theory, Technical Univ. of Denmark, September 1970.

29. Nielsen, E.D., and K. Pontoppidan, "Backfire Antennas with Dipole Elements," *IRE Trans. Antennas and Propagation,* 18, No. 3, May 1970, pp. 367–374.

29a. Baker, H.C., and A.H. La Grone, "Digital Computation of the Mutual Impedance Between Thin Dipoles," *IEEE Trans. Antennas and Propagation,* Vol AP-10, No. 2, March 1962, pp. 172–178.

30. Jordan, E.C., *"Electromagnetic Waves and Radiating Systems,"* Prentice-Hall, New Jersey, 1950.

31. Strom, J.A., and H.W. Ehrenspeck, "Backfire Antennas for SHF, UHF and VHF Bands," *AFCRL Report 63-0114,* April 1963.

31a. Longva, O.H., "Telemetry Antenna for NTNF Satellite Station in Tromse," *Report IR-92,* The Norwegian Institute of Technology, Norway, September 1967.

32. Strom, J.A., "A Dielectric-rod Backfire Antenna," *AFCRL Report 69-0347,* August 1969.

33. Loh, S.C., and J. Jacobsen, "The Radiation Characteristics of the Backfire Helical and Zig-Zag Antennas," *The Radio and Electronics Engineer,* 33, May 1967, pp. 317–321.

34. Hristov, H.D., and D. Taylor, "Rectangular Backfire Antenna with Dielectric SW Structure," *Electronics Letters,* Vol. 6, March 1972, pp. 163–165.

35. Hristov, H.D., and J.R. Urumov, "X-Band Backfire Antenna Arrays," *European Microwave Conference,* Hamburg, September 1975, pp. 112–115.

36. Hristov, H.D., and A. Kumar, "X-Band Waveguide-Fed Long Backfire Antenna with Dielectric SW Structure," Research Report, Dept. of Electrical and Electronic Engng., Queen Mary College, London University, August 1976.

37. Kumar, A., and H.D. Hristov, "Long Backfire Antenna with Dielectric SW Structure," *Int. Journal of Electronics,* Vol. 43, No. 3, 1977, pp. 309–312.

38. Kumar, A., "Backfire Antenna," Research Report, EETC Coventry, U.K., 1980.

39. Hristov, H.D., *Resonator-Type Antennas,* D.Sc. Thesis, Sofia, 1987, Available in Central Institute for Scientific and Technical Information (Tz.I.N.T.I.), (in Bulgarian).

40. Kumar, A., and H.D. Hristov, "Modified Backfire Antenna Lowers Cross-Polarization," *Microwave,* June 1977, pp. 70–73.

41. Simon, R.N., and D.T. Shahani, "Long Backfire Antenna with Insular Image Guide Feed for Dual Frequency Operation," *Electronics Letters,* 15, February 1979, No. 3, pp. 78–79.

42. Kumar, A., "Experimental Study of a Dielectric Rod Enclosed by a Waveguide for Use As a Feed," *Electronics Letters,* 12, No. 25, 1976, pp. 666–668.

43. Kumar, A., "Theoretical Analysis of a Long Backfire Antenna, Containing a Dielectric Structure," *IEE Proc.,* Pt. H-Microwaves, Optics and Acoustics, Vol. 2, No. 3, May 1978, pp. 91–95.

44. Barlow, H.M., and J. Brown, *Radio-Surface Waves,* Oxford, 1962.

45. Brown, J., and O. Spector, "The Radiating Properties of End-fire Aerials," *Proc. IEE,* January, 1957, pp. 27–34.

46. Kirov, N.M., Two-Reflector Confocal Antennas, Ph.D. Thesis, 1980, Available in Central Inst. for Scientific and Technical Information (Tz.I.N.T.I.), (in Bulgarian).

47. Hristov, H.D., and N.M. Kirov, "A Family of Two-Reflector Antennas with Identical Dielectic Primary Feeds," *Second Int. Conf. on Antennas and Propagation,* York, UK, April 1981, pp. 358–362.

47a. Ludwig, A.C., "The Definition of Cross-Polarization," *IEEE Trans. Antennas and Propagation,* Vol. AP-21, 1973, pp. 116–119.

48. Kun, R., *Microwave Antennas,* Leningrad, Sudostroene Publ. House, 1967 (in Russian).

49. Walter, C.H., *Traveling Wave Antennas,* Dover, N.Y., 1965.

50. Tereshin, O.N., "Reverse Electrodynamic Problem for an Impedance Antenna with a Special Profile," *Radiotechnika and Electronika,* Mocow, 8, No. 4, 1963, pp. 715–718, (in Russian).

51. Tereshin, O.N., and V.G. Goffman, "Synthesis of a Profiled Impedance Antenna," *Radiotek-nika,* Moscow, 24, No. 8, 1969, pp. 46–52, (in Russian).

52. Tereshin, O.N., B.M. Sedov, and A.F. Chaplin, *Synthesis of Slow-wave Antennas,* Moscow, Svjaz Publ. House, 1980 (in Russian).

53. Tereshin, O.N., and L.M. Kuznetzov, "Backfire Antennas with Modulated Phase Velocity in Dipole Slow-wave Structure," *Radiotechnika,* Moscow, Vol. 25, No. 12, 1970 (in Russian).

Author Index

SUBJECT INDEX

www.ingramcontent.com/pod-product-compliance
Lightning Source LLC
LaVergne TN
LVHW101454200926
848483LV00001B/125

* 9 7 8 0 8 9 0 0 6 3 3 4 7 *